Methods of Experimental Physics

VOLUME 23

NEUTRON SCATTERING

PART B

METHODS OF EXPERIMENTAL PHYSICS

Robert Celotta and Judah Levine, *Editors-in-Chief*

Founding Editors

L. MARTON
C. MARTON

Volume 23

Neutron Scattering

PART B

Edited by

David L. Price

Materials Science Division
Argonne National Laboratory
Argonne, Illinois

Kurt Sköld

Institute for Neutron Research
Uppsala University
Studsvik, Nyköping, Sweden

ACADEMIC PRESS, INC.
Harcourt Brace Jovanovich, Publishers

San Diego   New York   Berkeley   Boston
London   Sydney   Tokyo   Toronto

085769

COPYRIGHT © 1987 BY ACADEMIC PRESS, INC.
ALL RIGHTS RESERVED.
NO PART OF THIS PUBLICATION MAY BE REPRODUCED OR
TRANSMITTED IN ANY FORM OR BY ANY MEANS, ELECTRONIC
OR MECHANICAL, INCLUDING PHOTOCOPY, RECORDING, OR
ANY INFORMATION STORAGE AND RETRIEVAL SYSTEM, WITHOUT
PERMISSION IN WRITING FROM THE PUBLISHER.

ACADEMIC PRESS, INC.
1250 Sixth Avenue, San Diego, California 92101

United Kingdom Edition published by
ACADEMIC PRESS INC. (LONDON) LTD.
24–28 Oval Road, London NW1 7DX

Library of Congress Cataloging in Publication Data
(Revised for pt. B)

Neutron scattering.

(Methods of experimental physics ; v. 23)
Includes index.
1. Neutrons—Scattering.   2. Condensed matter.
I. Sköld, Kurt. II. Price, David L. (David Long),
Date   . III. Series.
QC793.5.N4628N496   1987   539.7'213   86–1128
ISBN 0–12–475969–6 (pt. B : alk. paper)

PRINTED IN THE UNITED STATES OF AMERICA

87 88 89 90   9 8 7 6 5 4 3 2 1

CONTENTS

PREFACE ix
WERNER SCHMATZ xiii
LIST OF SYMBOLS xv

8. Adsorbed Monolayers and Intercalated Compounds
by SUNIL K. SINHA

8.1. Introduction 1
8.2. Types of Adsorbate/Substrate Systems Studied
 by Neutron Diffraction 3
8.3. Diffraction from Two-Dimensional Structures 10
8.4. Studies of Phase Diagrams of Physisorbed
 Systems 16
8.5. Studies of Commensurate–Incommensurate
 Transitions 36
8.6. Studies of Incommensurate Two-Dimensional Lattice
 and Melting 42
8.7. Multilayer Studies 51
8.8. Inelastic Scattering from Adsorbed Monolayers . . . 53
8.9. Neutron Studies of Surface Magnetization 67
8.10. Neutron Studies of Intercalated Compounds 69
 References 79

9. Defects in Solids
by WERNER SCHMATZ

9.1. Introduction 85
9.2. Static Properties 96
9.3. Dynamic Properties 120
 References 126

10. Hydrogen in Metals
by Tasso Springer and Dieter Richter

10.1. Introduction: Hydrogen in Metals and Neutron Spectroscopy 131
10.2. Diffusion and Quasi-Elastic Scattering 132
10.3. Hydrogen Vibrations 153
10.4. Hydrogen Tunneling 179
10.5. Concluding Remarks 182
References 182

11. Fast Ion Conductors
by N. H. Andersen, K. N. Clausen, and J. K. Kjems

11.1. Introduction 187
11.2. Scattering Formalism 194
11.3. Model Systems 202
11.4. Diffraction Studies 205
11.5. Quasi-Elastic Diffuse Scattering 223
11.6. Inelastic Scattering 233
11.7. Conclusions 237
References 239

12. Glasses
by Kenji Suzuki

12.1. Introduction 243
12.2. Structure Factors and Pair Correlation Functions 245
12.3. Geometrical and Chemical Short-Range Structure 255
12.4. Structural Anisotrophy and Fluctuations in the Intermediate Range 270
12.5. Hydrogen Atoms as a Probe for Structure Characterization of Metallic Alloy Glasses 284
12.6. Atomic Vibrations by Inelastic Pulsed Neutron Scattering 293
References 300

13. Solid and Liquid Helium
by HENRY R. GLYDE AND ERIC C. SVENSSON

 13.1. Introduction 303
 13.2. Solid Helium 309
 13.3. Liquid ^{4}He 326
 13.4. Liquid ^{3}He 370
 13.5. Liquid ^{3}He–^{4}He Mixtures 385
   References 393

14. Classical Fluids
by PETER A. EGELSTAFF

 14.1. Introduction to Atomic and Simple Molecular Fluids 405
 14.2. Data Collection 414
 14.3. Data Analysis 419
 14.4. Inelasticity Corrections to Intensity Measurements . . . 428
 14.5. Evaluation of $S(Q)$ and $S(Q, \omega)$ 442
 14.6. Comparision of Experiment and Theory 455
   References 468

15. Ionic Solutions
by JOHN E. ENDERBY AND P. M. N. GULLIDGE

 15.1. Introduction 471
 15.2. The Method of Differences 473
 15.3. Experimental Aspects of the Method 475
 15.4. Some Results 477
 15.5. The Kinetics of Water Exchange 481
 15.6. Conclusions 487
   References 488

16. Colloidal Solutions
by Sow-Hsin Chen and Tsang-Lang Lin

16.1. Introduction	489
16.2. Small-Angle Neutron Scattering Technique	499
16.3. Determination of Intraparticle Structures	514
16.4. Determination of the Interparticle Structure	524
16.5. Conclusion	539
References	541

17. Liquid Crystals
by Jerzy A. Janik and Tormod Riste

17.1. Introduction	545
17.2. Single-Molecule Properties: Incoherent Scattering	553
17.3. Collective Properties: Coherent Scattering	567
17.4. Convective Instabilities	575
17.5. Concluding Remarks	582
References	582

Index	585
Contents of Volume 23, Part A	589
Contents of Volume 23, Part C	590

PREFACE

The neutron scattering technique for measuring the structure and dynamics of condensed matter has developed over the 50 years of the neutron's history into a widely used tool in physics, chemistry, biology, and materials science. Since the early diffraction studies in the 1940s and the first measurements of inelastic scattering in the 1950s, developments in experimental methods have greatly increased the sensitivity and range of applications of the technique. Thus, while the early measurements probed distances on the order of interatomic spacings (~ 3 Å) and times on the order of typical periods of lattice vibrations (~ 1 ps), the current range of neutron scattering experiments covers distances from 0.1 to 10,000 Å, and times from 10 fs to 1 μs. This has been achieved by expanding the range of neutron energies available to the experimenter from a few milli-electron-volts (at cold sources in research reactors) to several electron-volts (at pulsed spallation sources), and by using a variety of novel detection methods such as position-sensitive detectors and back-scattering and spin-echo techniques. As a result, the areas of investigation have expanded from the conventional crystal structures and lattice dynamics (and their magnetic analogs) of 30 years ago to high-resolution studies of the atomic spacings in amorphous thin films, biological structures on a cellular scale, unraveling of long chains of polymers, and transitions between energy levels in molecular solids.

Along with these developments, the community of neutron users has expanded and diversified. Whereas 30 years ago neutron scattering was practiced largely by solid-state physicists and crystallographers, the users of present-day centralized neutron facilities include chemists, biologists, ceramicists, and metallurgists, as well as physicists of diverse interests ranging from fundamental quantum mechanics to fractals and phase transitions. The neutron centers have developed from essentially in-house facilities at the national nuclear research laboratories into centralized facilities organized for use by the general scientific community at an international level. The pioneer of this mode of operation was the Institut Laue-Langevin in Grenoble, France, operated since 1972 by Britain, France, and Germany as a user-oriented facility for scientists from these and other countries. Similar modes of operation are now being established at other major reactor facilities like those at Brookhaven and Oak Ridge in the United States, and the pulsed spallation sources that have recently come into operation at Argonne in the United States, the KEK Laboratory in Japan, and the Rutherford Laboratory in Britain have been set up from the beginning with this mode of operation. The current population of users of these and other neutron facilities has been recently estimated* to be 500 in the United States, 1250 in Western Europe, and about 200 in Japan.

The aim of the present book is to describe the current state of the art of application of neutron scattering techniques in those scientific areas that are most active. The presentation is aimed primarily at professionals in different scientific disciplines, from graduate students to research scientists and university faculty members, who may be insufficiently aware of the range of opportunities provided by the neutron technique in their area of specialty. It does not present a systematic development of the theory, which may be found in excellent textbooks such as those of Lovesey or Squires, or a detailed hands-on manual of experimental methods, which in our opinion is best obtained directly from experienced practitioners at the neutron centers. It is rather our hope that this book will enable researchers in a particular area to identify aspects of their work in which the neutron scattering technique might contribute, conceive the important experiments to be done, assess what is required to carry them out, write a successful proposal for this purpose for one of the centralized user facilities, and carry out the experiments under the care and guidance of the appropriate instrument scientist. With this object in view, each chapter relating to a particular field of science has been written by a leading practitioner or practitioners of the application of the neutron methods in that field.

Volume 23, Part A, of this work starts out with a brief survey of the theoretical concepts of the technique and establishes the notation that will be used throughout the book. Chapters 2 and 3 review the fundamental hardware of neutron scattering, namely, sources and experimental methods, and Chapter 4 discusses fundamental physics applications in neutron optics. The remaining chapters of Part A treat various basic applications of neutron scattering to studies of the atomic structure and dynamics of materials. The Appendix contains a compilation of neutron scattering lengths and cross sections that are important in nearly all neutron scattering experiments.

Volume 23, Part B, contains surveys of the application of neutron scattering techniques to nonideal solids, such as solids with defects, two-dimensional solids and glasses, and to various classes of fluids. Finally, Volume 23, Part C, treats neutron scattering investigations of magnetic materials, solids undergoing phase transitions, and macromolecular and biological structures. In recognition of the expanding use of neutron scattering in technol-

* Current Status of Neutron-Scattering Research and Facilities in the United States (National Academy Press, Washington, D.C., 1984).

ogy, the last chapter in Part C is devoted to a survey of industrial applications.

We wish to thank the authors for taking time out of their busy schedules for contributing these chapters, Dr. R. Celotta for inviting us to undertake this work, and the staff of Academic Press for their encouragement and forbearance.

KURT SKÖLD
DAVID L. PRICE

WERNER SCHMATZ

During the preparation of this book, we were saddened by the news of the premature death of Werner Schmatz, a prominent member of the neutron scattering community and a contributing author to this volume (Chapter 9).

Werner Schmatz made important contributions in several areas of neutron scattering research and in materials science. Among these, his pioneering work on the study of disorder in metals by diffuse scattering and by small-angle scattering deserves special mention. In addition to his contribution to the scientific understanding of these systems, he also played a major role in the development of the neutron instrumentation used in such studies. For example, the state-of-the-art instruments for small-angle scattering and for diffuse scattering now in operation at the Institute Laue-Langevin in Grenoble are to a large degree based on concepts developed by Werner Schmatz. His contributions to materials science and to the experimental methods used in neutron scattering are of lasting importance, and he is greatly missed by colleagues and friends in the neutron scattering community.

We are grateful to Klaus Werner for completing and editing the chapter that Werner Schmatz contributed.

KURT SKÖLD
DAVID L. PRICE

LIST OF SYMBOLS

b	Bound scattering length
$\bar{b}$	Coherent scattering length
b^+	Scattering length for $I + \tfrac{1}{2}$ state
b^-	Scattering length for $I - \tfrac{1}{2}$ state
b_i	Incoherent scattering length
b_N	Spin-dependent scattering length
c	Velocity of light $= 2.9979 \times 10^{10}$ cm sec^{-1}
d	Mass density
d	Equilibrium position of atom in unit cell
$\mathbf{D}_\perp(\mathbf{Q})$	Magnetic interaction operator
$d\sigma/d\Omega$	Differential cross section
$d^2\sigma/d\Omega\, dE$	Double differential cross section
E_0, E_1	Incident, scattered energy
E	Energy lost by neutron $(E_0 - E_1)$
e	Charge on the electron $= 4.8033 \times 10^{-11}$ esu
$\mathbf{e}^j(\mathbf{q})$	Polarization vector of normal mode j [$\mathbf{e}_d^j(\mathbf{q})$ for non-Bravais crystal]
$F(\tau)$	Structure factor for unit cell
$f(\mathbf{Q})$	Form factor
$G(\mathbf{r}, t)$	Space–time correlation function $[G_d(\mathbf{r}, t) + G_s(\mathbf{r}, t)]$
$G_d(\mathbf{r}, t)$	"Distinct" space–time correlation function
$G_s(\mathbf{r}, t)$	"Self" space–time correlation function
$g(\mathbf{r})$	Pair distribution function $[\rho(\mathbf{r})/\rho_0]$
$\hbar$	Planck's constant$/2\pi = 1.0546 \times 10^{-27}$ erg sec
$I(\mathbf{Q}, t)$	Intermediate scattering function $[I_d(\mathbf{Q}, t) + I_s(\mathbf{Q}, t)]$
$I_d(\mathbf{Q}, t)$	Intermediate "distinct" scattering function
$I_s(\mathbf{Q}, t)$	Intermediate "self" scattering function
I	Angular momentum operator for nucleus
$\mathbf{k}_0, \mathbf{k}_1$	Incident, scattered wave vector
k_B	Boltzmann's constant $= 1.3807 \times 10^{-16}$ erg K^{-1}
l	Position of unit cell
M	Mass of atom
m	Mass of neutron $= 1.0087$ u
m_e	Mass of electron $= 9.1095 \times 10^{-28}$ g
N	Number of unit cells in crystal
N_A	Avagadro's number $= 6.0220 \times 10^{23}$
$n(r)$	Radial distribution function $[4\pi r^2 \rho(r)]$
Q	Scattering vector $(\mathbf{k}_0 - \mathbf{k}_1)$
q	Reduced wave vector $(\mathbf{Q} - \boldsymbol{\tau})$
r	Number of atoms in a unit cell
r_0	Classical electron radius $(e^2/m_e c^2) = 2.8179 \times 10^{-13}$ cm
S	Spin operator for ion or atom
$S(\mathbf{Q})$	Static structure factor $I(\mathbf{Q}, 0)$
$S_c(\mathbf{Q}, \omega)$	Coherent scattering function
$S_i(\mathbf{Q}, \omega)$	Incoherent scattering function
s	Spin operator for electron

u	Atomic mass unit $= 1.6606 \times 10^{-24}$ g
$\mathbf{u}^l$	Vibrational amplitude ($\mathbf{u}^l_d$ for non-Bravais crystal)
V	Volume of sample
v_0	Volume of unit cell
$2W$	Exponent of Debye–Waller factor (in cross section)
$Z(\omega)$	Density of phonon states
γ	Gyromagnetic ratio of neutron $= 1.9132$
Θ	Debye temperature
θ	Bragg angle
μ	Magnetic moment of ion or atom
μ_N	Nuclear magneton $(e\hbar/2m_pc) = 5.0508 \times 10^{-24}$ erg G^{-1}
μ_B	Bohr magneton $(e\hbar/2m_ec) = 9.2741 \times 10^{-21}$ erg G^{-1}
ρ_0	Average number density
$\rho(\mathbf{r})$	Pair density function $[G(\mathbf{r}, 0) - \delta(\mathbf{r})]$
σ	Bound total cross section (scattering plus absorption)
σ_c	Bound coherent scattering cross section
σ_i	Bound incoherent scattering cross section
σ_s	Bound scattering cross section ($\sigma_c + \sigma_i$)
$\tfrac{1}{2}\boldsymbol{\sigma}$	Spin operator for neutron
$\boldsymbol{\tau}$	Reciprocal lattice vector $\{2\pi[(h/a), (k/b), (l/c)]\}$
Φ	Neutron flux (n cm^{-2} sec^{-1})
ϕ	Scattering angle ($=2\theta$ for Bragg reflection)
χ	Susceptibility
$\chi(\mathbf{Q}, \omega)$	Generalized susceptibility
Ω	Solid angle
$\omega_j(\mathbf{q})$	Frequency of normal mode j

8. ADSORBED MONOLAYERS AND INTERCALATED COMPOUNDS

Sunil K. Sinha

Exxon Research and Engineering Company
Annandale, New Jersey

8.1. Introduction

The decade of the seventies saw an enormous increase of research in the area known as "surface physics" or "two-dimensional physics," and rapid growth in this area continues to this day. This interest has been fueled partly by the technological importance of understanding the properties of surfaces and thin films (particularly, the properties of hydrocarbons adsorbed on surfaces), partly by some of the novel features of physics in dimensions lower than three, and partly by the richness and variety of the phenomena available for study.

The most prevalent investigative tool for the study of surface structure has been low-energy electron diffraction (LEED), which has a venerable history dating back to the original experiments by Davisson and Germer that helped establish the wave nature of the electron. This is still the mainstay of surface structural investigations, since it is a probe that is specifically surface sensitive. The reason for this is the extremely strong interaction of low-energy electrons with matter, which presents bulk penetration by the electrons. However, this is also responsible for the biggest disadvantage of LEED, namely, the fact that the interaction of the electron with the surface is so strong that the Born approximation for scattering is not valid, so that detailed inferences regarding surface structure depend on a comparison of experiment and complicated theoretical computations that include involved multiple scattering corrections. The same is true of surface diffraction with beams of atomic particles, such as helium atoms. One would not at first think that the neutron would be useful as a probe of surfaces, since its relatively weak interaction with matter makes it ideal as a bulk probe, but in fact neutron diffraction was one of the earliest and most important tools in the study of physisorbed monolayers on solid surfaces. The validity of the Born approximation for the scattering of neutrons by atomic nuclei makes it possible to carry out detailed analyses of the diffraction pattern to obtain a large amount of information about the surface monolayer. The same is true for x-ray diffraction, which is also now being used for surface diffraction studies.

The biggest problem in such studies is to find ways to detect the weak diffraction signal from the surface and to prevent it from being overwhelmed by scattering from the bulk. (Bulk scattering is so many orders of magnitude larger than surface scattering that, under unfavorable conditions, even diffuse bulk scattering can overwhelm Bragg reflections from the surface.) There are two ways in which this may be accomplished. The first is to use materials that have very large specific surface areas (i.e., areas per gram of sample). Examples are crushed powders or very fine particles, such as platinum black, or materials that have very large internal surfaces such as porous silica or exfoliated graphite. The second is to use the technique of total external reflection, i.e., with the grazing angle that the incident radiation makes with the surface being less than the critical angle for total reflection. In this way, the beam does not penetrate the bulk, but is restricted to within a few angstroms of the surface region. In the case of high surface area samples, the available surface areas are of the order of square meters or greater so that one can see an appreciable "surface-specific" signal. In the case of total external reflection, one is generally dealing with a single (macroscopic) surface of area of the order of a square centimeter or less. Thus, with the relatively limited neutron fluxes available at even the high flux reactors today, neutron diffraction experiments from a single surface have not been carried out, although total reflection of neutrons from surfaces have recently been used to probe the depth dependence of magnetization at metallic surfaces, as will be discussed in more detail later in this chapter. Such studies have, however, been carried out with x-ray scattering, particularly at synchrotron sources,[1] owing to the much greater intensities available.

In spite of the above difficulties, neutron scattering has been a surprisingly successful probe for studying surface phenomena, particularly in physisorbed systems. To a large extent, this has been due to the existence of graphite substrates with large and homogeneous surface areas, which provided excellent model substrates with a low bulk signal in the regions of interest for the study of a variety of physisorption phenomena. This will be discussed in more detail in Section 8.2. Historically, neutron diffraction was one of the first methods used to study the details of the ordered structures, and of the two-dimensional phase diagrams for systems of molecules physisorbed on smooth surfaces, as in the pioneering experiments of the Brookhaven group on N_2 physisorbed on graphite.[2] These and neutron diffraction experiments on other graphite-substrate-based systems such as ^{3}He and ^{4}He, H_2 and D_2, CD_4 and other hydrocarbons, Ar, Ne, O_2, and several other adsorbates have been crucial to our understanding of the phenomena associated with two-dimensional ordering of molecules on smooth substrates. In addition, inelastic and quasi-elastic neutron scattering has yielded information on the dynamics of physisorbed molecules and atoms, which is not easily available by other means. In addition to the validity of the Born approximation for scattering, an advantage neutrons share with x rays in the study of physisorbed systems is that they are not significantly affected by the ambient vapor nor do they cause desorption from the substrate. This has restricted the use of LEED and atomic beam scattering in the case of physisorbed systems, although Fain and his collaborators have recently used low intensity electron beams with image intensifiers and carried out many successful LEED studies of several graphite-based physisorbed systems.[3] For purely structural (as opposed to dynamical) studies, the more recent use of synchrotron radiation for x-ray diffraction studies of adsorbed monolayers and thin films seems to have wider applicability than neutron diffraction, mainly for intensity reasons. A neutron diffractometer on a high-flux reactor can have an incident monochromatic neutron flux on the sample of at best $\sim 10^7$ n/cm^2/s while set for a wave vector resolution of ~ 0.01 Å^{-1} for the range of wave vectors of interest. On the other hand, an x-ray diffractometer installed on a wiggler beam line at a synchrotron light source can have an incident photon flux of $\sim 10^{13}$ photons/cm^2/s on the sample while set for a resolution of $\sim 5 \times 10^{-4}$ Å^{-1} for the wave vectors of interest.[4] The scattering cross section for x-ray photons per atom (which goes as Z^2, where Z is the atomic number) can go from one fiftieth of the corresponding neutron coherent scattering cross section to being several times larger. To some extent, this intensity advantage can be offset by the ability to use beams of much larger area in the neutron case and much thicker samples (since the attenuation length for neutrons is usually at least an order of magnitude larger than for x rays). Nevertheless, the advent of synchrotron radiation certainly makes neutrons less competitive for structural investigations except for special cases, such as that of light atoms or molecules, e.g., hydrogen, helium, or the lighter hydrocarbons (where the scattering from the adsorbed monolayer is weak compared to the bulk diffuse scattering from the substrate). In the case of magnetic structures of adsorbed monolayers (such as in the case of O_2 adsorbed on graphite[5]), or for *inelastic* scattering studies, however, neutrons provide unique information. We shall now describe how such studies are carried out and briefly describe the systems that have been studied.

8.2. Types of Adsorbate/Substrate Systems Studied by Neutron Diffraction

In general, a study of crystalline order on surfaces requires the surface to be fairly uniform and to correspond to only one type of crystallographic surface plane of the underlying substrate crystal structure. Thus, there are obvious problems with crushed powders of crystalline materials, where several crystallographic faces may be involved, or with surfaces which are inhomogeneous enough so that there is a nonuniform distribution of specific adsorption sites. There are, however, many forms of graphite possessing large specific surface areas, which when properly prepared and cleaned provide a suitably homogeneous and uniform surface. This surface is predominantly made up of the basal planes of graphite. Examples are the product known as Grafoil[6] or Papyex,[7] made from exfoliated graphite, which is prepared by intercalating graphite with $FeCl_3$, then heating to drive out the intercalant (which opens up the large internal surface area), and finally rolling it into thin sheets that partially orient the crystallographic basal planes (and thus the free surfaces) of the graphite crystallites parallel to the sheets. The surface area for adsorption available on Grafoil ranges from 20 to 40 m^2/g. The orientations of the crystallites are purely random in the basal plane (as in pyrolytic graphite) but the distribution of normals to the basal planes consists of an isotropic component and an oriented component with a broad mosaic distribution (full width at half-maximum being of the order of 30°) centered about the normal to the sheet. A grade of graphite known as UCAR-ZYX graphite,[6] which is prepared by exfoliating pyrolytic graphite crystals, possesses a much higher degree of basal plane orientation (full width at half-maximum being of the order of 17°) but a much smaller specific surface area (of the order of 2 m^2/g). From diffraction experiments on ordered adsorbate monolayers on these substrates, to be discussed later, the lateral size of coherent surface domains on Grafoil does not appear to be much larger than 100–200 Å,[8] whereas on ZYX graphite coherent surface domains of sizes ranging up to thousands of angstroms have been observed.[9] Figure 1 shows a schematic representation of exfoliated graphite that is partially oriented as Grafoil sheets. Completely unoriented graphite powders, such as graphon,[10]

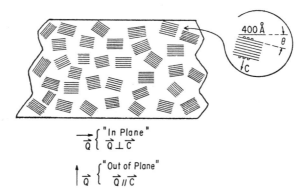

FIG. 1. Schematic representation of a Grafoil sheet.

Carbopack B,[11] or vermicular graphite,[12] have also been employed in monolayer studies.

The reason why graphite substrates have been taken as model systems is well illustrated by Figs. 2 and 3, which represent adsorption isotherms measured for methane on exfoliated graphite in a series of classic experiments by Thomy and Duval[13] several years before any diffraction measurements were carried out. The distinct steps in these isotherms represent the formation of successive distinct layers of adsorbate molecules and attest to the homogeneity of the substrate surface. The fine structure shown in Fig. 3 also shows the presence of distinct phase transitions with increasing coverage, which Thomy and Duval identified as transitions between two-dimensional vapor and two-dimensional liquid phases and two-dimensional liquid and two-dimensional solid phases. Graphite thus provides a convenient substrate to use for physisorbing a large variety of adsorbate atoms and molecules, ranging from the rare gas atoms, diatomic molecules (e.g., H_2, D_2, N_2, O_2), to the simple hydrocarbons, and for studying the rich variety of structural phases found in these systems at different coverages and temperatures. Neutron diffraction and other experiments have revealed ordered two-dimensional phases (i.e., two-dimensional crystalline phases) where the lattice structure of the monolayer is commensurate (or "registered") with the substrate (sometimes also known as an "epitaxial" phase), and also where it is incommensurate with the substrate lattice. These experiments have also

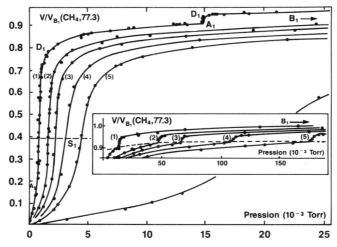

FIG. 2. Adsorption isotherms of methane of Grafoil as measured by Thomy and Duval.[13] The graph is plotted as a function of coverage (V/V_{B_1}) versus pressure and shows the formation of the first layer for temperatures of (1) 77.3 K, (2) 80.0 K, (3) 80.9 K, (4) 82.3 K, (5) 83.5 K, and (6) 90.1 K.

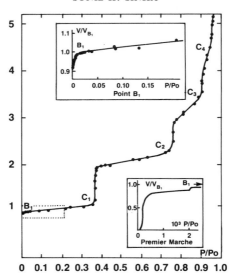

FIG. 3. Continuation of CH_4 isotherm at 77.73 K showing the formation of the second, third, etc., layers ($P_0 = 9.4$ mm). (From Thomy and Duvall.[13])

revealed two-dimensional solid phases where the molecules develop orientational order in addition to positional order, and two-dimensional liquid and vapor phases, where the molecules are positionally random, or where they form a lattice gas or liquid still epitaxial with the substrate. The structural details of these phases have been studied and also the phase transitions between them. The large variety of phenomena to be studied, coupled with the availability of relatively large specific area graphite substrate materials, has been the reason why neutron scattering studies of adsorbed monolayers has been almost exclusively restricted to experiments using graphite substrates. Efforts to develop new substrates for studying such phenomena have been slow in coming. A promising new substrate system that is currently being investigated is fumed MgO powder, which consists of tiny crystallites of MgO with cubic (100) faces obtained by deposition from MgO smoke. This may show interesting differences from experiments on graphite substrates because the substrate symmetry is now cubic rather than hexagonal as in the case of graphite substrates. As yet there is little published data on diffraction from these kinds of systems.[14]

We now discuss the methods used to carry out diffraction experiments on these kinds of samples. In the case of Grafoil or ZYX graphite, the substrate is initially baked (for several hours) at 1000°C in a vacuum furnace to remove strongly bonded chemical impurities introduced during the manufacturing process. While the predominant portion of the surface area is made of the basal plane of graphite on which most molecules are only weakly physisorbed (the adsorption energy per molecule arises due to the Van der Waals interaction with the substrate surface and is of the order of 50–200 K), the edges of the crystallites (which are faces normal to the basal plane) do have dangling bonds that strongly bind external molecules. However, it is a common assumption that these surfaces are probably always saturated (even after baking) and thus do not play a role in the physisorption process. After the initial high-temperature baking, any further cleaning of the substrate is achieved by baking at lower temperatures ($\sim 800°C$) for several hours in a vacuum furnace to remove water and physisorbed impurities. The substrate is then loaded in a glove box filled with dry nitrogen or argon gas into the neutron diffraction cell. In the case of Grafoil, the sheets are often cut into circular pieces that are then stacked in the sample cell. ZYX graphite comes in chips that are similarly stacked in the sample cell. Care must be taken to allow space for the ambient vapor to have access to all parts of the substrates for rapid equilibration. Since the sample cells are closed off during the experiment, care must also be taken not to have too large a volume for the ambient vapor, as the coverage will then not stay constant with temperature, especially at high temperatures and vapor pressures. In general, a small correction is always made for this slight loss of coverage with increasing temperature. An alternative is to keep the cell open but at constant vapor pressure during the experiment, while varying the temperature.

For neutron diffraction experiments, where the plane of scattering is generally horizontal, the usual geometry for the sample cell is to arrange to have the mean orientation of the basal planes *in* the scattering plane (i.e., also horizontal). For x-ray experiments with ZYX graphite, however, the basal planes of graphite are often arranged to be perpendicular to the plane of scattering, to minimize the beam attenuation through the sample.

The neutron diffraction experiments are then carried out in the usual manner on a powder diffractometer or a triple-axis spectrometer mounted with an analyzer crystal set for elastic scattering. The use of the latter largely eliminates the thermal diffuse background scattering from the substrate. Pyrolytic graphite monochromators are generally used for this purpose, with neutron wavelengths of either 2.36 Å and a pyrolytic graphite filter to remove higher order contamination of the beam, or >4 Å with a beryllium filter to remove higher orders. Because of the two-dimensional nature of the diffraction, as will be discussed in Section 8.3, the scattering is only weakly dependent on the component of **Q** normal to the adsorbing surface. Thus, the vertical resolution of the diffractometer can be profitably opened up, and the use of bent focusing monochromators and analyzers can significantly improve counting rates. As mentioned in the introduction, the wave vector resolution typically achieved in the longitudinal direction with these experiments is of the order of 0.01 Å^{-1} and may be measured by powder diffraction from a standard powder sample, such as Al_2O_3, which has diffraction peaks at wave vectors close to the peaks of interest in the sample.

The sample cell is usually made of thin-walled aluminum and mounted inside a cryostat on the spectrometer, with a capillary connecting it to an external gas-handling system. The gas of adsorbate molecules is introduced from a calibrated loading volume, while the sample is cold (~ 100 K), but not so cold as to allow the vapor to condense or freeze before it is adsorbed onto the substrate. Generally, before the experiment is carried out, some vapor pressure isotherms are run on the sample, one with a standard adsorbate (e.g., N_2) to calibrate the substrate surface area, and one with the actual sample adsorbate to calibrate the amount of gas introduced that would correspond to one monolayer. Figure 4 shows an example of such a vapor pressure isotherm for CD_4 on Grafoil. The amount of adsorbed gas is obtained by measuring the drop in pressure in the standard loading volume each time some gas is introduced into the sample cell. It should be remarked that there are two definitions of a "completed monolayer." For molecules that are adsorbed initially in a registered or commensurate phase (such as the ($\sqrt{3} \times \sqrt{3}$) structure to be discussed later), a second layer does not always form immediately after completion of the registered monolayer. Instead, the molecules "pack in" producing a (incommensurate) dense monolayer after which the second monolayer forms. The point of inflection (point A in

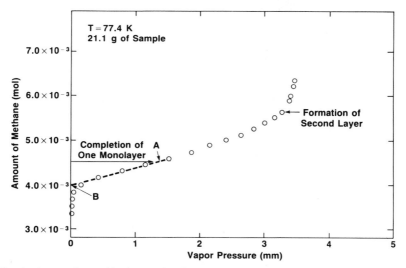

FIG. 4. An experimental isotherm taken for CD_4 on Grafoil as a calibration for coverage. The point A corresponds to the completed dense monolayer and the point B to the registered monolayer. (From Vora et al.[8])

Fig. 4) of the isotherm marks the completion of the dense monolayer, while the extrapolation of the slope in this region to zero vapor pressure (point B in Fig. 4) is conventionally taken to be the coverage corresponding to the completed registered monolayer. This has actually been checked with neutron diffraction measurements as will be discussed in Section 8.4. In any case, these two reference points in the isotherm provide one with a calibration of coverage for the system being studied.

The diffraction pattern from the monolayer is obtained by subtracting from the diffraction pattern for the loaded sample at the required coverage; that of the clean substrate at roughly the same temperature (see Fig. 5). The Bragg powder peaks from graphite are generally at larger Q than those from the monolayer, with the exception of the (002) peak. If the substrate was completely oriented, the peak would also not appear in the horizontal scattering plane, but owing to the partial misorientation of the graphite crystallites, it appears weakly but is nevertheless in general far bigger than the peaks from the monolayer. In principle, it should subtract out in the difference pattern, but quite often this peak causes problems owing to imperfect subtraction (due to counting statistics or slightly different attenuation corrections with and without the adsorbate molecules that are not always allowed for). In this manner, powder diffraction patterns corresponding to radial Q scans in the horizontal plane are taken as a function of coverage and temperature.

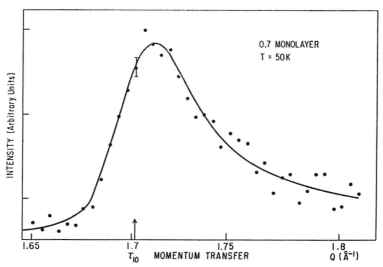

FIG. 5. A typical experimental neutron diffraction peak corresponding to the (10) peak of the ($\sqrt{3} \times \sqrt{3}$) registered phases of CD_4 on Grafoil. The solid curve shows a Warren-type line shape fit to the data. (From Vora et al.[8])

8.3. Diffraction from Two-Dimensional Structures

We now discuss some of the special considerations that apply to neutron (or x-ray) diffraction from two-dimensional structures. The original treatment of this problem was given in a classic paper by Warren over 40 years ago.[15]

Warren's paper applied to the case of diffraction from perfect two-dimensional crystallites of finite size with an orientation that was completely random in three-dimensional space. In order to use the result for analyzing diffraction from Grafoil or ZYX samples, Warren's formula has to be generalized in two ways. First, the result has to be rederived for crystallites that are not completely random in orientation but have a distribution of plane normals in space, and second, as we shall see, the structure factor of monolayers does not always correspond to perfect crystalline order. We shall sketch the derivation of the diffraction line shape from a set of two-dimensional systems with an arbitrary distribution of plane normals and with an arbitrary structure factor $S(\mathbf{Q})$, which is however assumed to be peaked in the vicinity of some $\mathbf{Q}_0$.

The principles of neutron diffraction are laid out in Chapter 1, Part A of this volume. The differential cross section for coherent elastic scattering of neutrons is given by

$$\frac{d\sigma}{d\Omega} = \frac{1}{N} \sum_{i,j} \bar{b}_i \bar{b}_j \langle e^{-i\mathbf{Q}\cdot\mathbf{R}_i} e^{i\mathbf{Q}\cdot\mathbf{R}_j} \rangle, \tag{8.1}$$

where $\bar{b}_i$ is the coherent scattering length associated with the nucleus with the position $\mathbf{R}_i$. The operators $\mathbf{R}_i$, $\mathbf{R}_j$ are assumed to be both evaluated at time $t = 0$ in the Heisenberg sense (see Chapter 1, Part A) inside the thermal expectation bracket, since we are assuming that all relevant energy transfers of the neutron have been integrated over for the given $\mathbf{Q}$. Here, N is the total number of atoms in the sample.

In the case where the system consists of molecules (of just one species) rather than independent atoms, and we assume that for the energies of relevance the molecules behave as rigid entities, we may replace Eq. (8.1) by

$$\frac{d\sigma}{d\Omega} = \bar{B}^2 |F(\mathbf{Q})|^2 S(\mathbf{Q}), \tag{8.2}$$

where $\bar{B}$ is the coherent scattering length of the molecule,

$$\bar{B} = \sum_{\nu} \bar{b}_\nu, \tag{8.3}$$

ν running over the nuclei in the molecule, and $F(\mathbf{Q})$ is the molecular form factor given by

$$F(\mathbf{Q}) = \frac{1}{B} \sum_\nu \langle b_\nu e^{-i\mathbf{Q}\cdot\mathbf{R}_\nu} \rangle, \tag{8.4}$$

where $\mathbf{R}_\nu$ measures the nuclear position relative to the molecular center of mass $\mathbf{R}_l$. The structure factor $S(\mathbf{Q})$ is given by

$$S(\mathbf{Q}) = \frac{1}{N} \sum_{l,l'} \langle e^{-i\mathbf{Q}\cdot\mathbf{R}_l} e^{i\mathbf{Q}\cdot\mathbf{R}_{l'}} \rangle. \tag{8.5}$$

Since $\mathbf{R}_l$, $\mathbf{R}_{l'}$ are all on a single plane (even though the individual $\mathbf{R}_\nu$ may not be), it is obvious that $S(\mathbf{Q})$ depends only on $\mathbf{Q}_\parallel$, the component of $\mathbf{Q}$ in the plane, i.e.,

$$S(\mathbf{Q}) = S(\mathbf{Q}_\parallel). \tag{8.6}$$

As in the case of the graphite substrates, let us assume that the crystallites have random orientations about the axis normal to this plane. Then the effective $S(\mathbf{Q})$ depends only on the magnitude of $\mathbf{Q}_\parallel$,

$$\tilde{S}(\mathbf{Q}_\parallel) = \frac{1}{2\pi} \int_0^{2\pi} d\psi \, S(\mathbf{Q}_\parallel, \psi), \tag{8.7}$$

where ψ is the angle made by $\mathbf{Q}_\parallel$ with respect to an arbitrary reference direction in the plane. We now have to average over the orientation distribution of the plane normals. Consider a crystallite plane whose normal is tilted with respect to the scattering plane (the x–y plane in Fig. 6) by an angle θ. We choose a new set of axes $x'y'z'$ such that the z' axis is normal to the crystallite plane and the y' axis lies in the lab-fixed y–z plane. The polar angle θ and the azimuthal angle ϕ of the crystallite plane normal with respect to the lab-fixed axes are shown in Fig. 6. We assume the neutron scattering vector $\mathbf{Q}$ to be along the y axis. Then the component of $\mathbf{Q}$ parallel to the crystallite plane is given by

$$Q_\parallel = |Q\hat{\mathbf{y}} - (\mathbf{Q}\cdot\hat{\mathbf{z}}')\cdot\hat{\mathbf{z}}'|, \tag{8.8}$$

where $\hat{\mathbf{y}}$ is a unit vector along the y axis, and $\hat{\mathbf{z}}'$ is a unit vector along the z' axis. Now,

$$\hat{\mathbf{z}}' = \sin\theta\cos\phi\hat{\mathbf{x}} + \sin\theta\sin\phi\hat{\mathbf{y}} + \cos\phi\hat{\mathbf{z}}. \tag{8.9}$$

Substituting in Eq. (8.8), we obtain

$$Q_\parallel = Q(1 - \sin^2\theta \sin^2\phi)^{1/2}. \tag{8.10}$$

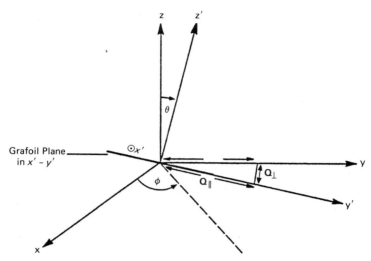

FIG. 6. Scattering geometry for a Grafoil crystallite with its c axis (z') tilted with respect to the lab z axis.

Let $P(\theta)$ be the probability distribution per unit solid angle of plane normals for tilt angle θ about the mean z axis, such that the fraction of plane normals within the solid angle $\sin\theta \, d\theta \, d\phi$ is $P(\theta) \sin\theta \, d\theta \, d\phi$. Then the cross section, from Eqs. (8.2), (8.6), (8.7), and (8.10), is given by

$$\frac{d\sigma}{d\Omega} = \bar{B}^2 |F(Q)|^2 \int_0^{2\pi} d\phi \int_0^{\pi} d\theta \sin\theta P(\theta) \tilde{S}[Q(1 - \sin^2\theta \sin^2\phi)^{1/2}]. \quad (8.11)$$

(We are here assuming that $F(\mathbf{Q})$ depends only on the magnitude of $\mathbf{Q}$.) Now we may manipulate this integral as

$$\frac{d\sigma}{d\Omega} = \bar{B}^2 |F(\mathbf{Q})|^2 \int_0^Q dQ' \int_0^{\pi} d\theta \sin\theta \, P(\theta)$$

$$\times \int_0^{2\pi} d\phi \, \delta[Q' - Q(1 - \sin^2\theta \sin^2\phi)^{1/2}] \tilde{S}(Q')$$

$$= \bar{B}^2 |F(Q)|^2 \int_0^Q dQ' \int_\beta^{\pi-\beta} d\theta \sin\theta \, P(\theta)$$

$$\times \frac{Q'}{Q^2[(Q'/Q)^2 - \cos^2\theta]^{1/2}} \frac{\tilde{S}(Q')}{[1 - (Q'/Q)^2]^{1/2}}, \quad (8.12)$$

where the last line is obtained by changing the variable of integration in the last integral from ϕ to $Q(1 - \sin^2\theta \sin^2\phi)^{1/2}$. In Eq. (8.12), β is given by

$$\beta = \sin^{-1}[1 - (Q'/Q)^2]^{1/2}. \tag{8.13}$$

We may write the result [Eq. (8.12)] in the general form

$$\frac{d\sigma}{d\Omega} = \bar{B}^2|F(Q)|^2 \int_0^\infty dQ'\, \tilde{S}(Q')L(Q, Q'), \tag{8.14}$$

where

$$L(Q, Q') \equiv \frac{Q'}{Q^2} \frac{\theta(Q - Q')}{[1 - (Q'/Q)^2]^{1/2}} \int_\beta^{\pi-\beta} d\theta \sin\theta\, P(\theta)[(Q'/Q)^2 - \cos^2\theta]^{-1/2}, \tag{8.15}$$

and $\theta(x)$ is the step function, which is unity for $x > 1$ and zero for $x \leq 1$. This is the line shape formula deduced by Dutta et al.[16] In the form of Eq. (8.14), the result is easily interpretable. The observed cross section is the folding of the true cross section with a function that represents the "smearing" of $Q_\parallel$ values due to the orientational distribution of the crystallites. This "smearing function" $L(Q, Q')$ depends only on this distribution, and hence may be computed for all Q, Q' of interest once this is known. It can then be folded with any assumed form of $\tilde{S}(Q')$ as in Eq. (8.14) and used to fit experimental line shapes. In practice, one must first fold $L(Q, Q')$ with the instrumental diffractometer resolution function $R(Q^* - Q)$, which for neutron diffractometers is usually taken to be a Gaussian function,

$$\frac{1}{\sqrt{2\pi}\,\sigma_R} e^{-(Q^*-Q)^2/2\sigma_R^2}.$$

The line shape one obtains by smearing with the function $L(Q, Q')$ has an asymmetric or "sawtooth" shape characteristic of diffraction from distributions of two-dimensional crystallites. This is because of the step function $\theta(Q - Q')$ in Eq. (8.15). Figure 7 shows how such an asymmetric line shape arises, for the case where $\tilde{S}(Q_\parallel)$ is from a two-dimensional powder average of crystallites with true long-range order. In such a case, in the vicinity of a particular two-dimensional powder peak,

$$\tilde{S}(Q_\parallel) \propto \delta(Q_\parallel - \tau), \tag{8.16}$$

where τ is the magnitude of the two-dimensional reciprocal lattice vector for the peak in question. Thus, $S(\mathbf{Q})$ is nonzero only on cylinders of radius τ normal to each crystallite plane. A powder diffraction scan corresponds to a radial scan of $\mathbf{Q}$ out from the origin in the lab-fixed coordinate system.

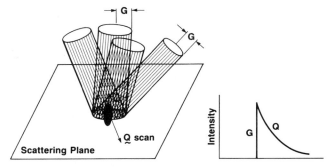

FIG. 7. Schematic showing how the asymmetric "sawtooth" line shape arises. The cylinders represent the locus of points in reciprocal space on which the two-dimensional Bragg scattering condition is satisfied for Grafoil-type substrates. The arrow radially outward from the origin represents the locus of a powder diffraction scan in reciprocal space.

From Fig. 7, it is obvious that the scan does not intersect the family of cylinders until $Q = \tau$, and thereafter one is continuously passing through a gradually decreasing density of such cylinder surfaces leading to the sawtooth line shape. In practice, finite resolution effects and finite crystallite coherence length effects smear out this further (see Fig. 5).

We now discuss the explicit forms for some of the above functions used in fitting experimental lineshapes. For monatomic systems such as helium, the form factor $F(\mathbf{Q}) = 1$ as is obvious from Eq. (8.4). For a diatomic molecule such as H_2 in a quantum $J = 0$ state, or one randomly oriented in three dimensions,

$$F(\mathbf{Q}) = \sin(Qr_0)/(Qr_0), \tag{8.17}$$

whereas for a diatomic molecule randomly oriented in the plane,

$$F(\mathbf{Q}) = J_0(Q_\parallel r_0), \tag{8.18}$$

where r_0 is the radius of the molecule. (In this case, the form factor must be included inside the integral of Eq. (8.14).) For a tetrahedral molecule (such as CD_4) with random orientations,

$$F(\mathbf{Q}) = \left[\bar{b}_C + 3\bar{b}_D \frac{\sin(Qr_0)}{(Qr_0)}\right] / [\bar{b}_C + 4\bar{b}_D], \tag{8.19}$$

where $\bar{b}_C$, $\bar{b}_D$ are the coherent scattering lengths of carbon and deuterium, respectively.

The distribution function $P(\theta)$ for the plane normals of the substrate crystallites was taken in the work of Kjems et al.[2] to be of the form

$$P(\theta) = P_0 + P_1 e^{-\theta^2/2\sigma_1^2}, \tag{8.20}$$

where the first term allows for a completely random component, and σ_1 is the standard deviation of the oriented distribution. However, although this form is commonly used it does not reproduce the large Q trailing edge of the diffraction peaks very well. Empirically, for both Grafoil and ZYX substrates, Dutta et al.[16] found that dividing the above form by (sin θ) gave a better fit to the line shapes. More recently, Dimon[17] found that a Lorentzian distribution, rather than the form given above, gives a satisfactory fit to the line shape. In principle, $P(\theta)$ may be obtained experimentally by carrying out a rocking curve with the substrate, although one should bear in mind that the bare substrate crystallites may not have quite the same orientational distribution as the adsorbed monolayers. The physical parameters of interest involved in $S(\mathbf{Q})$ that are extracted from line shape fitting are, however, dependent mainly on the *low Q* (leading) edge of the diffraction peaks and are fairly insensitive to the actual choice of $P(\theta)$.

For a two-dimensional crystallite of *finite* size, the theory of Warren[15] shows that the structure factor can be well approximated in the vicinity of a two-dimensional reciprocal lattice vector τ by

$$S(\mathbf{Q}) = \rho_0(L^2/a^2) \exp\{-|\mathbf{Q}_\parallel - \tau|^2 L^2/4\pi\}, \qquad (8.21)$$

where L is the size of the crystallite and a is the lattice constant. The factor (L^2/a^2) represents the number of unit cells in the crystallite, and must be changed appropriately if the lattice is not square [as assumed in Eq. (8.21)] but triangular, rectangular, etc.; ρ_0 is an amplitude factor related to the fraction of the sample that is in ordered crystallites. The amplitude of the Gaussian thus plays the role of a two-dimensional crystalline order parameter. Use of the form [Eq. (8.21)] for $S(\mathbf{Q})$ in Eqs. (8.7) and (8.14) for $P(\theta) = 1$ (completely random orientations) yields the original Warren formula. A modification of this result to approximately allow for a nonuniform $P(\theta)$ introduced by Kjems et al.[2] is not rigorously correct, but has been commonly used in analyzing neutron diffraction data. We shall refer to all line shapes based on Eq. (8.21), whether using the rigorous form of Eq. (8.14) or not, as "Warren line shapes," to be distinguished from the case where explicitly different forms of $S(\mathbf{Q})$ are used, such as Lorentzians, power laws, etc.

As the parameter $L \to \infty$ in Eq. (8.21), $S(\mathbf{Q})$ becomes a two-dimensional delta function. The theory of infinite two-dimensional crystals on smooth surfaces, to be discussed in Section 8.6, shows that this cannot be correct, as phonon fluctuations destroy this delta function and convert it to a power law singularity. Equation (8.21) is, however, valid in the case of the two-dimensional crystalline registered phases. At high temperatures, two-dimensional crystallites "melt" just as three-dimensional crystallites do, and Eq. (8.21) is not rigorously correct either. We shall discuss such studies of melting in Section 8.6. For the present, however, we shall discuss the use of the

Warren line shape as a way of obtaining qualitative (and semiquantitative) information about the nature of the two-dimensional phases studied. Thus, fitting to experimental diffraction peaks yields three parameters, from which the amount of crystalline order, the size of the ordered regions, and the lattice constant of the crystal may be determined for various temperatures and coverages. In the next section, we discuss how this has been used to obtain phase diagrams for a variety of physisorbed systems.

8.4. Studies of Phase Diagrams of Physisorbed Systems

Figure 8 illustrates schematically the underlying hexagonal network of carbon atoms on the basal plane of a graphite surface. In addition to the minimum in the potential along the Z axis just above the surface that corresponds to the adsorption energy, there will be lateral corrugations in the potential due to the adsorbate atom (molecule)–carbon atom interactions, although these are generally much smaller in magnitude (~ 50 K as opposed to several hundred degrees Kelvin). It is these corrugations in the lateral potential that can lead at low temperatures to the "locking in" or registry of the adsorbate lattice with the substrate. The competition between this interaction and the interaction between the adsorbate molecules themselves then leads to a rich variety of phases and commensurate/incommensurate (C/I) transitions. Quite often, the center of a carbon atom ring is a preferred adsorption site, but the adsorbed atoms or molecules have a mutual repulsion

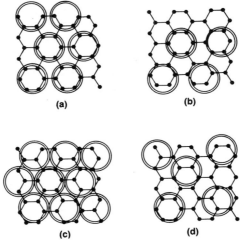

FIG. 8. Schematic showing the variety of phases of methane on the underlying hexagonal carbon rings of the graphite basal planes: (a) registered, (b) expanded, (c) compressed, and (d) fluid.

8. ADSORBED MONOLAYERS AND INTERCALATED COMPOUNDS

that makes it unfavorable for every such site to be occupied. A common situation is where the centers of next-nearest-neighbor carbon rings are occupied, leading to the registered phase shown in Fig. 8(a), known as the ($\sqrt{3} \times \sqrt{3}$) (R 30°) structure. The terminology arises because in this phase the adsorbate molecules form a triangular lattice with a lattice constant that is $\sqrt{3}$ times the lattice constant of graphite, while the a axis is rotated 30° from the graphite a axis. Figures 8(b) and 8(c) illustrate cases of incommensurate triangular lattice phases, where the lattice constant is greater than that of the ($\sqrt{3} \times \sqrt{3}$) phase (expanded incommensurate phase) or smaller (compressed incommensurate phase). Finally, Fig. 8(d) illustrates the case where the two-dimensional crystalline order is absent, and the molecules form a disordered structure on the substrate surface—an indication of a liquid or vapor-like phase. A special type of disorder (not illustrated here) is where the molecules are disordered but still confined to the adsorption sites. In this case we have a realization of a two-dimensional lattice liquid or lattice gas. Finally, at higher coverages, a bilayer begins to form and the diffraction line shapes change to reflect this. For cubic or rectangular symmetry substrates, as in the case of (100) or (110) faces of cubic metals, LEED studies have revealed registered adsorbed phases such as (2 × 2), (2 × 1), etc., but since very few neutron studies of such systems have been carried out, we shall restrict ourselves mainly to the case of hexagonal substrates.

Before reviewing the work done in this area, we shall illustrate in detail how two-dimensional phase diagrams are determined from neutron diffraction experiments with the particular case of CD_4 adsorbed on graphite,[8] which forms all the triangular ordered phases of the form illustrated in Fig. 8. The system CD_4/graphite was first studied by neutron diffraction by White and collaborators.[18] Their results are in good agreement with the later, more detailed studies described here. Figure 5 illustrates the (10) diffraction peak from a CD_4 monolayer on a Grafoil substrate. The asymmetric nature of the two-dimensional line shape as discussed in the previous section is apparent. The graphite substrate background has been subtracted. The smooth curve shows the result of fitting the "Warren line shape" discussed previously to this peak. The coverage is determined by the isotherm method discussion in Section 8.2. Figures 9–11 illustrate the behavior of the parameters for the CD_4 lattice constant, amplitude of the peak, and the coherence length L obtained by fitting to the (10) diffraction peak as a function of temperature for various coverages. The lattice constant shows the existence of a ($\sqrt{3} \times \sqrt{3}$) registered phase ($a = 4.26$ Å) at low temperatures up to a coverage that we denote as $n = 1.0$ (which corresponds quite well with the estimate of the completed registered monolayer coverage as determined by the isotherm method discussed in Section 8.2). The registered phase is replaced by an expanded incommensurate phase for $T \geq 50$ K. For $n > 1.0$, the lattice

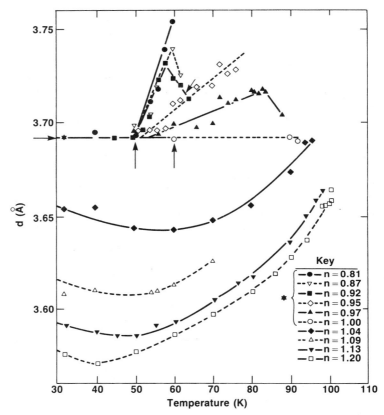

Fig. 9. Schematic showing d spacing of (10) reflection for CD_4/Grafoil as a function of temperature for various coverages; $n = 1$ corresponds to the completed registered monolayer. (From Vora et al.[8])

constant decreases at low temperatures indicating the presence of a compressed incommensurate phase, which with increasing temperature shows a minimum in the lattice constant before expanding toward the ($\sqrt{3} \times \sqrt{3}$) phase lattice constant just prior to melting. The peak amplitudes all decrease with increasing temperature, but at the lowest coverages, the temperature slope shows a break at the transition to the expanded phase and then another break at 60 K, identified as the melting temperature. At higher coverages ($n \geq 1.0$) there is a continuous decrease that becomes very rapid at the temperature at which the incommensurate (compressed) solid melts. The liquid phase is identified here in terms of a broad peak with a correspondingly small value of L.

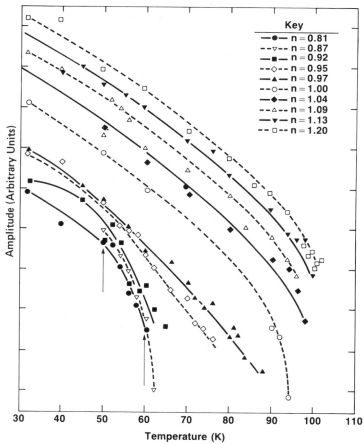

FIG. 10. Amplitude of (10) peak for CD_4/Grafoil as a function of temperature for various coverages. (From Vora et al.[8])

From a combined analysis of the data shown in Figs. 9–11, the phase diagram shown in Fig. 12 was deduced.[8] While the detailed arguments leading to this phase diagram will not be reproduced here, we shall discuss the consistency of this assignment with the results shown in Figs. 9–11. In Fig. 12 every point indicates a measured and fitted diffraction peak at the appropriate coverage and temperature. For coverages of $n \leq 0.93$, and $T \leq 50$ K, a registered ($\sqrt{3} \times \sqrt{3}$) solid phase was found that coexists with a low-density vapor phase (not visible in the neutron diffraction pattern). The pure low-density vapor phase must exist at very low coverages down to low temperatures, but this region was not investigated. At 50 K, the behavior of

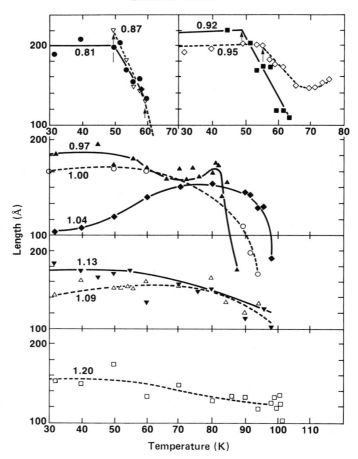

FIG. 11. Cluster size L in Warren line shape for (10) peak of CD_4/Grafoil as a function of temperature for various coverages. (From Vora et al.[8])

the lattice constant, amplitude factor A, and crystallite size L indicates a low-density vapor. Since this occurs on a vertical line in the coverage–temperature phase diagram, this was associated with the triple line for this phase transition where, according to the Gibbs phase rule, the registered solid (S_I), the expanded solid (S_{II}), and the vapor (V) are in equilibrium, and the transition is first order. The rapid decrease of both the amplitude and the size L with increasing temperature in this region points to decreasing islands of solid and increasing vapor content in accordance with the "lever rule" for coexisting phases. At 60 K, there is an abrupt decrease in both A and L and the peak is quite broadened, indicative of a fluid-like phase. Here, 60 K was

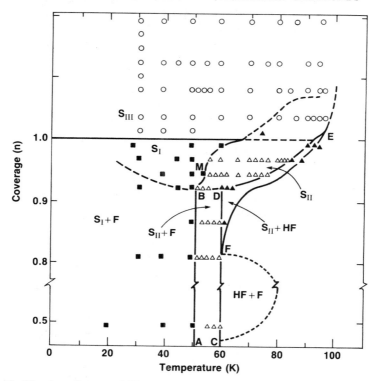

FIG. 12. The phase diagram of CD_4 on graphite, as determined by neutron diffraction. The points on the diagram represent measurements. The phases are denoted as S_I (registered solid), S_{II} (expanded solid), S_{III} (compressed solid), F (low-density fluid), and HF (high-density fluid). The dashed line represents a tentative line separating the compressed and expanded regions. (From Vora et al.[8])

accordingly identified with the triple line for first-order melting, where, according to the Gibbs phase rule, a solid (S_{II}), liquid, and vapor phase coexist. While neutron diffraction data cannot be used to investigate liquid–vapor coexistence, indirect information regarding a liquid–vapor coexistence region has been deduced for this system via quasi-elastic neutron scattering,[19] to be discussed in Section 8.8. For coverages $0.87 < n < 1.0$, the abrupt decrease of the size parameter L, which signals loss of the solid phase, is accompanied by an apparent contraction of the lattice constant back towards the registered value for higher temperatures. This has been interpreted as due to the coexistence of the expanded solid phase and a liquid-like phase (S_{II} + LQ) with an average interatomic spacing equal to that of the registered solid phase. These cannot be resolved in the experiment and so, as the liquidlike component increases, the effective peak position shifts back toward that of the registered phase, while the peak broadens overall (manifested as a decrease of L). It should be noted that a region of coexistence of a solid to a liquid phase at constant coverage generally implies a first-order melting transition, as in the case of solid–liquid coexistence at constant volume in three dimensions. At temperatures below this (S_{II} + LQ) phase, we have pure solid phases S_I and S_{II}. For $n \geq 1.0$, we have a pure compressed incommensurate solid (S_{III}) phase, which as stated above, appears not to have any phase transitions (the amplitude and L decreasing gradually and continuously with increasing temperature, while the lattice constant changes smoothly) until it melts at a temperature of ~ 100 K, at which point the peak broadens rapidly and L decreases. This phase diagram is in general supported by later thermodynamic,[20,21] NMR,[22] and neutron[23,24] measurements, although the registered–expanded solid phase transition line has more recently been determined to be 48 K and the melting temperature of the low coverage, solid phase to be 56 K. The general features of the original vapor pressure isotherms of Thomy and Duval[13] (see Fig. 2) for CH_4 on graphite can now be interpreted in terms of this phase diagram, e.g., the small steps can be identified with the solid–liquid (S_{II} + LQ) coexistence regions, while the vertical regions at low temperature and vapor pressure can be identified with the liquid–vapor (LQ + V) coexistence regions. We thus see how diffraction experiments can provide a great deal of detail regarding the phase diagram and the nature of the two-dimensional phases.

The "crystallite size" parameter L used in the Warren line shape is only qualitative, except in the registered phase. For describing the incommensurate solid structure factors and the melting process in more detail and in a quantitative fashion, fits involving a more complicated structure factor are necessary. These have been carried out for the case of CD_4 on graphite and the results are discussed in Section 8.6. We note for the moment, however, that the maximum L value obtained from diffraction experiments on Grafoil was only ~ 200 Å. This implies that the substrate is such that coherent regions for adsorbed monolayer crystallites cannot exceed this size. This is borne out by measurement of the intrinsic widths of the broad plane reflections from the empty Grafoil sample (after unfolding the instrumental resolution), which indicate the lateral size of the graphite crystallites themselves to be ~ 400 Å. ZYX graphite substrates have much larger surface coherence lengths (~ 5000 Å), larger than can be detected with typical neutron diffractometer resolutions, but which can be measured by high-resolution x-ray diffraction measurements at synchrotron sources.[9]

We now proceed to a general survey of the structural information obtained by studies of this sort on physisorbed systems. N_2 on Grafoil was, as mentioned in the Introduction, historically the first physisorbed monolayer system to be studied by neutron diffraction.[2] This was also one of the first

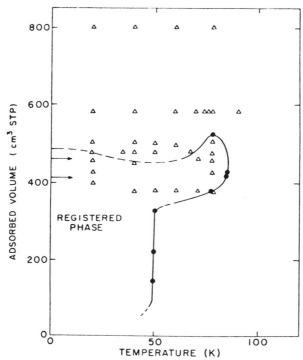

FIG. 13. Phase diagram of N_2 on graphite, as determined by neutron diffraction. (From Kjems et al.[2])

investigations of the structure of physisorbed monolayers, since it preceded the application of LEED techniques to such systems and also preceded x-ray diffraction studies of monolayers. Figure 13 shows the phase diagram deduced from these experiments. Below the critical coverage of $n = 1.0$, only a registered ($\sqrt{3} \times \sqrt{3}$) solid phase exists, which appears to coexist with a liquid phase at higher temperatures, implying a first-order solid-lattice liquid transition. For $n > 1.0$, there appears to be a coexistence (at low temperatures) of the registered phase and a dense incommensurate phase, which persists until $n = 1.09$ after which the pure dense phase exists. At higher temperatures, the dense phase appears to melt continuously (as in the case of CD_4/graphite) without going through a coexistence region. Kjems et al.[2] deduced from their line shapes that the dense phase consisted of more than a single monolayer. The "bilayer" deduced from their line shape fits (for details the reader is referred to the original paper) consisted of only a fractional statistical second layer that became uncorrelated with the first layer at higher temperatures. The authors assert that it is likely that the dense phase, however, does not exist as a pure monolayer phase at all. In a later study of N_2 on Grafoil by the Brookhaven group,[25] it was found that for temperatures below 30 K a weak additional peak indexable as the (21) reflection from the triangular lattice appeared, indicative of a doubling of the unit cell along both axes. [The (21) reflection has zero structure factor in the ($\sqrt{3} \times \sqrt{3}$) orientationally random phase.] This is shown in Fig. 14, and the temperature dependence of the (21) intensity is shown in Fig. 15. the same effect occurred in the dense phase as well, at almost the same temperature. This is an indication of orientational ordering of the molecules below 30 K due to their strong quadrupolar interactions, as was predicted theoretically.[26-28] Due to intensity limitations, however, the authors were unable to distinguish between the two-sublattice (2 × 1) herringbone structure and the four-sublattice (2 × 2) structure, both proposed theoretically. More recently Diehl et al.[29] have used LEED from N_2 on a single-crystal graphite surface to show that the orientionally ordered phase is indeed the (2 × 1) structure. Recent heat capacity measurements of Chan et al.[30] have confirmed all of these features of the phase diagram (see Fig. 16) and confirmed that both the melting transition from the ($\sqrt{3} \times \sqrt{3}$) phase and also the orientational order–disorder transition are first order. The general features of the phase diagram (in the case where only commensurate phases are involved) have been studied theoretically by Berker and Ostlund[31,32] using real space renormalization group techniques. The ($\sqrt{3} \times \sqrt{3}$) order–disorder transition is modeled via a three-state Potts model, because of the three possible sublattices involved in forming the registered phase on a graphite based plane. These methods do not work for incommensurate phases. This theory does, however, give a good

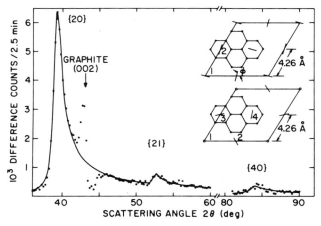

FIG. 14. Diffraction pattern from N_2 on Grafoil at 8 K in the ($\sqrt{3} \times \sqrt{3}$) phase. The (21) peak is the superlattice reflection arising from orientational ordering. (From Eckert et al.[25])

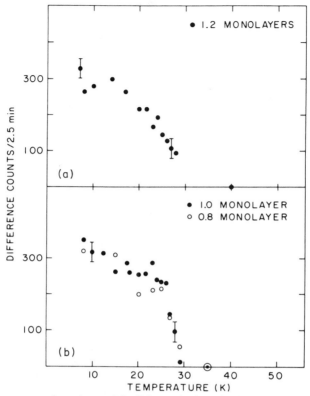

FIG. 15. Temperature dependence of the (21) superlattice reflection from the (a) dense-solid and (b) registered phases of N_2 on Grafoil. (From Eckert et al.[25])

quantitative accounting of the phase boundaries involving commensurate phases (e.g., in the case of He, N_2, or Kr on graphite).

The system O_2 on graphite is an interesting case of a more structurally complex system than most of the systems studied by neutron diffraction and is a case where early neutron diffraction results have been refined by higher resolution x-ray diffraction studies, and x-ray diffraction studies in turn have been refined by LEED studies. The first diffraction experiments on the system O_2/graphite were neutron diffraction experiments carried out on Grafoil substrates by McTague and Nielsen.[5] For $n < 1.6$ (where $n = 1.0$ is defined as the coverage for a completed ($\sqrt{3} \times \sqrt{3}$) registered monolayer), a single diffraction peak at $Q = 2.15$ Å^{-1} was observed, and they labeled this region the δ phase. For $n > 1.6$, two phases were observed. For $T > 11.9$ K, a single peak at 2.20 Å^{-1} was seen, while for $T < 11.9$ K, this peak split into two peaks centered at 2.17 Å^{-1} and 2.30 Å^{-1}. At low temperatures, an

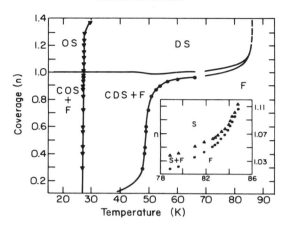

FIG. 16. Heat capacity and vapor pressure results for phase diagram of N_2 on graphite. COS, CDS, OS, DS, and F represent, respectively, commensurate (orientationally) ordered solid, commensurate (orientationally) disordered solid, orientationally ordered solid, orientationally disordered solid, and fluid. Solid lines represent first-order boundaries. Dashed lines are inferred to be continuous. The solid–fluid coexistence region is shown more clearly in the inset. (From Chan et al.[30])

additional peak at $Q = 1.14$ Å^{-1} was seen and ascribed to antiferromagnetic ordering of the two-dimensional O_2 lattice. The low-temperature phase (α-phase) was believed to have the same crystal and magnetic structure as the densest planes of bulk α-O_2. The high-temperature phase (β-phase) was inferred to have the same triangular structure as the densest planes of bulk β-O_2. The structure of the low-density δ phase was not solved by this experiment, although it was noticed that the single structural peak observed at 2.2 Å^{-1} was quite broad. Short-range antiferromagnetic correlations were also observed in this phase.

Subsequently, high-resolution x-ray diffraction experiments were carried out by the MIT group,[33] and extra peaks were found. In the δ phase, a structural peak was found at 1.58 Å^{-1}, and the broad peak at larger Q was found to be a double peak whose splitting was temperature dependent. It was concluded that the δ phase was a centered parallelogram structure, slightly distorted from the centered rectangular structure. For higher coverages, the x-ray data revealed that the two-peak structure seen by neutron diffraction was really a three-peak structure, with Q values of 2.14 Å^{-1}, 2.21 Å^{-1}, and 2.26 Å^{-1} at 15 K for $n = 2.7$, the position of the peaks being slightly temperature dependent. Thus, Heiney et al.[33] relabeled the β phase of McTague and Nielsen as the ζ phase. For $T < 11.9$ K, they find basic agreement with the α phase structure of McTague and Nielsen, although they label this the ε phase. The x-ray measurements did not establish a definitive assignment to the structure of the ζ phase, owing to intensity problems, overlapping peaks, and contamination with diffraction peaks of bulk α-O_2. The authors concluded, however, that it was not a simple triangular lattice, with two possible explanations of the structure: (1) triangular lattices with different lattice constants for successive layers and (2) modulation of the monolayer triangular lattice by interaction with the substrate or for other reasons. At higher temperatures only one peak remains (the η phase), which could be due to a vanishing of the modulation or melting of the second layer. Regions of coexistence between these phases and also with liquid-like phases were found, and it was concluded that the melting process was always first order.

More recently, however, LEED experiments on this system have been carried out[34] employing a single crystal substrate and have shown that the ζ phase really consists of two regions, the ζ_1 phase (12 K $< T <$ 18 K) and the ζ_2 phase (18 K $< T <$ 38 K). The first is a slightly distorted triangular lattice that may arise from molecular tilt considerations or coupling of the lattice to the magnetic long-range order (perhaps seen in the neutron diffraction data as short-range order owing to resolution problems with overlapping magnetic peaks). The ζ_2 phase is a pure triangular lattice. These authors ascribe the extra peaks seen in the x-ray diffraction data as due to contamination by bulk phases and a slightly different ζ' phase that exists on imperfect substrates. Thus, the O_2/graphite system is an example of a system sufficiently complicated that more refined results are obtained as one goes from relatively low-resolution, low-intensity measurements on polycrystalline substrates to high-resolution measurements, ultimately on single crystal substrates. It is also the only case where magnetic ordering in a monolayer has been seen directly via neutron diffration. The magnetic superlattice peak and intensity are consistent with a magnetic doubling of the unit cell along the same axis as in bulk α-O_2, with the moments confined to the plane.

The other diatomic molecule studied as an adsorbed monolayer on graphite is H_2.[35] One might think that the large incoherent cross section of hydrogen for neutrons and the small coherent scattering length would make such a system very hard to study by neutron diffraction. This turns out not to be the case for the following reason. The ground state of the H_2 molecule is the $J = 0$ state, where the antisymmetry of the proton wave functions requires the proton spins to be aligned antiparallel to each other. This causes the neutron scattering from one H_2 molecule to be perfectly coherent with that from another, provided the neutron cannot cause a spin-flip transition, which would involve having enough energy to excite the transition to the $J = 1$ state with an energy of 14.6 meV. This can be easily avoided by using cold neutrons. Neutron diffraction studies of D_2 monolayers has also been carried out.[35] The results show that for both H_2 and D_2 at low coverages and at low temperatures, the ($\sqrt{3} \times \sqrt{3}$) registered solid phase exists, as witnessed by the position of the (10) reflection from the triangular lattice. With increasing coverage, this monolayer squeezes in to form a dense incommensurate triangular lattice.

We turn now to studies of rare gas monolayers on graphite. Both isotopes of helium have been studied by heat capacity measurements and neutron diffraction. ^{4}He and ^{3}He and their mixtures adsorbed on graphite have been studied by heat capacity in the high-temperature, two-dimensional gas phase, where both quantum (band-structure) effects due to the lateral substrate potential and interaction effects have been identified,[36] and in the case of ^{4}He, Bretz[37] made one of the first careful thermodynamic studies of the order–disorder transition from the ($\sqrt{3} \times \sqrt{3}$) R 30° registered phase that exists at low temperatures. This appears to be a continuous second-order phase transition for which the critical exponent $\alpha = 0.36 \pm 0.02$ was obtained from the heat capacity measurements. As mentioned above, this commensurate order–disorder transition can be modeled by the three-state Potts model, for which renormalization group calculations[38] yield $\alpha = \frac{1}{3}$. For ^{4}He on graphite, for coverages up to a monolayer, neutron diffraction measurements by the Brookhaven group[39] and by Lauter[40] have confirmed the ($\sqrt{3} \times \sqrt{3}$) structure at low temperatures. The commensurate solid and the lattice liquid and gas are the only submonolayer phases of ^{4}He found for submonolayer coverages. Measurements of the second and higher layers are described in Section 8.9. It should be mentioned that heat capacity measurements have also been carried out for ^{4}He adsorbed on graphite on which a complete monolayer of krypton had already been adsorbed (krypton-plated graphite).[41] A peak in the heat capacity is assigned to an order–disorder transition from a (1 × 1) registered phase on the krypton two-dimensional lattice (every available site on this lattice occupied). This transition accordingly can be modeled by a two-dimensional Ising model (every site occupied or empty) for which theory gives $\alpha = 0$. This was also the value found experimentally, although no neutron diffraction measurements have yet been carried out on this system. ^{3}He on graphite exists as a dense incommensurate triangular lattice phase at low temperatures, as found by Lauter et al.[40] who studied coverages ranging from $n = 0.856$ to $n = 1.4$. The reciprocal lattice vector increases linearly with the square root of the coverage up to the dense monolayer completion, showing that all the ^{3}He atoms are homogeneously adsorbed into this phase.

Neon adsorbed on graphite has been studied by neutron diffraction by Wiechert et al.[42] Earlier heat capacity measurement by Huff and Dash[43] had established a phase diagram for this system with solid, liquid, vapor, solid and vapor, solid and liquid, and liquid and vapor coexistence regions, as for a classical van der Waals system. The neutron diffraction data confirmed this phase diagram and established that for most of the phase diagram, neon forms an incommensurate triangular lattice on graphite, while at low temperatures and up to a coverage of 0.94 (of the full dense monolayer), a commensurate ($\sqrt{7} \times \sqrt{7}$) R 19° solid phase was found. This is one of the few cases of a higher order commensurate structure found by neutron diffraction from an adsorbed monolayer on graphite. The ($\sqrt{7} \times \sqrt{7}$) R 19° structure contains four atoms in the unit cell and is formed by putting a neon atom at the center of a carbon ring on the surface and having the other three on saddle points between carbon atoms. Only the centers of fourth-nearest-neighbor carbon rings are occupied (see Fig. 20).

Argon on graphite also forms only a (dense) incommensurate solid monolayer phase on graphite from which it appears to melt continuously at higher temperatures.[44] Krypton and xenon on graphite (as well as argon) have been studied by x-ray diffraction.[9] In the submonolayer region krypton forms the ($\sqrt{3} \times \sqrt{3}$) R 30° solid phase from which it melts into a lattice liquid, while xenon always forms an (expanded) incommensurate solid phase, from which it melts into a fluid phase.

For absorbate atoms or molecules that form a close-packed structure in the three-dimensional solid phase (e.g., fcc crystals such as Ar, Kr, Ne, and Xe or hcp crystals such as ^{4}He, H_2, or D_2), the two-dimensional solid phase on graphite depends on how "mismatched" the nearest neighbor distance of the molecules in the densest three-dimensional solid planes (or of the equilibrium nearest neighbor distance given by the van der Waals potential) is with respect to the intermolecular spacing for a registered phase (such as the $\sqrt{3} \times \sqrt{3}$ R 30°) on graphite, and the attractive interaction between the adsorbate molecules and the graphite carbon atoms, compared to the forces between the adsorbate molecules. Figure 17 shows the trends for many of the systems studied so far. Note that molecules that have an equilibrium nearest-neighbor distance too large or too small compared with the registered value do not register except in the case of the lighter molecules where zero point motion expands the two-dimensional lattice naturally to the point where registry becomes possible.

Apart from methane, which has already been discussed, there have been several structural studies of adsorbed hydrocarbons on graphite. These include butane,[45] ethane,[46,47] benzene,[48,49] toluene,[48] and ethylene.[50] All of these studies have been carried out with deuterated compounds to increase the coherent Bragg scattering and decrease the incoherent background. These are more complex molecules than the ones discussed so far, and the exact manner in which they sit on the surface therefore requires more crystallographic data than was needed for the simpler systems. Quite often, as we shall see, information from LEED, Mössbauer, NMR, and particularly total energy calculations, is required to obtain the structure. This is because the

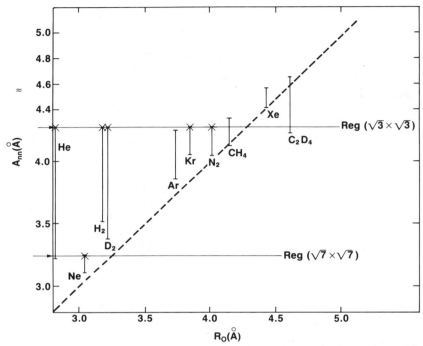

FIG. 17. Nearest neighbor distances for several simple adsorbate molecules as observed for monolayer and submonolayer coverages on graphite, plotted against nearest neighbor equilibrium distances obtained from their van der Waals radii. Crosses indicate the existence of a registered phase.

weakness and intrinsic width of the two-dimensional powder peaks make it quite difficult to uniquely determine the structure factors for the individual reflections. In practice, something analogous to profile refinement analysis (as done for bulk powder patterns) may often be carried out.

One of the first cases where the orientational configuration of molecules on a graphite surface was deduced from two-dimensional diffraction data is that of NO adsorbed on Papyex.[7] The authors were able to see several two-dimensional diffraction peaks at low temperatures. These were analyzed in terms of solid phases made up of lattices of N_2O_2 square molecular complexes. At low coverages and temperatures, these form a phase where the N_2O_2 units lie flat on the surface and form an oblique unit cell, labeled as the γ phase, while at higher coverges a different solid phase (β phase) was found with the N_2O_2 dimers standing on their sides in a rectangular unit cell. The cell dimensions and the angles of tilt of the N_2O_2 dimers relative to the $\vec{a}$ axis of the unit cell were determined in each case. At higher temperatures,

8. ADSORBED MONOLAYERS AND INTERCALATED COMPOUNDS

first-order melting was found via observation of coexiting solid and liquid phases for finite ranges of temperature at constant coverage.

In the case of butane (C_4D_{10}), neutron inelastic scattering studies of the molecular vibrational spectrum actually preceded structural studies and suggested that the carbon skeletal plane of the molecule was oriented parallel to the basal substrate plane. Figure 18 shows the diffraction pattern (in terms of difference counts) for C_4D_{10} on Grafoil together with a diffraction profile calculated from an assumed model, which best fit the data as a superposition of Warren line shapes.[45] The model puts the molecules in a parallelogram-shaped unit cell with orientational Euler angles and lattice constants as depicted in Fig. 19. (The angle β describing the tilt of the long axis away from the surface was found to be zero for the best fits.) The upper graph in Fig. 18 shows peaks at $Q > 5$ Å^{-1} not observed in the data, but the lower graph shows that this may be explained in terms of quite a reasonable Debye–Waller factor for the Bragg reflections. The lattice spacing suggested registry along only *one* direction with the graphite substrate. The general features of this structure (except this "quasi-registered" spacing) could be quantitatively explained by calculations involving only intermolecular interactions obtained by a superposition of empirical atom–atom potentials. A model including the interaction with the substrate, however, failed to account satisfactorily for the lattice spacing in the "registered" direction.

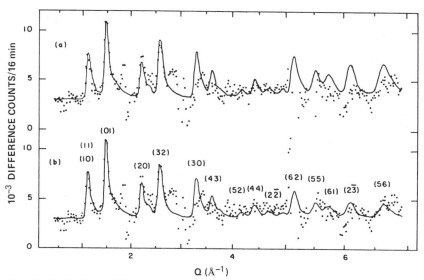

FIG. 18. Profile fits to diffraction pattern of 0.8 layer C_4D_{10} on Grafoil at 9 K ($\lambda = 1.19$ Å). The calculated profile (a) is based on the structure indicated in Fig. 19. The profile (b) is obtained on adding a Debye–Waller factor ($\langle u^2 \rangle = 0.02$ Å^2). (From Trott *et al.*[45])

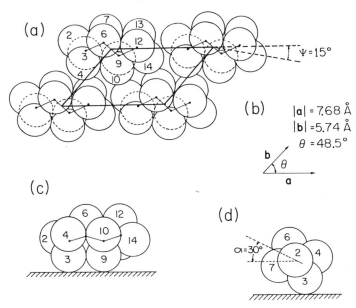

FIG. 19. Unit cell and molecular orientations of a butane submonolayer film deduced from neutron diffraction data. Circles depict approximate van der Waals radii of deuterium atoms. (From Trott et al.[45])

The diffraction pattern of deuterobenzene (C_6D_6) on Grafoil has been studied[48,49] and found to exhibit at submonolayer coverages and low temperatures three diffraction peaks, identified with a ($\sqrt{7} \times \sqrt{7}$) triangular lattice solid phase illustrated in Fig. 20, with the molecules lying flat on the surface. The diffraction data were not able to reveal that the orientation of the benzene molecule was parallel to the surface but not the orientation *on* the surface, which was determined from energy minimization calculations. d_8-Toluene[48,49] on Grafoil was also studied and found to progress through a commensurate (3 × 3) solid phase at low temperatures to an incommensurate solid before it melted. Both neutron diffraction[50] and x-ray diffraction[51] have been used to study ethylene on graphite and the structures thus obtained agree very well. From the neutron data, one finds that C_2D_4 exhibits three distinct solid phases. A low-density (LD) phase exists for $n \leq 0.83$, which possesses a triangular lattice incommensurate with the substrate, with a nearest neighbor distance of 4.65 Å. There is also a high-density (HD) phase that coexists with the LD phase at coverages above $n = 0.83$ and with the bulk solid at still higher coverages. This phase consists of a triangular lattice compressed about 9% relative to the LD phase. Because of the limited number of peaks studied, the detailed orientational structure

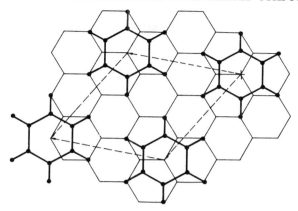

FIG. 20. A ($\sqrt{7} \times \sqrt{7}$) structure of C_6D_6 on basal planes of Graphite. The fact that the orientation of the molecules is parallel to the surface is deduced from the ratios of the (10) and (20) diffraction peak intensities. (From Monkenbusch and Stockmayer.[49])

of the molecules in these phases could not be determined from the diffraction data alone, but once again total energy calculations using semi-empirical atom–atom potentials indicated that the LD phase had the C=C bond parallel to surface and the plane of the molecule tilted 35° from the substrate surface plane. The model also predicts a herringbone type of orientational order, as in N_2/graphite, but the absence of the corresponding superlattice reflections indicates orientational disorder. As in the case of CD_4/graphite, this disagreement with predictions of classical total energy calculations is probably due to quantum rotational effects. From the calculations the HD phase arises from the ethylene molecules packing in and standing on end with the C=C bonds perpendicular to the surface. At a temperature of about 75 K, the HD lattice appears to suddenly expand in a first-order transition to a new (intermediate density, ID) phase, with the transition exhibiting hysteresis. The investigations did not reveal the nature of the ID phase, but indicated that the HD–ID transition appears to be correlated into the wetting transition (see Section 8.7), i.e., where the (bulk solid + monolayer) phase suddenly gives way to a bilayer phase. Deuteroethane (C_2D_6) also exhibits a large number of Bragg peaks in the powder diffraction pattern.[46,47] The measurements were carried out on a Papyex substrate that is very similar to Grafoil, discussed in Section 8.2. The results showed that for low temperatures, three different solid phases exist depending on coverage, with an S_1 phase existing from 0.4 to 0.8 monolayer, an S_2 phase that coexists with the S_1 phase for coverges up to about 0.95 monolayer, a narrow region of coverage of the pure S_2 phase, and then, for higher coverages, an S_3 phase that coexists with the S_2 phase. The structure of these phases has been investigated by the kind of profile fitting technique that was discussed in connection with butane. The S_1 phase was deduced to have one molecule per unit cell, and the orientation of the molecule relative to the graphite substrate was deduced from the observed intensities. The S_2 phase was assigned a unit cell containing two molecules. The only information obtained on the S_3 phase by neutron diffraction was that it appears to have a diffraction peak that can be indexed as the (10) peak of a $(\sqrt{3} \times \sqrt{3})$ structure.

More recently, a combination of the neutron diffraction results with the results of LEED experiments on ethane adsorbed on graphite have shown[52] that the above interpretations were incorrect, and that the correct structure for the S_1 phase is a herringbone structure with two molecules per unit cell, similar to that observed in N_2/graphite, the unit cell being a commensurate $(4 \times \sqrt{3})$ rectangular unit cell. The phase S_2 was found to be a $(10 \times 2\sqrt{3})$ rectangular structure with 12 molecules per unit cell. At higher temperatures (above 60 K) the phases S_1 and S_2 are replaced by intermediate phases (called I_1 and I_2) before the solid melts into a disordered phase. The structures of these phases have not been determined in detail, but I_1 appears to consist of an orientationally disordered 2×2 triangular lattice structure with only short-range order. Ethane/graphite thus provides a nice example of a complex orientationally ordered two-dimensional solid structure, the "crystallography" of which was carried out using a combination of neutron and LEED techniques rather than either one alone.

A combination of neutron diffraction, Mössbauer, and computer simulation techniques has been used to study the structure of submonolayer $Fe(CO)_5$ on graphite.[53] At 0.9 monolayer and low temperatures, an orientationally ordered phase is observed, indexable as a commensurate $(\sqrt{7} \times \sqrt{21})$ structure made up of two interpenetrating $(\sqrt{7} \times \sqrt{7})$ sublattices, with a herringbone arrangement of molecular axes. At higher temperatures the orientational order disappears, and the diffraction peaks are consistent with a nearly $(\sqrt{7} \times \sqrt{7})$ phase with the molecules freely rotating about an axis normal to the surface, which is consistent with the computer simulation. The Mössbauer data indicate an increase in molecular mobility at temperatures considerably below the orientational order–disorder transition.

CF_4 on graphite was first studied by Lauter et al.[54] using neutron diffraction. It provided the first case of a higher order commensurate phase on graphite other than the $(\sqrt{3} \times \sqrt{3})$ phase, namely, the (2×2) phase that is formed, for instance, by occupying the centers of third-nearest-neighbor carbon rings. Thus, the system can choose to order on one of four different sublattices, and consequently this system may provide a realization of the four-state Potts model, interesting to theorists because it is the marginal Potts model for critical behavior. (For higher state Potts models, the transitions are proven theoretically to be first order.) The neutron diffraction results up to 0.7 monolayer coverage and up to about 70 K showed the (10) and (11) peaks from the (2 × 2) triangular lattice. For higher coverages, the peaks became incommensurate and showed tails that the authors initially interpreted as satellites resulting from the modulation of the incommensurate monolayer structure by the underlying substrate. The authors also deduced a tentative phase diagram from their data. CF_4 has been subsequently investigated more thoroughly by both x rays[55,56] and further neutron diffraction measurements.[57] The later neutron data[57] indicated a rich phase diagram, with the existence of an incommensurate phase as well as the (2 × 2) commensurate phase and, at lower temperatures, a more complex solid phase characterized by three diffraction peaks centered at 1.465 $Å^{-1}$, 1.493 $Å^{-1}$, and 1.522 $Å^{-1}$, respectively. A complex domain wall model was invoked to explain the structure of this phase, but this must be for the moment regarded as tentative only. This "three-peak structure" is also seen with x rays.[55,56] A region of coexistence of a commensurate (2 × 2) peak and an incommensurate peak for coverages above that required to complete the (2 × 2) monolayer has been interpreted[55] as evidence for a striped phase (i.e., with parallel domain walls separating commensurate and incommensurate regions), but this does not seem consistent with more recent x-ray studies,[56] which do not show a fixed ratio for the intensities of these peaks as required by the striped domain model. All of these more recent data were carried out on ZYX graphite.

We may conclude this brief review of two-dimensional monolayer phase diagrams by noting that neutron diffraction has provided a great deal of information for the first time regarding the nature of the various phases. With the development of low-intensity LEED techniques and x-ray diffraction techniques to study physisorbed monolayer structures on single-crystal surfaces, neutron diffraction is not likely to be competitive in the future as an investigative tool for the majority of surface structures. However, in special cases such as that of very light atoms or molecules (^{3}He, ^{4}He, H_2, D_2, and many hydrocarbons) on surfaces of large area, neutrons have been and will continue to be extremely useful even for structural investigations. In the simplest cases, involving spherical molecules and commensurate or incommensurate triangular lattices, the complete phase diagram can be obtained uniquely and directly via neutron diffraction techniques alone, although this has been done only in a few cases. For nonspherical molecules, where orientationally ordered phases are involved, often a combination of techniques, e.g., neutron diffraction, LEED, Mössbauer, heat capacity, and computer calculations based on interatomic force models, have been found to be useful to obtain complete details of the phase diagram. We shall next discuss studies of the transitions between these phases such as the commensurate–incommensurate transition, the melting transition, and also studies of multilayer films.

8.5. Studies of Commensurate–Incommensurate Transitions

The commensurate–incommensurate (C–I) transition is caused by the competition between the interactions between the adsorbate molecules themselves and the interactions between the molecules and the substrate. If the latter is sufficiently strong, it will "lock" the interatomic spacing of the monolayer to that corresponding to, for example, the ($\sqrt{3} \times \sqrt{3}$) structure over some range of coverage and temperature, regardless of the spacing favored by the intermolecular forces. This will be true until the completion of the registered monolayer. All that will happen is that the "islands" of registered solid will grow at the expense of regions of vapor phase. If, however, the coverage is increased further, the tendency of the adsorbate molecules to still go into the first layer may cause the lattice structure of the monolayer to be compressed to accommodate the extra molecules. For a triangular lattice, the area density σ of molecules inside the two-dimensional solid islands is $\sim 1/a^2$, where a is the lattice constant, so we would expect σ to be constant with coverage n until the completion of a registered monolayer, and after that

$$\sigma = \sigma_R + \alpha(n - n_R) = \alpha n \tag{8.22}$$

where σ_R and n_R denote the area density and coverage for a completed registered monolayer, respectively. The constant α depends only on the geometry of the two-dimensional lattice and is equal to σ_R/n_R.

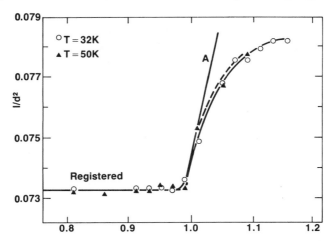

FIG. 21. Plot of $1/d^2$ versus (registered) monolayer fraction n (for CD_4/Grafoil). The straight line A represents the prediction of Eq. (8.22), if all the molecules continued to go onto the first monolayer after completion of the registered monolayer. (From Vora et al.[8])

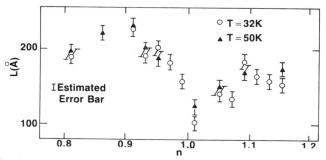

FIG. 22. Plot of the (Warren) coherence length L versus monolayer fraction n through the C-I transition for CD_4/Grafoil. (From Vora et al.[8])

Figure 21 shows the behavior of $1/d^2$ for CD_4 on Grafoil as a function of coverage as determined by Vora et al.[8] (d is the plane spacing for the (10) reflection). The behavior is as expected, except that the curve for higher n deviates from the straight line predicted by Eq. (8.22). This indicates that *all* the added molecules are not going into compressing the first layer, but probably starting the bilayer. Eventually, the lattice does not compress further, implying completion of the dense monolayer phase. Figure 22 shows the behavior of the size parameter L in the Warren line shape fits to the (10) peak as it goes through the C-I transition. There is a sudden dip in L at the transition, because the peak broadens considerably. This is illustrated in Fig. 23, which shows more recent data taken on CD_4 adsorbed on ZYX graphite through the C-I transition at 5 K.[58] The smooth curves are a fit representing a superposition of Warren line shape peaks at both the commensurate and incommensurate positions, with intensity shifting over from one to the other as coverage increases. The conclusion was that there was a coexistence region, and thus that the transition was first order. However, it should be borne in mind that this fit is not unique, and the line broadening can possibly be explained by other mechanisms to be discussed. Figure 24 illustrates the corresponding behavior for N_2 on Grafoil through the C-I transition at 20 K.[2] The behavior is very similar. For D_2, H_2 on Grafoil, Fig. 25 illustrates behavior similar to that in Fig. 21, except that during the compression of the monolayer, the increase in the reciprocal lattice vector τ is much more linear with $n^{1/2}$, and the break corresponding to the completion of the dense monolayer much sharper. Thus the molecules are going onto the first monolayer almost completely until the bilayer formation begins. Further increase of τ with coverage (seen for D_2 but not for H_2) indicates the compression of the first layer by the second. Similar behavior has been reported for neon on Grafoil for the $(\sqrt{7} \times \sqrt{7})$ R 19° C-I transition[42] and for ^{4}He on graphite.[39,40]

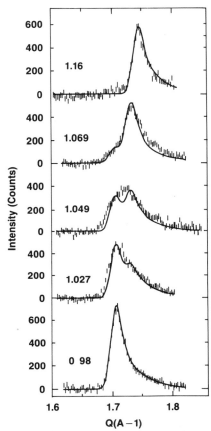

FIG. 23. Neutron diffraction measurement of the (10) peak from CD_4 adsorbed on UCAR-ZXY graphite through the C–I transition. The solid lines are Warren line shapes calculated from the superposition of a commensurate and an incommensurate peak. [From M. Nielsen, J. Als-Nielsen, and J. P. McTague, in "Ordering in Two Dimensions" (S. K. Sinha, ed.), p. 139. North-Holland Publ., New York, 1980.]

An interesting case is provided by CD_4/graphite, which is one of the few systems to break away from the $(\sqrt{3} \times \sqrt{3})$ registered phase into an expanded incommensurate phase at constant coverage before melting (see Fig. 12). The C–I transition does not appear to be accompanied by a sudden broadening of the peak and appears to be continuous. Thus, this case appears to be a realization of the "floating solid" phase discussed by Halperin and Nelson[59] in a renormalization group calculation of the phase diagram of an incommensurate solid on a substrate. According to Halperin and Nelson, a $(\sqrt{3} \times \sqrt{3})$ registering substrate potential should be "relevant" at all

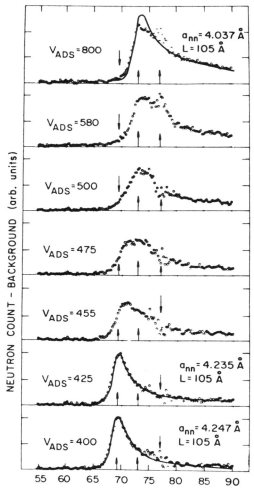

FIG. 24. Diffraction profiles from N_2 on Grafoil at 20 K at various coverages. The full lines indicate Warren lineshape fits. The arrows indicate the peak positions at the two extremes of coverage and the position of the 002 reflection from the graphite substrate. (From Kjems et al.[2])

temperatures, and thus one would always expect to find a range of coverage between the "expanded" and "compressed" solid phases (see Fig. 12) with the periodicity locked in at the registered value. However, measurements by Vora et al.[8] at 70 K (see Fig. 26) show an apparently continuous decrease in lattice spacing with increasing coverage right through the registered value. Thus, the registered phase is very narrow or else the expanded and compressed incommensurate solids are really a single phase.

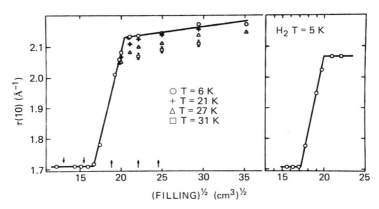

FIG. 25. The magnitude of the reciprocal lattice vector of the (10) reflection from monolayers of D_2 and H_2 plotted against $n^{1/2}$. (From Nielsen and Ellenson.[35])

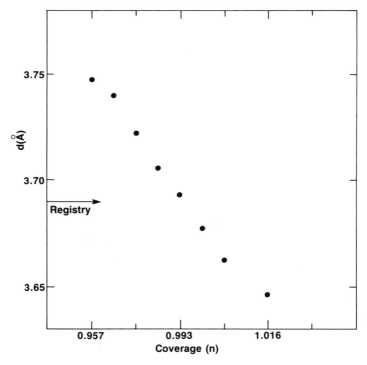

FIG. 26. The (10) peak plane spacing versus monolayer coverage fraction at 70 K through the expanded to compressed solid transition. No break in slope is evident at registry. (From Vora et al.[8] and Sinha et al.[58])

8. ADSORBED MONOLAYERS AND INTERCALATED COMPOUNDS 41

There has been a great deal of theoretical and experimental work done on the C-I transition in adsorbed monolayers over the last few years. Since experimental studies of the detailed nature of the transition require, in general, high intensities, high resolution, and often single-crystal surfaces, x-ray diffraction and LEED experiments have played the dominant role in these later studies, rather than neutrons, so we discuss these very briefly here. For the interested reader, several recent reviews are available.[60] The general theory (for $T = 0$ and one dimension) was put forward by Frank and Van der Merwe over 35 years ago.[61] This theory has been generalized for the case of two dimensions and finite temperatures by Bak et al.,[62] Pokrovskii and Talapov,[63] Shiba,[64] and others. The basic idea is that in the vicinity of the C-I transition domain walls or solitons appear, which contain excess atoms (or a lack of atoms). In the case where these have a positive wall intersection energy, parallel walls will appear, giving a striped domain structure, and for this case, the C-I transition is predicted to be continuous. In the opposite case, on a hexagonal substrate, a honeycomb network of domain walls of average hexagonal symmetry will appear, which theory predicts causes a first-order C-I transition.

One of the best studied C-I transitions in monolayer films is that of krypton absorbed on graphite. This has been studied by x-ray diffraction on ZYX substrates[9,65] and by LEED on a single-crystal substrate.[66] The earlier results[9,65,66] indicated a continuous C-I transition, but with no evidence of striped domains, and with a lattice parameter mismatch ε that appeared to behave as $(P - P_c)^\beta (\beta = 0.29 \pm 0.04)$ as a function of the ambient vapor pressure, which plays the role here of a chemical potential. This result could not be explained by any of the theories. More recently, several synchrotron studies of this transition have been carried out.[67–69] Nielsen et al.[67] found that at lower coverages and low temperatures there is a finite temperature region of coexistence of a commensurate and incommensurate peak, indicating a first-order C-I transition, as predicted theoretically.[60] At the higher coverages, where this transition occurs at high temperatures, the high-resolution x-ray studies[68,69] now indicate that a disordered phase (termed a reentrant fluid phase) intervenes between the commensurate and incommensurate phases. This phase joints continuously on to the higher temperature, regular fluid phase in the coverage-temperature phase diagram. Thus, the C-I transition is obviously a more complicated business than suspected previously. Various theoretical models exist[70] that predict an intervening disordered phase between the C and I phases, but the experiments cannot distinguish between these.[69]

An interesting aspect of the C-I transition has to do with the prediction of Novaco and McTague[71] that, close to the transition, the incommensurate layer could minimize its energy by rotating relative to the substrate lattice, thus minimizing the transverse strain energy. Obviously such an effect cannot be observed except on single-crystal substrates. It has been seen by Fain *et al.*[66] with LEED experiments on krypton on a single crystal graphite substrate. The rotation angle increases with increasing misfit. Very recently, high-resolution x-ray diffraction experiments for krypton on a single-crystal graphite surface[72] have shown that the rotation angle versus lattice mismatch parameter in the incommensurate phase quantitatively agrees with the theory of Shiba,[64] which puts in domain wall effects. At large values of the mismatch, the results agree asymptotically with the linear theory of Novaco and McTague.

8.6. Studies of the Incommensurate Two-Dimensional Lattice and Melting

One of the most remarkable facts about crystalline order in less than three dimensions is that, in principle, it does not exist! This was first pointed out by Landau and Peierls, many years ago. There exist general proofs that an infinite two-dimensional system with continuous symmetry (as is the case with a two-dimensional crystal on a smooth substrate) cannot have long-range order (LRO) except at $T = 0$.[73] The effect of phonon fluctuations causes the mean square displacement of an atom in the lattice to diverge logarithmically with the size of the crystal. This is in fact a very slow divergence, and one would not expect it normally to have much effect on a real experiment. However, the relevant point is the form of the structure factor $S(\mathbf{Q})$, and the effect of fluctuations upon it.

For an infinite three-dimensional crystal,

$$S(\mathbf{Q}) \sim \sum_{\tau} \delta(\mathbf{Q} - \tau)e^{-2W} + \text{inelastic part}, \tag{8.23}$$

where the τ are the reciprocal lattice vectors, e^{-2W} is the Debye–Waller factor, and the inelastic part is in general quite small compared to the Bragg part. For two dimensions, however, the theory[74-76] yields, for an infinite system

$$S(\mathbf{Q}) \sim \sum_{\tau} |\mathbf{Q}_\parallel - \tau|^{-[2-\eta(\tau)]}, \tag{8.24}$$

where $\mathbf{Q}_\parallel$ is the component of $\mathbf{Q}$ in the surface plane, and τ are the two-dimensional reciprocal lattice vectors. In other words, the delta functions of the Bragg reflections have been replaced by power-law singularities, as in a system at its critical point. Equation (8.24) corresponds to a pair-correlation function in real space that decays with the power law $r^{-\eta}$, which is a case of marginal (rather than true) LRO. The exponent η is a function of τ as well as temperature and goes as $\tau^2 T$ for low temperatures. Close to the melting point, the theory of Halperin and Nelson[59] predicts than η for the (10) reciprocal lattice point of a triangular lattice tends to a value η^*, which lies between $\frac{1}{4}$ and $\frac{1}{3}$, depending on the value of the elastic moduli of the two-dimensional crystal. As we have seen in Section 8.4, the crystallites on real substrates are limited to coherent regions of size ranging from 200 Å (for Grafoil) to about 5000 Å (for ZYX graphite). Thus, the question arises as to whether Eq. (8.24), which is valid for infinite two-dimensional crystallites, has any relevance to real experimental systems. This question has been recently resolved,[77] and it has been shown that for crystallites of finite size L, Eq. (8.24) should be replaced by

$$S(\mathbf{Q}) \sim \sum_\tau L^{(2-\eta)} \Gamma(1 - \eta/2) \Phi(1 - \eta/2; 1; -q^2 L^2/4\pi), \qquad (8.25)$$

where $\Phi(a; b; z)$ is a degenerate hypergeometric function also known as Kummer's function, and $\mathbf{q} \equiv \mathbf{Q}_\parallel - \tau$. In the limit $q^2 L^2 \ll 4\pi$,

$$\Phi(1 - \eta/2; 1; -q^2 L^2/4\pi) \simeq 1 - (1 - \eta/2)(q^2 L^2/4\pi), \qquad (8.26)$$

which may be interpreted as the first few terms of $S(q)$ for the Bragg peak of a finite crystallite of size L, as in the Warren approximation (see Eq. 8.20). In the limit $q^2 L^2 \gg 4\pi$,

$$\Phi(1 - \eta/2; 1; -q^2 L^2/4\pi) \simeq [\Gamma(\eta/2)]^{-1} [q^2 L^2/4\pi]^{(\eta/2 - 1)}, \qquad (8.27)$$

i.e., yielding the result of Eq. (8.24) as for the infinite crystal. Thus, the result is a smooth crossover from power-law behavior to the Warren form (for finite crystallites with perfect correlations) as a function of the dimensionless variable (qL). For diffraction line shapes obtained with an instrumental resolution function whose width in Q space is much larger than $(4\pi)^{1/2}/L$, it may be shown[58] that the observed line shape is indistinguishable from that for the infinite crystal, i.e., finite size effects on the power law structure factors are unobservable. This is the case for neutron diffraction experiments, but not for high-resolution, synchrotron x-ray experiments.

For the registered solids, the continuous symmetry is broken and perfect crystalline order can exist, and we thus expect the Warren line shape to be valid. This is borne out by both neutron and high-resolution synchrotron x-ray studies (see Fig. 27). In the incommensurate solid phase, however, to lowest order, the substrate potential averages out to zero, and we expect the behavior to resemble that of a two-dimensional solid on a smooth surface as discussed here. Figure 28 shows neutron diffraction results from CD_4 on ZXY graphite in the incommensurate dense solid phase, which shows the

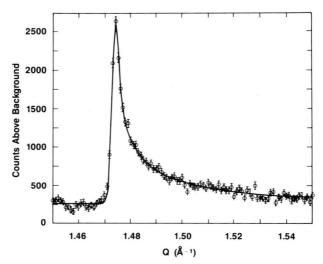

FIG. 27. High-resolution synchrotron x-ray profile of (10) peak for CF_4 on ZYX graphite in the commensurate (2×2) phase. The smooth line represents a Warren line shape fit to the profile. (From Nagler et al.[56])

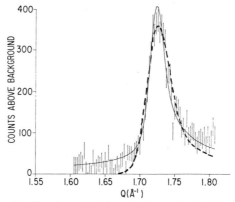

FIG. 28. Fit of power-law line shape to (10) peak shape obtained by neutron diffraction at 81.6 K from CD_4 on ZYX graphite in the dense incommensurate phase. The dashed line represents the best Warren line shape fit. The vertical bars represent statistical error bars centered on the experimental points. (From Dutta et al.[16])

presence of a "tail" at smaller Q that cannot be fitted with a Warren line shape but can be fitted well with a power-law line shape of the form of Eq. (8.24) by performing two-dimensional powder averaging and then using the procedure discussed in Section 8.3. Thus, the effect of two-dimensional fluctuations are quite apparent and show the dramatic difference between

Bragg scattering from two-dimensional as opposed to three-dimensional crystals. This system is the only one so far on which neutron diffraction has been used to do a detailed study of incommensurate solid diffraction line shapes. The results for $\eta(10)$ as a function of temperature are shown in Fig. 29. Although the uncertainties are large, $\eta(T)$ appears to show the correct behavior and to tend to a value close to 1/3 just prior to melting. Subsequently, synchrotron x-ray studies of power-law line shapes from the incommensurate solid phase of xenon on ZYX graphite have yielded more precise data on the incommensurate line shape. A fit to such a line shape using the Kummer function form (Eq. 8.25) is illustrated in Fig. 30.[17] For xenon on graphite, the limiting value η^* just before melting is 0.35 ± 0.02. Power-law line shapes have also been observed in thin (bilayer) liquid crystal films by x-ray scattering.[76] Since there is no orientational averaging to be done in this case, the (symmetric) appearance of "wings" on the sides of the peak is quite convincing evidence for the power-law form of $S(q)$.

We now discuss the process of melting from the incommensurate solid. In spite of the fact that the solid phase has only marginal LRO, a phase transition is still possible to a disordered or two-dimensional liquid phase. This was first shown by Kosterlitz and Thouless[78] for the case of the two-dimensional XY model, onto which a two-dimensional crystalline melting

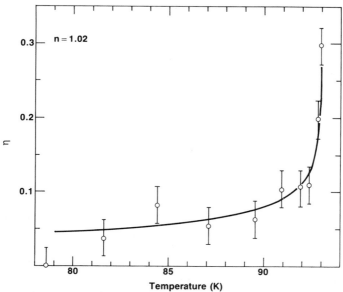

FIG. 29. The exponent η as measured from neutron diffraction line shape fits to CD_4 on graphite in the dense incommensurate phase as a function of temperature. The experimentally determined melting temperature is 95 K. (From Sinha et al.[58])

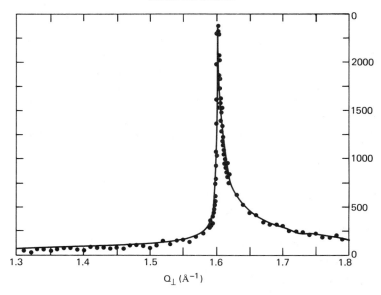

FIG. 30. High-resolution synchrotron x-ray profile of (10) peak for Xe/ZYX graphite in the incommensurate solid phase. The full line represents a Kummer function fit [see Eq. (8.25)]. (From Dimon.[17])

transition may be mapped. The melting process in such a theory is caused by the presence of topological defects, namely, dislocations in the two-dimensional crystal. These are defects where an extra row of atoms suddenly begins or ends at a point. At low temperatures, such dislocations are thermally activated in the form of bound dislocation pairs, but at a critical temperature T_{KT}, they suddenly unbind in a collective way and the presence of large numbers of free dislocations causes the shear modulus to go discontinuously to zero, and the crystal melts. Further elaborations of the theory were made by Halperin and Nelson[59] and Young.[79] Halperin and Nelson predicted that the melting process could proceed in two steps. The first would be to a phase where positional correlations were short ranged, i.e., exponentially decaying instead of power-law decaying, but where marginal bond orientational LRO was preserved, i.e., where the correlation function between the angles of lines connecting nearest-neighbor atoms decayed only with a power law. They named such a phase (in the case of a triangular lattice) a *hexatic* phase. At a slightly higher temperature, the bond-orientational order would also become short-ranged and a transition to an isotropic two-dimensional fluid was predicted. Both transitions were predicted to be continuous in these theories, although it was not ruled out that these could be preempted by first-order transitions.

These predictions have direct implications for the form of the structure factor through the melting transition and caused a great flurry of activity in the area of scattering studies from incommensurate monolayer films. For the hexatic phase, the prediction is that

$$S(\mathbf{Q}) \sim \xi^{2-\eta^*}/[1 + (\mathbf{Q}_\| - \tau)_L^2 \xi^2]^{1-\eta^*/2}, \tag{8.28}$$

where $(\mathbf{Q}_\| - \tau)_L$ is the radial component of $(\mathbf{Q}_\| - \tau)$ through the two-dimensional reciprocal lattice vector, and ξ is a correlation length. For the isotropic liquid phase,

$$S(\mathbf{Q}) \sim \xi^{2-\eta^*}/[1 + |\mathbf{Q}_\| - \tau|^2 \xi^2]^{1-\eta^*/2}. \tag{8.29}$$

The correlation length for the hexatic solid transition is predicted to diverge according to the law

$$\xi(T) = \xi_0 \exp(bt^{-\bar{\nu}}), \tag{8.30}$$

where ξ_0 and b are positive constants, $t = (T - T_{KT})/T_{KT}$, and $\bar{\nu}$ is a universal exponent with a value of ~ 0.37, Paradoxically, for constant coverage, first-order melting appears superficially to be more continuous, since there exists a finite temperature range of solid–liquid coexistence, as compared to second-order melting, where one goes directly (although continuously) from one phase to another. Figure 31 shows a series of (10) diffraction peaks taken for CD_4 on ZYX graphite for a coverage of $n = 1.09$ (i.e., in the dense incommensurate phase) as a function of temperature.[80] Because of the increase in vapor pressure in the closed sample cell with temperature, these do not all correspond exactly to the same coverage, but the corrections are small. The continuous broadening of the peak as the solid melts is evident. A Lorentzian line shape fit of the form given in Eq. (8.28) (taking the exponent as $\simeq 1$) was carried out to the line shape after two-dimensional powder averaging, using Eq. (8.14), and the result is shown in Fig. 31. The results are not accurate enough to make a meaningful fit to $\xi(T)$ of the form given in Eq. (8.30), and in any case this limiting behavior is believed theoretically to be valid only for very large ξ compared to the largest value observed in the experiments. Nevertheless, qualitative behavior reminiscent of a continuous phase transition is evident. As a consistency check, Fig. 32 shows the amplitude parameter of the peak plotted on a log–log scale versus ξ. According to Eq. (8.28) this should yield a straight line of slope $2 - \eta^*$. This is indeed observed and yields a value of $\eta^* = 0.32$ roughly consistent with η^* extracted from the solid-phase, power-law fits. By contrast, the same fits for a lower coverage of $n = 0.92$, where from the phase diagram a first-order transition involving solid–liquid coexistence for a finite range of temperature was observed, show a break in the straight line at the onset of the coexistence region as expected theoretically.[80] These neutron diffraction results were the

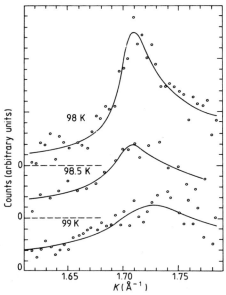

FIG. 31. Neutron diffraction line shapes at (incommensurate) (10) peak position for CD_4/Grafoil ($n = 1.10$) at different temperatures above the melting transition. The full lines represent Lorentzian line shape fits to the data. (From Sinha et al.[80])

first attempts to study in detail the diffraction line shape through the melting process, based on theories of incommensurate melting in two-dimensions. While they indicate qualitatively a dramatic difference in the melting processes of three-dimensional and two-dimensional crystals and are not inconsistent with the Kosterlitz-Thouless-Halperin-Nelson-Young (KTHNY) theories of melting, they cannot be used to rule out rigorously a narrow range of solid–liquid coexistence or an ultimate (weakly) first-order transition. In fact, this has proven to be extraordinarily hard to do to everybody's satisfaction, even with more recent studies of incommensurate melting with high-resolution synchrotron x-ray diffraction studies.[81-83] Part of the problem lies in the fact that here $\xi \gtrsim 200$ Å, and it is very difficult to distinguish between a power-law and a Lorentzian form of $S(\mathbf{Q})$, and thus between a pure highly correlated liquid phase and a superposition of solid and liquid. Thus, x-ray diffraction for xenon melting on ZYX graphite[81] at a coverge of $n = 1.2$ reveals a rapid but apparently continuous increase in the correlation length of the onset of the fluid–solid transition and a scaling of the amplitude of the structure factor with the $(2 - \eta^*)$ power of the correlation length, exactly as discussed above for the neutron diffraction results on CD_4/graphite but with much better resolution. Nevertheless, Abraham[84] has

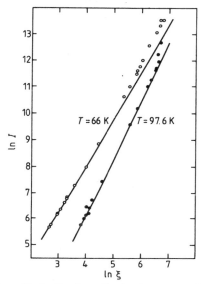

FIG. 32. Plot of ln I (where I is the fitted amplitude of the Lorentzian $S(\mathbf{Q})$ function used to fit the (10) line shapes) versus ln ξ (where ξ is the fitted coherence length parameter). The open circles are for $n = 0.92$ and the full circles for $n = 1.09$. (From Sinha et al.[80])

reproduced these results with computer simulations showing a mixture of solid and liquid coexisting through a first-order transition. Other computer simulation studies also yield first-order melting transitions[85] and the first-order melting picture is also lent some support by the work of Chui,[86] who considered a melting process that involved *lines* of dislocations, as on grain boundaries. Advocates of the dislocation-mediated melting theory point to the relatively short time scales involved in the computer simulation studies, and claim that the very slow movement of dislocations and the corresponding slow fluctuations may be misinterpreted at short time scales as a liquid–solid coexistence. It appears that there are certain levels of both experimental accuracy and practical computer simulations within which it appears difficult to distinguish clearly between these opposing viewpoints. Unfortunately, it appears very difficult at the present time to go beyond these limits. The existence of the hexatic phase is also controversial. Obviously, substrates such as Grafoil or ZYX graphite, where powder averages are taken in the substrate plane, cannot be used to say anything about bond orientational order. Recent experiments on single-crystal graphite substrates[83] indicate the presence of orientational correlations, where the angular spread of the peaks suddenly broadens above the temperature at which the positional correlations become short ranged. One cannot conclusively ascribe this to the

Halperin–Nelson mechanism as opposed to the effect of the hexagonal substrate orienting effects, however. In some respects, better systems for study appear to be provided by free-standing films of liquid crystals, where x-ray diffraction studies of films down to bilayer thicknesses have been carried out. Earlier studies of the liquid crystal $(\overline{14}S5)^{76}$ showed an abrupt hysteretic first-order transition for the crystalline B to smectic A melting transition. However, very recent measurements[87] on a bilayer film of $(460BC)$ indicate a continuous increase of the correlation length at melting indicative of a continuous transition, although direct evidence of bond orientational order was not seen. Thicker liquid crystal films do show hexatic ordering,[76] but this does not necessarily prove the case for the existence of the Halperin–Nelson phase in two dimensions. Nevertheless, these various experiments have demonstrated that the process of melting in two dimensions is quite different from that in bulk systems and that at least some of the ideas of the incommensurate melting theories have some validity, even if the "ideal" dislocation-mediated melting transition is preempted by other mechanisms.

It should be pointed out that earlier neutron diffraction experiments studying incommensurate melting have also been carried out for N_2 on Grafoil,[2,3]He on Grafoil,[40] and argon on Grafoil.[44] These studies were based on Warren line shape fits to the observed peaks and are thus only qualitative, but concluded that melting from these phases appeared to be continuous. Thus, Fig. 33 shows the rapid decrease of the coherence length L parameter for ^{3}He on Grafoil as the melting temperature is approached.[40] Interestingly enough, these data show loss of the two-dimensional crystallites at temperatures a degree or so below that at which the peak in the heat capacity

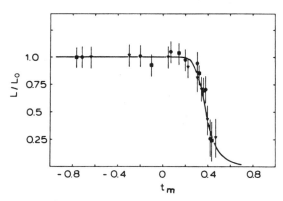

FIG. 33. Temperature dependence of the (Warren) coherence length L for various coverages of ^{3}He on Grafoil (normalized by its value $L_0 = 150$ Å at 1.6 K). The parameter t_m represents $(T_m - T_m)/T_m$ where T_m is the melting temperature. (From Lauter et al.[40])

is seen, consistent with theoretical predictions of the dislocation-mediated mechanism.[59,78]

There have been no detailed neutron diffraction studies of melting from registered phases, although there are many heat capacity studies, e.g., ^{4}He on graphite as discussed in Section 8.4. Melting from the $(\sqrt{3} \times \sqrt{3})$ R 30° phase of krypton on ZYX graphite was studied by x-ray diffraction[9,88] and a critical exponent $\beta = 0.09 \pm 0.03$ was extracted from the behavior of the order parameter at melting for $n \geq 0.92$ monolayers, assuming some finite-size temperature smearing of the transition. This is not very different from the value of $\beta = 0.072$ predicted for the multicritical point of the three-state Potts model.[89] However, recent synchrotron x-ray measurements indicate that the previous interpretation of the experimental results was incorrect and that the transition is actually first order.[69] For the melting of the (2×2) registered phase of CF_4 on graphite, recent x-ray results[56] also indicate a first-order melting process, with a liquid–solid coexistence region.

8.7. Multilayer Studies

Relatively few systematic diffraction studies have been carried out for n-layer films where $n = 2, 3,$ etc. Several interesting questions come to mind, e.g., whether the structure and spacing of the second layer is the same at the first, whether they melt at the same temperature, where true three-dimensional crystallite behavior is obtained, and so on. Some answers to these questions are now becoming available. Neutron diffraction has been carried out on the second layer of ^{4}He on Grafoil.[39,40] The results show that (1) the second layer compresses the first to a lattice constant about 25% smaller than the $(\sqrt{3} \times \sqrt{3})$ structure, and (2) that the second layer has a triangular structure but with an expanded nearest-neighbor distance (3.59 Å as opposed to 3.16 Å for the first layer; see Fig. 33) closer to the value in the basal plane of bulk hcp ^{4}He on the melting curve, which is 3.66 Å. The results also show that the second layer is incommensurate with the first and melts at a lower temperature than the first layer.[39] In CF_4 on Grafoil, Lauter et al.[54] have found that from $n = 1.9$ to $n = 4$, bulk CF_4 peaks appear in addition to the (10) and (11) monolayer peaks, which were shifted owing to slight compression of the two-dimensional lattice. This is an example of a partial wetting transition. Earlier, White and co-workers[18] studied ND_3 on graphon and found that the diffraction peaks seen at even the lowest coverages were characteristic of bulk crystallites, i.e., they did not show the characteristic asymmetric line shape associated with two-dimensional crystallites, as discussed in Section 8.3. As the temperature was raised, the intensity of these peaks decreased, and they disappeared rapidly at temperatures that were ascribed to the melting of these bulk crystallites. This melting temperature was about 155 K for a coverage equivalent to 0.5 "statistical monolayer" (based on an area/molecule of 16 Å^2 and a measured graphon surface area of 86 m^2/g), considerably lower than the bulk melting point for ND$_3$ of 199 K. At higher coverages this melting temperature rapidly approached the bulk melting temperature. As opposed to the case of CF$_4$/graphite, ND$_3$ has no two-dimensional solid phase, but apparently the bulk crystallites coexist with a dense fluid monolayer phase, inferred by these authors from quasi-elastic scattering experiments (see Section 8.8).

The wetting transition has also been studied for ethylene on graphite by both neutron[50] and x-ray diffraction.[51] In both cases, symmetric Bragg peaks characteristic of bulk ethylene crystallites appear for $n > 1$ and coexist with the two-dimensional crystallite peaks ($n = 1$ is defined as the coverage for a completed $\sqrt{3} \times \sqrt{3}$ monolayer on the graphite, as before). The relative intensities of the bulk peaks indicated a nonrandom orientation of the bulk crystallites, probably due to the orienting effects of the substrate. The intensities of the bulk peaks increased linearly as a function of coverage from zero at a coverage of $n = 1.05$. The x-ray diffraction results, carried out with a rotating anode x-ray source, showed that for the higher coverages, raising the temperature caused the bulk peak intensity to suddenly change in intensity at 75 K, while the underlying two-dimensional diffraction peak shape changed from one characteristic of a monolayer to that of a bilayer. At still higher coverages, a further third bulk-layer transition took place at 98.4 K (see Fig. 34) although the layers now show broad line shapes characteristic of a fluid, and finally a fourth bulk-layer transition was inferred at

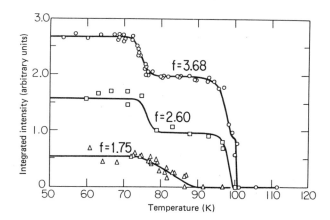

FIG. 34. Integrated intensity of bulk ethylene peaks on ZYX graphite from x-ray diffraction for several coverages as a function of temperature. Solid lines are guides to the eye. The intensity scale is normalized to 2.7 for $n = 3.68$ at low temperatures. (From Sutton et al.[51])

101 K, close to the bulk triple-point temperature. Such transitions are theoretically expected[90] to be first order, and the authors ascribed the broadness of the observed transitions to a distribution of chemical potentials at adsorption sites.

There appears to be tremendous potential in this relatively unexplored area for further diffraction and inelastic and quasi-elastic scattering studies to study exactly how bulk crystals are built up layer by layer from monolayers, how they melt as a function of layer thickness, and so forth.

8.8. Inelastic Scattering from Adsorbed Monolayers

Inelastic neutron scattering (INS) and quasi-elastic neutron spectroscopy (QENS) can be used to probe the dynamics of adsorbate molecules, an advantage that x-rays do not share. Again, because of signal limitations, one is restricted to large area substrates. Since these experiments are not carried out on single crystals, they are most efficiently done on time-of-flight spectrometers, where large solid angles are available and several values of Q can be simultaneously measured. However, triple-axis spectrometers have also been used for such experiments. Inelastic neutron spectroscopy has been used to study low-frequency tunnelling modes of adsorbed molecules, translational and rotational diffusion on the surface, collective vibrational modes of the two-dimensional lattice, and internal vibrational modes of the adsorbed molecules.

For molecules containing hydrogen, the large incoherent scattering cross section (~ 80 b) of the proton often allows one to neglect scattering from the other nuclei. Thus, the partial differential cross section, as discussed in Part A, Chapter 1, may be written as

$$\frac{d^2\sigma}{d\Omega\, dE} = N \frac{k_1}{k_0} \frac{\sigma_i}{4\pi} S_i(\mathbf{Q}, E), \qquad (8.31)$$

where N is the total number of hydrogen atoms, σ_i the incoherent cross section per hydrogen atom, $\mathbf{k}_1$ and $\mathbf{k}_0$ are the scattered and incident neutron wave vectors, respectively and $\mathbf{Q}$ and E the wave vector and energy transfer of the neutron, respectively. $S_i(\mathbf{Q}, E)$ is the space and time Fourier transform of the positional autocorrelation function of the hydrogen atoms and is related to the power spectrum of their dynamics. For translational diffusion on a smooth surface,

$$S_i^{\text{tr}}(\mathbf{Q}, E) \sim \frac{\hbar}{\pi} \frac{D_{\text{tr}} Q_\parallel^2}{[\hbar D_{\text{tr}} Q_\parallel^2]^2 + E^2}, \qquad (8.32)$$

where $Q_\parallel$ is the component of Q parallel to the surface, and D_{tr} is the translational diffusion coefficient of the molecule. In principle, diffusion in two dimensions is theoretically predicted to be different to that in three dimensions, so that D_{tr} actually diverges. This behavior is observed in molecular dynamics simulations of diffusion on surfaces[91] for sufficiently low coverages. What this means is that the actual form of $S_i(Q, E)$ given by the hydrodynamic expression (8.32) is not strictly correct, but the deviations from it are probably subtle and beyond the capability of detection at current experimental resolutions. For higher coverages, molecular dynamics simulations yield agreement with Eq. (8.32).

In addition to translational diffusion, molecules on surfaces will undergo rotational diffusion (or even free rotations). The motions, if they take place independently of translational motion, cause a folding in energy of the appropriate $S_i(Q, E)$, because the appropriate time-dependent autocorrelation functions factor out. In the case of rotational diffusion about a single axis,[92]

$$S_i^{\text{rot}}(Q, E) \sim J_0^2(Qa \sin \theta) \delta(E) + \frac{2}{\pi} \sum_{n=1}^{\infty} J_n^2(Qa \sin \theta) \frac{\hbar n^2 D_{\text{rot}}}{(\hbar n^2 D_{\text{rot}})^2 + E^2}, \tag{8.33}$$

where a is the radius of rotation, θ the angle between Q and a, the J's are Bessel functions, and D_{rot} the rotational diffusion constant. For diffusion on the surface of a sphere,

$$S_i^{\text{rot}}(Q, E) \sim j_0^2(Qa)\delta(E) + \frac{1}{\pi} \sum_{l=1}^{\infty} (2l + 1)j_l^2(Qa) \frac{\hbar l(l + 1)D_{\text{rot}}}{(\hbar l(l + 1)D_{\text{rot}})^2 + E^2}, \tag{8.34}$$

where a is the radius of the sphere. For $Qa \ll 1$, the higher order Bessel functions are small, and thus one probes mainly the translational diffusion. In this case, the quasi-elastic energy spectrum possesses broad wings and broadens with $Q_\parallel^2$. In order to extract D_{tr} or D_{rot} accurately from quasi-elastic spectra measured for monolayers on substrates such as Grafoil, one must first subtract the empty substrate background (which may sometimes be a little tricky owing to attenuation by the adsorbate molecules or multiple scattering effects), perform an appropriate weighted average of Eqs. (8.32) or (8.33) allowing for the orientational distribution of the substrate surfaces, as in the case of diffraction (Section 8.3), and finally fold with the instrumental energy resolution function, before fitting the experimental spectra.

Coulomb et al.[19] carried out a QENS study of CH_4 adsorbed on Grafoil at various coverages and temperatures. Their initial analysis did not properly account for the orientational distribution of the substrate surfaces, but their more recent work[93] does take this into account. Rotational diffusional effects were ignored. The results indicate, as expected, high values for D_{tr} (22×10^{-4} cm^2 s^{-1}) at low coverages ($n = 0.45$), decreasing as the coverage increases, and a rapid decrease in D_{tr} as one goes from the two-dimensional liquid phase to the two-dimensional solid phase. An interesting feature is that for 60 K $< T <$ 70 K, there is a range of coverage over which D_{tr} is almost constant. This was ascribed to a liquid–vapor coexistence region (unobservable by diffraction techniques), where the vapor scattering was essentially unobservable, so that one would be probing only the quasi-elastic scattering from the liquid phase at that temperature. Toxvaerd[91] carried out some molecular dynamics computer simulations for this system and found qualitative agreement with the experiment for $n \simeq 0.85$. More recently, Grier et al.[92] have carried out a QENS study of the system ethylene (C$_2$H$_4$) on isotropic exfoliated graphite. Their results are shown in Fig. 35. For temperatures above 30 K (in the two-dimensional solid phase), they observed quasi-elastic broadening characteristic of rotational diffusion about both an axis normal to the substrate surface and an axis in the surface. The fitted values yielded $D_{rot} = 3.3 \times 10^{11}$ s^{-1} for the former, and $D_{rot} = 2.5 \times 10^{10}$ s^{-1} for the latter. The intensity of the elastic component as a function of Q roughly obeyed the Bessel function laws given in Eqs. (8.33) and (8.34) and fell somewhat between them. In the liquid phase at 90 K, they observed translational diffusional broadening of the elastic peak (the rotational part being so broad as to be essentially flat), which obeyed the expected Q^2

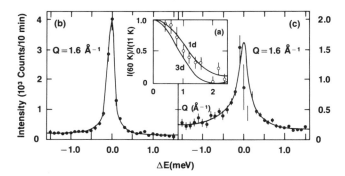

FIG. 35. (a) Q dependence of the elastic scattering from the LD solid phase of ethylene at 60 K normalized to the scattering at 11 K. The solid lines are intensity profiles calculated for one-dimensional and three-dimensional rotational diffusion. (b) Quasi-elastic scattering from the LD solid phase at 60 K. The solid line was calculated assuming two-dimensional rotational diffusion with D_{rot}(C–C) = 3.3×10^{11} s^{-1} and $D_{rot}(\perp) = 2.5 \times 10^{10}$ s^{-1}. (c) Quasi-elastic scattering from the liquid phase at 90 K. The solid line was calculated assuming translational diffusion with $D_{tr} = 2.0 \times 10^{-15}$ cm^2 s^{-1}. The double-ended horizontal line indicates the spectrometer resolution. (From Grier et al.[92])

behavior. From their data they deduced $D_{tr} = 2 \times 10^{-5} \, \text{cm}^2 \, \text{s}^{-1}$ at 90 K. These results are in reasonably good agreement with molecular dynamics simulations of C_2H_4 on a smooth surface of Nosé and Klein.[94]

It would be of great interest to study in detail the onset of the diffusion process at the melting transition, since unlike the case of three-dimensional crystals, this transition can be continuous or almost continuous, as discussed in the previous section. Thus, for instance, if the hexatic phase really exists, one would expect a loss of positional order not accompanied by rapid diffusion, since only the (relatively few) unbound dislocations are as yet mobile in this phase, but on the transition to the isotropic liquid, one would expect a rapid increase in diffusion rates. For the case of melting from the registered phase to the lattice liquid, one would expect the process to take place via jump diffusion between filled and empty lattice sites on the substrate. However, detailed studies of this kind would be likely to involve very high resolution inelastic spectroscopy, such as neutron spin echo, which has not as yet been applied to this problem.

With regard to INS study of phonons and molecular vibrations, we point out that the technique is complementary is some sense to electron energy loss spectroscopy (EELS), atomic scattering spectroscopy, and Raman or infrared spectroscopy. Nevertheless, INS has its special strengths, namely, its extreme sensitivity to hydrogen modes, the simplicity of the interaction of the neutron with the vibrational modes (which make it easier to compare observed spectra quantitatively with those derived from model calculations), and its applicability to bulk or opaque samples and to samples inside cells where conditions of pressure or temperature may be changed at will. The study of catalytic reactions on surfaces, for instance, appears to be an area where INS has great potential and, as we shall see, this has recently begun to be exploited. In the case of vibrational scattering, we have

$$S_i^{\text{vib}}(\mathbf{Q}, E) = e^{-2W}\{n(E) + 1\} \frac{\hbar^2 Q^2}{2E} \sum_{\nu \mathbf{q} j} P_\nu \left| \frac{\mathbf{Q} \cdot \mathbf{e}_\nu^j(\mathbf{q})}{M_\nu} \right|^2 \delta[E - \hbar\omega_j(\mathbf{q})], \tag{8.35}$$

where $n(E)$ is the Bose–Einstein population factor for vibrations of energy E, neutron energy loss processes only are considered, M_ν is the mass of the νth molecule in the unit cell containing P_ν hydrogen atoms, $\mathbf{e}_\nu^j(\mathbf{q})$ is the eigenvector of the νth molecule for the mode characterized by wave vector $\mathbf{q}$, branch index j, and frequency $\omega_j(\mathbf{q})$. Equation (8.35) is easily modified for the case of molecular vibrations from a set of independent molecules. In this case ν simply runs over the set of protons in the molecule, and the factor P_ν is absent. In Eq. (8.35) we have assumed purely translational motion of the molecules as rigid units. In some cases Eq. (8.35) may have to be modified to include torsional modes and their coupling to the translational modes.

8. ADSORBED MONOLAYERS AND INTERCALATED COMPOUNDS

In the absence of the latter, and for the case of monatomic two-dimensional lattices (e.g., triangular lattices), one may see from Eq. (8.35) that

$$S_i^{\text{vib}}(\mathbf{Q}, E) = e^{-2W}\{n(E) + 1\}\frac{\hbar^2 Q^2}{2ME}\{Q_{\parallel}^2 Z_{\parallel}(E) + Q_{\perp}^2 Z_{\perp}(E)\}, \quad (8.36)$$

where $Z_{\parallel}(E)$ and $Z_{\perp}(E)$ are, respectively, the phonon densities of states for vibrations parallel and perpendicular to the surface. In principle, measurements of the vibrational spectrum should be carried out at low temperatures to avoid seeing effects due to diffusion, particularly in the low-frequency regime. Various instrumental corrections, such as the variation of spectrometer efficiency with energy transfer, must also be taken into account. Measurement of this type have been carried out by Lauter et al.[95] for CH_4 on ZYX graphite, by Grier et al.[92] for C_2H_4 on graphite, and by Monkenbusch and Stockmayer[49] for C_6H_6 (benzene) on Grafoil. The latter authors have carried out a detailed calculation of the vibrational spectrum assuming the (registered) structure deduced from their neutron diffraction measurements of C_6D_6 on Grafoil (see Section 8.4) and potential functions in the literature for the H–H, C–H, C–C, and hydrocarbon-graphite substrate interactions, and used Eq. (8.35) to calculate $S_i^{\text{vib}}(\mathbf{Q}, E)$. The results are in qualitative agreement with experiment for the energy region 3–16 meV but not in the low-frequency regime, which the authors ascribe to neglect of the dynamical adsorbate-substrate coupling.

One of the earliest neutron scattering studies of the phonon spectrum of a two-dimensional lattice was made by Taub et al.[44] on the argon/graphite system. The isotope ^{36}Ar has an extremely large coherent neutron scattering cross section (77.8 b). The measurement of the phonon spectrum is thus somewhat complicated by coherence effects. In place of Eq. (8.35), one has

$$S_c^{\text{vib}}(\mathbf{Q}, E) = e^{-2W}\{n(E) + 1\}\frac{\hbar^2 Q^2}{2ME}\left\langle \sum_j |\mathbf{Q}\cdot\mathbf{e}^j(\mathbf{q})|^2 \delta(E - \hbar\omega_j(\mathbf{q}))\right\rangle, \quad (8.37)$$

where the average is taken over all values of $\mathbf{q}$ and orientations of $\mathbf{e}^j(\mathbf{q})$ for a given $\mathbf{Q}$, allowing for the properly weighted distribution of two-dimensional microcrystallites in the sample. The method is analogous to measurements of phonon spectra by coherent scattering from powder samples.[96,97] In the case of Grafoil, however, one is not dealing with a completely isotropic system. In principle, averaging over spectra for many different values of $\mathbf{Q}$ yields a result like that in Eq. (8.36) from which $Z_{\parallel}(E)$ and $Z_{\perp}(E)$ can be extracted. Taub et al.[44] carried out their experiments on a triple-axis spectrometer, on which measurements at only a few values of Q are practicable. Accordingly, these authors analyzed their data by modeling the

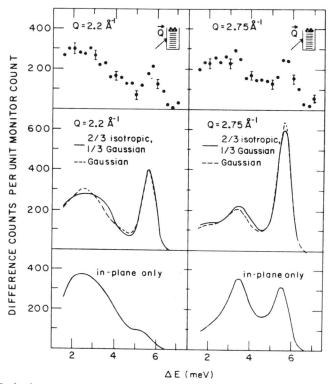

FIG. 36. Inelastic spectra from argon monolayers at 5 K observed with scattering vector **Q** at a 45° angle to the plane of the graphite foils. The solid circles at the top are experimental difference data. Plotted underneath are spectra computed using Eq. (8.37) assuming a partly ordered–partly isotropic orientational distribution of graphite crystalline normals (solid line) and a distribution with only the partly ordered component (dashed line). The curves at the bottom represent the calculated contribution to the observed spectrum. (From Taub et al.[44])

phonon spectrum of the triangular two-dimensional lattice using a nearest-neighbor force model based on a Lennard–Jones potential (obtained from gas-phase virial coefficient studies), numerically carrying out the averages in Eq. (8.37), and comparing with the observed spectra. No interaction with the substrate was assumed. Calculations were carried out for three cases: for **Q** parallel to the surface, for **Q** normal to the surface, and for **Q** at 45° to the surface. Qualitative agreement was obtained with the experimental data and with the observed variation with Q. Figure 36 shows the observed and calculated spectra for two different values of **Q** oriented at 45° to the mean basal plane direction. Note that this direction of **Q** picks up both in-plane and surface-normal modes of vibration. At higher temperatures, in the liquid phase, the spectra broaden and become monotonically decreasing with frequency, similar to those of classical three-dimensional liquids. Similar results have also been seen in CH_4/graphite.[95]

There are several interesting aspects of the lattice dynamics of adsorbed monolayers that have not yet received detailed experimental study. One has to do with the appearance of a gap in the acoustic phonon spectrum at zero wave vector when the monolayer goes into registry on the substrate (assumed to be rigid). This is due to the breakdown of lateral translational invariance in this case and hence the disappearance of the "Goldstone" modes. For a "weakly registered" monolayer, this gap may become quite small, and it has been shown[58] that this can indirectly manifest itself in the static structure factor $S(\mathbf{Q})$ in the vicinity of a Bragg peak. For such a case, we may write the phonon dispersion relation at small $\mathbf{q}$ as

$$\omega^2 = \Delta^2 + c^2 q^2. \qquad (8.38)$$

The Gaussian form (Eq. 8.21) for $S(\mathbf{Q})$ associated with the Bragg peak of a finite two-dimensional crystallite develops an addition Lorentzian component due to the phonon scattering. This component grows and takes over the Bragg peak to yield the power-law form of $S(\mathbf{Q})$ in the limit of zero gap. (We note that it is the vanishing of the phonon frequencies at $\mathbf{q} \to 0$ that causes the divergence of the mean squared displacement in the infinite-crystal limit.) From the width of the Lorentzian component, one may deduce the value of the ratio (Δ/c). Such an effect has been seen in CD_4/graphite in the $(\sqrt{3} \times \sqrt{3}) R\, 30°$ phase just below the transition to the expanded phase.[58] Another interesting question is that of dynamical coupling to the substrate. Model calculations of these effects have been carried out by Cleveland et al.[98]

Perhaps the highest resolution inelastic neutron scattering work done on monolayers is the work of Newbery et al.[99] on CH_4 adsorbed on graphite. In an elegant experiment carried out using 10 Å incident neutrons and 20 μeV energy resolution, they obtained the low-frequency spectrum shown in Fig. 37. At low temperature, several peaks can be seen, which they ascribe to transitions between quantum states split by tunneling through rotational barriers about axes parallel to the surface. A model for the rotational barrier, with an adjustable barrier height, is able to give good agreement with the positions and intensities of the peaks in the observed spectrum, not all of which were totally resolved. These distinct tunneling transitions were not observable for $T \gtrsim 20$ K indicating that by 20 K the molecule is almost freely rotationally diffusing about all three axes.

We now discuss an interesting application of INS to the case of excitations in thin films of superfluid helium on graphite. Third[100] and fourth[101] sound measurements and heat capacity measurements[102] in few atomic layer films of ^{4}He indicated a smaller superfluid fraction in these films relative to the

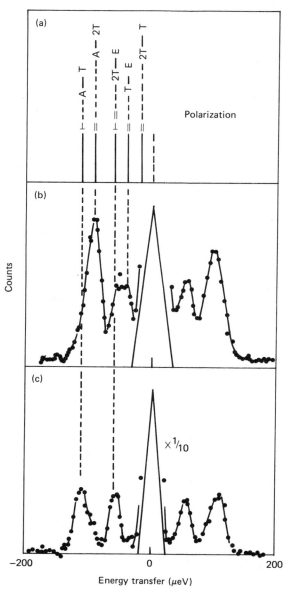

FIG. 37. Rotational tunneling spectra of CH_4 on Papyex at 4 K and for a coverage of 0.7 monolayers. The background has been subtracted: (a) assignation of the five transitions, (b) **Q** parallel to the surface, and (c) **Q** perpendicular to the surface. Resolution (full width at half-maximum) = 20 µeV. (From Newbery et al.[99])

8. ADSORBED MONOLAYERS AND INTERCALATED COMPOUNDS

bulk superfluid and also a smaller roton gap. Neutron scattering studies have been carried out by the Brookhaven group[103, 104] and by a group at the Institut Laue-Langerin (ILL) at Grenoble.[105] These results (carried out for films from three layers to 20 layers thick) show that there are basically two types of excitations in these films (see Fig. 38). The first is a roton excitation essentially unchanged in energy relative to the roton excitations in bulk superfluid ^{4}He. This is thus identified as a "bulk" roton, and its intensity decreases linearly to zero for about three layers coverage (see Fig. 39). Since structural studies of ^{4}He films indicate that the first two layers are solid-like at these temperatures, this means that this bulk excitation vanishes when only one monolayer of liquid remains. The linewidth increases dramatically as the

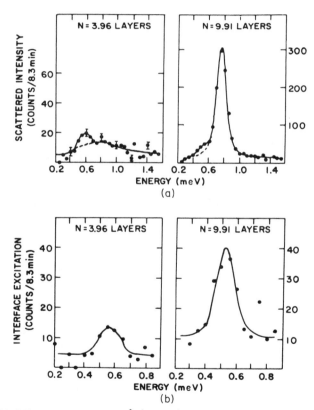

FIG. 38. (a) INS spectra at $Q = 2.0$ Å^{-1} from ^{4}He films adsorbed on Graphon (multiple-scattering and unfilled cell backgrounds have been subtracted); $T \leq 1.32$ K. The solid lines are Lorentzians fitted to the bulk-like roton peak at 0.7 meV and the peak at 0.54 meV identified as an interfacial excitation. (b) INS spectra in the neighborhood of 0.54 meV with background of bulk-like roton scattering subtracted. (From Thomlinson et al.[104])

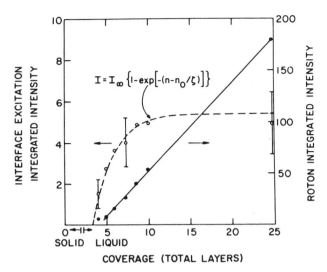

FIG. 39. Coverage dependence of the integrated intensity of the bulk roton scattering (solid circles) and the interfacial excitation scattering (open circles). Here ζ and n_0 were taken as 2.3 and 3.3 layers, respectively. (From Thomlinson et al.[104])

number of layers is decreased, indicating that the roton lifetimes are limited by the boundary scattering at the film surfaces. The other excitation is a lower energy, relatively dispersionless excitation (see Fig. 38) at 0.54 meV. The coverage dependence of the intensity of this excitation is very different and consistent (Fig. 39) with a mode originating in the first liquid layer and having a penetration depth of only 2.3 layers into the bulk. This mode has been called a "surface roton"[104] as its energy is close to that of rotons in bulk ^{4}He of density equal to that in the first liquid layer. It may, however, be a more general dynamical mode at interfaces, since there is evidence that it is also seen in normal two-dimensional liquid monolayer films, such as neon on graphite.[106]

Taub and collaborators[107, 108] have made a series of detailed INS studies of the dynamics of short-chain hydrocarbons, such as butane (C_4H_{10}), pentane (C_5H_{12}), and hexane (C_6H_{14}) on graphite surfaces. Figure 40 shows the inelastic neutron spectrum (obtained by time-of-flight techniques) for butane adsorbed on a Carbopack B substrate (having a surface area of 80 m^2/g and no preferred orientations) for several different temperatures. At the lower temperatures, several distinct peaks are seen in the spectrum. By a series of model force constant calculations, based on empirical atom–atom potentials, these peaks can be identified with the intramolecular torsions of the end CH_3 groups (A, B), the internal CH_2–CH_2 torsion (C), and three surface-specific modes not found in the bulk solid, namely, two librational

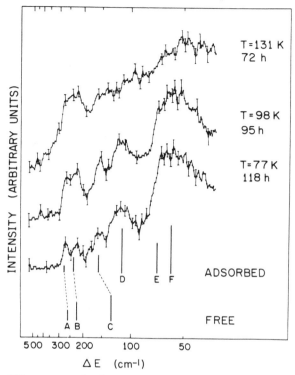

FIG. 40. The INS spectrum of a butane monolayer adsorbed on Carbopack B for three temperatures. In each case the background from the substrate has been subtracted. The vertical bars labeled ADSORBED represent the monolayer spectrum calculated from the force constant model referred to in the text. The spectrum calculated for the intramolecular modes of the free molecule is shown for comparison. (From Wang et al.[108])

modes about axes parallel to the graphite surface (D, E) and a "bouncing" mode of the whole molecule normal to the surface (F). The frequencies predicted by the model as shown agree fairly well with the observed peaks. Note that the internal modes are slightly shifted from those of the free molecule due to the presence of the surface. The configuration of the butane molecule relative to the surface was obtained from neutron diffraction studies of C_4D_{10} on graphite (referred to in Section 8.4) with the carbon skeletal plane tilted 30° from the graphite surface. Reasonable agreement was similarly obtained for the case of hexane, but not for pentane, where the orientation of the molecules on the surface are not as well known.

INS studies have also been made on chemisorbed systems, e.g., hydrogen adsorbed on catalytic surfaces such as nickel and platinum. Cavanagh et al.[109] carried out a triple-axis spectrometer study of the INS spectrum of

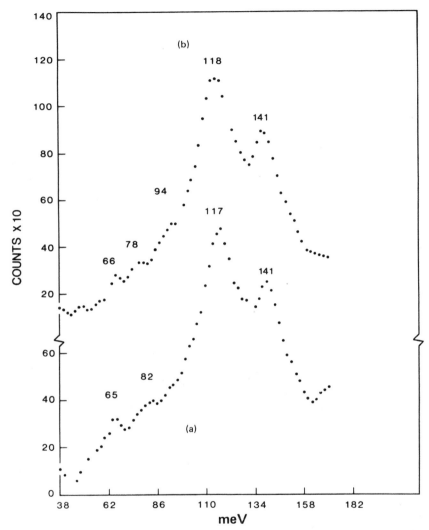

FIG. 41. (a) INS spectrum of saturation coverage of H_2 on Raney nickel adsorbed at 150 K; (b) INS spectrum from same sample annealed at 300 K. Both spectra recorded at 80 K. (From Cavanagh et al.[109])

hydrogen adsorbed on Raney nickel, a fine powder of nickel–aluminium alloy (with Al < 10 at. %) having a surface area of 14 m²/g. The results are shown in Fig. 41 for the case of hydrogen adsorbed at 150 K and 300 K. The two prominent peaks at 118 V and 141 meV were identified with vibrations of hydrogen chemisorbed at a threefold symmetric site on the nickel (111) face. This was deduced on the basis of model calculations assuming the Ni—H bond length and the Ni-H force constant as adjustable parameters. Eigenvectors and eigenfrequencies were then calculated and the parameters adjusted to give agreement for the two frequencies and their 2 : 1 intensity ratio in the scattering. It was found that assumption of a fourfold symmetric site gave unrealistic values of these parameters, but the values found for the threefold site ($d_{Ni-H} = 1.88 \pm 0.04$ Å, and $k = 0.58 \pm 0.3$ mdyn/Å) are reasonable, the bond length being in agreement with LEED studies for hydrogen on nickel (111). Thus, the Raney nickel microcrystallites are dominated by (111) faces. The features in the lower energy part of the spectra are dependent on the temperature at which the hydrogen was absorbed and, while not definitely accounted for, are ascribed by the authors to metastable hydrogen binding sites at low temperatures.

In a subsequent experiment,[110] these authors studied INS from equal amounts of CO and H absorbed on Raney nickel. They found that if the CO is introduced at 80 K on the already hydrogen-covered substrate, there is almost no perturbation of the hydrogen vibrational spectrum (the CO scattering is negligible compared to that of hydrogen here), i.e., the CO does not affect the hydrogen threefold absorption sites or the Ni-H force constants. However, on warming the system to room temperature and recooling to 80 K, a considerably changed spectrum resulted indicating strong perturbations of the hydrogen sites or a new hydrogeneous species on the surface, as yet unidentified. These authors also carried out INS studies on this system after subjecting it to reactive flow conditions at 300 K (where very little CO conversion was chemically detected) and at 450 K (where 50% of the CO was broken down into hydrocarbons). The similarity of the spectra after these two different reactive flow treatments indicate no change in the adsorbed layer even during CO-hydrocarbon conversion. Experiments were also carried out for acetylene and ethylene adsorbed on platinum black at 120 K[111] (see Fig. 42). Fairly well-defined peaks were found in the spectrum, and the results were in general in good agreement with those obtained from electron energy loss spectroscopy (EELS) for these molecules on platinum (111) surfaces. Again force-field calculations for eigenvectors and eigenfrequencies for assumed configurations of these molecules on threefold sites on the platinum surface were able to yield reasonable agreement with the INS spectra, as shown in Fig. 42 for the case of acetylene on platinum black. Figure 43 shows the configurations for the models that gave good fits to the data. The third configuration is for ethylidyne (C-CH$_3$ groups) on the surface, for which the authors find some evidence as a new hydrocarbon product that appears when the acetylene or ethylene is annealed on the surface at 300 K and some additional hydrogen introduced. The authors emphasize, however, that their models are not unique and other possibilities

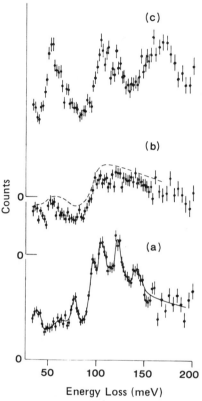

FIG. 42. INS spectrum of acetylene on platinum black at 150 K: (a) spectrum for acetylene adsorbed at 120 K; (b) annealed to 300 K (dashed line indicates changes observed when one hydrogen per C_2 species is introduced); (c) annealed to 300 K with 400 Torr of H_2 introduced. (From Cavanagh et al.[111])

for fitting the data exist. In general, a correct picture of what is happening on the surface must be pieced together by an evaluation of INS spectra, comparison with EELS or optical spectral data, knowledge of the general nature of hydrocarbon–metal bonds and some educated guesswork. It is encouraging to see that detailed evaluations of this kind are beginning to be made of INS spectra of hydrocarbons on metal surfaces. The potential for general studies of hydrocarbons on catalysts with INS (particularly with the development of the spallation neutron sources and their ability to study large-energy transfers with relative ease) appears to be very exciting. Spectra of molecules adsorbed in microporous materials such as zeolites should also be very interesting, particularly from the point of view of the importance of such processes in various technological applications.

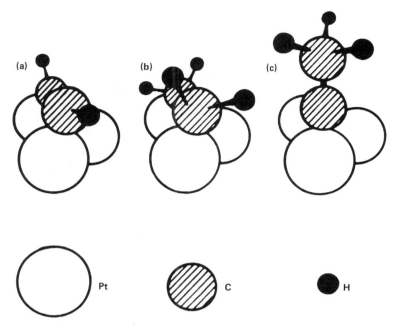

FIG. 43. Model geometries for adsorbed hydrogenous species on platinum (a) acetylene, (b) ethylene, and (c) ethylidene. (From Cavanagh et al.[111])

8.9. Neutron Studies of Surface Magnetization

Neutron diffraction from magnetic monolayers has been only rarely studied (an exception being the case of O_2 on graphite as discussed in Section 8.4). Recently, Felcher and collaborators[112,113] at Argonne National Laboratory (ANL) have developed a technique for studying the depth dependence of surface magnetization or magnetic flux at crystal surfaces, by using total reflection of neutrons. This technique studies the profile of magnetization normal to the surfaces only, rather than lateral surface structure of the magnetization. This effect has been applied to the study of the penetration depth for magnetic flux into superconducting niobium films.

Imagine a solid surface with an applied field H parallel to the surface, which decays with a characteristic length Λ into the depth of the material. The z dependence of the magnetic induction is given by

$$B(z) = He^{-z/\Lambda}. \tag{8.39}$$

The refractive index of the material for neutrons will depend both on the nuclear scattering length and density and the magnetic induction (in the absence of direct magnetic scattering by atomic moments) and will thus be given by[113]

$$n^{\pm}(z) = 1 - \frac{\lambda^2}{2\pi}\left[\frac{b}{V} \mp \frac{2\pi\mu_n m}{h^2} H(1 - e^{-z/\Lambda})\right], \quad (8.40)$$

where λ is the neutron wavelength, μ_n the neutron magnetic moment, and m the neutron mass. The $\pm$ signs refer to the case of neutron spins polarized parallel ($+$) or antiparallel ($-$) to the field H; b is the nuclear scattering length, and V the volume per nucleus.

The reflectivity for polarized neutrons at the surface as a function of the angle of incidence θ_i and the neutron wavelength λ can in principle be calculated from Eq. (8.40) but not in closed analytic form. However, an approximate expression, valid to first order in H and away from the critical wavelength, λ_c, is[112]

$$R^{\pm} = |r_1|^2\left[1 \mp \frac{2\theta_i}{P_0}\frac{CH}{B/V}\frac{1}{1 + (4\pi P_0 \Lambda/\lambda)^2}\right], \quad (8.41)$$

where $r_i = (P_0 - \theta_i)/(P_0 + \theta_i)$, the reflectivity in the absence of the magnetic terms, $P_0 = (\theta_i^2 - (\lambda^2/\pi)b/V)^{1/2}$, and $C = 2\pi\mu_n m/h^2$. Calculations based on Eq. (8.41) show that the reflectivity of the surface is sufficiently spin dependent to yield a reasonable sensitivity to Λ in the range of a few hundred angstroms. The experiment thus consisted of allowing a beam of polarized neutrons (parallel or antiparallel to the surface of the niobium film) from the pulsed neutron source at ANL fall on the surface at a mean incident angle of 0.34° and measuring the flipping ratio R^+/R^- as a function of the neutron wavelength λ in the range 3.8 Å $< \lambda <$ 8 Å. An example is shown in Fig. 44 for a niobium film at 4.6 K and with an applied field of 500 Oe. The lines drawn through the data are numerical solutions for the flipping ratio based on Eq. (8.40) assuming different values of Λ. The deduced value of Λ was 430 ± 40 Å, which is consistent with other measurements carried out at different fields and mean angles of incidence. In the calculations, corrections for finite angular divergence of the beam, surface roughness, and the effects of a finite niobium film thickness had to be made and checked from measurements made for the same film *above* T_c, where there are no penetration depth effects. At 7 K, Λ was measured to be 600 ± 80 Å. The increase is as expected, since Λ must diverge at T_c. These values of Λ are consistent with estimates made from other studies and theoretical calculations and provide one of the first direct measurements of the penetration depth for a superconductor. No doubt this technique will be further exploited to study the depth dependence of surface magnetism, the occurrence of magnetically inactive surface layers, and so on. Thus, it should provide a useful complementary technique to polarized LEED and other surface magnetic probes.[114]

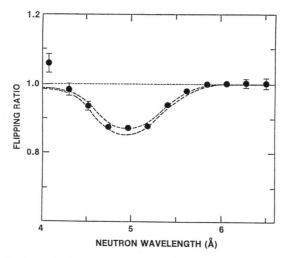

Fig. 44. The flipping ratio for a niobium film at 4.6 K and 500 Oe as a function of neutron wavelength; $\theta_i = 0.34 \pm 0.02°$. The lines drawn are calculated (in ascending order) for $\Lambda = 380$ Å and 480 Å. (From Felcher et al.[112])

8.10. Neutron Studies of Intercalated Compounds

The literature on neutron investigations of layered and intercalated compounds is already so vast that to reasonably cover the subject would require a separate chapter. In this section, we shall present a brief review of the salient features of those studies that bear closely on the quasi-two-dimensional behavior of intercalant molecules in such systems and are thus related to the studies of the adsorbed monolayer systems discussed previously. In this sense, the intercalated compound is a system that offers the "macroscopic" surface areas that are needed to study monolayer behavior, although, as we shall see, three-dimensional behavior often intrudes in many of these studies. For general reviews on layered and intercalated compounds, the reader is referred to Refs. 115–117.

Intercalation compounds are formed by starting with a host material that has a highly anisotropic layered structure, such as graphite or the transition metal dichalcogenides, and inserting (under conditions of high temperature and vapor pressure) atomic or molecular layers of a different chemical species, called the intercalant, between the layers of host atoms. Technological interest in these materials lies in the ability to control their electrical and thermal conductivity by suitable intercalation, the anisotropy of their transport properties, the appearance of superconductivity in many of these systems at low temperatures, and the large in-plane ionic conductivity of some of these systems.

The strong electron–phonon interaction in the pure layered transition metal dichalcogenides leads to charge-density-wave-type instabilities, and much neutron diffraction and inelastic neutron scattering work (as well as x-ray and electron diffraction) was carried out on such systems over 10 years ago and has been reviewed in several articles.[118,119] Two-dimensional and quasi-two-dimensional behavior of intercalant molecules has mainly been studied in graphite intercalation compounds (GICs), and it is with such studies that we shall concern ourselves in this section, although again we shall concentrate on neutron as opposed to x-ray studies, both of which have been carried out extensively over the last few years.

The most common method of intercalation is the so-called two-zone vapor transport method, where the graphite is kept at a high temperature T_g (~ 200–$500°C$) in a crucible connected by a tube to another crucible containing the intercalant at a slightly lower temperature T_i. The difference ($T_g - T_i$) controls the phase of the intercalate, known as the *stage*, A GIC is said to be "stage-n" if n graphite layers separate the intercalant layers (Fig. 45). The lower ($T_g - T_i$) is, the lower is the stage of the GIC produced. Mixed stages or even disordered stages can also be formed. Other methods for preparing GICs, such as liquid intercalation, electrochemical methods, and cointercalation techniques, also exist.[115,120,121] Although single-crystal graphite may be intercalated and x-ray diffraction studies have been carried out on such samples, neutron studies have been confined to intercalated highly oriented pyrolytic graphite (HOPG) crystals, where the graphite basal planes of the crystallites are randomly oriented, but possess a common c axis with a small mosaic spread. The situation is thus analogous to the case of UCAR-ZYX substrates discussed in Section 8.2. Since most GICs are stable at room temperature, but highly reactive in air, the samples are usually prepared and then transferred to a glove box, where they are sealed in thin-walled aluminum containers under a slight pressure of a nonreactive vapor, such as helium. These are them mounted in the usual way on a neutron spectrometer for the measurements.

From the point of view of diffraction studies, the majority of structural investigations have been carried out with x rays. Neutrons offer no particular advantage except possibly in the case of intercalants containing appreciable hydrogen or for high-pressure studies, where the increased penetrability of neutrons through cell windows can be an advantage. The most common structural investigation carried out on GICs is that of the staging phenomenon. For ($00l$) reflections (along the c^* axis), it is obvious that the study of staging reduces to a one-dimensional diffraction problem, namely, measuring the Fourier transform of scattering density normal to the c-axis. More general (hk) reflections are required to obtain the ordering within the layers.

8. ADSORBED MONOLAYERS AND INTERCALATED COMPOUNDS 71

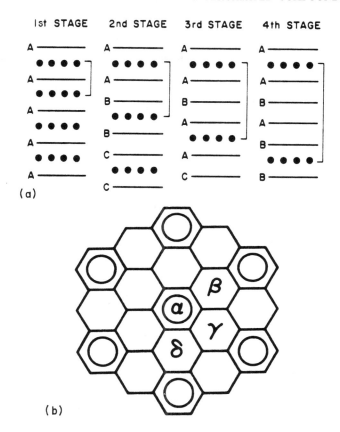

FIG. 45. (a) Stacking sequence of carbon (——) and alkali (···) layers in stages 1 to 4 compounds. (b) In plane 2 × 2 structure of potassium in C_8K; α, β, γ, and δ denote the positions of alkali atoms in subsequent layers.

The lateral ordering within the layers is usually correlated three-dimensionally, at least for the lower stages. The arrangement of the intercalant molecules is often on a triangular lattice, but the stacking arrangement shifts its basis to each of four possible sublattices as one moves from layer to layer (see Fig. 45). In addition, the host graphite lattice, which consists of hexagonal rings of carbon atoms (Fig. 8) and normally has a stacking sequence ABAB..., where the layer B is shifted laterally relative to layer A by a vector Δ_G, may also change its stacking sequence under intercalation. Thus, for instance, x-ray diffraction data show that stage-2 $C_{24}Rb$ has a stacking sequence of the form $A\alpha AB\beta BC\gamma CA...$,[122] where the capital Arabic letters refer to the graphite layers and the Greek letters to the intercalant layers. (The graphite layer C is shifted by $2\Delta_G$ relative to layer A.) As in the case of adsorbed monolayers, the intercalant layers can be commensurate with the graphite lattice, forming $(\sqrt{3} \times \sqrt{3})$ R $10°$, (2×2), or $(\sqrt{7} \times \sqrt{7})$ R $19.1°$ structures (depending on the size of the intercalant) or incommensurate with the graphite lattice (usually at higher temperatures) or even liquid-like. For the higher stage compounds, where the intercalant layers are widely separated, the ordering in the layers can become essentially two-dimensional. Thus, GICs offer a rich variety of systems in which to study the effect of dimensionality on the structure and dynamics of condensed matter.

Let us briefly discuss how diffraction studies yield information about the structure. For the case of the commensurate structures, the unit can be defined from the positions of the Bragg reflections. (For HOPG samples, only the magnitudes of $\mathbf{Q}_\parallel$, the component of $\mathbf{Q}$ for the reflections parallel to the graphite basal planes, are available.) Since the reflections are three dimensional, and there is no large misorientation of the basal planes in this case, the detailed line shape analysis in Section 8.3 is not necessary here. The stacking sequence of layers can be determined by fitting structure factors for the Bragg reflections based on any assumed stacking sequence and noting the agreement with experiment. Thus, for instance, from a study of the $(h0l)$ reflections from an intercalated HOPG C_8Rb sample at 290 K, Ellenson et al.[123] were able to obtain the stacking sequence $A\alpha A\beta A\gamma A\delta\ldots$.

An early set of neutron diffraction investigations was carried out by Watanabe et al.[124] on stage-2 $C_{24}K$ in powder form, with D_2 adsorbed to form $C_{24}K(D_2)_2$. From a study of $(00l)$ reflections, they concluded that the D_2 was adsorbed within the alkali atom layers, but no further information regarding the structure could be obtained. More recently, Zabel et al.[125] have carried out a neutron diffraction investigation of $C_{24}Rb(D_2)_{0.2}$ in intercalated pyrolytic graphite and studied the $(hk0)$, $(10l)$, and $11l)$ type reflections of the alkali metal superlattice. The graphite lattice does not contribute to these reflections. Assuming that the D_2 molecules go into the interstitial sites indicated in Fig. 46, the structure factor may be written

$$F(Q) = [1 + 2\cos 2\pi/3(h + k + l)]$$
$$\times [b_A + C_{D_2}b_{D_2}(-1)^{h+k}2\cos \pi/3(h - k)], \qquad (8.42)$$

where b_A and b_{D_2} are the coherent scattering lengths of the alkali metal atoms and the D_2 molecules, respectively, and C_{D_2} is the average D_2 occupation number of the triangular interstitial sites. Debye–Waller factors have been neglected in Eq. (8.42). The observed intensities correspond well with the predictions of Eq. (8.42), with C_{D_2} obtained as 0.21, in good agreement with the calibrated D_2 loading. Thus, the co-planar structure of D_2 and alkali intercalant atoms was confirmed.

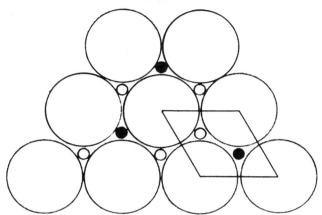

FIG. 46. Alkali close-packed structure showing triangular interstitial sites that may be randomly occupied (full circles) or unoccupied (open circles) by D_2 molecules. Each alkali metal unit cell contains two interstitial sites, yielding a maximum concentration of $D_2/Rb = 2$ when completely filled. (From Zabel et al.[125])

If the intercalant ordering within the layer becomes incommensurate with the graphite, the only reflections where the graphite and intercalant atoms are coherent with each other are the $(00l)$ reflections. For general (hkl) reflections, the reflections separate into those from the intercalant lattice and those from the graphite lattice. This was observed in a neutron diffraction experiment of Suzuki et al.[126] in stage-2 $C_{24}Rb$ for temperatures between 106 and 165 K. The positions of the observed Bragg peaks are shown in Fig. 47. The extra peaks due to rubidium show values of $Q_\parallel$ equal to 1.20 Å^{-1}, 2.08 Å^{-1}, and 2.40 Å^{-1}, which are indexable in terms of a triangular lattice of rubidium atoms in the planes that is incommensurate with that of the graphite lattice. In addition, an extra set of peaks with $Q_\parallel = 1.78$ Å^{-1} is observed. This may be explained in terms of a modulation of the incommensurate rubidium lattice by the graphite host lattice via periodic strains. Let us consider an alkali atom whose undistorted position is $\mathbf{R}_m$ (m is an index running over all the alkali atoms). Due to the interactions with the graphite lattice, its position is changed to $\mathbf{R}_m + \mathbf{u}_m$. Since $\mathbf{u}_m$ arises due to a potential with the in-plane periodicity of the graphite lattice, it may be Fourier decomposed as

$$\mathbf{u}_m = \sum_{\tau_\parallel} \mathbf{U}_{\tau_\parallel} \exp(i\tau_\parallel \cdot \mathbf{R}_m), \qquad (8.43)$$

where $\tau_\parallel$ are the set of graphite reciprocal lattice vectors in the basal plane, and the $\mathbf{U}_{\tau_\parallel}$ are a corresponding set of basal plane vectors. The diffracted

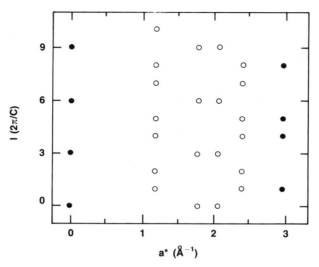

FIG. 47. Schematic representation of the reciprocal lattice space for the more intense Bragg peaks from $C_{24}Rb$ between 106 K and 165 K ($c = 27.9$ Å). The open circles are rubidium reflections only. The closed circles are mixed graphite and rubidium reflections. (From Suzuki et al.[126])

intensity from the alkali atoms is

$$I(\mathbf{Q}) \propto \left| \sum_m \exp[i\mathbf{Q} \cdot (\mathbf{R}_m + \mathbf{u}_m)] \right|^2. \quad (8.44)$$

If the $\mathbf{U}_{\tau_\parallel}$ are small, we may expand this to first order in these quantities as

$$I(\mathbf{Q}) \propto \left| \sum_m \exp(i\mathbf{Q} \cdot \mathbf{R}_m) \left[1 + i \sum_{\tau_\parallel} (\mathbf{Q} \cdot \mathbf{U}_{\tau_\parallel}) \exp(it_\parallel \cdot \mathbf{R}_m) \right] \right|^2. \quad (8.45)$$

We may now use Bloch's theorem and the fact that the $\mathbf{R}_m$ are incommensurate with the $\tau_\parallel$ to obtain from Eq. (8.45),

$$i(Q) \propto \sum_{\tau_A} \left[\delta(\mathbf{Q} - \tau_A) + \sum_{\tau_\parallel} (\mathbf{Q} \cdot \mathbf{U}_{\tau_\parallel})^2 \delta(\mathbf{Q} - \tau_A - \tau_\parallel) \right], \quad (8.46)$$

where the τ_A are the reciprocal lattice vectors of the alkali atom superlattice. The second terms shows that there will arise satellite peaks in the alkali atom reflections displaced from the τ_A by the most important basal-plane reciprocal lattice vectors of the graphite lattice. The actual magnitude $|\mathbf{Q}|$ of this satellite position will then depend on the relative orientation of the graphite and alkali atom lattices as well as their spacings. From the observed position of 1.78 Å^{-1} of the superlattice reflections, Suzuki et al. deduced that the rotation angle between the two lattices was 9.2°. However, a direct verification of this is not possible without single-crystal diffraction data. The intensities of the satellite reflections were, however, much larger than a small value of $(\mathbf{Q} \cdot \mathbf{U}_{\tau_\parallel})^2$ would lead one to expect, indicating that the linearized approximation made in Eq. (8.45) may be breaking down. Similar satellites due to modulation effects have been observed by x-ray diffraction from potassium-intercalated graphite at low temperatures,[127,128] and in incommensurate adsorbed monolayer films by x-ray and LEED studies.[66,60] At higher temperatures, these modulation effects disappear due to the decreased "susceptibility" of the "soft" lattice with respect to potential fluctuations from the "hard" lattice (substrate or host lattice). Many detailed studies of the order–disorder ("melting") transition in the intercalant lattice have been carried out[127-129]—mainly by x-ray diffraction, however—so we shall not discuss them here in detail. An interesting aspect is that in stage-2 (incommensurate) $C_{24}Cs$, the buildup of interlayer correlations never truly achieves long-range order, the correlation length along the c axis never exceeding 40 Å,[128] while the correlation length in the planes increases continuously to about 400 Å at 100 K. This may be a case of a quasi-two-dimensional incommensurate solid developing, although detailed line shape studies of the development of the power-law structure factor have not yet been done and probably must await high-resolution synchrotron x-ray studies. For the case of the incommensurate phase of $C_{24}Rb$, the neutron studies also show a short correlation length of ~30 Å along the c axis at 130 K.

Several INS studies of the lattice dynamics of pure graphite and alkali-intercalated GICs have been carried out. For $L(00q)$ modes, i.e., longitudinal modes propagating along the c axis, the problem reduces to the textbook case of the dynamics of a linear chain with a distribution of masses and force constants along it, representing the planes of graphite and intercalant atoms, respectively. In the extended-zone scheme, accordingly, one expects to see for an n-stage GIC [corresponding to a repeat distance of $(n + 1)$ layers along the c axis], $(n + 1)$ longitudinal modes. Figure 48 indicates results for the dispersion of the $L(00q)$ modes in the heavy alkali metal intercalated compounds for various stages, as measured by Zabel and Magerl.[130] The splittings at the zone center and zone boundaries that develop are due to the mass and force-constant differences between the layers. The dispersion curves indicate the existence of effective long-range interactions between the layers, but can be fitted quite well with a simple one-dimensional shell model where a massless "shell" is introduced associated with the alkali layer and coupled to the alkali layers and graphite layers via short-range force constants, in addition to the graphite–graphite layer force constant. This indicates that electronic polarization effects of the intercalant layers may be responsible for transmitting long-range forces in the c axis direction.

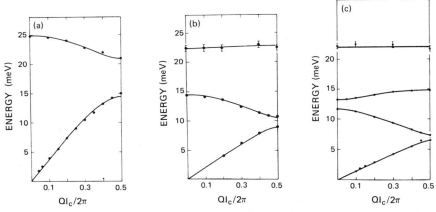

FIG. 48. Measured dispersion curves for L(001) modes in potassium-intercalated graphite compounds (a) C_8K, (b) $C_{24}K$, and (c) $C_{36}K$ at 296 K. Full lines are best fits with one-dimensional shell models. (From Zabel and Mager.[130])

In the absence of large single crystals, a rigorous measurement of dispersion curves for other modes is not possible. However, in an earlier study of phonons in pure HOPG, Nicklow et al.[131] noticed that the dispersion curves for modes with **q** in the basal plane and polarization along the c axis (the so-called layer bending or "rippling" modes in the basal plane) were almost isotropic throughout the Brillouin zone. Thus, by staying within the first zone, it is possible to measure these dispersion curves, and this has also been done for the GICs.[132,133] Some problems arise at small **q** owing to the mosaic spread of the HOPG samples, which ranged from 1.5° to 3°. Figure 49 shows results for stage-2 $C_{24}K$ and $C_{24}Rb$. Of interest is the pronounced curvature of the TA1 branch in the $(q00)$ direction. For "ripplon" modes within a single isolated layer, it has long been known[134] that the dispersion is of the form $\omega \sim q^2$ for small q. Introducing a coupling to the other layers introduces a shear modulus C_{44} and thus for small q, we have

$$\omega^2 \approx (C_{44}/\rho)q^2 + Bq^4 + \cdots. \qquad (8.47)$$

The neutron results indicate that the interlayer coupling, represented by the first term, is considerably reduced in the intercalated compounds relative to pure graphite, particularly in the stage-2 compounds. This has interesting implications for the Bragg reflection line shapes of these compounds, which have not yet been explored in detail. The higher branches have the form of the graphitic TA1 mode hybridizing with rather flat Einstein-like alkali metal optic modes. There have been several theoretical treatments of the lattice dynamics of these compounds in an attempt to account for the INS, Raman

8. ADSORBED MONOLAYERS AND INTERCALATED COMPOUNDS

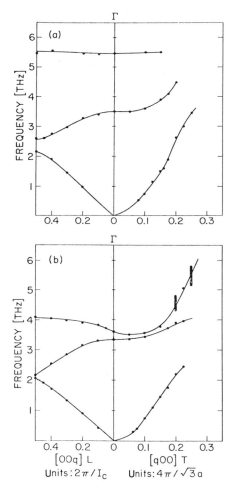

FIG. 49. Measured dispersion curves for [001]L and [100]TD⊥ modes in (a) $C_{24}K$ and (b) $C_{24}Rb$ at room temperature. (From Zabel et al.[133])

scattering, and elastic constant results.[135-137] Stage-1 C_6Li (where the Li atoms form a commensurate ($\sqrt{3} \times \sqrt{3}$) structure with the simple stacking sequence AαAα...) and stage-2 $C_{12}Li$ have also been studied by INS.[138,139] The basal plane TA1 modes still show upward curvature (q^2 behavior) but C_{44} is much larger, indicating a strong Li–C interaction, presumably due to increased ionicity of the lithium layers.

Surprisingly, at low temperatures, the effect of in-plane ordering of the alkali metal atoms does not seem to induce any splittings at the zone

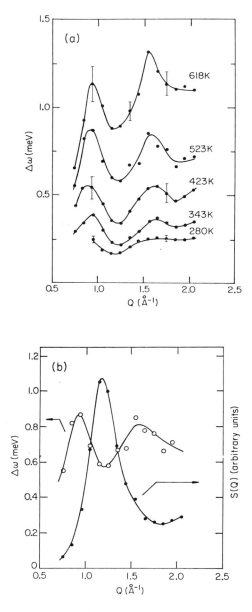

FIG. 50. (a) Energy widths (FWHM) of the Lorentzian component of the scattering law $S(\mathbf{Q}, \omega)$ of $C_{24}Rb$ as a function of Q for several temperatures. (b) The Q dependence of the energy width of $C_{24}Rb$ at 523 K is compared with the structure factor $S(Q)$ [obtained by integrating $S(\mathbf{Q}, \omega)$ with respect to ω]. (From Zabel et al.[142])

boundaries of this superlattice structure in the basal plane TA1 modes. Splittings of the lowest acoustic branch are seen in both the L($00q$) and TA1 ($q00$) branches in $C_{24}K$,[140] but this persists above the alkali order–disorder transition and is ascribed to hybridization with a low-lying, Einstein-like, in-band mode. Such an effect has also been seen in $C_{24}Rb$.[141] Its origin is not completely clear at the present time. A detailed high-resolution study of the phonons in the incommensurate phase would be fascinating from the point of view of questions relating to phasons, etc., but have not yet been carried out, presumably because single-crystal experiments are not possible.

Recently, Zabel et al.[142] have succeeded in studying the in-plane diffusional motion of alkali atoms in the disordered phase in $C_{24}Rb$ and $C_{24}Cs$ via QENS. Figure 50 shows the measured energy widths of the quasi-elastic lines from $C_{24}Rb$ as a function of neutron momentum transfer Q at various temperatures. It should be noted that the scattering is predominantly coherent quasi-elastic scattering, as contrasted to the incoherent QENS discussed in Section 8.8, which measures the self-correlation function of the diffusing particles. No rigorous theory yet exists to interpret coherent QENS, although computer simulations may provide some help with this in the near future, and the results have been analyzed in terms of a rather simplified theory of de Gennes narrowing,[143] in which

$$\Delta\omega = 2DQ^2/S(Q), \qquad (8.48)$$

$S(Q)$ being the structure factor of the (liquid-like) alkali atoms. The dips in $\Delta\omega$ in Fig. 50 certainly correlate quite well with the peaks in $S(Q)$ for $C_{24}Rb$. Using Eq. (8.48), the authors were able to extract values for D for rubidium and cesium in stage-2 intercalated compounds. These were of the order of $\sim 5 \times 10^{-5}$ cm^2/s at 523 K and much smaller than values observed in the corresponding bulk liquid alkali metals at this temperature. The results indicate dense and highly viscous fluids. An interesting further feature is the existence of phonon-like excitations at large wave vectors, in contrast to the situation in bulk liquids. Thus, a high degree of local atomic correlations must be present in the two-dimensional liquid.

Acknowledgments

I would like to express my gratitude to M. Chan, P. Dimon, S. Fain, D. Moncton, L. Passell, J. J. Rush, H. Taub, and H. Zabel for providing information extremely helpful to me in the course of writing this review.

References

1. W. C. Marra, P. H. Fuoss, and P. Eisenberger, *Phys. Rev. Lett.* **49**, 1169 (1982).
2. J. K. Kjems, L. Passell, H. Taub, and J. G. Dash, *Phys. Rev. Lett.* **33**, 724 (1974); J. K. Kjems, L. Passell, H. Taub, J. G. Dash, and A. D. Novaco, *Phys. Rev. B* **13**, 1446 (1976).

3. S. C. Fain, Jr., in "Chemistry and Physics of Solid Surfaces IV" (R. Vanselow and R. Howe, eds.), p. 203. Springer, New York, 1982; S. C. Fain, Jr., M. F. Toney, and R. D. Diehl, *Proc. Int. Vac. Congr., 9th, Int. Conf. Solid Surf., 5th* (J. L. deSegovia, ed.), p. 129. Imprenta Moderna, Madrid, 1983.
4. D. E. Moncton and G. S. Brown, *Nucl. Instrum. Methods* **208**, 579 (1983).
5. J. P. McTague and M. Nielsen, *Phys. Rev. Lett.* **37**, 596 (1976); M. Nielsen and J. P. McTague, *Phys. Rev. B* **19**, 3096 (1979).
6. Grafoil and UCAR-ZYX are the trade names of products marketed by the Union Carbide Corp., Carbon Prod. Div., 270 Park Ave., New York, N.Y. UCAR-ZYX is made by the same company.
7. J. Suzanne, J. P. Coulomb, M. Bienfait, J. Matecki, A. Thomy, B. Croset, and C. Marti, *Phys. Rev. Lett.* **41**, 760 (1978).
8. P. Vora, S. K. Sinha, and R. K. Crawford, *Phys. Rev. Lett.* **43**, 704 (1979); P. Vora, Ph.D. Thesis, Univ. of Illinois, Chicago Circle, 1980.
9. R. J. Birgeneau, E. M. Hammons, P. Heiney, P. W. Stephens, and P. M. Horn, in "Ordering in Two Dimensions" (S. K. Sinha, ed.), p. 29. North-Holland Publ., New York, 1980.
10. Graphon is made by the Cabot Corp., Billerica, MA 01821. It is characterized in M. H. Folley, W. D. Schaffer, and W. R. Smith, *J. Phys. Chem.* **57**, 469 (1953).
11. Made by Supelco, Inc., Bellefonte, PA 16823.
12. Made by Le Carbon Lorraine, 45 Rue des Acacois, F-75821, Paris, France.
13. A. Thomy and X. Duval, *J. Chim. Phys.* **67**, 287 (1970); **67**, 1101 (1970).
14. J. L. Jordan, J. P. McTague, L. Passell, and J. B. Hastings, *Bull. Am. Phys. Soc.* **28**, 870 (1983).
15. B. E. Warren, *Phys. Rev.* **59**, 693 (1941).
16. P. Dutta, S. K. Sinha, P. Vora, M. Nielsen, L. Passell, and M. Bretz, in "Ordering in Two Dimensions" (S. K. Sinha, ed.), p. 169. North-Holland Publ., New York, 1980.
17. P. Dimon, Ph.D. Thesis, Univ. of Chicago, Chicago, Illinois, 1984; P. Dimon, P. M. Horn, M. Sutton, R. J. Birgeneau, and D. E. Moncton, *Phys. Rev. B* **31**, 437 (1985).
18. J. W. White, R. K. Thomas, T. Trewern, I. Marlow, and G. Bomchil, *Surf. Sci.* **76**, 13 (1978).
19. J. P. Coulomb, M. Bienfait, and P. Thorel, *Phys. Rev. Lett.* **42**, 733 (1979).
20. M. Chan, private communication; also to be published.
21. R. Marx and E. F. Wassermann, *Solid State Commun.* **40**, 959 (1981); *Surf. Sci.* **117**, 267 (1982).
22. J. H. Quateman and M. Bretz, *Phys. Rev. B* **29**, 1159 (1984). The phase diagram given here agrees with that in Ref. 16 if allowance is made for the fact that the monolayers referred to here refers to completed *dense* monolayers, as opposed to completed registered monolayers.
23. A. Glachant, J. P. Coulomb, M. Bienfait, P. Thorel, C. Marti, and J. G. Dash, in "Ordering in Two Dimensions" (S. K. Sinha, ed.), p. 203. North-Holland Publ., New York, 1980.
24. T. Rayment, R. K. Thomas, G. Bomchil, and J. W. White, *Mol. Phys.* **43**, 601 (1981).
25. J. Eckert, W. D. Ellenson, J. B. Hastings, and L. Passell, *Phys. Rev. Lett.* **43**, 1339 (1979).
26. C. R. Fuselier, N. S. Gillis, and J. C. Raich, *Solid State Commun.* **25**, 747 (1978).
27. A. B. Harris and A. J. Berlinsky, *Can. J. Phys.* **57**, 1852 (1979).
28. S. F. O'Shea and M. L. Klein, *Chem. Phys. Lett.* **66**, 381 (1979).
29. R. D. Diehl, M. F. Toney, and S. C. Fain, Jr., *Phys. Rev. Lett.* **48**, 177 (1982); see cf also R. D. Diehl and S. C. Fain, Jr., *Surf. Sci.* **125**, 116 (1983).
30. M. H. W. Chan, A. D. Migone, K. D. Miner, and Z. R. Li, *Phys. Rev. B* **30**, 2681 (1984).

31. A. N. Berker, *in* "Ordering in Two Dimensions" (S. K. Sinha, ed.), p. 9. North-Holland Publ., New York, 1980.
32. A. N. Berker and S. Ostlund, *J. Phys. C* **12**, 4691 (1979).
33. P. A. Heiney, P. W. Stephens, S. G. J. Mochrie, J. Akimitsu, and R. J. Birgeneau, *Surf. Sci.* **125**, 539 (1983).
34. M. F. Toney and S. C. Fain, Jr., *Phys. Rev. B* **30**, 1115 (1984).
35. M. Nielsen and W. Ellenson, *Proc. Int. Conf. Low Temp. Phys., 14th* (M. Krusis and M. Vuorio, eds.), Vol. 4, p. 437. North-Holland Publ., Amsterdam, 1975.
36. W. E. Carlos, M. W. Cole, S. Rauber, G. Vidali, A. F. Silva-Moreira, J. L. Codona, and D. L. Goodstein, *in* "Ordering in Two Dimensions" (S. K. Sinha, ed.), p. 263. North-Holland Publ., New York, 1980.
37. M. Bretz, *Phys. Rev. Lett.* **38**, 501 (1971).
38. B. Nienhuis, A. N. Berker, E. K. Riedel, and M. Schick, *Phys. Rev. Lett.* **43**, 737 (1979).
39. K. Carneiro, L. Passell, W. Thomlinson, and H. Taub, *Phys. Rev. B* **24**, 1170 (1981).
40. H. J. Lauter, H. Wiechert, and R. Feile, *in* "Ordering in Two Dimensions" (S. K. Sinha, ed.), p. 291. North-Holland Publ., New York, 1980.
41. M. J. Tejwani, O. Ferreira, and O. E. Vilches, *Phys. Rev. Lett.* **44**, 152 (1980).
42. H. Wiechert, C. Tiby, and H. J. Lauter, *Physica B + C (Amsterdam)* **108B + C**, 785 (1981).
43. G. B. Huff and J. G. Dash, *J. Low Temp. Phys.* **24**, 155 (1976).
44. H. Taub, K. Carneiro, J. K. Kjems, L. Passell, and J. P. McTague, *Phys. Rev. B* **16**, 4551 (1977).
45. G. J. Trott, H. Taub, F. Y. Hansen, and H. R. Danner, *Chem. Phys. Lett.* **78**, 504 (1981); G. J. Trott, Ph.D. Thesis, Univ. of Missouri, Columbia, 1981.
46. H. Taub, G. J. Trott, F. Y. Hansen, H. R. Danner, J. P. Coulomb, J. P. Biberian, J. Suzanne, and A. Thomy, *in* "Ordering in Two Dimensions" (S. K. Sinha, ed.), p. 91. North-Holland Publ., New York, 1980.
47. J. P. Coulomb, J. P. Biberian, J. Suzanne, A. Thomy, G. J. Trott, H. Taub, H. R. Danner, and F. Y. Hansen, *Phys. Rev. Lett.* **43**, 1878 (1979).
48. M. Monkenbusch and R. Stockmayer, *in* "Ordering in Two Dimensions" (S. K. Sinha, ed.), p. 223. North-Holland Publ., New York, 1980.
49. M. Monkenbusch and R. Stockmayer, *Ber. Bunsenges. Phys. Chem.* **84**, 808 (1980).
50. S. K. Satija, L. Passell, J. Eckert, W. Ellenson, and H. Patterson, *Phys. Rev. Lett.* **51**, 411 (1983).
51. M. Sutton, S. G. J. Mochrie, and R. J. Birgeneau, *Phys. Rev. Lett.* **51**, 407 (1983).
52. J. Suzanne, J. L. Seguin, H. Taub, and J. Biberian, *Surf. Sci.* **125**, 153 (1983).
53. R. Wang, H. Taub, H. Schechter, R. Brener, J. Suzanne, and F. Y. Hansen, *Phys. Rev. B* **27**, 5864 (1983).
54. H. J. Lauter, B. Croset, C. Marti, and P. Thorel, *in* "Ordering in Two Dimensions" (S. K. Sinha, ed.), p. 211. North-Holland Publ., New York, 1980.
55. K. Kjaer, M. Nielsen, J. Bohr, H. J. Lauter, and J. P. McTague, *Phys. Rev. B* **26**, 5168 (1983).
56. S. Nagler, P. M. Horn, S. K. Sinha, P. Dutta, and D. Moncton, to be published.
57. B. Croset, C. Marti, P. Thorel, and H. J. Lauter, *J. Phys. (Orsay, Fr.)* **43**, 1659 (1982).
58. S. K. Sinha, P. Dutta, M. Nielsen, P. Vora, M. Bretz, and L. Passell, to be published.
59. B. I. Halperin and D. R. Nelson, *Phys. Rev. Lett.* **41**, 121 (1978); **41**, 519 (1978); D. R. Nelson and B. I. Halperin, *Phys. Rev. B* **19**, 2457 (1979).
60. P. Bak, *Rep. Prog. Phys.* **45**, 587 (1982); J. Villain, *in* "Ordering in Two Dimensions" (S. K. Sinha, ed.), p. 123. North-Holland Publ., New York, 1980.
61. F. C. Frank and J. H. Van der Merwe, *Proc. R. Soc. London* **198**, 205 (1949).
62. P. Bak, D. Mukamel, J. Villain, and K. Wentowska, *Phys. Rev. B* **19**, 1610 (1979).

63. V. L. Pokrovskii and A. L. Talapov, *Phys. Rev. Lett.* **42**, 66 (1979).
64. H. Shiba, *J. Phys. Soc. Jpn.* **48**, 211 (1980).
65. P. W. Stephens, P. Heiney, R. J. Birgeneau, and P. M. Horn, *Phys. Rev. Lett.* **43**, 47 (1979).
66. S. C. Fain, Jr., M. D. Chinn, and R. D. Diehl, *Phys. Rev. B* **21**, 4170 (1980).
67. M. Nielsen, J. Als-Nielsen, J. Bohr, and J. P. McTague, *Phys. Rev. Lett.* **47**, 582 (1981).
68. D. E. Moncton, P. W. Stephens, R. J. Birgeneau, P. M. Horn, and G. S. Brown, *Phys. Rev. Lett.* **46**, 1533 (1981).
69. E. D. Specht, M. Sutton, R. J. Birgeneau, D. E. Moncton, and P. M. Horn, *Phys. Rev. B* **30**, 1589 (1984).
70. S. N. Coppersmith, D. S. Fisher, B. I. Halperin, P. A. Lee, and W. F. Brinkman, *Phys. Rev. B* **25**, 349 (1982); D. A. Huse and M. E. Fisher, *Phys. Rev. Lett.* **49**, 793 (1982).
71. A. D. Novaco and J. P. McTague, *Phys. Rev. Lett.* **38**, 1286 (1977).
72. K. L. D'Amico, D. E. Moncton, E. D. Specht, R. J. Birgeneau, S. E. Nagler, and P. M. Horn, *Phys. Rev. Lett.* **53**, 2230 (1984).
73. L. D. Landau and E. M. Lifshitz, "Statistical Physics," p. 478. Pergamon, Oxford, 1969; R. Peierls, *Helv. Phys. Acta* **7**, Suppl. II, 81 (1934); N. D. Mermin, *Phys. Rev.* **176**, 250 (1968).
74. B. Jancovici, *Phys. Rev. Lett.* **19**, 20 (1967).
75. Y. Imry and L. Gunther, *Phys. Rev. B* **3**, 3939 (1971); Y. Imry, *CRC Crit. Rev. Solid State Sci.* **8**, 157 (1979).
76. D. E. Moncton and R. Pindak, in "Ordering in Two Dimensions" (S. K. Sinha, ed.), p. 83. North-Holland Publ., New York, 1980; D. E. Moncton, R. Pindak, S. C. Davey, and G. S. Brown, *Phys. Rev. Lett.* **49**, 1865 (1982).
77. P. Dutta and S. K. Sinha, *Phys. Rev. Lett.* **47**, 50 (1981).
78. J. M. Kosterlitz and D. J. Thouless, *J. Phys. C* **5**, L124 (1972); **6**, 1181 (1973).
79. A. P. Young, *Phys. Rev. B* **19**, 1855 (1979).
80. S. K. Sinha, P. Vora, P. Dutta, and L. Passell, *J. Phys. C* **15**, L275 (1982).
81. P. A. Heiney, P. W. Stephens, R. J. Birgeneau, P. M. Horn, and D. E. Moncton, *Phys. Rev. B* **23**, 6416 (1983).
82. J. P. McTague, J. Als-Nielsen, J. Bohr, and M. Nielsen, *Phys. Rev. B* **25**, 7765 (1982).
83. T. F. Rosenbaum, S. E. Nagler, P. M. Horn, and R. Clarke, *Phys. Rev. Lett.* **50**, 1791 (1983).
84. F. F. Abraham, *Phys. Rev. B* **29**, 2606 (1984).
85. R. Kalia, P. Vashishta, and F. W. de Leeuw, *Phys. Rev. B* **23**, 4794 (1981).
86. S. T. Chui, in "Melting, Localization and Chaos" (R. K. Kalia and P. Vashishta, eds.), p. 29. North-Holland Publ., New York, 1982.
87. S. C. Davey, J. Budai, J. W. Goodby, R. Pindak, and D. E. Moncton, *Phys. Rev. Lett.* **53**, 2129 (1984).
88. P. M. Horn, R. J. Birgeneau, P. Heiney, and E. M. Hammonds, *Phys. Rev. Lett.* **41**, 961 (1978).
89. S. Ostlund and A. N. Berker, *Phys. Rev. Lett.* **42**, 843 (1979); *J. Phys. C* **12**, 4961 (1979).
90. R. Pandit, M. Schick, and M. Wortis, *Phys. Rev. B* **26**, 5112 (1982).
91. S. Toxvaerd, *Phys. Rev. Lett.* **43**, 529 (1979).
92. B. H. Grier, L. Passell, J. Eckert, H. Patterson, D. Richter, and R. J. Rollefson, *Phys. Rev. Lett.* **53**, 814 (1984).
93. J. P. Coulomb, M. Bienfait, and P. Thorel, *J. Phys. (Orsay, Fr.)* **42**, 293 (1981).
94. S. Nosé and M. L. Klein, *Phys. Rev. Lett.* **53**, 818 (1984).
95. H. J. Lauter, S. K. Sinha, and P. Obermeyer, to be published.

8. ADSORBED MONOLAYERS AND INTERCALATED COMPOUNDS

96. B. P. Schweiss, B. Renker, E. Schneider, and W. Reichardt, *in* "Superconductivity in d- and f-band Metals" (D. H. Douglass, ed.), p. 189. Plenum, New York, 1976.
97. S. D. Bader, S. K. Sinha, and R. N. Shelton, *in* "Superconductivity in d- and f-band Metals" (D. H. Douglass, ed.), p. 209. Plenum, New York, 1976.
98. C. L. Cleveland, C. S. Brown, and U. Landman, *in* "Ordering in Two Dimensions" (S. K. Sinha, ed.), p. 207. North-Holland Publ., New York, 1980.
99. M. W. Newbery, T. Rayment, M. V. Smalley, R. K. Thomas, and J. W. White, *Chem. Phys. Lett.* **59**, 461 (1978); M. V. Smalley, A. Huller, R. K. Thomas, and J. W. White, *Mol. Phys.* **44**, 533 (1981).
100. J. E. Rutledge, W. L. McMilland, J. M. Mochel, and T. E. Washburn, *Phys. Rev. B* **18**, 2155 (1978).
101. D. J. Bishop, J. M. Parpia, and J. D. Reppy, *in* "Low Temperature Physics, LT-14" (M. Krusius and M. Vusrio, eds.), Vol. 1, p. 380. Am. Elsevier, New York, 1975.
102. D. F. Brewer, A. J. Symonds, and A. L. Thompson, *Phys. Rev. Lett.* **15**, 182 (1965).
103. K. Carneiro, W. D. Ellenson, L. Passell, J. P. McTague, and H. Taub, *Phys. Rev. Lett.* **37**, 1695 (1976).
104. W. Tomlinson, J. A. Tarvin, and L. Passell, *Phys Rev. Lett.* **44**, 266 (1980).
105. H. J. Lauter, H. Wiechert, and C. Tiby, *Physica B + C (Amsterdam)* **107B + C**, 239 (1981).
106. H. J. Lauter, private communication.
107. H. Taub, H. R. Danner, Y. P. Sharma, H. L. McMurry, and R. M. Brugger, *Phys. Rev. Lett.* **39**, 215 (1977).
108. R. Wang, H. R. Danner, and H. Taub, *in* "Ordering in Two Dimensions" (S. K. Sinha, ed.), p. 219. North-Holland Publ., New York, 1980.
109. R. R. Cavanagh, R. D. Kelley, and J. J. Rush, *J. Chem. Phys.* **77**, 1540 (1982).
110. R. D. Kelley, R. R. Cavanagh, and J. J. Rush, *J. Catal.* **83**, 464 (1983).
111. R. R. Cavanagh, J. J. Rush, R. D. Kelley, and T. J. Udovic, *J. Chem. Phys.* **80**, 3478 (1984).
112. G. P. Felcher, R. T. Kampwirth, K. E. Gray, and R. Felici, *Phys. Rev. Lett.* **52**, 1539 (1984).
113. G. P. Felcher, *Phys. Rev. B* **24**, 1595 (1981).
114. G. P. Felcher, S. D. Bader, R. J. Celotta, D. T. Pierce, and G. C. Wang, *in* "Ordering in Two Dimensions" (S. K. Sinha, ed.), p. 107. North-Holland Publ., New York, 1980.
115. M. S. Dresselhaus and G. Dresselhaus, *Adv. Phys.* **30**, 139 (1981).
116. *Proceedings of the International Conference on Graphite Intercalation Compounds, 3rd, Port à Mousson, France; Synth. Met.* **8** (1983).
117. J. E. Fischer, *Physica B + C (Amsterdam)* **99B + C**, 383 (1980).
118. F. J. DiSalvo, *in* "Electron–Phonon Interactions and Phase Transitions" (T. Riste, ed.), p. 107. Plenum, New York, 1977.
119. N. Wakabayashi, *in* "Electrons and Phonons in Layered Crystal Structures" (T. J. Wieting and M. Schluter, eds.), p. 409. Reidel, Dordrecht, Netherlands, 1979.
120. L. B. Ebert, *Annu. Rev. Mater. Sci.* **6**, 181 (1976).
121. A. Hérold, *in* "Physics and Chemistry of Materials with Layered Structures" (F. Lévy, ed.), Vol. 6, p. 323. Reidel, Dordrecht, Netherlands, 1979.
122. G. S. Parry, D. E. Nixon, K. M. Lester, and B. C. Levene, *J. Phys. C* **2**, 2156 (1969).
123. W. D. Ellenson, D. Semmingsen, D. Guerard, D. G. Onn, and J. E. Fischer, *Mater. Sci. Eng.* **31**, 137 (1977).
124. K. Watanabe, M. Soma, T. Onishi, and K. Tamaru, *Proc. R. Soc. London, Ser. A* **333**, 51 (1973).
125. H. Zabel, J. J. Rush, and A. Magerl, *Synth. Met.* **7**, 251 (1983).
126. M. Suzuki, H. Ikeda, H. Suematsu, Y. Endoh, H. Shiba, and M. T. Hutchings, *J. Phys. Soc. Jpn.* **49**, 671 (1980).

127. H. Zabel, in "Ordering in Two Dimensions" (S. K. Sinha, ed.), p. 61. North-Holland Publ., New York, 1980.
128. R. Clarke, Ref. 127, p. 53.
129. R. Clarke, N. Caswell, and S. A. Solin, *Phys. Rev. Lett.* **42**, 61 (1979).
130. H. Zabel and A. Magerl, *Phys. Rev. B* **25**, 2463 (1982).
131. R. Nicklow, N. Wakabayashi, and H. G. Smith, *Phys. Rev. B* **5**, 4951 (1972).
132. W. A. Kamitakahara, N. Wada, and S. A. Solin, *Solid State Commun.*
133. H. Zabel, W. A. Kamitakahara, and R. M. Nicklow, *Phys. Rev. B* **26**, 5919 (1982).
134. Lord Rayleigh, "Theory of Sound," Vol. 1, p. 352. Macmillan, New York, 1984.
135. C. Horie, M. Maeda, and Y. Kuramoto, *Physica (Amsterdam)* **99**, 430 (1980).
136. S. Y. Leung, G. Dresselhaus, and M. S. Dresselhaus, *Phys. Rev. B* **24**, 6083 (1981).
137. R. Al-Jishi and G. Dresselhaus, *Phys. Rev. B* **26**, 4514 (1982).
138. H. Zabel, A. Magerl, and J. J. Rush, *Phys. Rev. B* **27**, 3930 (1983).
139. K. C. Woo, W. A. Kamitakahara, D. P. DiVincenzo, D. S. Robinson, H. Mentwoy, J. W. Milliken, and J. E. Fischer, *Phys. Rev. Lett.* **50**, 1982 (1983).
140. A. Magerl, H. Zabel, and J. J. Rush, *Synth. Met.* **7**, 339 (1983).
141. S. Funahashi, T. Kondow, and M. Iizumi, *Solid State Commun.* **44**, 1515 (1982).
142. H. Zabel, A. Magerl, A. J. Dianoux, and J. J. Rush, *Phys. Rev. Lett.* **50**, 2094 (1983).
143. P. D. de Gennes, *Physica (Amsterdam)* **25** (1959).

9. DEFECTS IN SOLIDS

Werner Schmatz*

Kernforschungszentrum Karlsruhe
Institut für Nukleare Festkörperphysik
D-7500 Karlsruhe, Federal Republic of Germany

9.1. Introduction

How much have neutrons contributed to a deeper understanding of disordered solids? As a complementary tool to many other experimental techniques, neutron scattering has indeed given a broad variety of specific answers for the structure and dynamics of disordered solids. Depending on the exact definition of this chapter's title, there are between 10 and 20 review articles and between 200 and 1000 original papers concerned with the topic. The condensation of this huge amount of information into a single chapter for a new book results in a short description of the principal ideas and the general trends without the possibility of accounting for all the beautiful details of the individual scientists' efforts.

The first task is to define the frame. A possible way is to introduce, into the ideal periodic lattice, defects with increasingly complex disorder. We start with point defects as substitutional atoms, vacancies, and interstitials in dilute concentration. There are two ways toward higher complexity: (1) the defects can be considered on an atomic scale, but they are present at high concentrations and (2) instead of simple point defects, we have defect agglomerates but still in dilute concentration. Finally, the most complex case of disorder in a crystal—still keeping the periodic structure—is realized by defect agglomerates at high concentrations.

We will follow approximately in Section 9.2 the sequence of Krivoglaz[1] in the first part of his book "Theory of X-Ray and Thermal Neutron Scattering by Real Crystals." For understanding the details of the various scattering phenomena this textbook is an important base, especially for elastic nuclear scattering.

---

*Professor Werner Schmatz died unexpectedly on November 13, 1986. I will always remember him. It has been a special honor for me to edit and complete this chapter. If there is anything overlooked or not as clearly expressed as it would have been by the author himself, I ask the reader's forgiveness—Klaus Werner, March 1987.

With neutrons we can look at either the structure or the dynamics of a disordered system by elastic or inelastic scattering, and these are further subdivided into nuclear or magnetic scattering. Although, in terms of physics, in a thorough investigation all four types of neutron scattering might be applied to one individual disordered solid, in practice most original papers emphasize one of the four topics. Furthermore, there are many more studies on static than on dynamical properties and practically no case exists where a disordered system has been thoroughly investigated by nuclear *and* magnetic as well as by elastic *and* inelastic scattering. A further complexity arises from the many different materials, e.g., metals, ionic crystals, molecular crystals, covalent crystals, and so on. Thus, type of disorder, type of scattering, and type of material span a three-dimensional space with $4 \times 4 \times n$ boxes, some of them loaded with numerous investigations, some of them still empty today.

Faced with the problem of describing at least part of the work in a one-dimensional sequence, we present in Section 9.2 static properties, i.e., elastic scattering, in a sequence of increasing complexity (9.2.1) and after that two sections (9.2.2 and 9.2.3) describing very recent developments. Dynamic properties (9.3) are discussed only shortly, with main emphasis on quasi-elastic scattering. Magnetic scattering is always described after nuclear scattering, and no attempt is made to present the different type of materials in sequence, quite in accordance with the widespread and sometimes fleeting interests of the scientists. For the newcomer a few introductory sections with elementary explanations are included.

9.1.1. Scattering Functions

The connecting link between experiment and theory is

$$\frac{d^2\sigma}{d\Omega\, d\omega} = \frac{k_1}{k_0} \hat{S}(\mathbf{Q}, \omega), \tag{9.1}$$

where $d^2\sigma/d\Omega\, d\omega$ is the double differential scattering cross section evaluated from the scattering experiment by absolute calibration, subtraction of background, and proper corrections for multiple scattering;[2] $\mathbf{k}_1$ and $\mathbf{k}_0$ are the wave vectors of the scattered and the incident beam, respectively; $\hbar\mathbf{Q} = \hbar(\mathbf{k}_0 - \mathbf{k}_1)$ is the momentum transfer and $\hbar\omega$ is the energy transfer. The general scattering function $\hat{S}$ is a sum of individual scattering functions $S_{\nu,\mathrm{i}}$ and $S_{\nu\mu}$:

$$\hat{S}(\mathbf{Q}, \omega) = \sum_{\nu=1}^{\nu_0} \frac{\sigma_{\mathrm{i},\nu}}{4\pi} S_{\nu,\mathrm{i}}(\mathbf{Q}, \omega) + \sum_{\nu\mu=1}^{\nu_0} b_\nu b_\mu S_{\nu\mu}(\mathbf{Q}, \omega). \tag{9.2}$$

The sums are extended over all ν_0 individual elements in the scattering sample, with b_ν the coherent scattering length and $\sigma_{\mathrm{i},\nu}$ the incoherent scattering cross section[2] of the individual element v. The b_ν are normally sufficiently well known, whereas the values for σ_i are much less accurate, especially if they are small. A comprehensive listing of scattering lengths and cross sections is given in Appendix (Part A). For defect scattering the inaccuracies in σ_i and sometimes in b can be a very limiting factor, and improvement would be desirable. If isotopically enriched elements are used, the values of σ_i and b of the special isotope mixture have to be used for Eq. (9.2). In cases where different isotopes behave structurally and dynamically different, they have to be considered separately. Up to now this seems to be important only for hydrogen and deuterium or hydrogen–deuterium mixtures.

The functions $S_{\nu,i}$ and $S_{\nu\mu}$ can be expressed in terms of the $(\mathbf{Q}, \omega)$, Fourier transforms $\tilde{G}(\mathbf{Q}, \omega)$ of the van Hove space-, and time-correlation functions:[2]

$$S_{\nu,i}(\mathbf{Q}, \omega) = c_\nu \tilde{G}_{\nu,s}(\mathbf{Q}, \omega) \tag{9.3}$$

$$S_{\nu\mu}(\mathbf{Q}, \omega) = c_\nu c_\mu \tilde{G}_{\nu\mu}(\mathbf{Q}, \omega). \tag{9.4}$$

Here $G_{\nu,s}(\mathbf{r}, t)$ is the van Hove self-correlation function for element v and $G_{\nu\mu}(\mathbf{r}, t)$ is the van Hove pair correlation function combining $\nu\mu$ pairs with $\nu\nu$ pairs included in the definition. The concentrations c_ν are introduced for convenience in dealing with disordered crystals; the sum of all c_ν equals 1.

For defect scattering work Eqs. (9.2)–(9.4) prove to be very convenient, especially with a proper idea about the correlation functions, even if this is only a classical one as a starting point.[3] Disregarding, at first, diffusion of atoms, both $G_{\nu,s}$ and $G_{\nu\mu}$ have finite asymptotic values for $t \to \infty$ and the Fourier transforms $\tilde{G}_{\nu,s}$ and $\tilde{G}_{\nu\mu}$ contain $\delta(\omega)$ contributions [$\delta(\omega)$ is the dirac function with $\int \delta(\omega)\, d\omega = 1$]. The integral over ω at $\omega = 0$ is the elastic scattering cross section to be evaluated with Eqs. (9.1)–(9.4) or directly written by

$$(d\sigma/d\Omega)_{\text{el}} = (d\sigma/d\Omega)_{\text{el},i} + (d\sigma/d\Omega)_{\text{el},c}, \tag{9.5}$$

with

$$\left(\frac{d\sigma}{d\Omega}\right)_{\text{el},i} = \sum_{\nu=1}^{\nu_0} c_\nu \frac{\sigma^0_{i,\nu}}{4\pi} = \sum_{\nu=1}^{\nu_0} c_\nu \frac{\sigma_{i,\nu}}{4\pi} \overline{e^{-2W_\nu}} \tag{9.6}$$

and

$$\left(\frac{d\sigma}{d\Omega}\right)_{\text{el},c} = \frac{1}{N}\left|\sum_{n=1}^{N} b^0_n e^{i\mathbf{Q}\cdot\mathbf{r}_n}\right|^2 = \frac{1}{N}\left|\overline{\sum_{n=1}^{N} b_n e^{-W_n} e^{i\mathbf{Q}\cdot\mathbf{r}_n}}\right|^2. \tag{9.7}$$

The sum over N atoms in Eq. (9.7) may be considered as a sum over a sufficiently large microcrystal to contain all the average information on the defect structure obtainable by a scattering experiment. Also an average over many sufficiently large-sized microcrystals may be a proper starting point to evaluate Eq. (9.7). (For a more detailed discussion of this special point see

Refs. 1–5.) Scattering from defects like dislocations or grain boundaries is better discussed by introducing a density of scattering lengths (see Section 9.2.3).

The elastic incoherent scattering cross section [Eq. (9.6)], in accordance with its name, contains no information on the relative positions of the atoms. In favorable cases the determination of the Debye–Waller factor via Eq. (9.6) is an important tool. Exceptional in this sense is hydrogen in a host lattice like niobium with a nearly vanishing incoherent scattering cross section. For defect scattering, however, the coherent elastic scattering obtained after correction for the incoherent elastic scattering is of far greater interest. It should be noted that many elements have rather large values for σ_i, which makes the determination of the coherent elastic scattering fairly difficult, at least at dilute concentrations. There would, for instance, be no chance to measure diffuse scattering from 0.1 at. % Cu in vanadium.

We will not go into a detailed discussion of inelastic scattering within this chapter, but we have to be aware of the inelastic scattering contribution to the experimentally measured elastic scattering cross section from an experimental viewpoint. Though we have the ability with neutrons to separate elastic and inelastic scattering, the "elastic" experiment is always performed with a finite resolution $\hbar\delta\omega$. Let us assume $\hbar\delta\omega \leq 1$ meV. Then most of the vibrational spectrum is outside of this energy window. The remainder are multiphonon processes and, near reciprocal lattice points, acoustic phonons, because $\hbar\omega_{ph} \to 0$ for $q \to 0$ with $q = \mathbf{Q} - \tau_{hkl}$, where τ_{hkl} is a reciprocal lattice vector. This is one factor to be considered, while another is the diffusion of the atoms themselves. As long as the residence time τ of the atoms at a lattice or at an interstitial site is considerably longer than $1/\delta\omega$, the neutron sees the scattering object at its instantaneous site and the atom contributes to the scattering intensity with the "scattering amplitude" of $b_n \exp(-W_n) \exp(i\mathbf{Q} \cdot \mathbf{r}_n)$ for coherent scattering and with $\sqrt{\sigma_i/4\pi} \exp(-W_n)$ for incoherent scattering. The factorization of the Debye–Waller factor is correct if the majority of the vibrational motions contain frequencies larger than $\delta\omega$. [The factor $\exp(i\mathbf{Q} \cdot \mathbf{r}_n)$ is the positional phase factor for coherent scattering with $\mathbf{r}_n$ measured for all atoms from an arbitrary origin in real space.] For hydrogen or deuterium in metals, the residence time at high temperatures may well be shorter than $\hbar/(1 \text{ meV})$, forcing us in such cases or other similar ones to more detailed considerations. With high-resolution neutron spectrometers (back scattering and spin-echo preferably), $\hbar\delta\omega$ values of 1 μeV and smaller can be achieved, and the broadening of the "elastic" scattering into quasi-elastic scattering by diffusion of the atoms can be observed as discussed in part in Section 9.3.1.

Magnetic scattering and interference of magnetic scattering with nuclear scattering is a powerful domain for neutrons because of the lack of other similarly informative scattering processes for bulk materials. In a scattering experiment provided with polarization analysis we have not one but four distinctly different double differential scattering cross sections:

$$\frac{d^2\sigma}{d\Omega\, d\omega} \to \frac{d^2\sigma^{\uparrow\uparrow}}{d\Omega\, d\omega}, \frac{d^2\sigma^{\uparrow\downarrow}}{d\Omega\, d\omega}, \frac{d^2\sigma^{\downarrow\uparrow}}{d\Omega\, d\omega}, \frac{d^2\sigma^{\downarrow\downarrow}}{d\Omega\, d\omega}, \quad (9.8)$$

of which two are spin-flip and the other non-spin-flip processes. A neutron beam polarized parallel or antiparallel to a given axis at the sample position changes its polarization when spin flip occurs. (The possibility that the scattering system turns the direction of the polarization is, in the Born approximation, only to be considered for an inequivalent quantity of left- and right-hand helical magnetic moment arrangements,[3,6] a case that has never been studied, at least in defect scattering.) Several additional aspects are important.

(1) Nuclear spin incoherent scattering also gives spin flips.
(2) For samples with a finite-bulk magnetization M_B, the polarization p_0 of the incident beam has to be parallel to M_B, otherwise the polarization vectors p_0 and p_1 would turn around M_B with fixed angle and scattering would take place under "uncontrolled" conditions.
(3) A similar situation occurs for a macroscopic sample with $M_B = 0$, which has large magnetic domains inside.
(4) A single-domain antiferromagnetic crystal ($M_B = 0$ also for the microcrystals) has distinctly different scattering cross sections for the colinear magnetic moment axis parallel or perpendicular to the neutron beam polarization p_0 (Fig. 1).

For nuclear coherent plus magnetic scattering from rigid-spin, magnetic-moment systems, we can write for the scattering cross section[3,6]

$$\frac{d^2\sigma}{d\Omega\, d\omega} = \frac{1}{N}\frac{k_1}{k_0} \cdot \frac{1}{2\pi} \int dt\, e^{-i\omega t} \sum_{m,n}^{N} e^{-i\mathbf{Q}\cdot\mathbf{r}_m(0)} e^{i\cdot\mathbf{r}_n(t)} \sum_{v=1}^{8} T_v, \quad (9.9)$$

$$\mathbf{p}_1 \frac{d^2\sigma}{d\Omega\, d\omega} = \frac{1}{N}\frac{k_1}{k_0} \cdot \frac{1}{2\pi} \int dt\, e^{-i\omega t} \sum_{m,n}^{N} e^{-i\mathbf{Q}\cdot\mathbf{r}_m(0)} e^{i\cdot\mathbf{r}_n(t)} \sum_{v=1}^{8} \mathbf{P}_v, \quad (9.10)$$

where the values of T_v and $\mathbf{P}_v$ are given in Table I and $\mathbf{r}_m(0)$ and $\mathbf{r}_n(t)$ are time-dependent Heisenberg operators, which for elastic scattering reduce to $\exp(-\mathbf{Q}\cdot\mathbf{r}_m)\exp(-W_m)$ and $\exp(-\mathbf{Q}\cdot\mathbf{r}_n)\exp(-W_n)$, respectively. The

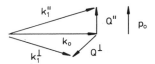

FIG. 1. Geometrical sketch of a scattering experiment with momentum transfer parallel ($Q^{\parallel}$) and perpendicular ($Q^{\perp}$) to the neutron beam polarization (p_0).

TABLE I. Values of T_v and P_v for Different v's

v	T_v	P_v
1	$a_m a_n$	0
2	0	$a_m a_n \mathbf{p}_0$
3	0	$a_m m_n \langle \mathbf{S}_n^\perp \rangle$
4	0	$a_n m_m^* \langle \mathbf{S}_m^\perp \rangle$
5	$a_m m_n \langle \mathbf{S}_n^\perp \rangle \cdot \mathbf{p}_0$	$i a_m m_n \langle \mathbf{S}_m^\perp \rangle \cdot \mathbf{p}_0$
6	$a_n m_m^* \langle \mathbf{S}_n^\perp \rangle \cdot \mathbf{p}_0$	$-a_n m_m^* \langle \mathbf{S}_m^\perp \rangle \cdot \mathbf{p}_0$
7	$m_m^* m_n \langle \mathbf{S}_m^\perp(0) \cdot \mathbf{S}_n^\perp(t) \rangle$	$i m_m^* m_n \langle \mathbf{S}_m^\perp(0) \cdot \mathbf{S}_n^\perp(t) \rangle$
8	$i m_m^* m_n \langle \mathbf{S}_m^\perp(0) \cdot \mathbf{S}_n^\perp(t) \rangle \cdot \mathbf{p}_0$	$m_m^* m_n \{ \langle \mathbf{S}_m^\perp(0)(\mathbf{S}_n^\perp(t) \cdot \mathbf{p}_0) \rangle$ $+ \langle (\mathbf{S}_m^\perp(0) \cdot \mathbf{p}_0) \mathbf{S}_n^\perp(t) \rangle - \mathbf{p}_0 \langle \mathbf{S}_m^\perp(0) \cdot \mathbf{S}_n^\perp(t) \rangle \}$

functions $\mathbf{S}_m$ and $\mathbf{S}_n$ are the total electron spin operators at the atoms m and n, respectively; $\mathbf{S}_m^\perp$ and $\mathbf{S}_n^\perp$ are the components of those vectors perpendicular to the scattering vector $\mathbf{Q}$. The m_n are defined as $m_n = r_0 |\gamma| F_n(Q)$ with $|\gamma| = 1.91$, $r_0 = 2.8 \times 10^{-13}$ cm, and the form factor $F_n(Q \to 0) = 1$. The use of Eqs. (9.9) and (9.10) in combination with Table I has the advantage of keeping an open mind for all combinations of S-vector arrangements relative to $\mathbf{p}_0$ and $\mathbf{Q}$. Extension of Eqs. (9.9) and (9.10) to non-spin-only magnetism can be found in textbooks.[7] The polarization vector of the scattered beam $\mathbf{p}_1$ is the ratio of Eq. (9.10) to Eq. (9.9), whereby some care is necessary with respect to domain averaging. For the final polarization $\mathbf{p}_1$ parallel or anti-parallel to $\mathbf{p}_0$, in addition to the elements T_2, T_3, T_4, and $\mathbf{P}_1$ the elements T_8, $\mathbf{P}_5$, $\mathbf{P}_6$, and $\mathbf{P}_7$ also vanish,[3] and we obtain

$$\frac{d\sigma^{\uparrow\uparrow}}{d\Omega \, d\omega} = Q(T_1 + T_5 + T_6 + T_{7z}), \quad (9.11)$$

$$\frac{d\sigma^{\uparrow\downarrow}}{d\Omega \, d\omega} = Q(T_{7x} + T_{7y}), \quad (9.12)$$

$$\frac{d\sigma^{\downarrow\uparrow}}{d\Omega \, d\omega} = Q(T_{7x} + T_{7y}), \quad (9.13)$$

$$\frac{d\sigma^{\downarrow\downarrow}}{d\Omega \, d\omega} = Q(T_1 - T_5 - T_6 - T_{7z}), \quad (9.14)$$

with the Q operator including all factors, integrations, and lattice summations of Eq. (9.9). Here, T_{7x}, T_{7y}, and T_{7z} are short notations of the respective parts of the scalar product in T_7 (e.g., $S_{n,x}^\perp S_{n,x}^\perp \to T_{7x}$), and z is the direction of $\mathbf{p}_0$ with $\mathbf{p}_0 \parallel \mathbf{p}_1 = 1$ for Eqs. (9.11) to (9.14). The scattering cross sections for polarized beams ($0 \leq p_0 \leq 1$) without polarization analysis and for unpolarized beams are obtained by the appropriate summations.

Having in Eqs. (9.8)–(9.14) and Table I a sufficient basis for simple explanations, we should also mention the concept of quasi-elastic scattering. We can already obtain quite easily an impression of what happens for a single magnetic ion in a nonmagnetic matrix. In the absence of an external magnetic field, by coupling to the conduction electrons the average value $\langle S_n \rangle$ is zero and the only term remaining is $\langle S_m^\perp(0) S_m^\perp(t) \rangle$, which in the simplest picture decays exponentially with a relaxation time τ. The Fourier transform of the time-dependent expectation value is then a Lorentzian with full width at half-maximum (FWHM) Γ, where $\Gamma/2 = 1/\tau$. In case of the so-called Korringa relaxation, Γ is proportional to the absolute temperature T with a proportionality factor α varying from 10^{-1} to 0.4 for different systems ($\hbar\Gamma/2 = \alpha \cdot k_B T$). Thus, in many cases the relaxation is sufficiently fast that the quasi-elastic scattering exceeds the typical $\hbar\delta\omega$ window of about 1 meV. This fact should be kept firmly in mind.

9.1.2. Types of Defects

Single-point defects (magnetic or nonmagnetic) like vacancies, self-interstitials, foreign interstitials, and substitutional atoms (Fig. 2) are of interest with respect to their local structural and dynamical properties and also with respect to the impact they have on the surrounding matrix. Except for a few favorable hosts like beryllium, aluminum, niobium, or lead, which have very small σ_i values, neutrons are not a sensitive tool for very small defect concentrations ($c < 10^{-3}$). In general, higher concentrations—ranging from 10^{-2} to 10^{-1}—must be used, and then defect interactions can be important and must be accounted for by a concentration series. In nuclear scattering the defect positions, the lattice distortion around the defect, and the dynamical properties of the defect are the main quantities of interest. Magnetic scattering of substitutional atoms in ferro-, ferri-, and antiferromagnetic hosts reveals the magnetic moment at the defect site and the local magnetic disturbance of the surrounding matrix. The other extreme is represented by a magnetic substitutional atom in a nonmagnetic matrix (e.g., iron in gold), where mainly the spin dynamics are of interest. A complete combined study of nuclear and magnetic scattering for dilute systems, including all relevant properties, is not known to the author.

For high point-defect concentrations, considerable difficulties arise in the

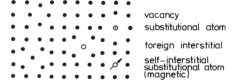

FIG. 2. Different types of single-point defects.

FIG. 3. Binary system showing tendency for (a) clustering or (b) short-range ordering.

interpretation, cancelling in part the advantage of easier access to experiments. As in x-ray scattering work, most experiments with neutrons have been performed for binary solid solutions. In the homogeneous one-phase regime above miscibility gaps or above concentration–temperature fields with long-range ordering, substitutional solid solutions exhibit clustering or short-range order (Fig. 3) with corresponding scattering patterns. From high-temperature data (normally obtainable with quenched samples), conclusions about effective pair interaction potentials V_{AA}, V_{AB}, and V_{BB} can be obtained. Also the kinetics following a temperature change and the increase of concentration fluctuations on approaching the one-phase boundary in the c, T-phase diagram are of interest with regard to equilibrium and nonequilibrium approaches in statistical mechanics. Although up to now most investigations have been performed on metal alloys and metal–hydrogen systems, the number of experiments performed on other materials, such as nonstoichiometric refractory compounds (e.g., NbC_{1-x}, Fig. 4), superionic conductors, metallic glasses, and molecular crystals, has increased in recent years.

Nuclear and magnetic scattering experiments on highly concentrated magnetic metallic alloys, binary or pseudobinary compounds, and also on binary or pseudobinary insulating magnetic compounds, have been performed in numerous cases. Either the formation of magnetic moments for varying concentration [as in ferromagnetic metals (Fig. 5)] or the mutual

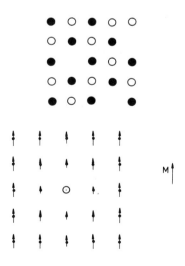

FIG. 4. Arrangement of atoms in the nonstoichiometric refractory compound NbC_{1-x}; Nb (●) and C (○).

FIG. 5. Variation of magnetic moment of a ferromagnetic metal in the neighborhood of a nonmagnetic defect.

9. DEFECTS IN SOLIDS

FIG. 6. Mutual arrangement of magnetic moments in a spin-glass system.

arrangement of magnetic moment direction [as in spin-glass systems (Fig. 6)] are of interest, again on a local atomic scale. Examples for both will be presented in Sections 9.2.1.2, 9.2.2, and 9.2.3.

Concentration fluctuations in binary alloys above the decomposition or ordering temperature and magnetic moment couplings in spin glasses above the freezing temperature tend to increase their average characteristic wavelength to higher values on lowering the temperature. This is one possible way to get long-wavelength fluctuations (50 Å to 1000 Å). Below the respective "critical" temperatures, spinodal decomposition, nucleation and growth, long-range ordering with domain formation, and magnetic-moment cluster freezing take place with quite different kinetics, leading to further long-wavelength fluctuations or large-sized inhomogeneities (50 Å to a few thousand angstroms). Neutron small-angle scattering has proved to be a very powerful method for the investigation of this type of phenomena as well as for a series of other large-sized defects like voids, dislocations, incoherent precipitates, dislocation loops, damage zones produced by neutrons or other fast particle irradiation, and many more. Because large-sized inhomogeneities play an essential role in the deeper understanding of materials in technological applications (e.g., fatigue and creep), the interest in neutron small-angle scattering grew rapidly, and the list of publications with very different subjects is extremely long. Only an outline of some basic principles and a few specific examples will be given here; for the interested reader there are quite a number of review articles.[8-10] Both nuclear and magnetic scattering have been applied widely in small-angle scattering. A broader extension to polarized neutron might be desirable. Inelastic small-angle scattering is restricted to a few cases of doubt and to a few real investigations of very slow processes in spin-glass systems as discussed briefly in 9.2.3. Excitation spectroscopy seems to be reasonable for magnetic systems. For atomic rearrangements on a large scale real-time experiments [Eq. (9.10] are probably more promising.

9.1.3. Instrumentation

No attempt is made to describe experimental techniques in detail. Nearly all diffractometers and spectrometers[12] installed at the approximately 60

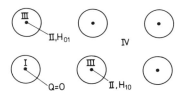

FIG. 7. Subdivision of the reciprocal space into four regions as explained in the text.

neutron sources in the world are used in part for studying disorder in solids, some of them especially suitable for this purpose like special time-of-flight spectrometers or small-angle scattering devices. There are estimated to be about 500 instruments all over the world with very different performance, flux, wavelength range, resolution, and production rates. A rough grouping is into diffractometers (no energy analysis of the scattered beam) and spectrometers (with energy analysis).

For theory and experiment, it is useful to subdivide reciprocal space into four regions (Fig. 7).[4] Region I, comprising $|\mathbf{Q}|$ values smaller than about one-fifth of the nearest reciprocal lattice point, is the small-angle scattering region. A noteworthy point is that small-angle scattering cross sections are mostly measured without energy analysis. Thus the energy-integrated cross section contains contributions to $\hat{S}(\mathbf{Q}, \omega)$ from much higher $\mathbf{Q}$ values than the $\mathbf{Q}$ values of actual interest in the experiment. Further, special care has to be taken for spin waves and for incident neutron velocities comparable to acoustic sound velocities (see Refs. 9 and 12). Another important point is the transition from the simple Born approximation to multiple scattering within the extended scattering center. This may occur for particle sizes larger than 1000 Å assuming a neutron wavelength $\lambda_0 = 10$ Å, as described in detail by Schmatz.[4] Region II in Fig. 7 consists of the small volumes around reciprocal lattice points with $|\mathbf{q}|$ values ($q = \mathbf{Q} - \tau_{hkl}$) as small as in region I. Scattering in this region arises from long-wavelength fluctuations in density and composition and from static and dynamic distortion. The calculation of the elastic scattering cross section in region II (Huang scattering) is given by Dederichs[13] and Krivoglaz.[1] Experiments in region II require high resolution both in $\mathbf{q}$ and ω. The high resolution in ω is necessary to separate acoustic phonons from elastic scattering because $\hbar\omega_{ph} \sim q$ in this region (ω_{ph} is the phonon frequency). A simple consideration[5] shows that, even for typical Huang scattering cross sections with $\hat{S}(\tau_{hkl} + \mathbf{q}) \sim 1/q^2$, an experiment at $q \cong 0.05$ Å^{-1} is already very difficult, and with decreasing q the intensity drops proportional to q^3 because of unavoidable constraints by resolution conditions. Furthermore, one must bear in mind that mosaic spread resulting in a smearing of the reciprocal lattice point perpendicular to τ_{hkl} raises difficulties for experiments with small $\mathbf{q}$ vectors and $\omega = 0$ or small ω values. Region III in Fig. 7 is rather convenient for both elastic and inelastic scattering. Elastic scattering in this region is sensitive to short-range local atomic arrangement and distortion. The experimental technique is most recently described by Bauer[14] and Schmatz[5] for elastic nuclear scattering and shortly reviewed by Hicks[15] for elastic magnetic scattering. For inelastic nuclear scattering in region III and in region II with high ω values, we can refer to the extensive general literature and to Windsor.[11]

A general remark for planning experiments should be given. Maier-Leibnitz[16] has emphasized the connection of the counting rate at a detector $\Delta Z_1(\mathbf{k}_0 \to \mathbf{k}_1)$ for a neutron transition from $\mathbf{k}_0$ to $\mathbf{k}_1$ in a scattering experiment with the resolution elements $\mathbf{k}_0$ and $\Delta \mathbf{k}_1$ defined by the primary and secondary part of a spectrometer. It is

$$\Delta Z_1(\mathbf{k}_0 \to \mathbf{k}_1) \sim p(\mathbf{k}_0) \Delta \mathbf{k}_0 \Delta \mathbf{k}_1. \qquad (9.15)$$

The function $p(\mathbf{k}_0)$ is the momentum space density of neutrons in the source, which is simply proportional to $\exp(-E_0/k_B T)/T^2$ with T the effective neutron temperature. There is an enormous drop of $p(\mathbf{k}_0)$ from $\lambda = 10$ Å to a λ value of 0.7 Å, for instance, of approximately three or four orders of magnitude. This is the major reason why cold neutrons have extended the application of neutrons so dramatically and why experiments with hot neutrons—though very important for large $\mathbf{Q}$ and ω transfers—are much more tedious and by far less numerous. A second advantage for cold neutrons is the neutron guide tube technique, which allows the installation of a large number of instruments at *one* cold source.

A large amount of diffraction studies of disordered crystals or materials has been performed with conventional neutron diffractometers, i.e., without energy analysis. The scattering data in such cases are often interpreted in

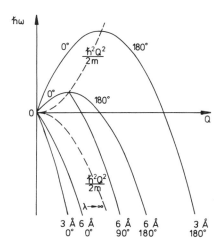

FIG. 8. Energy transfer as a function of momentum transfer for different scattering angles and different incident neutron wavelengths.

terms of elastic scattering. Because of the curved integration path (see Fig. 8), compared to x rays that integrate on a vertical line, one has to be careful in comparing x-ray data with neutron data or neutron data obtained with different wavelengths among themselves. For high concentrations of defects, the corrections are normally small but not negligible.

The ($\mathbf{Q}$, ω, $\Delta\mathbf{Q}$, $\Delta\omega$) space achievable today is respectable. Also, Q values between 10^{-4} Å^{-1} and 20 Å^{-1} and $\hbar\omega$ values between 0.02 μeV and 300 meV are obtainable. The high-flux reactor at the Institut Laue Langevin (ILL) Grenoble, with its many and often specialized instruments, is a demonstration of a carefully diversified instrumentation designed to cover very different needs from very different disciplines. The topic presented in this chapter also benefitted from these capabilities, and the extent of experimental results available today compared to 1973 is certainly overwhelming.[3]

Worldwide, the progress in magnetic neutron scattering, especially at its highest level, i.e., with polarization analysis, seems still to be underdeveloped. Polarized beams—monochromatic and with broad wavelength distribution—are available at many places. However, polarization analysis with ($\mathbf{Q}$, ω) tracing is still restricted to a few triple-axis spectrometers. Only two multidetector devices with polarization analysis are known to the author: the instrument at Lucas Heights, Australia, using the iron-foil transmission technique,[17] and the supermirror analyzer upgraded diffuse scattering spectrometer D7 at the ILL.[18] Polarization analysis is, in all cases of scattering work, performed for the polarization direction parallel to the incident beam polarization, so that spin-flip and non-spin-flip scattering are separated as described in the famous paper by Moon et al.[19]

In the interpretation of scattering data with various models there is a principal argument one should always be aware of. The scattering function is the product AA^* of a scattering amplitude A: the information on scattering phase is lost. In addition $\hat{S}$ is a sample average of small volumes $1/\Delta\mathbf{Q}$ within the sample. Thus a measured scattering function can fit data from distinctly different atomic arrangements in both $\mathbf{r}$ and t space very well in all details. In other words, a given correlation function $G(\mathbf{r}, t)$ is not uniquely related to a single model.

9.2. Static Properties

9.2.1. General Types of Disorder

9.2.1.1. Isolated Point Defects.
For host lattices with small incoherent scattering cross sections, such as aluminum, niobium, lead, and others, defect scattering cross sections at concentrations $c > 1\%$ can be measured quite well in region III of Fig. 7. For substitutional defects, the low-concentration-

limit scattering cross section for all **Q** values except at $\mathbf{Q} = \tau_{hkl}$ is

$$\frac{d\sigma}{d\Omega} = c|(b_D^0 - b_H^0) + b_H^0 i\mathbf{Q} \cdot \tilde{\mathbf{s}}(\mathbf{Q})|^2, \qquad (9.16)$$

with

$$\phi(\mathbf{Q})\tilde{\mathbf{s}}(\mathbf{Q}) = \tilde{f}(\mathbf{Q}), \qquad (9.17)$$

where b_D^0 and b_H^0 are the coherent scattering lengths (including the Debye-Waller factors) of the substitutional and the host atom, respectively; $\tilde{\mathbf{s}}$ is the Fourier transform of the lattice displacement originating from an individual substitutional atom; $\phi(\mathbf{Q})$ the dynamical matrix of the host lattice; and $\tilde{f}(\mathbf{Q})$ a fictitious quantity. It is the Fourier transform of the so-called Matsubara-Kanzaki (MK) forces. The real space **f**'s are the forces that in an ideal lattice around a normal atom cause exactly the same displacements as the substitutional atom. Data evaluation according to Eqs. (9.16) and (9.17) in terms of MK forces reveals the symmetry of the defect and the number of neighboring shells that are directly affected by the local electronic disturbance of the substitutional atom. Systems analyzed by this method are Pb-Bi, Al-Mn, Al-Cu, Al-Mg, Al-Zn, and Nb-D. For Pb-Bi and Al-Mn nonsymmetric MK forces are claimed from the analysis of the scattering data. It would be of interest to prove these results by other methods as Huang scattering, channeling, and internal friction. Also, it was clearly demonstrated for the systems Al-Cu, Al-Mg, Al-Zn, and Nb-D that MK forces applied to nearest neighbors only are not sufficient to explain the scattering data. This, however, was to be expected for metal hosts. The next step beyond the simple MK force parametrization is to calculate the **f**'s by cluster or at least by pseudopotential theory. The only attempt into this direction is reported by Werner et al.[20] and by Solt and Werner[21] for Al-Mg; a rather satisfying result is shown in Fig. 9.

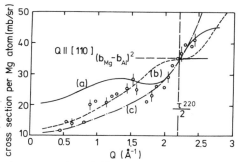

FIG. 9. Diffuse elastic cross section of dilute AlMg measured with the "Spectrometer for Diffuse Neutron Scattering" of the KFA Jülich. (a) Pseudopotential theory, (b) nearest neighbor Kanzaki forces, and (c) Kanzaki forces, fitted to the slope at $\tau^{220}/2$.

We conclude that diffuse elastic neutron scattering by point defects (substitutional atoms, vacancies, interstitials, and foreign interstitials) in dilute concentration can reveal the symmetry of defects and give information on the extension of the local electronic disturbance, and we refer the more deeply interested reader to the reviews by Bauer,[14] Schmatz,[5] and Kostorz.[10] A broader application of the method can be expected as soon as theory can offer more detailed calculations. X rays may well be competitive or complementary to neutrons in this special area of research as demonstrated by Haubold[22] and by Haubold and Martinson,[23] with the excellent results they obtained for electron-irradiated aluminum and copper, respectively.

The lattice sum $\tilde{s}(\mathbf{Q})$ is periodic in reciprocal space, and near reciprocal lattice points $\tilde{s}(\mathbf{Q}) = \tilde{s}(\mathbf{q} + \tau_{hkl}) = \tilde{s}(\mathbf{q})$ is proportional to $1/q$, which for sufficiently small q values results in a $1/q^2$ dependence of $d\sigma/d\Omega$. This so-called Huang scattering, along with the asymmetry, which arises for larger q values by interference of the first and the second term in the bracket of Eq. (9.5) [because $\tilde{s}(\mathbf{q}) = -\tilde{s}(-\mathbf{q})$], has proved to be an important tool for the analysis of point defects and point-defect clusters. For small q values, $\tilde{f}(\mathbf{Q})$ is related directly to the tensor of elastic dipole forces by

$$\tilde{f}(\mathbf{q}) = i\mathbf{Pq} \qquad (9.18)$$

with

$$P_{\alpha\beta} = \sum_n f_\alpha^n r_\beta^n. \qquad (9.19)$$

The matrices $\boldsymbol{\phi}(\mathbf{Q}) = \boldsymbol{\phi}(\mathbf{q} + \tau_{hkl}) = \boldsymbol{\phi}(\mathbf{q})$ can be expressed rather simply by a combination of elastic constants with a linear q dependence. Extensive use has been made of the simplification of the scattering cross section by x-ray scatterers in the study of point defects and clusters, which is reviewed by Kostorz.[10] An example of Huang neutron scattering (HNS) on NbH obtained by Burkel et al.[24] is shown in Fig. 10. Because the inelastic scattering by phonons can be separated in neutron scattering—provided a high-resolution

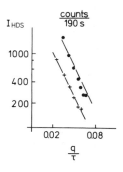

FIG. 10. Intensity of Huang neutron scattering on NbH versus reduced wave vector.

instrument is available—Huang scattering with neutrons can be performed at high temperatures, where the solubility limit is higher, which is especially important for metal–hydrogen systems. Other systems studied by HNS are neutron irradiated MgO and LiF[24] and Au in Pb (see 9.2.2).

Nearly 20 years ago a remarkable effort was undertaken by the Harwell research group to measure the magnetic disturbance due to a large variety of nonmagnetic and magnetic substitutional atoms in ferromagnetic host lattices (iron, cobalt, nickel). The magnetic scattering was separated by $\mathbf{M} \perp \mathbf{Q}$ and $\mathbf{M} \parallel \mathbf{Q}$ experiments on magnetized samples ($\mathbf{M}$ = magnetization) with the well-known Glopper instrument. The Lucas Heights (Hicks) and the Oak Ridge (Cable) groups have contributed considerably to the same topic, partly in cooperation. The subject is reviewed by Low,[25] Schmatz,[3] and Hicks.[15] The magnetic disturbance in the host lattice can be of long range, as for instance for a series of substitutional atoms in nickel (with a diameter up to 20 Å, e.g., NiV), or can be restricted to the impurity site (e.g., NiFe). The magnetic moment at the impurity site can either be zero within experimental limits (e.g., NiCu or NiCr with impurity site magnetic moments of $0 \pm 0.01\,\mu_B$) and can be parallel (e.g., NiMn with $2.4 \pm 0.1\,\mu_B$ at the Mn sites) or antiparallel (e.g., CoMn with $-1.0 \pm 0.7\,\mu_B$) in orientation. For a particular system, the determination of the configuration requires the knowledge of the change of the average magnetic moment $d\mu/dc$ due to alloying. Apart from a possible extension of the early work by adding other substances to the list, or by use of polarized neutrons or in special cases by searching for canted moments with polarization analysis, the major hindrance for further studies seems to be the lack of theory. Considerable efforts are presently underway to better understand the ground state and the excitations (spin waves, electron transition) of pure ferromagnetic materials (mainly iron and nickel). This may also have a theoretical impact on the impurity problem. Recent work on MnCu and MnPd and on CrFe related to the topic above will be discussed in Sections 9.2.1.2 and 9.2.2, respectively.

9.2.1.2. Short-Range Order and Clustering. For binary alloys with nonrandom local atomic arrangements, as indicated in Fig. 3, it is convenient[1] to introduce a position-dependent occupation number c_n with $c_n = 1$, if a B atom occupies the lattice site R_n and $c_n = 0$ otherwise. The elastic coherent scattering cross section is then[1,3]

$$\left(\frac{d\sigma}{d\Omega}\right)_{c,\,el} = |\tilde{c}(\mathbf{Q})|^2 |b_B^0 - b_A^0|^2 \qquad (9.20)$$

with

$$\tilde{c}(\mathbf{Q}) = \frac{1}{\sqrt{N}} \sum_{n=1} c_n e^{i\mathbf{Q}\cdot\mathbf{R}_n}. \qquad (9.21)$$

For a random distribution, the squared Fourier transform of the concentration fluctuations $|\tilde{c}(\mathbf{Q})|^2$ is $c(1 - c)$ with c equal to the average concentration of B atoms.

At very high temperatures, the entropy term in the free energy drives the system to near-random distribution and the deviations from this can be expressed by[1,3,14,26]

$$|\tilde{c}(\mathbf{Q})|^2 = \frac{c(1 - c)}{1 + [c(1 - c)\,\Delta\tilde{V}(\mathbf{Q})]/k_B T} \qquad (9.22)$$

with

$$\Delta\tilde{V}(\mathbf{Q}) = \sum_{n=1}^{\infty} [V_{AA}(\mathbf{R}_n) + V_{BB}(\mathbf{R}_n) - 2V_{AB}(\mathbf{R}_n)]e^{i\mathbf{Q}\cdot\mathbf{R}_n}. \qquad (9.23)$$

Equation (9.23) starts to be incorrect at small deviations from the random value ($\cong 50\%$). Moss and Clapp[27] have pointed out an easy fundamental correction. Quite generally it is

$$\int |\tilde{c}(\mathbf{Q})|^2\, d\mathbf{Q} = c(1 - c)\Omega_B, \qquad (9.24)$$

with Ω_B the Brillouin-zone volume. This gives, as a first correction, a renormalization for the use of Eq. (9.22). For approximations better than this, we refer to the results obtained by statistical mechanics by de Fontaine.[27]

According to Eqs. (9.20)–(9.23), the elastic coherent scattering cross section would be periodic in reciprocal space apart from the Debye–Waller factor. But this holds only for systems with negligible local lattice distortions. Assuming linear superposition of lattice distortions centered on single sites, and also relatively small distortions, a possible first-order correction of Eq. (9.20) is [1,3]

$$\left(\frac{d\sigma}{d\Omega}\right)_{c,\,el} = |\tilde{c}(\mathbf{Q})|^2|(b_B^0 - b_A^0) + \bar{b}^0 i\mathbf{Q}\cdot\tilde{\mathbf{s}}(\mathbf{Q})|^2, \qquad (9.25)$$

where $\bar{b}^0 = cb_B^0 + (1 - c)b_A^0$ and $\tilde{\mathbf{s}}$ can be expressed in terms of the individual distortions arising from A and B atoms but in practice is assumed to be an adjustable parameter. According to experimental results, one essential feature for $\tilde{\mathbf{s}}$ remains the proportionality to $1/q$ for small q-values. The distortion contribution in Eq. (9.25) increases with Q and distorts the periodicity of $(d\sigma/d\Omega)_{c,\,el}$ in reciprocal space. Borie and Sparks[28] developed a method to separate $|\tilde{c}(\mathbf{Q})|^2$ in the presence of distortions for a data set measured in a sufficient number of Wigner–Seitz cells of the reciprocal lattice. As reviewed by Hayakawa and Cohen,[29] Bauer,[14] and Kostorz,[10] this method has proved to be very successful for a series of x-ray scattering experiments. For x rays the Borie–Sparks method also has the advantage of separating the thermal diffuse scattering.

In neutron scattering studies the method has been applied by Lefebvre et al.[30,31] for Ni_3Fe and in a modified version by Moisy-Maurice et al. for $TiC_{0.67}$ and $NbC_{0.73}$.[32] In the last mentioned case (NaCl structure), the metal atoms form a periodic lattice, and the "binary" mixture is on the anion sublattice with carbon atoms and vacancies. A typical result is shown in Fig. 11 for experimental data and distortion-corrected scattering cross sections. Single-crystal measurements are rather rare in x-ray and especially in neutron scattering. From the many investigations on polycrystalline materials—reviewed by Bauer,[15] Kostorz,[11] and Chen et al.[33]—the studies on $Al_{0.6}Zn_{0.4}$[34] and $Cu_{1-x}Ni_x$[35–38] will be explained in more detail, because they are quite demonstrative for other studies also. Both systems show clustering. For $Al_{0.6}Zn_{0.4}$ the concentration fluctuations above the "incoherent" critical temperature $T_{c,i}$, between the "incoherent" and the coherent critical temperature, and below the coherent critical temperature $T_{c,c}$, have been

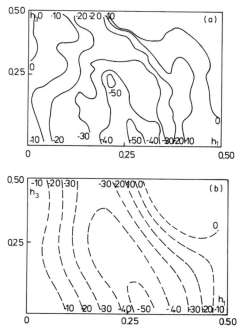

FIG. 11. Application of the Borie–Sparks method: (a) original experimental data and (b) distortion-corrected data.

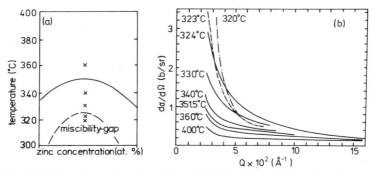

FIG. 12. SANS by concentration fluctuations in Al-Zn: (a) part of the phase diagram; solid and dashed lines are boundaries for incoherent precipitations and spinodal decomposition, respectively, and (b) SANS data.

measured (see Fig. 12). Above $T_{c,i}$ thermal-equilibrium concentration fluctuations have been observed, and between $T_{c,i}$ and $T_{c,c}$ it was possible to observe the lattice coherent concentration fluctuations sufficiently long before they disintegrated into lattice incoherent precipitates. Below $T_{c,c}$ immediate decomposition was observed as expected.

The system $Ni_{1-x}Cu_x$ has many favorable and interesting aspects. (1) The lattice distortions are nearly negligible. (2) The critical temperature for decomposition is low according to calculations.[37] It is between 360 K and 600 K at a concentration of 65 at. % Ni. (3) Whereas the system is not suitable for diffuse x-ray scattering, by using proper isotope mixtures diffuse neutron scattering can be greatly increased. (With a $^{65}Cu_{0.435} - {}^{62}Ni_{0.565}$ alloy, a null matrix, i.e., zero Bragg scattering, is obtained.[35]) (4) By adding copper to nickel, a considerable disturbance to the magnetization of the surrounding nickel matrix occurs, which is in no way simply additive with increasing copper concentration. (5) The Curie temperature of $Ni_{1-x}Cu_x$ drops from the nickel value to zero at the critical concentration $x = 0.477$.

The nuclear scattering measured in a temperature range from 340 to 700°C was interpreted in terms of Warren–Cowley short-range order parameters α_l[38–40] by

$$\left(\frac{d\sigma}{d\Omega}\right)_{c,\,el} = \langle|\tilde{c}(\mathbf{Q})|^2\rangle (b_B^0 - b_A^0)^2 = c(1-c)(b_B^0 - b_A^0)^2 \sum_{l=1}^{\infty} \alpha_l e^{i\mathbf{Q}\cdot\mathbf{R}_l}, \quad (9.26)$$

and from this an interatomic interaction potential was obtained (Fig. 13), which fits reasonably well to a Friedel oscillation. For a fit over all temperatures and concentrations, an additional triplet interaction term of 5 meV was necessary, indicating that metals cannot be simply discussed in terms of pair interaction potentials.

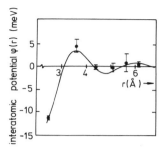

FIG. 13. Interatomic potential as a function of r for NiCu.

Another interesting aspect observed in the Cu-Ni alloys is the kinetics after quenching from high temperature and subsequent thermal annealing.[41] An example is shown in Fig. 14 for a Cu-Ni null-matrix alloy quenched from 700°C and subsequently annealed at 450°C. Apart from some care necessary for very short annealing times—due to excess vacancies, which restrict the data evaluation to Q values smaller than 0.8 Å^{-1}—there is excellent agreement with Cook's theory of the kinetics of clustering as the calculated solid lines in Fig. 14 show.

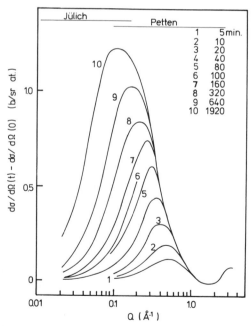

FIG. 14. The various stages of the relaxation of the null matrix at 450°C, as measured at Petten and at Jülich.

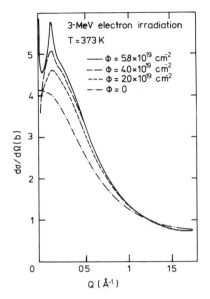

FIG. 15. Neutron small-angle scattering and diffuse SRO scattering of the null matrix, measured after the preparatory thermal treatment (6-h annealing at 870 K and slow furnace cooling) and subsequent isothermal electron irradiation. The SANS data collected at the FRJ-2 extend from $Q = 0.01 \text{ Å}^{-1}$ up to $Q = 0.22 \text{ Å}^{-1}$, the diffuse-scattering data obtained at BER II extend from $Q = 0.18 \text{ Å}^{-1}$ upwards.

Wagner et al.[37] have performed neutron scattering studies on an electron-irradiated ^{62}Ni$_{0.414}$^{65}Cu$_{0.586}$ alloy. A typical result is shown in Fig. 15. With increasing electron dose, a peak develops at approximately 0.15 to 0.20 Å^{-1}. At intermediate electron dose, the forward scattering even drops below the original value of the 6 h, 870 K annealed and slow-furnace-cooled sample. The interpretation of this result as an irradiation-induced spinodal decomposition seems rather convincing. (The rise at small Q values after a $\Phi = 5.8 \times 10^{19}$ cm^{-2} dose might be an indication of incoherent precipitates.) From the experiments discussed above, the authors deduced a lower limit for the critical temperature of about 400 K, in good agreement with the range estimated by Radelaar from the high-temperature experiments.

As described in detail by Hicks,[15] alloying copper with nickel causes magnetic disturbances at the nickel sites, which are in no way additive in the sense of the Marshall model.[43] The most detailed experimental information has been obtained by Medina and Cable[44] with polarized neutrons. The vector $\mathbf{Q}$ was perpendicular to the externally applied field $\mathbf{H}_{\text{ext}}$ and the neutron polarization $\mathbf{p}_0 \parallel \mathbf{H}_{\text{ext}}$ nearly one. The results for three concentrations are shown in Fig. 16. The upper curves (black dots) are proportional to the nuclear-magnetic interference term obtained from the difference data for $p_0 = \pm 1$. The steep increase toward $Q = 0$ is partly caused by clustering. Therefore the ratio of nuclear-magnetic interference scattering to nuclear scattering is calculated, which results in the lower curves. These curves

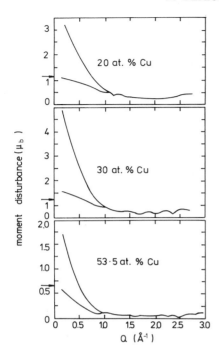

Fig. 16. The polarization-dependent cross section (expressed in terms of moment disturbance) for Ni–Cu alloys after Medina and Cable (1976).

would be constant in the case of strictly atomic-site-defined magnetic moments. This evidently does not hold, and the curves give an indication of the spatial extent of the disturbance caused by the copper atoms, which reduces the nickel moments. From the slope of the curves, one can see further that the average range of this disturbance is roughly a factor two larger for the 53.5 at. % Cu alloy than for the 20 at. % Cu alloy, with root-mean-square extensions in real space of about 4 Å and 2 Å, respectively. Medina and Cable[44] conclude from their data that there is no magnetic moment at the copper sites, and that the average magnetic moment at the nickel sites

TABLE II. Magnetic Moments and Concentration Derivatives for Ni–Cu Alloys[a]

c	H (kOe)	$\langle\mu_{Ni}\rangle - \langle\mu_{Cu}\rangle$[b]	$\langle\mu\rangle/(1-c)$	$M(0)/S(0)$	$d\langle\mu\rangle/dc$	$\langle\mu_{Cu}\rangle$
0.198	25	0.478 (5)	0.486 (3)	−1.125 (10)	−1.140 (10)	0.006 (13)
0.296	25	0.413 (6)	0.397 (6)	−1.128 (10)	−1.120 (10)	−0.011 (12)
0.525	10	0.091 (2)	0.103 (5)	−0.65 (2)	−0.66 (2)	0.006 (7)
0.525	57	0.106 (2)	0.114 (6)	−0.72 (2)	−0.79 (6)	0.004 (7)

[a] From Medina and Cable.[44]
[b] Multiple-scattering uncertainty of 0.015 μ_B must be added to the statistical error quoted.

decreases with increasing copper concentration. Furthermore, they get excellent agreement for the general equation

$$C\frac{(d\sigma/d\Omega)_{\text{nuc-mag}}}{(d\sigma/d\Omega)_{\text{nuc}}} = \frac{d\langle\mu\rangle}{dc}, \tag{9.27}$$

which holds for a local disturbance model. This actually means that localization occurs within $1/Q_{\min}$ where $Q_{\min}$ is the smallest Q value in the experiment. Some values obtained by Medina and Cable[44] are listed in Table II.

For concentrated $Ni_{1-x}Mn_x$ alloys ($x = 0.05, 0.10, 0.15,$ and 0.20) Cable and Child[45] measured the polarization-dependent scattering cross section

$$\left(\frac{d\sigma}{d\Omega}\right)^{+-}_{c,\,el} = |\tilde{c}(\mathbf{Q})|^2(\Delta b \pm \Delta p)^2 \tag{9.28}$$

at temperatures ranging from 42 to 298 K with an applied field of 20 kOe and $\mathbf{H}_{\text{ext}} \perp \mathbf{Q}$. For low concentrations ($x = 0.05$), the nuclear and the nuclear-magnetic interference scattering showed the same behavior, whereas for high concentrations, magnetic and nuclear short-range order differ, i.e., Eq. (9.28) fails. However, contrary to the situation in the Cu–Ni system, it is not the disturbance at the nickel sites that is responsible for this discrepancy but the major disturbance occuring at the manganese sites. From their scattering and magnetization data, the authors deduce temperature- and concentration-dependent local-site magnetic moments, which are given in part in Table III. Clearly, the average magnetic moment at the manganese sites decreases considerably with concentration. The authors tried several models with colinear magnetic moments and antiferromagnetic coupling, i.e., some of the manganese moments are antiparallel to $\mathbf{H}_{\text{ext}}$ due to a sufficient number of manganese near neighbors. None of their models has been able to explain the detailed results. As Hicks[15] proposes, the next step would be to check with a polarization analysis experiment ($\mathbf{Q} \parallel \mathbf{M}$) whether some of the manganese moments are canted, i.e., noncolinear.

TABLE III. Magnetic Moments and Relative Magnetizations at Elevated Temperatures[a]

At. % Mn	T (K)	$\frac{T}{T_c}$	$\sigma(T)$	$\mu_{\text{Mn}}(T)$	$\mu_{\text{Ni}}(T)$	$\frac{\mu_{\text{Mn}}(T)}{\mu_{\text{Mn}}(0)}$	$\frac{\mu_{\text{Ni}}(T)}{\mu_{\text{Ni}}(0)}$
5	188	0.33	0.69	3.05 ± 0.13	0.566	0.87 ± 0.05	0.97 ± 0.03
5	298	0.52	0.64	2.63 ± 0.14	0.535	0.75 ± 0.05	0.92 ± 0.03
10	298	0.61	0.61	2.17 ± 0.07	0.436	0.68 ± 0.03	0.83 ± 0.03
15	298	0.75	0.44	1.34 ± 0.05	0.281	0.63 ± 0.03	0.61 ± 0.03
20	96	0.32	0.43 ±0.01	0.98 ± 0.04	0.292 ±0.013	0.89 ± 0.05	0.97 ± 0.06

[a] From Cable and Child.[45]

A very elegant method of separating diffuse elastic nuclear scattering has been used by Davis and Hicks.[46] With the Longpol instrument at Lucas Heights,[17] they measured the defect scattering from a 10 at. % Cu γ-MnCu and a 10 at. % Pd γ-MnPd alloy by polarization analysis. (γ-Mn is antiferromagnetic.) The magnetic disturbances in γ-MnCu occur predominantly on first nearest neighbors, while in γ-MnPd the disturbance is a maximum at second nearest neighbors. Within an accuracy of $\pm 0.2 \mu_B$, there appears, in both cases, to be no magnetic moment at the impurity site itself. The nuclear defect scattering cross section measured for MnCu is shown in Fig. 17a. Clearly, some clustering occurs for copper in γ-Mn. The magnetic scattering cross section (Fig. 17b) is roughly a factor four smaller than the nuclear one. In order to get the magnetic cross section, the experiment was performed with $\mathbf{p}_0$ parallel to $\mathbf{Q}$. In the polycrystalline multidomain sample, all domains with the sublattice magnetization perpendicular to $\mathbf{Q}$ will scatter magnetically, but for them $\mathbf{p}_0$ is perpendicular to the sublattice magnetization. Therefore the magnetic scattering will occur with spin flip, whereas for nuclear disorder scattering no spin flip occurs. From the spin-flip scattering cross sections, the disturbances on the neighboring shells have been calculated. In both cases (MnCu and MnPd), the integral numbers [$(-4.3 \pm 0.6) \mu_B$ for copper in γ-Mn and $(+3.1 \pm 0.3) \mu_B$ for palladium in γ-Mn] agree well with the numbers obtained from the variation of the Bragg intensity at the (001) magnetic Bragg peak.

Up to now we have considered ferro- or antiferromagnetic host lattices with dilute or high-defect concentrations. The other extreme is to start with nonmagnetic hosts. Adding magnetic impurities with moderate concentrations ($x = 0.1$–0.2) leads in many cases to so-called spin-glass systems. For

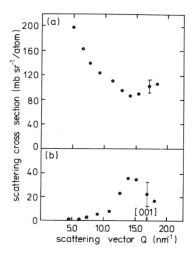

FIG. 17. Scattering cross section for MnCu as a function of Q: (a) nuclear defect scattering and (b) magnetic scattering. (From Davis and Hicks.[46])

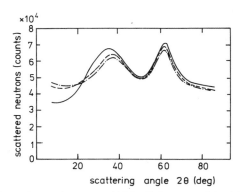

FIG. 18. Scattering at the system Al$_2$Mn$_3$Si$_3$O$_{12}$ as explained in Ref. 47.

such systems at high temperatures, the thermal excitation "decouples" the mutual magnetic interaction and no significant correlation in the scattering pattern occurs. In many cases (e.g., manganese in copper or iron in gold), moreover, the relaxation is fast enough to be observable by medium- or high-resolution spectroscopy. Lowering the temperature leads to preferential either ferromagnetic or antiferromagnetic coupling, while in the first-mentioned case a maximum occurs in the neutron small-angle scattering at the spin-glass freezing temperature, as discussed in some detail in Section 9.2.3. An example of antiferromagnetic coupling far above the freezing temperature is given for the amorphous insulating glass Al$_2$Mn$_3$Si$_3$O$_{12}$, which simultaneously is an "antiferromagnetic" spin-glass system (Fig. 18, reference 47). For the temperature-dependent relaxation in this system, see Section 9.3.1. There are many other spin-glass systems with preferential antiferromagnetic coupling.[48]

9.2.1.3. Rotational Disorder. Rotational disorder in solids leads to a wide variety of fascinating phenomena. At high temperatures, spherical molecules such as NH$_4^+$ in NH$_4$Br or CH$_4$ in fcc solid methane have a random distibution of their orientation, analogous to the randomness of magnetic moment orientation far above the transition temperature. On cooling, orientational correlations occur, which can be studied by the Q dependence of the integrated quasi-elastic scattering. By analogy with the situation for diffusion or magnetic moment relaxation, the $\Delta\omega$ window must be broad enough to ensure $\Delta\omega\,\tau_R > 1$, where τ_R is the orientational jump time. Figure 19 shows the contour map for integrated quasi-elastic scattering from CD$_4$ at a temperature of 1.3 T_c. As there are only two different orientations in the fcc lattice for the tetrahedral CD$_4$ molecule, the scattering can be treated similarly to a binary alloy with two different form factors dependent on the direction of **Q**. As $Q \to 0$, the form factor of the molecule takes the

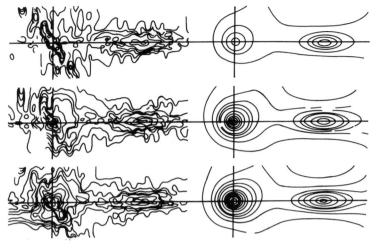

FIG. 19. Contour map for integrated quasi-elastic scattering from CD_4.

same value for the two orientations, and the scattering due to rotational disorder vanishes. Lattice distortions are of minor importance for nearly spherical molecules like CD_4 compared with metallic alloys. Though there are numerous investigations[49–51] on the dynamics (rotational, librational, and vibrational modes) of molecular crystals, very little work has been done in terms of **Q**-dependent mapping of diffuse scattering with different $\Delta\omega$ windows. This would be quite interesting from the viewpoint of disorder phenomena, not only for monomolecular crystals like CD_4 or NH_4Br but also for binary or pseudobinary crystal systems, where by analogy with the phenomena in magnetic materials two extremes can be considered: replacement of spherical molecules by aspherical ones like NH_4Br in KBr or replacement of a spherical molecule by spherical ones like argon in CD_4.

9.2.1.4. Long-Wavelength (Small-q) Fluctuations. By means of the long-range elastic strain field, point defects induce long-wavelength fluctuations with regard to the exact periodicity of an ideal lattice. For an elastic isotropic medium, $\tilde{s}(\mathbf{q})$ is independent of the **q** direction in the case where the defect itself has cubic symmetry. Then the product $\mathbf{Q}\tilde{s}(\mathbf{Q})$ near a reciprocal lattice point τ_{hkl} is simply $(\tau_{hkl}/q)\cos\alpha$, where α is the angle between **q** and τ_{hkl}. (For τ_{000} it is $Q = q$ and the second term in Eq. (9.16) is simply a constant for a cubic defect in an isotropic medium.) For an elastic anisotropic medium and an anisotropic defect as well as for a cubic defect in an elastic anisotropic medium, the small q distortion term in Eq. (9.16)—i.e., $\tau_{hkl} \cdot \boldsymbol{\phi} \cdot (\mathbf{q})\mathbf{P} \cdot \mathbf{q}$—shows occasionally strong anisotropy in its dependence on the **q** direction compared with the simple $\cos\alpha$ dependence described above.[1, 13, 24] In most

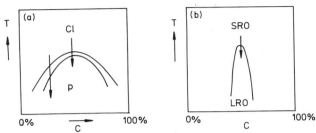

FIG. 20. Temperature via concentration maps, displaying the regions of (a) clustering-precipitated and (b) short-range orders–long-range orders.

cases the scattering pattern for an anisotropic defect has to be averaged with respect to equivalent orientations. Nevertheless, the broad variety of τ vectors and τ versus q orientations allows a sufficient number of data sets to discriminate between a considerable number of defect models on the basis of Huang scattering. Up to now neutrons have been used only in a few cases, as described in Section 9.2.1.1 and 9.2.2.

For point defects at high concentration (mainly studied for substitutional atoms and foreign interstitials), a second type of long-wavelength fluctuation arises from lattice-site occupation on the pathway from short-range order to long-range order or from clustering to spinodal decomposition or precipitation, as sketched schematically in Fig. 20. As long as the size of the long-range order domains, or the size of the precipitates or the longest wavelength in spinodal decomposition, is smaller than the primary extinction length or the mosaic crystal size, we can still use Eq. (9.24), i.e., the Brillouin-zone integrated scattering remains constant. Thus the narrowing of the scattering pattern at the positions of superlattice reflections, as in Fig. 20, or their narrowing around reciprocal lattice points (including $\tau = 0$), increases the peak intensities $s(q)$ roughly in proportion to $1/q_c^3$, where $1/q_c$ is the average size of the inhomogenieity. Thus long-range order domains can mostly be observed quite easily, even in powder diffraction, both with x rays and with neutrons. An example of single-crystal work on the Pd/D system is shown in Fig. 21. Such studies reveal details in local atomic arrangement in the

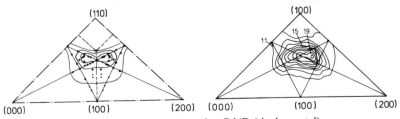

FIG. 21. Intensity lines measured at Pd/D (single crystal).

course of the transition to long-range order.[52] The complicated structure of the scattering pattern suggests influence of electrons near the Fermi surface on the interaction potential. The integral over all superlattice reflections within one Brillouin zone—as given by Eq. (9.24)—is an indication of the amount of material transformed.

Precipitation or spinodal decomposition can be observed at $\tau = 0$ and $\tau_{hkl} \neq 0$. The small-angle scattering regime ($\tau = 0$) is in most cases much more favorable, especially for polycrstalline materials. At the Bragg positions, additional scattering arises from the wings of the Bragg peaks and the thermal diffuse scattering. In addition the lattice strains due to the precipitates have to be accounted for. Moreover, small-angle scattering allows much easier adjustment to different Q ranges and Q resolution. For numerous examples we refer to Refs. 8–10. A few general aspects should be mentioned however. (1) Parallel to the measurement of $|\tilde{c}(\mathbf{Q})|^2$ around $\tau = 0$, it would be desirable in some cases to have $|\tilde{c}(\mathbf{Q})|^2$ also for the whole Brillouin zone, in order to check Eq. (9.24) as completely as possible. (If the integrated intensity were higher than $c(1 - c)$, parasitic scattering, for instance, could be detected in this way.) (2) Absolute scattering cross sections should be determined whenever possible. (3) The anisotropy of $\hat{S}(\mathbf{Q})$, especially for single crystals, but also in the case of textured polycrystalline materials, should be determined. (4) The usual textbook interpretation in terms of single-particle (sphere, ellipsoid, platelet) scattering functions should be considered as only a zeroth-order approach. Highly desirable—to be applied whenever possible—are additional experimental methods, especially transmission electron microscopy, in order to have a proper guide for particle shape, size, and orientational distribution. (5) For high Q values compared to the reciprocal size of the smallest particles present in sufficient number to produce significant scattering, the Porod law $S(Q) \sim Q^{-4}$ should be considered only an approximation, because at very high Q values the scattering function merges into the diffuse scattering, due to the local atomic fluctuations in the precipitates and in the host, and does not go to zero. (6) Finally the interparticle interference scattering has to be included in the interpretation. The interference function (Fig. 22) is similar to the structure factor of a liquid for highly concentrated alloys. Because of less dense packing the first maximum is not very high, but for small Q values its value can be far below 1 (the value for random packing). A typical small-angle scattering pattern is shown in Fig. 23.[53,54] The anisotropic maxima are due to interparticle interference. The maxima are along the [001] direction and are probably caused by correlations due to long-range elastic interactions.

For coherent precipitates there is a further complication due to the second term in Eq. (9.25). For dilute concentrations this additional component can be evaluated by analogy with the scattering from isolated point defects.

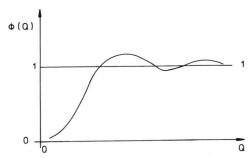

FIG. 22. Interference function $\phi(Q)$ versus Q for a highly concentrated alloy.

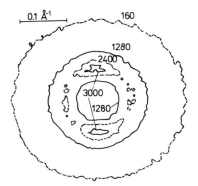

FIG. 23. Small-angle neutron scattering pattern from a highly concentrated alloy.

However, one must keep in mind that the strain originating from an atom in a precipitate is not the same as the strain for an isolated substitutional impurity. For high concentrations the same ambiguity in interpretation of $\bar{\bar{s}}$ arises as discussed in Section 9.2.1.2. Probably because of the complexity of this problem, there seems to be no thorough experimental investigation. Following Krivoglaz,[1] the theoretical treatment seems not to be an obstacle because elastic continuum theory can be used. There are, however, other problems: Large-sized coherent precipitates release their strain by production of dislocations. Moreover, precipitate grain boundary interactions may alter the result. For x rays, where Huang scattering can be traced very close to reciprocal vectors and thus $\bar{\bar{s}}$ can be observed rather well for precipitates also, the question of ambiguity may be less severe. For neutron small-angle scattering, only the proper corrections as $Q \to 0$ due to the $\bar{\bar{s}}$ term are of interest.

The use of the theoretical framework of Eqs. (9.16)–(9.26) for long-wave concentration fluctuations requires a periodic reference lattice [see Eqs (9.1), (9.3), and (9.5)]. Incoherent precipitates, dislocations, and grain boundaries are large-sized inhomogeneities which must be treated differently. For $\tau = 0$

($Q = q$), the best concept is to introduce a density of scattering lengths $\rho(\mathbf{r})$ by

$$\rho(\mathbf{r}) = \frac{1}{V_s} \sum_{i=1}^{N_s} b_i, \qquad (9.29)$$

where N_s atoms are in the microscopic volume V_s. Because neutron scattering is a bulk method averaging over many inhomogeneities, the $\rho(\mathbf{r})$ concept is very reasonable for model calculations. Local variation of $\rho(\mathbf{r})$ is due to density *and* compositional fluctuations. The scattering cross section is given simply by the squared Fourier transform of $\rho(\mathbf{r})$ for a volume $V \gg V_s$ divided by N—the number of atoms in V. It is no problem to calculate $d\sigma/d\Omega$ with a given $\rho(\mathbf{r})$. The difficulty is to have the necessary metallurgical and other background information concerning the types of large-sized inhomogeneities to take into consideration in calculating the scattering cross section. Some illuminating examples will be given in Section 9.2.4. The recent reviews by Kostorz[8-10] will guide the interested reader into this rapidly developing field. Without doubt, this is a discipline where neutron scattering lies very close to technology.

While for nuclear small-angle scattering, x-ray methods (because of recent improvements in sources and equipment) are partly competitive, neutrons are a unique tool for the investigation of long-wavelength magnetic fluctuations. "Simple" cases are (1) critical scattering above the transition temperature;[55] (2) magnetic precipitates in nonmagnetic materials, e.g., Refs. 56–58; (3) nonmagnetic precipitates in magnetic materials, e.g., Refs. 59–62; (4) long-wavelength ferromagnetic clusters in spin glasses or at the percolation limit, where the spin-glass state is in competition with magnetic ordering; (5) helical magnetic structures in reentrant superconductors; and many other. The word "simple" in no way implies that the physics involved are simple. It merely means that the scattering cross sections to be used are sufficiently well described in the Born approximation in accordance with Eqs. (9.8)–(9.14), or similarly, after proper introduction of the magnetic scattering amplitudes, with the extensions given in Eqs. (9.16)–(9.26). An exception to the Born approximation may occur for critical scattering very near the Curie temperature. From this general concept, one immediately concludes that, for instance, there should be magnetic Huang scattering, a subject that has up to now never been investigated.

Long-wavelength magnetic fluctuations occur also in a homogeneous background of material, as in the cases of vortex lattices in type II superconductors or magnetic domains separated by domain walls in ferro-, ferri-, or antiferromagnetic materials. (Domain walls in antiferromagnets can carry a two-dimensional apparently ferromagnetic layer.) Also, magnetic materials with long-wavelength inhomogeneities in density or composition—either coherent (precipitates, LRO domains) or incoherent (precipitates, dislocations, grain boundaries)—are inhomogeneous also with respect to the local induction **B(r)**. Using the scattering operator $\boldsymbol{\mu}_n \cdot \mathbf{B(r)}$ ($\boldsymbol{\mu}_n$ = neutron magnetic moment), either with local magnetic moments or, for small-angle scattering, with a definition similar to that used for the scattering length density in Eq. (9.29), with proper models all scattering cross sections can be calculated. Use of polarized neutrons and polarization analysis, together with the appropriate choice of orientation of the **Q** vector relative to an externally applied field, gives a wide variety of experimental conditions not fully exploited at present. Schmatz,[3,4] Hicks,[15] and Kostorz[8-10] have reviewed part of the subject concerning the possibilities for studying long-wavelength fluctuations.

9.2.2. Specific Systems with Static Disorder

From the many very different investigations in recent years a few illustrative examples are now described. (1) For the explanation of the fast diffusion of gold in lead, information about the structural state of the gold impurities is crucial. X-ray lattice parameter determinations indicated that gold dissolves as a single substitutional atom in lead,[63] while there are other indications of the formation of Au_4 interstitial clusters.[64] Huang scattering is very sensitive to the formation of clusters, giving at small q values four times as much scattering intensity in the case of Au_4 clusters as for single interstitial gold atoms at the same total gold concentration. X-ray Huang scattering[65] indicated single-site interstitials, but a neutron scattering measurement was necessary to ensure that the x-ray result[65] was not a surface effect. A neutron Huang scattering experiment[66] was performed at the (600) Bragg reflection of a lead single crystal with 864 ppm gold at 10 K after rapid quenching from 220°C to liquid nitrogen temperature. The neutron Huang diffuse-scattering intensity showed the usual $1/q^2$ dependence, and by comparing the intensity with the lattice parameter change it was proved unambiguously that gold in lead is in a single interstitial site.[66]

(2) As already investigated in many x-ray scattering experiments, copper, when dissolved in low concentrations (a few at. %) in aluminum, precipitates in so-called Guinier–Preston zones (Fig. 24) over periods of 1–10 days at

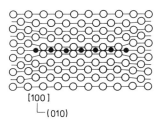

FIG. 24. Guinier–Preston zones in the dilute system AlCu: Al (○) and Cu (●).

temperatures slightly above room temperature. By applying an external stress (8×10^5 dyn/cm^2) parallel to the [100] cylinder axis of an aluminum single crystal containing 1.8 at. % copper resulting in a strain of 0.2%, it was possible at various aging temperatures to overpopulate the Guinier–Preston zones with the normal parallel to the cylinder axis[67] (sample diameter, 8 mm; height, 8 mm). From the diffuse scattering pattern obtained with neutrons, the authors deduced the population factor for the Guinier–Preston zones with normals parallel to the cylinder axis. At all aging temperatures this value is considerably above the 33.3% random value. The 60°C aged sample was used for a lattice dynamical study as described in Section 9.3.3. The evaluation of the population factor is based on a proper derivation of the scattering cross section, Eq. (9.25), where both $|\tilde{c}(\mathbf{Q})|^2$ and $\bar{\bar{s}}(\mathbf{Q})$ must be used. The experiment demonstrates the advantage of using neutrons to investigate both structure and dynamics in the same sample.

(3) Neutrons are an especially informative probe for metals, because many light-scattering methods fail for them. This may explain why defect scattering with neutrons is somewhat concentrated toward metals. There is, however, an increasing tendency toward nonmetals. An especially informative result is shown in Fig. 25, with contour maps of the elastic diffuse scattering of $Ba_{1-x}La_xF_{2+x}$ obtained with a triple-axis spectrometer[68] for $x = 0.209$ and $x = 0.492$. In accordance with other investigations[69] on the related intrinsic superionic conductors CaF_2, $SrCl_2$, and PbF_2, the authors[68] interpreted their result in terms of so-called 222 clusters. This cluster consists of two F^- anion interstitials with the two trivalent metal ions at the nearest-neighbor position and the two nearest-neighbor ions on regular sites

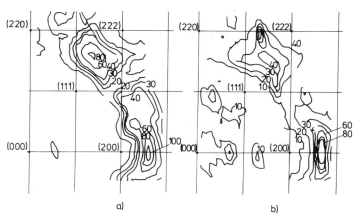

FIG. 25. Contour maps of the diffuse elastic scattering of the system $Ba_{1-x}La_xF_{2+x}$ measured with a neutron triple-axis spectrometer: (a) $x = 0.209$ and (b) $x = 0.492$.

displaced from the interstitial anions. This model requires a three-parameter fit, which with reasonable numbers reproduces very well the scattering pattern of Fig. 25a. For the $x = 0.492$ crystal, extra contour lines appear, which have the character of broad superlattice reflections [e.g., at $(\frac{4}{3}, 2, 2)$ in Fig. 25]. These extra contours can be explained by ordered arrays of the 222 clusters with a correlation length of about 30–40 Å. The satellite peak (2.69, 0, 0) showed nearly a 100% decrease in the temperature range from 400–800°C, which must be related to the fact that the ionic conductivity starts to increase anomalously at about 600°C. This indicates that the dissolution of the ordered array of 222 clusters and of the clusters themselves is essential for the high-temperature conductivity in this system. A high-temperature contour map would be desirable in this case.

(4) Among the vast variety of magnetically ordered host materials, chromium is a very special one. Pure chromium has a Néel temperature of 310 K, which is lowered by adding iron impurities with a simultaneous reduction of the apparent chromium magnetic moment. An important question concerning the local behavior of iron is thus raised. With the D5 instrument at the ILL[70] and by comparing spin-flip to non-spin-flip scattering with the polarization vector perpendicular and parallel to the **Q** vector (compare the experiment on MnCu in Section 9.2.1.2), an attempt was made to obtain information on the local magnetic moment at the iron impurity site by diffuse scattering on polycrystalline materials. The signal obtained was about (0.2 ± 0.2) of the value of iron in iron. The authors claim that iron in antiferromagnetic chromium is either a very long-range static defect, or if its magnetic moment is local, its magnitude is considerably reduced.

9.2.3. Special Studies in Small-Angle Scattering

Historically,[71] neutron small-angle scattering was started with extremely weak scatterers, e.g., at dislocations of plastically deformed copper single crystals. Successful results were obtained because, in contrast to x-ray scattering, double Bragg scattering could be avoided. (Count rates were about 1/min.) The early encouraging results stimulated the construction of better instruments at better sources, at first in Jülich and Saclay, and later at the ILL in Grenoble (D11 and D17). In addition, the first results on type II superconductors, biological materials, and polymers had considerable impact.[71] The technique is now employed worldwide, and the literature can no longer be summarized on a few pages. Hence, in the following, two topics have been selected: one representative of the sensitivity of the technique and the other of its importance in a new research field.

9.2.3.1. *Studies with Very Small Scattering Cross Sections.* As early as 1967, at a 4-MW swimming-pool reactor without a cold source, Herget[72] made a complete study of anisotropic dislocation scattering from plastically deformed copper single crystals by using photographic techniques, in which the individual neutrons were seen as spots with a counting efficiency of 20%. Though the neutron intensity at the D11 (ILL) is roughly a factor 1000 more, the number of investigations on defects with very small scattering cross sections is not large, probably because of the beam-time competition. Nevertheless, some of the studies are at the extreme limit, even beyond the level of Herget's experiment. Kettunen et al.[73] investigated void formation in fatigued copper single crystals with the stress axis parallel to $\langle 111 \rangle$. After subtracting the background and also the dislocation scattering, there was a tail extending to large Q values (Fig. 26), which was attributed to voids of 20-Å diameter and a volume fraction of only 10^{-5}. The dislocation density created by fatigue was 4×10^{10} cm^{-3} for a sample fatigued to 80% of its expected lifetime. The material used was pure to one part in 10^5, and the surfaces were carefully polished to reduce surface scattering. A special heat treatment lowered the background of the unfatigued sample. A rather extensive program on cavity formation in polycrystalline copper was performed by Weertman and colleagues[74, 75] for fatigue, fatigue and subsequent creep, and for creep alone. Data were taken in a Q range between 0.0015 and 0.085 Å^{-1} for various fatigue temperatures, cycles, and stress amplitudes on extremely carefully prepared samples as in Kettunen et al.[73] The SANS data gave definite conclusions with regard to various theoretical models of nucleation and growth of cavities. The sensitivity limit claimed by the authors is a volume fraction of 5×10^{-7}. (For particles of about 300 Å the sensitivity limit is better than for 20 Å particles, see Schmatz.[5]) Data from the creep experiments are shown in Fig. 27. The important result was that no incubation time could be detected. A fatigue study in polycrystalline nickel was interpreted in terms of voids ranging in size from approximately 100 to 1400 Å in diameter.[76] Slight discrepancies are evident on comparing

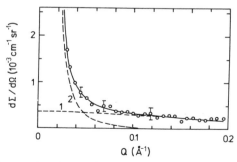

FIG. 26. Small-angle scattering picture of a fatigued copper single crystal (stress axis parallel to $\langle 111 \rangle$).

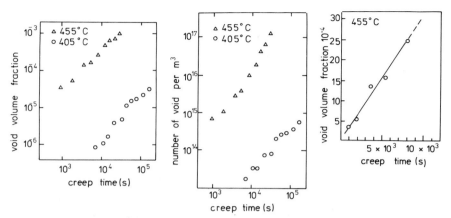

FIG. 27. Results from creep experiments on polycrystalline copper.

the results with high-voltage electron microscopy. Another study of Cu 1.8 at. % Co single crystals was aimed at controlling the influence of fatigue on 54-Å diameter lattice coherent precipitates.[77] No change of the shape of the scattering curves and only a slight reduction of the overall intensity (about 10%) was observed, indicating that most of the material is unaffected by dislocation rearrangements during fatigue.

The Reading–Oxford group (Stewart, Mitchell, and co-workers) have approached the limits of present-day SANS methods with studies on silicon and GaAs. For 10 Å neutrons and small scattering angles, silicon single crystals are ideal "zero" scattering samples. Newman et al.[78] produced ellipsoidal oxygen precipitates with an average diameter of 200 Å to 500 Å in Czochralski-grown, dislocation-free single crystals by annealing at temperatures between 650 and 1050°C. The oxygen concentration in their material was 10^{-5}. Etch-pit counting and infrared absorption measurements at 9 μm confirmed the result for the number density of precipitates deduced from SANS. The group reported a somewhat mysterious observation for silicon doped with carbon and oxygen in concentrations for 10^{-5} to 10^{-4} but without special annealing treatment. They observed for $\mathbf{k}_0 \parallel \langle 111 \rangle$ an anisotropic starlike diffuse scattering pattern with an intensity maximum at $Q \cong 0.1 \text{ Å}^{-1}$ and with a scattering cross section of 2–10 mb/sr/silicon atom.[79] An astonishing observation was that the scattering disappeared on cooling to 100 K. They then used D17 in a TOF mode of operation and were able to show that, if the diffuse scattering is inelastic at all, the energy transfer is less than 2 meV.[80] The puzzle remains because 2 meV < 8 meV ($\cong$ 100 K). The group was faced with similar difficulties in the course of their investigations of GaAs.[81] Here, part of the unexplained scattering could be removed

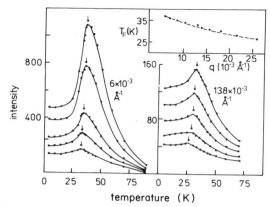

FIG. 28. Forward scattering intensity as a function of temperature for a series of q values for AuFe alloys containing (a) 10 at. % Fe and (b) 13 at. % Fe. The arrows mark the temperatures T_f of the discontinuities in the scattering. These are plotted versus q in the insets.

by avoiding any type of scattering from the cadmium slits, a precaution entirely unnecessary for most "normal" SANS experiments.

9.2.3.2. Ferromagnetic Spin-Glass Systems. These systems show a typical temperature-dependent, small-angle scattering pattern (Fig. 28, Ref. 82). With decreasing temperature, the directional dynamics of individual iron magnetic moments slow down (approximately in proportion to T) and in a concentration range with dominant ferromagnetic interactions, dynamical ferromagnetic clusters develop, increasing in size with decreasing temperature. This can be deduced from Fig. 28 by plotting the intensity versus Q curves for different temperatures. Below a freezing temperature T_f, the intensity decreases again and the average size becomes smaller. The T_f value is slightly dependent on Q, but for the smallest Q value, it corresponds well to the T_f value measured as a cusp in the low-frequency a.c. susceptibility. It is generally assumed, and proven in part by experiments, that starting at T_f the dynamical clusters freeze into static ones. Thus for $T > T_f$, the scattering curves give the ω-integrated generalized susceptibility, whereas for $T < T_f$ the static part has to be subtracted, which requires a high-resolution experiment. Though neutron scattering work started soon after Mydosh's discovery of spin-glass systems, there is much more work necessary for a thorough understanding. In particular, for a proper calculation of the dynamics with correct interaction potentials, the chemical short-range order must be determined and more must be known about the magnetic pair-interaction potentials.[83] Increasing the magnetic impurity concentration above a critical value, e.g., for iron in gold above 16 at. %, one finds systems with a spectacular behavior. On cooling there is at first a transition from the paramagnetic state to slight ferromagnetic ordering, which on further cooling develops a $T > T_f$ spin-glass behavior, and then finally condenses for $T < T_f$ as a normal spin glass, as concluded by Murani[84] from his results obtained with SANS and magnetic Bragg scattering.

9.3. Dynamic Properties

9.3.1. Quasi-Elastic Scattering

When rapid diffusion is present, the elastic scattering broadens into quasi-elastic scattering. Apart from the many studies on hydrogen and deuterium in metals (see, e.g., Refs. 85 and 86) and on superionic conductors,[87] there are only a few on disordered crystals. In Fig. 29 the quasi-elastic linewidth (FWHM) for the three principal directions in sodium, measured at 269.2 K, are shown. The data agree well with a model of monovacancy diffusion with nearest-neighbor jumps. It would be desirable to measure directly the vacancy diffusion by coherent scattering as a counterpart. However, thermal equilibrium concentrations of vacancies of monoatomic metals are so small that, to the author's knowledge, no experiment has ever been performed to observe vacancies in integrated quasi-elastic scattering. Better chances probably exist for nonstoichiometric compounds with vacancies (such as NbC_{1-x}) at high temperatures, where diffusion is not as fast as in superionic conductors but still fast enough for a high-resolution experiment. Probably there is no chance of observing formation and dissolution of defect clusters by excitation spectroscopy. Real-time observations seem to be more appropriate, representing one of the possible applications of pulsed neutron sources.

The inelastic scattering cross section of mutually independent local magnetic moments **m** can be written

$$\frac{d^2\sigma}{d\Omega\,d\omega} = \frac{1}{4}(\gamma r_e)^2 \frac{k_1}{k_0} \frac{2}{3} \frac{1}{\mu_B^2} \cdot \frac{1}{2\pi} \int dt\, e^{i\omega t} c \langle \mathbf{m}(0) \cdot \mathbf{m}(t) \rangle. \tag{9.30}$$

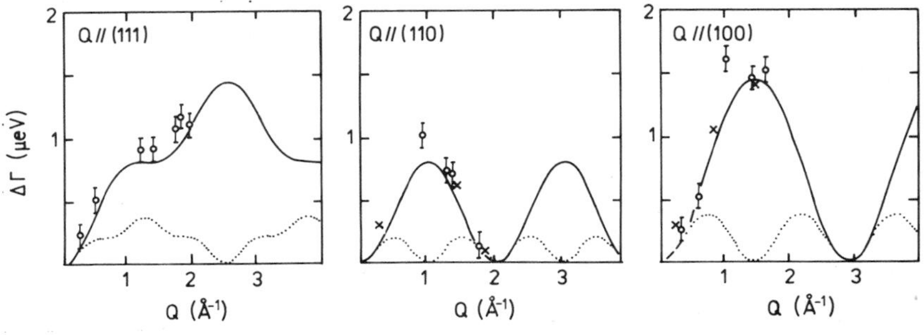

FIG. 29. The quasi-elastic linewidth (FWHM) $\Delta\Gamma$ for sodium as a function of Q parallel to the three main directions.

Coupling to conduction electrons in metals or to phonons in insulators drives the expectation value of $\langle \mathbf{m}(0) \cdot \mathbf{m}(t) \rangle$ to zero for $t \to \infty$. With a single relaxation time τ, a Lorentzian in ω results with a full width at half-maximum $\Gamma/2 = 1/\tau$. The temperature dependence of $\Gamma/2$ characterizes the coupling mechanisms, though it is not fully understood for intermediate valence, Kondo, and heavy fermion systems. For rather dilute iron in gold and manganese in copper, $\Gamma/2$ is proportional to T as expected in metals (Fig. 30, references 90 and 91). For 680 ppm Fe in Cu, a typical Kondo system, Loewenhaupt[92] observed a crossover with decreasing temperature, i.e., at $T > T_K$ the value of $\Gamma/2$ was smaller than $k_B T$, and at values $T < T_K$ the value of $\Gamma/2$ was larger than $k_B T$. For $\Gamma/2 \cong k_B T$ and $> k_B T$ the Lorentzian in ω has to be multiplied with the detailed-balance factor $\hbar\omega/[1 - \exp(-\hbar\omega/k_B T)]$, resulting in a purely excitational spectrum for $T \to 0$ and $\Gamma/2$ finite, which in the literature is still understood as "quasi-elastic" scattering. In such cases, at $T = 0$ the excitation spectrum rises linearly with ω at $\omega = 0$. No "elastic" magnetic scattering remains for $\Delta\omega \to 0$, which is in accordance with the understanding that the iron magnetic moment is compensated by the conduction electrons in the Kondo ground state.

In earlier experiments it was believed that a separation of the local moment and the compensation cloud had been seen in polycrystalline AlMn, as cited in Schmatz.[3] This, however, turned out to be invalid, probably because of an incorrect concentration determination, as explained by Bauer[15] using new data on AlMn single crystals. Another quite different behavior for $\Gamma/2$ is reported by Holland-Moritz and Prager.[93] For $Tm_{0.05}La_{0.95}Se$ measured in comparison to the reference substance LaSe, the quasi-elastic linewidth first

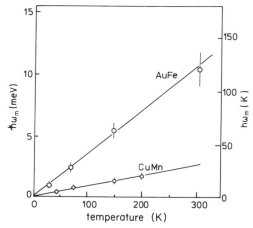

FIG. 30. Relaxating linewidth $\hbar\omega_m$ for quasi-elastic scattering of Fe in Au and Mn in Cu as a function of temperature.

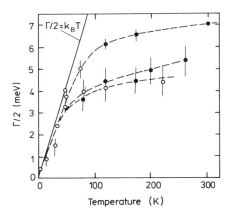

FIG. 31. Temperature dependence of the quasi-elastic linewidths for TmSe, $TmSe_{0.85}Te_{0.15}$, and $Tm_{0.05}La_{0.95}Se$ (from upper to lower curves).

decreases more slowly than $\sim T$ and then decreases nearly linearly and in coincidence with the $\Gamma/2 = k_B T$ line (Fig. 31). With slightly different values, the same result was obtained for the stoichiometric compounds TmSe and $TmSe_{0.85}Te_{0.15}$ (Holland-Moritz[94]), demonstrating that the RKKY interaction of 4f electrons via conduction electrons is not the major mechanism for this particular behavior in the stoichiometric compounds TmSe and $TmSe_{0.85}Te_{0.15}$.

In order to illustrate the wide variety of $\Gamma(T)$ dependences possible, we quote finally a result obtained for the insulating amorphous spin glass $Al_2Mn_3Si_3O_{12}$. The relaxation time τ over a temperature range from 294 to 3.0 K follows an Arrhenius law $\tau = \tau_\infty \cdot \exp(E_a/k_B T)$ with E_a an activation energy as shown by the Γ versus $1/T$ plot in Fig. 32.[95] For the five decades that were covered, the use of six different spectrometers at Karlsruhe and Grenoble was necessary. The lowest values of Γ (black dots in Fig. 32) are measured around the freezing temperature ($T_g = 3.2$ K) with T_g as glass temperature with a Q value of 1.3 Å^{-1}, where the main contribution is due to antiferromagnetically coupled spin pairs. It is astonishing that the high-temperature single-site behavior can be extrapolated down to the freezing temperature.

An example in the realm of band magnetism completes the list. Burke et al.[96] have analyzed by time-of-flight spectroscopy the short-lived local spin fluctuations (localized paramagnons) on the nickel impurity by substraction of the TOF spectra of PdNi and pure palladium. At 14 K they obtained for Q values smaller than 1.2 Å^{-1} approximately a $\Gamma \sim Q^2$ law with $\Gamma = 2$ meV at 1 Å^{-1} with the ω-integrated scattering around $\omega = 0$ following approximately the form factor. At higher temperatures (50 and 95 K) Γ becomes significantly larger. The $\Gamma \sim Q^2$ behavior is quite an interesting result:

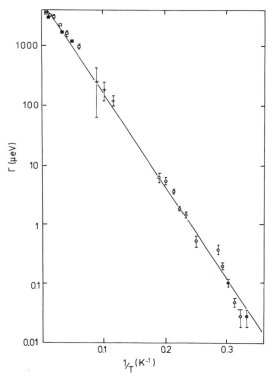

FIG. 32. Measured dependency of the relaxation rate Γ as a function of the inverse temperature.

it means that the local paramagnon diffuses slowly from the impurity site as in a diffusion jump process.

Now let us consider correlation functions $\langle \mathbf{m}_i(0) \cdot \mathbf{m}_j(t) \rangle$ with $i = j$ and $i \neq j$. One of the first substances in which the relevant quasi-elastic scattering was measured was MnF_2 with $T > T_N$. Here, $\Gamma/2$ increases proportionally to Q^2 and eventually reaches a limiting value. This means that a local instantaneous magnetic moment "diffuses" according to the normal diffusion equation. Murani[97] measured $\Gamma(Q)$ dependence in Ag–Mn spin-glasses as a function of the concentration down to 0.35 Å^{-1}. For 1 and 2 at. % Mn, $\Gamma(Q)$ was constant. For 5 and 10%, a decrease of Γ with decreasing Q was observable below 0.7 Å^{-1}, in fairly good accord with moment-to-moment interactions in which a magnetic moment decrease parallel to an arbitrary axis creates a magnetic moment at another nearby site. These measurements were performed at room temperature, where no appreciable long-range correlations exist. For low temperatures but still above T_f, a single relaxation time approximation is no longer justified, as

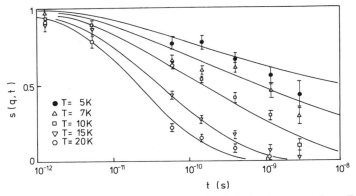

FIG. 33. The Fourier transform $s(q, t)$ versus time for the spin-glass system $La_{0.75}Er_{0.25}Al_2$, measured with the spin-echo spectrometer at the ILL.

shown by Mezei[98] with the neutron spin-echo spectrometer. A typical result[98] is shown in Fig. 33 for the spin-glass system $La_{0.75}Er_{0.25}Al_2$. The Fourier transform $S(\mathbf{Q}, t) = S(\mathbf{Q}, \omega)$ was directly determined by this method. A weighted distribution of relaxation times according to

$$S(t) = \frac{1}{E_m} \int_0^{E_m} e^{-t/\tau_0} \exp(E/kT) \, dE, \qquad (9.31)$$

with $E_m = 80$ K and $\tau_0 = 4.5 \times 10^{-12}$ s can be used for explaining the temperature and time dependence. An additional Q-dependent plot would be of great interest in regard to understanding temperature-dependent dynamics for various extensions in $\mathbf{r}$ space. We are faced with similar questions for the vast variety of quasi-elastic linewidth measurements on rare-earth ions in metals and metal compounds. No Q dependence of Γ is reported even for concentrated systems, as if the local magnetic moment relaxation extends to a widely expanded region of the conduction electrons.

9.3.2. Localized and Resonance Modes

Compared with the considerable amount of data obtained in metal–hydrogen systems,[85,86] investigations on other systems have not been pushed very strongly in recent years. Most of the subject is reviewed by Nicklow.[99,100]

For dilute concentrations of heavy or light substitutional atoms, the clear-cut picture of localized and resonant modes has been proved, with the general trend that for chemical similarity of substitutional and host-lattice atoms (such as aluminum in copper or gold in copper) some predictions of the low-concentration, mass-defect theory are fulfilled fairly well. The next step would be to analyze the results with respect to the additional effect of changes in the force constants *and* to compare this with first principles calculations. No such attempt is known to the author. As Nicklow emphasizes, there is the additional difficulty that the dilute concentration limit from a theoretical point of view seems to be rather low (about 1 to 5%), whereas most experiments claimed to be in the dilute limit have been performed in the concentration range from approximately 3 to 12%, where defect interactions may considerably change the $c \to 0$ result. On the other hand, it is extremely difficult to get precise results for concentrations as low as 1%. Such low-concentration experiments have been performed for dumbbell configuration defects such as 0.37% CN-molecules in KCl and lower than 0.1% dumbbell self-interstitials in low-temperature-irradiated aluminium and copper. The low-lying vibrational modes of CN in KCl gave a well-resolved resonance splitting at 10 K with an energy transfer of 2 meV, which at higher temperatures (40 K = 3.3 meV) runs into the usual resonance damping with a single phonon line shifted and broadened at the resonance position. Considerable effort was necessary for the experiments with reactor-irradiated copper and aluminum, but no definite result emerged to compare with theoretical expectations. For instance, for aluminum a significant change of the slope of the dispersion curve was observed, but no resonance-like structure (Fig. 34, reference 101). It might be that in both cases the vibrational modes of self-interstitials have to be calculated on a much more fundamental basis than in the hitherto applied "spring" models. On the other hand, defect interactions cannot be excluded for the 0.8% Frenkel defect concentration in aluminum. (For copper the concentration was 40 ppm.)

A nice example of resonance modes is obtained in Al 1.8 at. % Cu with copper precipitated into Guinier–Preston zones preferentially perpendicular

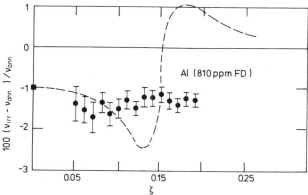

FIG. 34. Expected resonance mode and measured data on irradiated Al (containing 810 ppm Frenkel defects).

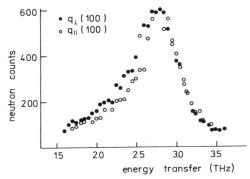

FIG. 35. Resonance modes in Al containing 1.8 at. % Cu for $q_\|[100]$ and $q_\perp[100]$.

to stress direction [100] (see Section 9.2.2).[67] As Fig. 35 shows, there is more resonance damping obtained in the transverse acoustic phonons with **q** ∥ [100] than for the direction **q** perpendicular to it (**q** ∥ 010). This is because for **q** ∥ [100] more Guinier–Preston zones are seen in full phase coherence. Rotation of **q** should reveal further phenomena related to the reciprocal planar extension of the Guinier–Preston zones $1/\phi$, where ϕ is the average diameter of the Guinier–Preston zones.

There are many efforts aimed at a better description of the lattice dynamics of high defect-concentrated systems. The general features are reasonably well understood, but precise first-principles theories describing the phonon shifts, the damping, and the extra modes are missing. Recent cluster calculations with a pseudopotential approach may be more quantitative, at least for simple metal alloys. It is very tedious to get nearly first-principles data for complex ordered substances (e.g., A15 metals or refractory compounds). Disordered systems are far more complicated than this!

References

1. M. A. Krivoglaz, "Theory of X-ray and Thermal Neutron Scattering by Real Crystals" (S. C. Moss, transl. ed.). Plenum, New York, 1969. (Second Ed., in Russ., 1983.)
2. D. L. Price and K. Sköld, this volume, Part A, Chapter 1.
3. W. Schmatz, in "Treatise on Materials Science and Technology" (H. Herman, ed.), Vol. 3, p. 105. Academic Press, New York, 1973.
4. W. Schmatz, in "Neutron Diffraction" (H. Dachs, ed.), Chap. 5. Springer-Verlag, Berlin and New York, 1978.
5. W. Schmatz, *Methods Exp. Phys.* **21**, 147 (1983).
6. I. I. Gurevich and L. V. Tarasov, "Low-Energy Neutron Physics" (R. I. Sharp and J. Chomet, transl. eds.). North-Holland Publ., Amsterdam, 1968.
7. W. Marshall and S. W. Lovesey, "Theory of Thermal Neutron Scattering." Oxford, Univ. Press, London and New York, 1971.
8. G. Kostorz, in "Neutron Scattering" (G. Kostorz, ed.), p. 227. Academic Press, New York, 1979.

9. DEFECTS IN SOLIDS

9. G. Kostorz, in "Small Angle X-ray Scattering" (O. Glatter and O. Kratky, eds.), p. 467. Academic Press, London, 1982.
10. G. Kostorz, in "Physical Metallurgy" (R. W. Cahn and P. Haasen, eds.), p. 794. Elsevier, Amsterdam, 1983.
11. C. G. Windsor, *Methods Exp. Phys.* **23A**, Chap. 3 (1986).
12. W. Schmatz, *Proc. Conf. Neutron Scattering, Gatlinburg, Tenn.* p. 1037 (1976).
13. P. H. Dederichs, *J. Phys. F* **3**, 471 (1973).
14. G. S. Bauer, in "Neutron Scattering" (G. Kostorz, ed.), p. 291. Academic Press, New York, 1979.
15. T. J. Hicks, in "Neutron Scattering" (G. Kostorz, ed.), p. 337. Academic Press, New York, 1979.
16. H. Maier-Leibnitz, *Nukleonik* **8**, 61 (1966).
17. S. J. Campbell, N. Ahmed, T. J. Hicks, F. R. Ebdon, and D. A. Wheeler, *J. Phys. E* **7**, 195 (1974).
18. O. Schärpf and W. Just, in press.
19. R. M. Moon, T. Riste, and W. Koehler, *Phys. Rev.* **181**, 920 (1969).
20. K. Werner, W. Schmatz, G. S. Bauer, E. Seitz, H. J. Fenzl, and A. Baratoff, *J. Phys.* **F8**, L207 (1978).
21. G. Solt and K. Werner, *Phys. Rev. B* **24**, 817 (1981).
22. H.-G. Haubold, *J. Appl. Crystallogr.* **8**, 175 (1975).
23. H.-G. Haubold and D. Martinson, *J. Nucl. Instrum. Methods* **69/70**, 644 (1978).
24. E. Burkel, B. v. Guerard, W. Metzger, J. Peisl, and C. M. E. Zeyen, *J. Phys. B* **25**, 227 (1979).
25. G. G. Low, *Proc. Phys. Soc., London* **92**, 938 (1967).
26. D. de Fontaine, *Solid State Phys.* **34**, 73 (1979).
27. S. C. Moss and P. C. Clapp, *Phys. Rev.* **171**, 764 (1968).
28. B. Borie and C. J. Sparks, *Acta Crystallogr., Sect. A* **A27**, 198 (1971).
29. M. Hayakawa and J. B. Cohen, *Acta Crystallogr., Sect. A* **A31**, 635 (1975).
30. S. F. Lefebvre, F. Bley, M. Bessiére, M. Fayard, M. Roth, and J. B. Cohen, *Acta Crystallogr., Sect. A* **A36**, 1 (1980).
31. S. F. Lefebvre, F. Bley, M. Fayard, and M. Roth, *Acta Metall.* **29**, 749 (1981).
32. V. Moisy-Maurice, C. H. de Novion, A. N. Christensen, and W. Just, *Solid State Commun.* **39**, 661 (1981).
33. H. Chen, R. J. Comstock, and J. B. Cohen, *Annu. Rev. Mater Sci.* **9**, 51 (1979).
34. D. Schwahn and W. Schmatz, *Acta Metall.* **26**, 1571 (1978).
35. J. Vrijen and C. van Dijk, in "Fluctuations, Instabilities and Phase Transitions" (T. Riste, ed.), p. 43. Plenum, New York, 1975.
36. J. Vrijen and S. Radelaar, *Phys. Rev. B* **17**, 409 (1978).
37. W. Wagner, R. Poerschke, A. Axmann, and D. Schwahn, *Phys. Rev. B* **21**, 3087 (1980).
38. J. Vrijen, Ph.D. Thesis, Rijks Univ., Utrecht, 1977.
39. J. Vrijen, C. van Dijk, and S. Radelaar, *Proc. Conf. Neutron Scattering, Gatlinburg, Tenn.* p. 92 (1976).
40. J. Vrijen, C. van Dijk, and S. Radelaar, Rep. RCN-76-062 (1976).
41. J. Vrijen, S. Radelaar, and D. Schwahn, *J. Phys. (Orsay, Fr.)* **C7**, 347 (1977).
42. H. E. Cook, *Mater. Sci. Eng.* **25**, 127 (1976).
43. W. Marshall, *J. Phys. C* **1**, 88 (1968).
44. R. A. Medina and J. W. Cable, *Phys. Rev. B* **15**, 1539 (1977).
45. J. W. Cable and H. R. Child, *Phys. Rev. B* **10**, 4607 (1974).
46. J. R. Davis and T. J. Hicks, *J. Phys. F* **7**, 2153 (1977).
47. W. Nägele, K. Knorr, W. Prandl, P. Convert, and J. L. Bueroz, *J. Phys. C* **11**, 3295 (1978).

48. F. Bruss, K. Baberschke, M. Loewenkaupt, and H. Scheuen, *Solid State Commun.* **32**, 135 (1979).
49. M. Bee, *J. Chim. Phys. Chim. Biol.* **82**, 205 (1985).
50. J. Howard and T. C. Waddington, in "Advances in Infrared and Raman Spectroscopy" (R. J. H. Clark and R. E. Hester, eds.), Vol. 7, p. 86. Heyden, Philadelphia, Pennsylvania, 1980.
51. W. Press, *Springer Tracts Mod. Phys.* **29** (1981).
52. R. Beddoe and K. Werner, in preparation.
53. E. Bubeck, Diplome Thesis, Univ. Stuttgart, 1980.
54. G. Kostorz, *Proc. Yamada Conf. VI Neutron Scattering Condensed Matter, 1982* p. 389. North-Holland Publ., Amsterdam, 1983.
55. J. Als-Nielsen, *Phase Transitions Crit. Phenom.* **5a**, 88 (1976).
56. M. Ernst, J. Schelten, and W. Schmatz, *Phys. Status Solid.* **7**, 469 (1971).
57. G. Abersfelder, K. Naack, K. Stierstadt, J. Schelten, and W. Schmatz, *Philos. Mag. B* **41**(5), 519 (1980).
58. P. Kournettas, K. Stierstadt, and D. Schwahn, *Philos. Mag. B* **51**(4), 381 (1985).
59. D. Schwahn, H. Ullmaier, J. Schelten, and W. Kesternich, *Acta Med.* **31**(12), 2003 (1983).
60. D. Schwahn, D. Pachur, and J. Schelten, Jül-1543, ISSN 0 366-0885 (1978).
61. D. Schwahn, W. Kesternich, S. Spooner, H. Schröder, H. Ullmaier, and J. Schelten, *Proc. ILL Workshop*, 197 (1985).
62. R. Beddoe, P. Haasen, and G. Kostortz, in "Decomposition of Alloys, the Early Stages" (P. Haasen et al., eds.), p. 233. Pergamon, Oxford, 1983.
63. W. R. Warburton, *Phys. Rev. B* **11**, 4945 (1975).
64. W. R. Warburton and S. C. Moss, *Phys. Rev. Lett.* **42**, 1694 (1979).
65. T. Bolze, Diplomarbeit, Univ. Munich.
66. T. Bolze, H. Metzger, J. Peisl, and S. C. Moss, Annex to ILL-Rep., p. 296. Inst. Laue Langevin, Grenoble (1980).
67. S. Kawarazaki, N. Kunitomi, R. M. Nicklow, and H. G. Smith, *Proc. Yamada Conf. VI Neutron Scattering Condensed Matter, 1982* p. 401. North-Holland Publ., Amsterdam, 1983.
68. J. K. Kjems, N. H. Anderson, J. Schoonman, and K. Clausen, *Proc. Yamada Conf. IV Neutron Scattering Condensed Matter, 1982* p. 357. North-Holland Publ., Amsterdam, 1983.
69. K. Clausen, W. Hayes, M. T. Hutchings, J. K. Kjems, P. Schnatrel, and C. Smith, *Solid State Ionics* **5**, 589 (1981).
70. S. K. Burke, J. R. Davis, P. W. Mitchell, and J. G. Booth, Annex to Annu. ILL-Rep., p. 283. Inst. Laue Langevin, Grenoble (1980).
71. W. Schmatz, *Conf. Ser.—Inst. Phys.* No. 64, p. 301 (1983).
72. P. Herget, Ph.D Thesis, TU Munich, 1968; see also W. Schmatz, *Riv. Nuovo Cimento* **5**, 398 (1975).
73. P. O. Kettunen, T. Lepisto, G. Kostorz, and G. Götz, *Acta Metall.* **29**, 969 (1981).
74. R. Page, J. R. Weertman, and M. Roth, *Acta Metall.* **30**, 1357 (1982).
75. R. Page and J. R. Weertman, *Scr. Metall.* (1980).
76. M. H. Yoo, J. C. Ogle, B. S. Borie, E. H. Lee, and R. W. Hendricks, *Acta Metall.* **30**, 1733 (1982).
77. D. Steiner, R. Beddoe, V. Gerold, G. Kostorz, and R. Schmelzer, *Scr. Metall.*, in press.
78. R. C. Newman, R. J. Stewart, S. Messoloras, and B. Pike, submitted.
79. E. W. J. Mitchell, M. Dusic, S. Messoloras, and R. J. Stewart, Annex to Annu. ILL-Rep., p. 300. Inst. Laue Langevin, Grenoble (1980).

80. E. W. J. Mitchell, R. J. Stewart, M. Dusic, and S. Messoloras, Annex to Annu. ILL-Rep., p. 258. Inst. Laue Langevin, Grenoble (1982).
81. E. W. J. Mitchell, R. J. Stewart, S. Messoloras, and P. Raig, Annex to Annu. ILL-Rep., p. 228. Inst. Laue Langevin, Grenoble (1981).
82. A. P. Murani, *Phys. Rev. Lett.* **37**, 450 (1976).
83. J. Mydosh, private communication.
84. A. P. Murani, *Solid State Commun.* **34**, 705 (1980).
85. K. Sköld, M. H. Mueller, and T. O. Brun, *in* "Treatise on Materials Science and Technology" (G. Kostorz, ed.), Vol. 15, p. 381. Academic Press, New York, 1979.
86. T. Springer and D. Richter, this volume.
87. N. H. Anderson, K. N. Clausen, and J. K. Kjems, this volume, Chapt. 11.
88. M. Ait-Salem, T. Springer, A. Heidemann, and B. Alefeld, *Philos. Mag.* **39**, 797 (1979).
89. G. Göltz, A. Heidemann, H. Mehrer, A. Seeger, and D. Wolf, *Philos. Mag., Part A* **41**, 723 (1980).
90. H. Scheuer, M. Loewenhaupt, and W. Schmatz, *Physica B + C (Amsterdam)* **86/88B + C**, 842 (1972).
91. H. Scheuer, private communication.
92. M. Loewenhaupt, *J. Magn. Magn. Mater.* **2**, 99 (1976).
93. E. Holland-Moritz and M. Prager, *J. Magn. Mater.* **31–34**, 395 (1983).
94. E. Holland-Moritz, *J. Magn. Magn. Mater.* **38**, 253 (1983).
95. W. Nägele, W. Prandl, and P. v. Blanckenhagen, Annex to Annu. ILL-Rep., p. 285. Inst. Laue Langevin, Grenoble (1980).
96. S. K. Burke, B. D. Rainford, E. J. Lindley, and O. Moze, *J. Magn. Magn. Mater.*
97. A. P. Murani, Annex to Annu. ILL-Rep., p. 284. Inst. Laue Langevin, Grenoble (1980).
98. F. Mezei, *Proc. Yamada Conf. VI Neutron Scattering Condensed Matter, 1982.* North-Holland Publ., Amsterdam, 1983.
99. R. M. Nicklow, *in* "Treatise on Materials Science and Technology" (G. Kostorz, ed.), Vol. 15, p. 191. Academic Press, New York, 1979.
100. R. M. Nicklow, *Methods Exp. Phys.* **21**, 172 (1983).
101. K. Boenig, G. S. Bauer, H. J. Fenzl, R. Scherm, and W. Kaiser, *Phys. Rev. Lett.* **38**, 852 (1977).

10. HYDROGEN IN METALS

Tasso Springer and Dieter Richter*

Institut für Festkörperforschung
Kernforschungsanlage Jülich GmbH
D-5170 Jülich
Federal Republic of Germany

10.1. Introduction: Hydrogen in Metals and Neutron Spectroscopy

Research on hydrogen–metal systems is strongly stimulated by its relation to applied problems, such as energy storage, fusion technology, and hydrogen embrittlement. In addition, due to the small mass of the hydrogen, there is a great variety of interesting features for basic solid-state physics, such as fast hydrogen diffusion, large isotope effects, hydrogen tunneling in metals, and hydrogen influence on superconductivity. Also, the close conceptual relation between hydrogen and muons in metals should be emphasized.

Among the experiments probing hydrogen on an atomistic scale, neutron scattering plays a special role. The high incoherent scattering cross section of the proton, 79.7 b, compared with ~2–10 b for most other elements, allows convenient observation by neutron spectroscopy. Deuterium in metals has also been studied and has an appreciable coherent contribution ($\sigma_c = 5.4$ b, $\sigma_i = 3.3$ b), as does tritium ($\sigma_c = 3.1$ b).

This chapter deals with neutron scattering in terms of these main items: quasi-elastic spectra to investigate the diffusive step of hydrogen and also to determine diffusion constants; more complicated effects like trapping by impurities; the problem of high concentration hydrides and deuterides; localized vibrations and their higher harmonics and anharmonicity; phonons in metal deuterides; localized modes of trapped hydrogen in order to obtain information on the type of the site and tunneling in the vicinity of traps. The review tries to summarize work in the field of neutron spectroscopy essentially from the period from 1977 to 1983, sometimes referring to older papers,

---

* Present address: Institute Laue-Langevin, 156 X Centre de tri, 38042 Grenoble Cedex, France.

if it seems necessary. For earlier work we refer to Sköld,[1] Springer,[2] and Richter.[3] Diffraction is not treated in this chapter.[4] The effect of diffuse scattering in disordered hydrides is dealt with in Chapter 9 of this volume. During recent years, important experimental progress has been made, in particular with respect to quasi-elastic scattering due to the improved resolution available with back-scattering spectrometers and on the higher harmonics of localized modes, which have become accessible in particular as a result of reactor hot-source instruments and instruments on spallation sources.

10.2. Diffusion and Quasi-Elastic Scattering

10.2.1. Short Outline of the Theory for Dilute Hydrides

Incoherent neutron scattering on a diffusing hydrogen atom causes a bell-shaped "quasi-elastic line," centered at energy transfer $\hbar\omega = 0$. The width Γ and the shape of this line as a function of momentum transfer $\hbar\mathbf{Q}$ yield information on the diffusive process.[†] As a typical example, Fig. 1 shows quasi-elastic spectra[5] for a $NbH_{0.02}$ single crystal well separated from the phonon contributions. So far, most scattering experiments of this kind were interpreted in terms of a simple jump model (Chudley–Elliott model).[6] This is based on a number of simplifications. First, the diffusion occurs over well-localized interstitial sites interconnected by a set of jump vectors $\mathbf{d}$. The mean residence time τ of the hydrogen on a certain site is assumed to be large compared to the time $\tau_j \simeq d(2M_H/kT)^{1/2} \simeq 10^{-13}$ s needed for the motion from one site to the other. The jump probability to any of the n adjacent sites is $1/n\tau$. Second, subsequent jumps are uncorrelated; jumps and oscillations are uncorrelated as well. Third, the hydrogen concentration is small, so that blocking and other interaction effects can be neglected.

For a simple Bravais interstitial lattice, as for hydrogen in fcc palladium, a single rate equation is obtained for the probability of occupancy at a site $\mathbf{r}$

$$\partial P(\mathbf{r}, t)/\partial t = (1/n\tau) \sum_{k=1}^{n} [P(\mathbf{r} + \mathbf{d}_k, t) - P(\mathbf{r}, t)]. \qquad (10.1)$$

Here, n is the number of adjacent sites to the site $\mathbf{r}(k = 1, ..., n)$. The self-correlation function is then

$$G_s(\mathbf{r}, t) = P(\mathbf{r}, t), \quad \text{if} \quad P(\mathbf{r}, 0) = \delta(\mathbf{r}), \qquad (10.2)$$

[†] If the diffusion is frozen in, the line becomes a delta function at $\hbar\omega = 0$, in analogy to the Mössbauer line in gamma resonance scattering; only momentum is transferred to the lattice, but no energy.

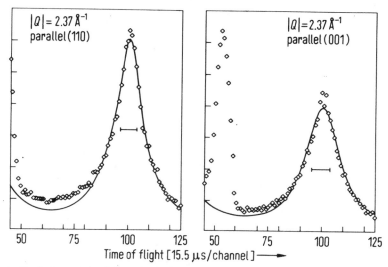

FIG. 1. Typical quasi-elastic lines for hydrogen diffusing in α-NbH$_{0.02}$ at 580 K, as measured by time-of-flight spectroscopy. Solid lines shows fit with theory (see Section 10.2.2). (From Lottner et al.[5])

where the oscillatory part is left out. The resulting incoherent scattering function is then found to be a Lorentzian:

$$S_i(\mathbf{Q}, \omega) = \frac{f(\mathbf{Q})/\pi}{\omega^2 + f^2(\mathbf{Q})} \qquad (10.3)$$

with a half-width at half-maximum (HWHM)

$$\Gamma = f(\mathbf{Q}) = \frac{1}{n\tau} \sum_{k=1}^{n} [1 - \exp(-i\mathbf{Q} \cdot \mathbf{d}_k)]. \qquad (10.4)$$

For an interstitial lattice that is not of the Bravais type, there is a finite number of inequivalent interstitial sublattices ($\nu = 6$ for the tetrahedral sites in a bcc lattice), and one obtains a system of ν rate equations, namely,[7]

$$\frac{\partial P_i(\mathbf{r}, t)}{\partial t} = \frac{1}{n\tau} \sum_{jk} [P_j(\mathbf{r} + \mathbf{d}_{ijk}) - P_i(\mathbf{r}, t)], \qquad i = 1, \ldots, \nu. \quad (10.5)$$

The jump vector $\mathbf{d}_{ijk}$ connects a site of a sublattice with a certain local symmetry $i = 1, \ldots, \nu$ with an adjacent site of local symmetry $j = 1, \ldots, \nu$, where $k = 1, \ldots, n$ labels a site in a sublattice j. The sum goes over all sites k and sublattices j, which are nearest neighbors. Here, $P_i(\mathbf{r}, t)$ is the probability of finding the hydrogen at time t at the site of type i in a lattice cell **r**. The initial condition in this case is, for a certain sublattice,

$$P_i(\mathbf{r}, 0) = \delta(\mathbf{r}) \quad \text{for} \quad i = j,$$

$$P_i(\mathbf{r}, 0) = 0 \quad \text{for} \quad i \neq j. \quad (10.6)$$

This yields G_s as the average over all starting sites

$$G_s(\mathbf{r}, t) = \frac{1}{\nu} \sum_{j=1}^{\nu} P^j(\mathbf{r}, t) \quad \text{with} \quad P^j = \sum_i P_i, \quad (10.7)$$

assuming, for simplicity, energetically equivalent sublattices. The system of rate equations can again be solved by Fourier transformation,[7] and one obtains ν eigenvalues. Correspondingly, the incoherent scattering function for diffusion is a superposition of ν Lorentzians with **Q**-dependent intensities and widths. Under certain circumstances, it may be advantageous to replace the rate equation approach by a random-walk method, which can be written as a series expansion in terms of jump sequences,[8,9] which, of course, gives the same results.

If **Q** approaches a reciprocal lattice vector of the interstitial lattice, at least one of the eigenvalues (widths) approaches zero, and a delta function appears in the spectrum. Qualitatively, this can be understood as the diffraction of

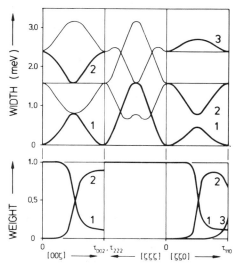

FIG. 2. Calculated half-width (eigenvalues) and partial quasi-elastic intensities (weights or eigenvectors) for hydrogen diffusion over tetrahedral sites in a bcc lattice. Abscissa: reduced scattering vector. The numbers correspond to the different eigenvalues. Weight ≥ 0.5% (———) and weight ≤ 0.5% (———). (From Lottner et al.[5])

the neutron wave by a *single* proton: by diffusion it spreads over a finite region of the interstitial lattice during the "coherency time" of the neutron wave $\hbar/\Delta E$ (ΔE is the spectral width of the beam).[10] This was used to study site occupancies.[11,12] As an example, Fig. 2 shows calculated eigenvalues and partial quasi-elastic intensities for a bcc lattice.

For $Q \to 0$ there is always one eigenvalue proportional to Q^2, and the corresponding partial line intensity dominates. For any kind of diffusion in a cubic lattice, one gets for the half-width at half-maximum

$$\Gamma = Q^2 D \qquad (10.8)$$

for $Qd <$ or $\ll 1$. Here, D is the single-particle (tracer), self-diffusion constant (see also Section 10.2.5). In the case of sublattices that differ strongly with respect to their binding energy or in the presence of traps (impurities, deviations from stoichiometry, etc.), it may be necessary to replace this criterion by a different and more restrictive one (see Section 10.2.3). For noncubic systems Eq. (10.8) is to be replaced by a tensor product.

10.2.2. Experiments on the Diffusive Step in Dilute Hydrides

For hydrogen in fcc palladium, the first experimental results by Sköld and Nelin[13] gave the verification of the Chudley–Elliott model [Eqs. (10.1)–(10.4)], and the observed consistency is very satisfactory,[13,14] assuming octahedral interstitial sites; also more recent experiments by Anderson et al.[15] confirm the first results. On the other hand, striking discrepancies were observed by Rowe et al.[16] for hydrogen in bcc metals, namely, α-TaH$_x$ and, more recently, by Lottner et al. for hydrogen in α-NbH$_x$:[5,12] At elevated temperatures $\Gamma(\mathbf{Q})$ turned out to be considerably smaller than calculated for diffusion over tetrahedral and octahedral sites (tetrahedral sites are assumed to be the correct location of the interstitials). This can be seen from Fig. 3 for the case of α-NbH$_{0.02}$. On the other hand, at room temperature the Chudley–Elliott theory with T sites works quite well.

Generally, this observation can be related to a larger effective jump distance, and the theory was modified by Lottner et al.[5] introducing jump sequences over tetrahedral sites in this way: the hydrogen atom alternates between a mobile "state," where it performs repeated jumps between adjacent tetrahedral sites, with a rate $1/4\tau_1$, during a period τ_e on the average, and an "immobile" state where it stays for a mean time τ_r. The model generalizes a double-jump model, where sequences of only two rapid jumps were introduced (this leads to 16 jump types, including jumps to second and third neighbors and backward jumps). The simplified double-jump model already describes the results very well, and the longer jump

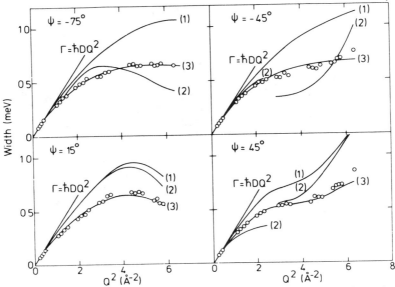

FIG. 3. Quasi-elastic width (effective half-width at half-maximum, not an eigenvalue) as obtained from the quasi-elastic spectra for α-NbH$_{0.02}$ (580 K) as a function of Q^2 (see Fig. 1). Here, ψ is the angle $\mathbf{k}_0$ versus [110] in the scattering plane. Solid curves: (1) simple tetrahedral jumps; (2) simple tetrahedral jumps including jumps to second neighbors; and (3) double jumps (see Section 10.2.2). (Reprinted with permission from Lottner et al.[12] Copyright 1979, Pergamon Press, Ltd.)

sequences do not lead to relevant improvements. Figure 3 also demonstrates how the simple next-neighbor model fails, whereas the multiple-jump model gives very good agreement for various crystal orientations. Typical jump sequences were determined to be of the order of four jumps within the time τ_e at 580 K. An alternative concept was the introduction of additional jumps to second neighbors (instead of *double* jumps).[17] This leads to an apparent agreement of the Q dependence of Γ. However, an incorrect value of τ or D was obtained, and the model cannot be maintained.

We interpret the physical origin of double- or multiple-jump sequences in the following way. The hydrogen interstitial causes a strong distortion of the surrounding host lattice with a tetragonal axis of symmetry (Fig. 4). This distortion field has been evaluated by means of diffuse neutron scattering experiments[18] (see Chapter 9). Obviously, a jump from a tetrahedral site to an adjacent one rotates this axis by 90° and is related to a rearrangement of the surrounding lattice atoms. This process requires a certain relaxation time τ_{rel}. The barrier for leaving the unrelaxed site is much lower (by 100 meV as an order of magnitude) than for a fully relaxed site (Fig. 5). Consequently,

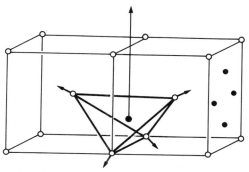

FIG. 4. Host lattice distortion. Arrows show displacement of the atoms, caused by a hydrogen atom on a tetrahedral site (solid circle). Niobium is shown by the open circles. Vertical arrow shows the tetragonal axis of strain field. A diffusive jump leads to a rotation of the axis by 90°. (From Bauer et al.[18])

the probability of leaving the site *before* it has relaxed increases with raising the temperature (we assume that the change of τ_{rel} with temperature is small). As a consequence, rapid jump sequences become increasingly probable at elevated temperatures. These considerations were stimulated by, and are related to, the observation of a dispersive behavior of the $T_1 A[110]$ acoustic branch in the hydrides of Nb[19] and Ta,[20] which corresponds to the shear modulus $c' = \frac{1}{2}(c_{11} - c_{12})$ in the long wavelength limit. It has also been theoretically shown that the $T_1 A[110]$ mode is connected with a strongly enhanced hydrogen amplitude, which enhances the neutron scattering intensity.[21] The $T_1 A[110]$ mode is able to couple to the diffusive jumps that create a deformation with the symmetry of this mode (see Section 10.3.1). As a conjecture, and using arguments from small-polaron hopping,[22] we

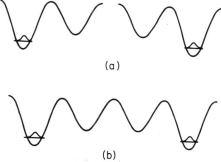

FIG. 5. (a) Hydrogen atom jumps with full host-lattice relaxation. (b) Second jump occurs before relaxation takes place thus leading to a jump sequence.

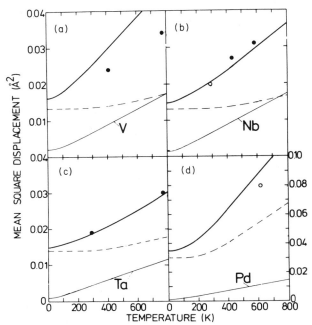

FIG. 6. Mean square amplitude of hydrogen $\langle u^2 \rangle$ in (a) vanadium, (b) niobium, (c) tantalum, and (d) palladium from the integrated intensity of the quasi-elastic line. Harmonic theory (———); contribution of optic (localized) modes only (----); Nb, Ta, and V (●) (from Ref. 5); Nb (○) (from Ref. 24); and Pd (from Ref. 14); thin lines $\langle u^2 \rangle$ for host metal atom (———). (From Schober and Lottner.[23])

assume that the relaxation for a jump is of the order of

$$\tau_{rel} \simeq 1/\Delta \nu \simeq 10^{-12} \text{ s}. \qquad (10.9)$$

Here, $\Delta \nu \simeq 10^{-12} \text{ s}^{-1}$ is the width of the vibrational band corresponding to these modes.[21] Therefore, τ_{rel} is sufficiently long to justify nonrelaxed jumps.

Finally, we turn briefly to the *intensity* of the quasi-elastic line, $I_0(\mathbf{Q})$. This is defined as an integral over the line intensity from $\omega = -\infty$ to $+\infty$ where an extrapolation by the best possible model into the region of the superimposed phonon spectrum is necessary. The intensity $I_0(\mathbf{Q})$ can be described by a Debye–Waller factor for the hydrogen vibrating on its interstitial site, with a mean-square hydrogen amplitude $\langle u^2 \rangle$ during its rest time, namely,

$$I_0(\mathbf{Q}) = \exp(-Q^2 \langle u^2 \rangle), \qquad (10.10)$$

assuming isotropic vibrations for simplicity. Here, $\langle u^2(T) \rangle$ is normally found to have values in the range of 0.02–0.06 Å^2. Figure 6 compares measured and theoretical values. The contributions due to the band modes and the optic (or localized) modes were calculated by Schober *et al.* in the harmonic approximation,[23] and $\langle u^2 \rangle$ was found as $\langle u^2 \rangle = \langle u^2 \rangle_{band} + \langle u^2 \rangle_{optical}$.

There is good agreement with the experimental values from the initial slope of $\ln I_0$ versus Q^2. Also, the surprisingly high mean-square amplitude observed by Sköld *et al.* for hydrogen in palladium[13,14] is confirmed by these calculations. In this case the optic modes contribute strongly because of their relatively low frequencies. Deviations from a simple Debye–Waller factor appear for hydrogen in the bcc vanadium above 700 K for $Q \gtrsim 1.5$ Å^{-1}, where the negative slope of $\ln I_0(Q)$ versus Q^2 start to increase. The reason could be that τ_0 approaches the order of magnitude of τ_j, the jump time, which is expected to be about 10^{-13} s. This causes a "transfer" of quasi-elastic intensity into the region of vibrational energies and thus leads to an apparent reduction of $I_0(\mathbf{Q})$.

In general, for the investigated bcc hydrides, as well as for the palladium hydride, we expect a rather strong localization of the hydrogen caused by the elastic polarization or relaxation of the surrounding host matrix. This is confirmed by the sharp localized optic modes (see Section 10.3). Obviously the mean residence time of about 10^{-12} s is sufficiently large compared to the vibrational period $2\pi/\omega_0$ of these modes. We may point out briefly the qualitative difference from fast ionic diffusion in superionic conductors. Here the diffusion is similar to that in a classical liquid, namely, a viscous motion, where the vibrational modes are completely overdamped. The other extreme is observed for the diffusion of the muon, where, due to its very small mass, a strong delocalization can be observed.[3]

It would be interesting to study metal hydrides at very high temperatures, where diffusion may approach the behavior of a free gas with a mean-free path $\tilde{d}$ on the order of a few lattice distances, and D would approach $\approx \tilde{d} v_{th}/3$.

Finally, we refer briefly to measurements published recently, dealing with yttrium hydride[25] on a powder sample, where inequivalent sites and different residence times were evaluated.

10.2.3. Diffusion with Trapping Impurities

Interstitial atoms, as nitrogen or oxygen in a metal, as well as substitutional metal atoms in niobium or tantalum, are known to induce "traps" for the dissolved hydrogen. This can be concluded from various observations, such as the increase of the hydrogen solubility; the appearance of an additional Snoek relaxation maximum, apparently caused by jumps of the trapped hydrogen close to the impurity atom; the decrease of the self-diffusion

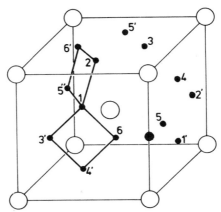

FIG. 7. Nitrogen atom (large solid circle) in the bcc niobium lattice. Open circles are niobium and closed circles are tetrahedral sites. The numbers characterize the six sublattices of the interstitial sites; 5' and 6' are possible tetrahedral sites where the hydrogen atom is trapped by the nitrogen (see Section 10.4.2).

constant; and the occurrence of modified localized modes and tunneling states, as described in Sections 10.3.3.1 and 10.4. Figure 7 indicates the possible geometry of an impurity atom in a bcc lattice and the surrounding sites that are able to trap hydrogen.

To calculate the quasi-elastic scattering for diffusion in the presence of an impurity, a simple "two-state model" was introduced by Richter and Springer.[26] Alternatively, the hydrogen spends a time τ_0 on the average on a trapping site, and then, during a period τ_1, diffuses in the lattice. The "free" diffusion is described approximately by the diffusion constant $D^{(f)}$ of the undisturbed or impurity-free lattice. This leads to the typical *anticrossing* shown schematically in Fig. 8. At small Q, a single Lorentzian is the dominating fraction of the quasi-elastic intensity. Its width and the self-diffusion constant are reduced by the trapping, namely,

$$\Gamma = Q^2 D = Q^2 D^{(f)} \tau_1 / (\tau_0 + \tau_1). \qquad (10.11)$$

At larger Q, however, *two* quasi-elastic components appear. One has a width of approximately $1/\tau_0$; the other has a width close to that expected for the free diffusion. With increasing temperature, the first contribution obviously decreases. The "crossover" occurs near

$$Q^2 \simeq 1/D^{(f)} \tau_0. \qquad (10.12)$$

Only below this value, the width yields the "true" (macroscopic) diffusion constant D. We refer to this point in Section 10.2.4.

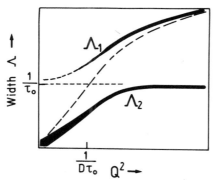

FIG. 8. Eigenvalues (partial widths) for diffusion with traps. The weights (quasi-elastic intensities) of the components are indicated by the thickness of the curves for the two-state model, indicating the "anticrossing" (schematic).

For a distribution of inequivalent traps with different trapping times τ_0, a bundle of eigenvalue branches appears in Fig. 8 and the slope of Γ versus Q^2 reaches the macroscopic value D if τ_0 in Eq. (10.12) is identified with the largest existing value of τ_0. If there are very "deep" traps (e.g., $\tau_0 > 10^{-7}$ s), quasi-elastic scattering may never experimentally reach the corresponding asymptotic region and the "true" value of D. A general treatment of the incoherent quasi-elastic scattering for a random trap distribution has been formulated by Kehr et al.,[27,28] which yields the two-step model as a special case.

A series of experiments[26] was carried out on niobium doped with a few tenths of an atomic percent of nitrogen and hydrogen. The results were interpreted in this model, and the measured spectra reveal clearly the existence of the trapped state by a narrow component whose width is about $1/\tau_0$, and whose intensity decreases with increasing temperature. In addition, there is a much broader component from the "free state" (Fig. 8).

The trapping rate $1/\tau_1$ was calculated from first principles in terms of a simple elastic interaction mode. Figure 9 presents the results for the characteristic times as obtained from the quasi-elastic experiments for $Q = 0.15$–1 Å^{-1}. Only the narrow component was measured with a back-scattering spectrometer. The weight of the unobserved broad component was obtained from the intensity of this line and also from additional low-resolution experiments.[29] At larger Q, the general trapping model[27,28] was successfully applied. There was some controversy regarding the question whether diffusion in the free state is properly described by $D^{(f)}$.[30] The consistent description of all experimental data and also quantitative arguments[31] makes us believe that the model is certainly applicable in the temperature region studied, where the deformation energy in the lattice sufficiently far from the impurities is small compared to kT.

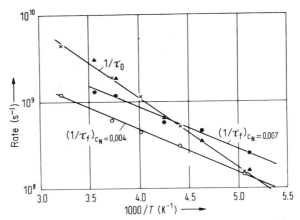

FIG. 9. Mean escape rate $1/\tau_0$ from the traps and mean trapping rate $1/\tau_1$ as functions of temperature T, for hydrogen diffusion in NbN_cH_y. Here, $y = 0.4$ at. %; c is the trap (nitrogen) concentration. (From Richter and Springer.[26])

10.2.4. Self-Diffusion Constants, in Particular for Alloys

The asymptotic quasi-elastic width $\Gamma = DQ^2$ in Eq. (10.8) has been frequently used to determine the (single-particle) self-diffusion constant. The resolution of back-scattering spectrometers (in Jülich and Grenoble), with a resolution of a few tenths of a µeV, allow D measurements between $\sim 10^{-8}$

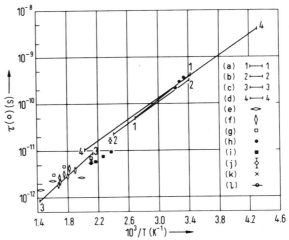

FIG. 10. Mean residence time $\tau(0)$ extrapolated to concentration $c = 0$ [see Eq. (10.13)] from quasi-elastic spectra (QENS) and from other methods as a function of temperature T. For indicated letters, see Table I. (From Anderson et al.[34])

and $\sim 10^{-6}$ cm^2/s. Conventional time-of-flight instruments are restricted to the region of $\sim 10^{-6}$ cm^2/s and higher.

In the past, accurate and reliable values of $D(T)$ were obtained for simple metal hydrides such as α-NbH$_x$,[5,32] α-TaH$_x$,[33] and for PdH$_x$.[34] Results for palladium are summarized in Fig. 10 and Table I and include alternative methods. All data are normalized to a concentration $c \to 0$ and presented in terms of the mean residence time $\tau(c = 0)$, using the relation (Section 10.2.5)

$$D(c) = f_c(c)d^2/6\tau(c), \qquad (10.13)$$

where $\tau(c) = \tau(0)/(1 - c)$, s is the jump distance, $(1 - c)$ the "blocking correction," and f_c the correlation factor. The mutual agreement is quite satisfactory.

TABLE I. Self-Diffusion Constant D in PdH$_c$[a]

Fig. 10[b]	Phase	Method	Reference[c]
(a)	β	Permeation	1
(b)	—	NMR spin echo	2
(c)	α	Permeation	3, 1
(d)	α	Gorsky effect	4
(e)	α	QENS[d]	5
(f)	β	QENS	6
(g)	α	QENS[e]	7
(h)	β	QENS[e]	8
(i)	β	QENS	9
(j)	β D	QENS[e]	9
(k)	β	QENS[e]	9
(l)	β	QENS[e]	9

[a] Comparison of results from different methods (extrapolated to $c = 0$).
[b] Letters correspond to Fig. 10.
[c] References
 1. G. Bohmholdt and E. Wicke, *Z. Phys. Chem. (Wiesbaden)* **56**, 133 (1967).
 2. P. P. Davis, E. F. W. Seymour, D. Zamir, W. D. Williams, and R. M. Cotts, *J. Less-Common Met.* **49**, 159 (1976).
 3. W. D. Davis, *KAPL Rep.* No. 1227 (1954).
 4. J. Völkl, G. Wollenweber, K. H. Klatt, and G. Alefeld, *Z. Naturforsch. A* **26A**, 922 (1971).
 5. K. Sköld and G. Nelin, *J. Phys. Chem. Solids* **28**, 2367 (1967).
 6. G. Nelin and K. Sköld, *J. Phys. Chem. Solids* **36**, 1175 (1975).
 7. C. J. Carlile and D. K. Ross, *Solid State Commun.* **15**, 1923 (1974).
 8. I. S. Anderson, D. K. Ross, and C. J. Carlile, in "IAEA Symposium on Neutron Inelastic Scattering," IAEA, Vienna, 2(8), 427 (1977).
 9. I. S. Anderson, Ph.D. Thesis, Univ. of Birmingham, 1978.

[d] QENS—quasi-elastic neutron scattering.
[e] Single-crystal sample.

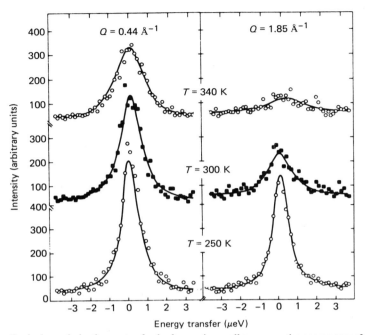

FIG. 11. Typical quasi-elastic spectra for hydrogen in an alloy versus the energy transfer for $Ti_{0.8}Zr_{0.2}CrMnH_3$ in the µeV range, as measured with a back-scattering spectrometer. Solid lines show Lorentzian convoluted with the resolution spectrum. (From Hempelmann et al.[35] Copyright of the Institute of Physics.)

More recently, similar experiments were carried out for the hydrides of certain alloys, many of them being of interest as storage materials. As an example of such measurements, where diffusion is much slower than in a pure metal, typical quasi-elastic spectra in the µeV range and the corresponding width curves are shown in Figs. 11 and 12, respectively, for the "prealloy" $Ti_{0.8}Zr_{0.2}CrMnH_3$ (Hempelmann et al.[35]). It should be realized that multiple scattering in the sample (~80% transmission) is an important source of systematic errors; corrections by a Monte Carlo program from Harwell were applied.[36] A survey of the available data together with alternative methods is quoted in Tables II and III.[†] The very high value quoted for $LaNi_5H_6$ is obviously wrong. The quasi-elastic line may disguise a broad component caused by a rapid "local" motion that is not separated from the diffusive part because of insufficient spectrometer resolution.

[†] Of the figures in the tables, the 300 K values are the most reliable; D_0 and activation energies must be treated with caution because of the relatively small T range covered in most of the measurements.

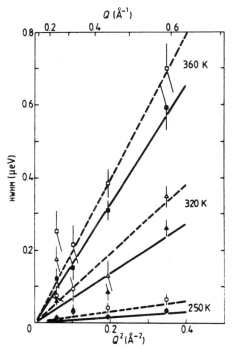

FIG. 12. The widths from Fig. 11, uncorected (○) and corrected (●) for multiple scattering. (From Hempelmann et al.[35] Copyright of the Institute of Physics.)

A striking effect observed for $LaNi_5H_x$ is an inverse concentration dependence of D: the diffusion constant increases as the concentration c increases[37] (Fig. 13). This can be explained by the coexistence of two nonequivalent sites in the lattice. The energetically deeper sites saturate first, and the remaining hydrogen atoms are then able to diffuse faster because the trapping probability at the deeper sites is diminished. This effect compensates the blocking factor $(1 - c)$, which tends to reduce D as c increases.

Another interesting effect should be mentioned that was found in $LaNi_5H_6$[37] and also in $Ti_{1.2}Mn_{1.8}H_3$,[38] namely, the appearance of a very broad component in the spectrum. This leads to a decrease of the quasielastic intensity with increasing Q inside the resolution window, much steeper than explicable by a Debye–Waller factor. This can be interpreted by two possible mechanisms:[38] either (1) a *backjump mechanism*, which occurs at high concentration ($c \simeq 1$) (Section 10.2.5), or (2) *structural effects*, that is, "local hopping" in the vicinity of an energetically deep site surrounded by shallow sites. A model was worked out by Hempelmann et al.,[38] which takes

TABLE II. Hydrogen Diffusion Coefficients in Different Intermetallic Hydrides[a]

Specimen	D_0 (cm^2/s)	u (meV)	D_{300K} (cm^2/s)	Reference[b]
PdH$_{x\to 0}$	5.3×10^{-3}	236	5.7×10^{-7}	1
NbH$_{x\to 0, T>250K}$	3.6×10^{-4}	108	5.5×10^{-6}	2
Ti$_2$NiH$_2$	2.0×10^{-4}	345	3.2×10^{-10}	3
TiFeH	7.2×10^{-4}	500	2.9×10^{-12}	4
TiFeH	4.2×10^{-7}	330	1.2×10^{-12}	5
LaNi$_5$H$_6$	3.2×10^{-4}	249	2.1×10^{-8}	6
LaNi$_5$H$_6$	2.1×10^{-3}	275	5.0×10^{-8}	7
TiCr$_{1.8}$H$_{2.6}$		270		8
Ti$_{1.2}$Mn$_{1.8}$H$_3$	5.9×10^{-4}	225	9.8×10^{-8}	9
Ti$_{0.8}$Zr$_{0.2}$CrMnH$_3$	3.1×10^{-4}	220	6.2×10^{-8}	10
Mg$_2$NiH$_{0.3}$	6.7×10^{-5}	280	1.3×10^{-9}	11

[a] For comparison, the values for dilute PdH$_x$ and NbH$_x$ are also included.
[b] References
1. E. Wicke, H. Brodowsky, and H. Zückner, *Top. Appl. Phys.* **29**, 73 (1978).
2. J. Völkl and G. Alefeld, in "Diffusion in Solids" (A. S. Nowick and J. J. Burton, eds.). Academic Press, London, 1975; J. Völkl and G. Alefeld, *Top. Appl. Phys.* **28**, 321 (1978).
3. J. Töpler, E. Lebsanft, and R. Schätzler, *J. Phys. F* **8**, L25 (1978).
4. E. Lebsanft, D. Richter, and J. Töpler, *J. Phys. F* **9**, 1057 (1979).
5. R. C. Bowman and W. E. Tadlock, *Solid State Commun.* **32**, 313 (1979).
6. T. K. Halstead, N. A. Abood, and K. H. J. Buschow, *Solid State Commun.* **19**, 425 (1976).
7. D. Richter, R. Hempelmann, and L. A. Vinhas, *J. Less-Common Met.* **88**, 353 (1982).
8. R. C. Bowman and J. R. Johnson, *J. Less-Common Met.* **73**, 254 (1980).
9. R. Hempelmann, D. Richter, and A. Heidemann, *J. Less-Common Met.* **88**, 343 (1982).
10. R. Hempelmann, D. Richter, R. Pugliesi, and L. A. Vinhas, *J. Phys. F* **13**, 59 (1983).
11. J. Töpler, H. Buchner, H. Säufferer, K. Knorr, and I. J. Prandl, *J. Less-Common Met.* **88**, 397 (1982).

into account four different time scales: (1) the jump rate $1/\tau_f$ over free sites; (2) a trapping rate in deep sites $1/\tau_1$; (3) a trap escape rate $1/\tau_0$ from the deep sites (see Section 10.2.3); and (4) a rate $1/\tau_b$ corresponding to the fast local motion already mentioned. Figure 14 shows the width curves and the related characteristic times with their T dependence. Also, in other experiments[39,40] such a local motion was observed and postulated. Experiments with poor resolution tend to mix the broad component with the true "diffusive" line. Obviously this always leads to an exaggerated value of D.

As pointed out earlier [Eq. (10.12)], traps interferring with free diffusion cause an anticrossing behavior (Fig. 8), and the true diffusion constant can only be found if $Q^2 < 1/D\tau_0$. Using the values quoted by the authors,[38] one obtains $Q \le 0.2$ Å^{-1}, consistent with the range where D was evaluated.

The experiments described originally started with the aim of determining D values, but they also led to an interesting insight into the diffusion

TABLE III. Literature Data for the Hydrogen Diffusion Coefficient in LaNi$_5$H$_6$

D_0 (cm^2/s)	u (meV)	$D_{300\text{K}}$ (cm^2/s)	Method	Reference[b]
		120×10^{-8}	QNS	1
0.6×10^{-4}	110	85×10^{-8}	QNS	2
8.8×10^{-5}	242	0.8×10^{-8}	QNS	3
2.1×10^{-3}	275	5.0×10^{-8}	QNS	4
1.5×10^{-4}	218	3.2×10^{-8}	NMR-CW	5
3.2×10^{-4}	249	2.1×10^{-8}	NMR-T$_{1\rho}$	6
8.0×10^{-4a}	230	11.0×10^{-8}	NMR-T$_1$	7
1.6×10^{-3}	300	1.5×10^{-8}	NMR-T$_2$	8
1.4×10^{-1}	412	1.4×10^{-8}	NMR-APFG	
	206		NMR-T$_1$	9
	302			10
	218		NMR-T$_1$	

[a] Calculated by Lebsanft[3] using the jump distance of Halstead.[5]
[b] References
1. P. Fischer, A. Furrer, G. Busch, and L. Schlapbach, *Helv. Phys. Acta* **50**, 421 (1977).
2. D. Noréus, L. G. Olsson, and U. Dahlborg, *Chem. Phys. Lett.* **67**, 432 (1979).
3. E. Lebsanft, D. Richter, and J. Töpler, *Z. Phys. Chem. (Wiesbaden)* **116**, 175 (1979).
4. D. Richter, R. Hempelmann, and L. A. Vinhas, *J. Less-Common Met.* **88**, 353 (1982).
5. T. K. Halstead, *J. Solid State Chem.* **11**, 114 (1974).
6. T. K. Halstead, N. A. Abood, and K. H. J. Buschow, *Solid State Commun.* **19**, 425 (1976).
7. Y. F. Khodosov, A. I. Linnik, G. F. Kobzenko, and V. G. Ivanchenko, *Fiz. Met. Metalloved.* **44**, 433 (1977); *Phys. Met. Metallogr. (Engl. Transl.)* **44**, 178 (1978).
8. R. C. Bowman, D. M. Gruen, and M. H. Mendelsohn, *Solid State Commun.* **32**, 501 (1979).
9. R. F. Karlicek and I. J. Lowe, *J. Less-Common Met.* **73**, 219 (1980).
10. H. Chang, I. J. Lowe, and R. J. Karlicek, in "Nuclear and Electron Resonance Spectroscopies Applied to Materials Science" (E. N. Kaufmann and G. K. Shenoy, eds.). North-Holland Publ., Amsterdam, 1981.

mechanism, and the simple trap model from Section 10.2.3 proved to be useful for the interpretation of the data.

The diffusion constants of hydrogen can be related to the reaction rate constant $K_r = 1/\tau_r$ for absorption or desorption in a storage powder. For grains of size R we obtain approximately

$$K_r \approx \pi^2 D/R^2. \tag{10.14}$$

For Ti$_{0.8}$Zr$_{1.2}$CrMnH$_3$[35] with $D = 6 \cdot 10^{-8}$ cm^2/s from quasi-elastic scattering and $R = 0.5\,\mu$m, this yields $K_r = 400$ s^{-1}, if the absorption and desorption were entirely diffusion controlled. Actually, the measured rate is only 0.1 s^{-1}. Consequently, the reaction rate is retarded by surface penetration effects and not determined by the relatively fast bulk diffusion.

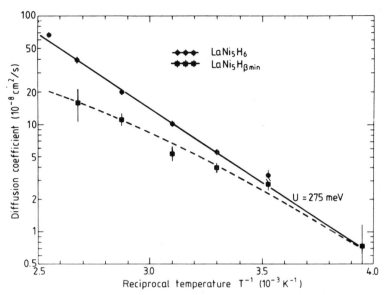

FIG. 13. Self-diffusion constant from quasi-elastic scattering on LaNi$_5$H$_x$ for high concentration ($x = 6$) and low concentration ($x = 5$–6). (From Richter et al.[37])

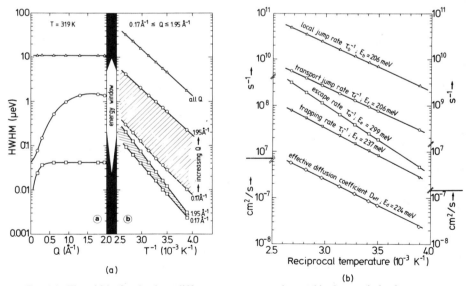

FIG. 14. The widths for the three different components observed in the quasi-elastic spectrum for hydrogen in Ti$_{1.2}$Mn$_{1.8}$H$_3$, showing (a) Q and T dependence and (b) the corresponding characteristic times. Arrow shows resolution window of back-scattering spectrometer. (From Hempelmann et al.[38])

10. HYDROGEN IN METALS

The opposite case occurs for TiFeH$_x$,[41,42] where the self-diffusion is very slow. Here the reaction is diffusion controlled. However, it turns out that if K_r is calculated from Eq. (10.14), the resulting value is still smaller than measured, using the actual grain size for R. This was understood after inspecting metallographic pictures that show many microcracks in the grains generated during the absorption process. The cracks partially bypass the diffusion path, and the value of R inserted into Eq. (10.14) is too large. Actually, the distance between the microcracks is much larger than $1/Q$. Consequently, they do not affect D as measured from neutron scattering that averages over volumes $\lesssim 1/Q \approx 10$ Å.

Finally, we point out that the diffusion during absorption or desorption is influenced by the driving force related to the hydrogen density gradient (see the end of Section 10.2.5). Therefore, the chemical diffusion constant $D_{ch}(c)$ comes into play instead of D. Rigorously, the problem is nonlinear and very complex, and Eq. (10.14) is only a rough approximation.

10.2.5. Nondilute Systems: Incoherent and Coherent Cases

In Section 10.2.1, incoherent quasi-elastic scattering for hydrogen diffusion was derived in the limit of zero concentration. For finite, but not too high, average occupancy of the *available* sites, a mean-field approach is a useful, although rough, approximation, namely, reducing the jump probability from $1/\tau$ to $(1 - c)/\tau$, where $(1 - c)$ is the "blocking factor." It implies that the only interaction between two hydrogen atoms is the mutual site exclusion. Consequently, the shape of the incoherent scattering function in Eq. (10.3) would remain unchanged with $f(\mathbf{Q})$ being replaced by $(1 - c)f(\mathbf{Q})$.

At higher concentrations the situation is much more complicated. In this case, experimental results are practically nonexistent, and in this section we mainly quote theoretical papers rather than measurements. The theoretical approaches still postulate mutual exclusion as the main interaction.[34,43–46] In this case, correlations between successive jumps occur, which can be easily explained for the case of high concentration: if the hydrogen atom has jumped from a site j to a site $j + 1$, the probability for a jump back to j is obviously higher than the mean value because, initially, site j is certainly unoccupied and available for a return jump. Only for long times site j will be filled by *other* hydrogen atoms, and the return jump probability approaches its average value $(1 - c)/\tau$. For $1 - c = c_v \ll 1$ (vacancy assisted diffusion), one gets in particular

$$D_c = c_v D(c = 0) f_c, \quad (10.15)$$

where the correlation factor f_c is well known for the sc, bcc, and fcc lattice.[47–49] The full scattering function $S_i(\mathbf{Q}, \omega)$ in Fig. 15 presents results

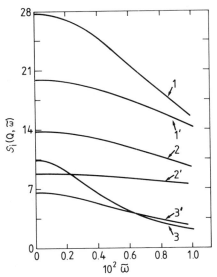

FIG. 15. Calculated quasi-elastic width for nondilute systems: 1 = fcc; 2 = bcc; 3 = sc lattice. Theory of Tahir-Kheli and Elliott,[45] which agrees also with Kehr et al.[44] Here, 1', 2', and 3' show the mean field approach. The pairs of curves are normalized; therefore, a lower value at $\bar{\omega} = 0$ is related to a broader wing at large $\bar{\omega}$; $\bar{\omega}$ is the energy transfer in units of the empty lattice jump rate $1/\tau(0)z$. (From Tahir-Kheli and Elliott.[45])

from a general approach for Q at the zone boundary,[45] in order to demonstrate the deviations from the mean-field approach, which lead mainly to broad wings of the quasi-elastic spectra. A semi-analytical approach as developed by Kehr et al.[44] reduces the problem to the knowledge of waiting time distributions $\psi_k(t)$, where $\psi_k(t)\,dt$ is the joint probability that a hydrogen atom has carried out no jump until a time t, and jumps in the time interval between t and $t + dt$;[49] the mean field approach gives $\psi_k(t) = \exp\{-tn(1 - c)/\tau\}$, where n/τ is the "attempted jump rate." The index k labels the type of jump (e.g., backward jump $k = 0$; there are various classes of forward jumps that differ for different orientations of the considered jump relative to the previous one). For a fcc lattice, these functions were calculated by a Monte Carlo method. For the sake of simplicity only the "memory" of the diffusing particle to its previous jump is taken into consideration, neglecting correlations to jumps more distant in the past. With these Monte Carlo generated functions, the self-correlation functions and the incoherent scattering function can be calculated, and the results agree very well with those from Tahir-Kheli and Elliott[45] in Fig. 15. For completeness, the series expansion quoted in Section 10.2.1 should also be mentioned here; it has been successfully applied[43] to problems of nondilute diffusion.

To judge the value and versatility of the various approaches in general, one has to analyze to what extent more realistic assumptions can be included for the mutual interaction (e.g., the concept of an excluded volume surrounding the diffusing particle) based on realistic computing times. So far, no systematic experiments have been carried out to test existing theories, except in the region of small Q.

Coherent quasi-elastic scattering could be investigated experimentally for deuterium in metals, correcting for its incoherent contribution. Theoretically, the treatment is simpler than for incoherent scattering, because there is no distinction between the atom under consideration and the other atoms. For a simple Bravais lattice the rate equations read (Ross and Wilson;[43] Kutner[50])

$$\frac{\partial}{\partial t} P(\mathbf{r}, t) = \frac{1}{n\tau} \sum_{\mathbf{r}'} [P(\mathbf{r}', \hat{\mathbf{r}}, t) - P(\mathbf{r}, \hat{\mathbf{r}}', t)]. \qquad (10.16)$$

Here, $P(\mathbf{r}', \hat{\mathbf{r}}, t)$ is the joint probability of finding a diffusing particle at $\mathbf{r}'$ and *no* particle at $\hat{\mathbf{r}}$ at time t. The sum extends over the adjacent sites $\mathbf{r}'$ of $\mathbf{r}$. Obviously, the sum of the complementary probabilities yields the probability $P(\mathbf{r}, t)$ of finding a particle at $\mathbf{r}$ for a time t, namely,

$$P(\mathbf{r}', \hat{\mathbf{r}}, t) + P(\mathbf{r}', \mathbf{r}, t) = P(\mathbf{r}', t)$$

$$P(\mathbf{r}, \hat{\mathbf{r}}', t) + P(\mathbf{r}, \mathbf{r}', t) = P(\mathbf{r}, t) \qquad (10.17)$$

which leads to

$$\frac{\partial}{\partial t} P(\mathbf{r}, t) = \frac{1}{n\tau} \sum_{\mathbf{r}'} [P(\mathbf{r}', t) - P(\mathbf{r}, t)]. \qquad (10.18)$$

This agrees with the Chudley–Elliott equation for the single-particle correlation function. However, instead of Eq. (10.1) the initial condition for the calculation of $S_c(\mathbf{Q}, \omega)$ is

$$P(\mathbf{r}, 0) = \begin{cases} 1 & \text{for} \quad \mathbf{r} = 0 \\ c & \text{for} \quad \mathbf{r} \neq 0, \end{cases} \qquad (10.19)$$

where c is the average occupancy of the available sites. The scattering function per site is then

$$S_c(\mathbf{Q}, \omega) = c(1 - c) \frac{f(\mathbf{Q})/\pi}{\omega^2 + f^2(\mathbf{Q})} + c^2 \delta(\omega) \times [\text{Bragg lines}]. \qquad (10.20)$$

The quasi-elastic width does not depend on the concentration; the line intensity is proportional to $c(1 - c)$, the well-known Laue factor for a random lattice (cf. the convolution approximation). For small Q one obtains, in analogy to incoherent scattering, the quasi-elastic width

$$\Gamma_c = (d^2/n\tau)Q^2 = D_{ch}Q^2. \qquad (10.21)$$

This result holds generally for coherent scattering on a collectively diffusing isotropic system.

To understand why we introduced the *chemical* or collective diffusion constant D_{ch}, we recall the next relations. The collective particle current is

$$j = -cM(\partial\mu/\partial c)\nabla n(\mathbf{r}) = -D_{ch}\nabla n(\mathbf{r}), \qquad (10.22)$$

where $n(\mathbf{r})$ is the particle density, and B the single particle mobility is related to D_{ch} by $D_{ch} = (\partial\mu/\partial c)cB$. In the dilute case $\partial\mu/\partial c = kT/c$ and $D_{ch} = BkT$. For an interaction-free lattice cell gas, the free energy is $F = kTc\ln c + kT(1-c)\ln(1-c)$, which yields

$$D_{ch} = D/(1-c). \qquad (10.23)$$

Therefore, in this model the blocking factor $1/(1-c)$ in D_{ch} cancels, and thus leads to Eq. (10.21).

Figure 16 compares the quantities discussed above, as measured for tantalum hydride,[33] namely, the chemical diffusion constants D_{ch} and D. The (normalized) chemical factor

$$F_{ch} = (\partial\mu/\partial c)/(\partial\mu/\partial c)_{c\to 0} = (\partial\mu/\partial c)/(kT/c) \qquad (10.24)$$

was determined by the Gorski relaxation method.[51] Obviously, for lower concentrations, D and D_{ch} agree quite well. However, at high concentration, there is a strong difference between both curves, and one expects

$$D/D_{ch} = f_c/F_{ch}, \qquad (10.25)$$

where f_c is the single-particle correlation factor introduced before (for $c \to 0$ both $f_c \to 1$ and $F_{ch} \to 1$). The dashed line shows $D' = D_{ch}/F_{ch} = Df_c$, where f_c is close to unity (0.6–0.8 for $c \to 1$). The remaining discrepancy includes f_c and, as well, the systematic errors of the experimental quantities involved.

An interesting effect was observed for the niobium–hydrogen system at the critical concentration (~ 33 at. %). In this case $\partial\mu/\partial c$ has a singularity at the critical temperature T_c on top of the miscibility region. This leads to[51,52]

$$D_{ch} = D(T - T_c)/T \qquad (10.26)$$

where the entropy contribution (see above) has been left out for simplicity. Consequently, D_{ch} reveals a spectacular slowing down whereas, as expected, D from incoherent quasi-elastic scattering behaves completely regularly in the critical region.[51] So far, no systematic studies exist of the quasi-elastic width for coherent scattering in an extended Q region.

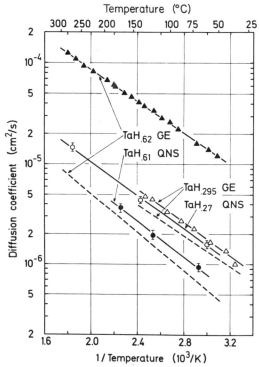

FIG. 16. Chemical and self-diffusion constants for α-TaH$_x$: D from quasi-elastic scattering (QNS); D_h from Gorski effect (GE). Dashed line is D' (see text). (From Potzel et al.[33])

10.3. Hydrogen Vibrations

While in the previous sections we were concerned with hydrogen jumps between different sites, we now consider the vibrational motions that the hydrogen atom exhibits during its rest time at a particular interstitial site. They can be discussed in terms of two classes of motion: (1) the band modes, where the hydrogen isotopes move according to the displacements of the host atoms during their acoustic vibrations, and more or less mirror the host density of states; and (2) the local or optic vibrations where hydrogen vibrates against its metal neighbors with frequencies typically a few times higher than those of the host vibrations.

The influence of hydrogen loading on the acoustic host vibrations has been studied widely.[1] In particular the consequences of changes in the electronic structure of the host on the acoustic phonons were investigated.[53] Here, we focus on the recently detected quasioptic modes within the acoustic band in

NbH and TaH[54,55] and on their counterpart at low hydrogen concentrations, where Lottner et al.[21] reported resonantlike enhancements of the hydrogen amplitude for certain lattice modes.

The main emphasis in this section, however, will be placed on the discussion of the high-frequency hydrogen vibrations. At low hydrogen concentration, or small direct H-H interaction, the hydrogen can be considered as a three-dimensional Einstein oscillator in a cage formed by the slowly vibrating metal atoms. A determination of these local vibrations yields information on the hydrogen metal forces and therefore on the hydrogen potential. We surveyed recent investigations of local modes in bcc refractory metals, including measurements on the three different isotopes and the higher harmonics that yielded detailed information on the hydrogen potential in these materials. The theoretical interpretation of these results is still on a semi-empirical level and will be reviewed briefly.

For high hydrogen concentration, the dispersion of the optic hydrogen modes also reveals information on direct H-H interactions. While the direct forces between interstitial hydrogen atoms in bcc metals appear to be weak,[56] they are important in fcc structures and were investigated extensively for the case of PdH_x and for some rare earth dihydrides (for a review see Sköld[1]). There are some further investigations on Ce hydrides[57-59] and on rare earth trihydrides,[60-62] which all show considerable dispersion of the optic modes but do not reveal any new features. We therefore omit the discussion of these results due to space limitations.

The point symmetry of the hydrogen site determines the degeneracy of the local hydrogen modes, e.g., two different fundamentals with an inversed sequence of degeneracy exist for hydrogen on tetrahedral (T) or octahedral (O) sites in bcc structures. Sites of lower symmetry lead to three different fundamentals, while for cubic surroundings like that of T and O sites in fcc structures three-fold degeneracy results. This relation between point symmetry and local vibrations has been used in order to obtain information on hydrogen sites by inelastic neutron scattering, where diffraction results were not sensitive enough. We examine successful applications in the field of hydrogen trapping[63,64] and that of glassy metals.[65]

Though the discovery of superconductivity[66] in PdH together with the anomalous isotope effect[67] in T_c dates back ten years, its explanation is still controversial and attracts scientific interest. We surveyed recent investigations on the anharmonicity of the optic PdH phonons by a measurement of higher harmonics.[68] The superconducting properties of α-ZrH_x and α-TiH_x are discussed in relation to their local hydrogen vibrations.[69]

Finally, hydrogen tunneling between adjacent potential minima has been observed in $NbO_{0.013}H_{0.016}$.[70] We devote the last section to a discussion of this new feature in the dynamics of hydrogen in metals.

10.3.1. Acoustic Phonons

Hydrogen loading affects the acoustic phonons of the host in various ways. First, the addition of hydrogen atoms changes the electronic structure of the host metal. According to band structure calculations by Switendick[71] and others, upon hydrogenation there is a general tendency in d-electron metals to pull high-energy states well below the Fermi level. The newly introduced electrons partly occupy those states and are partly added to the d band at the Fermi level. As a consequence the shape of the Fermi surface as well as the binding strength is expected to change. In niobium, for example, with increasing hydrogen concentration, the first results in a gradual removal of all phonon anomalies of pure niobium,[53,55] until the latter causes significant modifications of the metal–metal force constants surrounding a hydrogen atom.[23] Second, at high hydrogen concentrations the hydrogen interstitials have the tendency to order on their sublattice. Even in the disordered liquid-like phases important short-range order phenomena have been observed.[72] These ordering phenomena strongly affect the symmetry of the effective force field between the host atoms and change the dispersion relations significantly.[54,55,73,74] Third, the introduction of hydrogen expands the lattice and as a consequence a general tendency toward softening of the force constants should be present.

The acoustic phonons of the metal lattice are probed best by coherent inelastic neutron scattering. The double differential one-phonon coherent cross section is given by[75]

$$\frac{d^2\sigma}{d\Omega\, d\omega} = \frac{k_1}{k_0}\frac{(2\pi)^3}{2v_0} \sum_\tau \sum_{j\mathbf{q}} g_j(\mathbf{Q})[n(\omega_j) + 1 \pm \tfrac{1}{2}]\delta[\omega \pm \omega_j(q)]\delta(\mathbf{Q} \pm \mathbf{q} - \tau),$$

(10.27)

where $g_j(\mathbf{Q})$ is the phonon form factor

$$g_j(\mathbf{Q}) = \frac{1}{\omega_j(\mathbf{q})}\left|\sum_d b_d \exp[-W_d(Q)]\exp(i\mathbf{Q}\cdot\mathbf{d})\mathbf{Q}\cdot\frac{\mathbf{e}_d^j(\mathbf{q})}{\sqrt{M_d}}\right|^2.$$ (10.28)

In this equation b_d and M_d are the scattering length and mass of atom d in the unit cell with the volume v_0, $n(\omega) = [\exp(\hbar\omega/kT) - 1]^{-1}$ is the Bose–Einstein factor, $W_d(Q)$ is the Debye–Waller exponent of atom d and $\mathbf{d}$ is its position vector in the cell, and $\mathbf{e}_d^j$ is the eigenvector of the displacements of the mode corresponding to $\omega_j(\mathbf{q})$. Energy and momentum conservation yield

$$\hbar\omega_j(q) = \hbar^2/2m(k_0^2 - k_1^2)$$ (10.29)

and

$$\mathbf{Q} = \mathbf{k}_0 - \mathbf{k}_1 = \tau \pm \mathbf{q},$$ (10.30)

where τ is a reciprocal lattice vector and q is the momentum of the phonon in the Brillouin zone. Dispersion curves are usually measured keeping q constant and varying the energy transfer to the samples. Peaks in $d^2\sigma/d\Omega\, d\omega$ will occur for $\omega = \omega_j(q)$.

In the following discussion, we survey the main topics of interest, restrict ourselves to an outline of experiments on the relation between hydrogen ordering effects and dispersion relations, and elaborate on recent attention that has been paid to hydrogen band modes at low hydrogen concentrations, where local changes of the host force constants are of importance.

Recently Shapiro et al.[54] reported the observation of flat opticlike excitations within the acoustic band of niobium. Speculations on their origin included the suggestion of hydrogen tunneling states as a cause for the modes and initiated further extensive investigations.[55, 73, 74] Figure 17 shows the

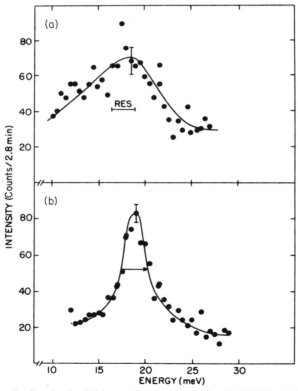

FIG. 17. Flat optic vibration in niobium at the zone center $Q = (2, 2, 2)$ for $NbD_{0.85}$ showing the change in linewidth between the (a) α' phase at 433 K and (b) β phase at 220 K. (From Shapiro et al.[55])

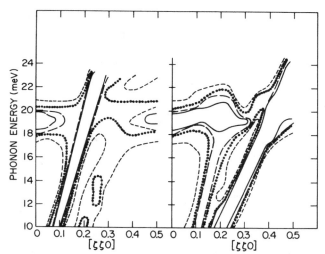

FIG. 18. Equal intensity contours along the $[\xi, \xi, 0]$ direction of β-NbD$_{0.85}$ at 220 K. The intensity scale is given as 200 (———); 150 (– – –); 100 (·····); and 50 (----). (From Shapiro et al.[55])

dispersionless excitation measured at the zone center $\mathbf{Q} = (2, 2, 2)$ in both the ordered β and the liquid-like α' phase of NbD$_{0.85}$. Though considerably broader in the α' phase, the excitation is well defined in both phases and appears at the same energy of about 19 meV. It is characterized by these features: (1) temperature-dependent measurements showed that its intensity follows the Bose–Einstein factor; (2) experiments on NbH$_{0.82}$ revealed the absence of any isotope effect in the frequency; (3) in TaD$_{0.78}$ a similar excitation appears around 15 meV, approximately lowered by the square root of the mass ratio of the host atoms; and (4) while in the α' phase the intensity does not change throughout the Brillouin zone, there are large intensity variations in the β phase.

This can be seen best from intensity contours. They show the measured intensities corrected for resolution volume, thermal occupation, and the form factor $Q^2 e^{-2W(Q)}/\omega$ and are a measure of the weighted eigenvector coefficients in Eq. (10.28). As an example we show the contours in $[\xi, \xi, 0]$ direction (Fig. 18). Most prominent is the band of intensity around 19 meV, which is most intense at the zone center. Under transverse polarization, the longitudinal mode is also seen, indicating strong deviations from cubic symmetry. Finally, there appears to be some coupling between the 19 meV mode and the acoustic phonons.

Burkel et al.[73] reported a more detailed study of the excitations in the $[\xi, \xi, 0]$ direction of α'-NbD$_x$ ($0.57 \leq x \leq 0.77$) and found similar results as

Shapiro et al.[55] in the β phase. In particular there is strong evidence for an anticrossing between the [$\xi, \xi, 0$]-LA mode and the 19 meV excitation. Furthermore, a softening of the LA zone boundary phonon at the N point was observed with increasing deuterium concentration. It goes along with anomalies in the shear constant $c' = (c_{11} - c_{12})/2$ and may indicate the beginning of a lattice instability. For the β phase, Shapiro et al.[55] report on anomalous broadening of the same mode near the N point.

Generally the lattice distortions in the ordered hydrides are small and *cannot* be responsible for large deviations of the phonon eigenvectors from cubic symmetry. However, the interaction between metal and hydrogen atoms is very strong. Therefore, at low frequencies it mediates an effective force between adjacent metal atoms. If hydrogen ordering now occurs in other than cubic symmetries, this ordering creates an effective force field between the host atoms that strongly deviates from the near cubic symmetry of the host lattice. In addition hydrogen ordering results in a multidomain structure of the crystal, further complicating the situation.

Based on these considerations Magerl et al.[74] gave a lattice dynamical explanation of the flat modes: as a consequence of the orthorhombic force field in β-NbD$_x$, certain cubic zone-edge modes appear as optic niobium modes along the principal cubic directions. In addition the multidomain structure enhances this effect. Quantitatively, the detailed calculations reveal that nearest- or next-nearest-neighbor force constants are changed strongly by the presence of hydrogen or deuterium and that in the main directions an excitation around 20 meV nearly always possesses a nonvanishing structure factor. The agreement between model calculations and experiments was improved if the actual nonstoichiometry of the β phase was considered. Namely, as a result of the strong Nb–D force constant a missing deuterium atom affects the effective Nb–Nb coupling strongly and changes the resulting spectrum considerably.

Besides the general tendency towards the 19 meV excitation, the calculations also reproduce details like the apparent hybridization of the acoustic modes and the 19 meV feature. Thereby, its large phonon structure factor is almost entirely due to the vibrations of the deuterium atoms. From the eigenvectors of the deuterium atoms, it was concluded that the deuterium amplitudes in the flat mode are around 1.5–2 times as large as in other modes near the same energy. Thus, one can speak of a resonant-like enhancement of the deuterium amplitudes for certain lattice modes. Similar observations were also reported at low hydrogen concentrations[21] and will be discussed in this section.

Finally, considering the indications of short-range order also in the α' phase, where preferred occupation in [110] planes was observed,[72] the above lattice dynamical explanation of the 19 meV excitation for β-NbH(D) may

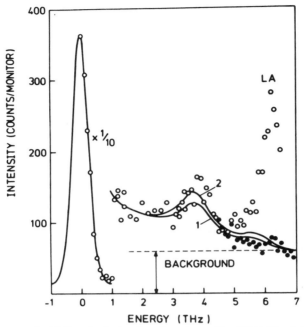

FIG. 19. Spectrum for NbH$_{0.05}$ in the band mode region at $T = 57°C$. Line 1 shows the one-parameter model and line 2 the six-parameter model. The LA peak originates from the host lattice phonons. $Q = (0, 0, 2.9)$ (○): $Q = (0, 0, 2.1)$ (●); the temperature is 57°C. (From Lottner et al.[21])

also hold for α'-NbH(D). Its appearance in other systems like TaD$_{0.78}$[55] and LaNi$_5$D$_6$[76,77] suggests that it may be typical for the lattice dynamics of hydrides, where hydrogen ordering occurs with symmetries lower than that of the empty host metal.

At low hydrogen concentrations the hydrogen was supposed to move according to the host vibrations and to mirror the nearly unchanged host density of states. Only recently Lottner et al.[21] reported a strong peak in the inelastic intensity of NbH$_{0.05}$ around the 16 meV, where the transverse acoustic modes have a van Hove singularity (Fig. 19). This peak cannot simply be explained by protons mirroring the host density of states. The authors suggest an interpretation of this feature as a resonant-like proton mode within the acoustic band. They calculated the local proton density of states $Z_i^H(\omega)$ by the Greens function technique. Two models were considered: model (1) included only longitudinal nearest-neighbor force constants f between the niobium and the hydrogen. They were adjusted such as to reproduce the local hydrogen vibrations correctly. In this case the local spectrum can be given analytically:

$$Z_i^H(\omega) = (2/\pi)M^H\omega \, \text{Im}\{M^H\omega^2 - b_i f/[1 + \hat{g}_j(\omega)]\}^{-1}. \quad (10.31)$$

In this equation the $\hat{g}_j(\omega)$ are appropriate combinations of elements of the Greens function and the b_i are geometrical factors. In the region of the band modes $M^H\omega^2$ is small and we get

$$Z_i^H(\omega) = \frac{2}{\pi}M^H\omega \frac{1}{b_i} \text{Im} \, \hat{g}_i(\omega). \quad (10.32)$$

In Eq. (10.32) $Z_i^H(\omega)$ is independent of the Nb–H coupling and depends solely on geometry. Thus the hydrogen mirrors the host density of states, but due to $\hat{g}_i(\omega)$ the weighting may differ for different vibrations. This can be seen from Curve (1) in Fig. 19, which shows a preference of the TA modes as observed experimentally. Model (2) refines model (1) and allows for changes in the force constants between the surrounding niobium atoms as suggested by band structure calculations and in agreement with the model calculations at high hydrogen concentrations. The six parameters were fitted in order to reproduce the local mode frequencies, the elastic double-force tensor, the change of the elastic constants upon alloying with hydrogen and the observed peak in the density of states. The result is represented by line (2) in Fig. 19 and leads to a better agreement with the experiments than model (1). We emphasize that both models yield a strong peak near the van Hove singularity of the TA modes, but they predict only low intensity in the region of the LA cutoff.

The resonant-like enhancement of the hydrogen amplitudes in the region of the TA zone boundary modes results mainly from the strong longitudinal Nb–H(D) force constant that the hydrogen avoids to strain. It is enhanced by local changes of the host force constants. Thus, the low lying flat modes at high hydrogen concentrations and the peak in the band modes have the same origin.

The large hydrogen amplitude also means that for certain TA modes the lattice exerts a large pull on the hydrogen atom and may constitute the first experimental evidence for lattice activated jump processes as proposed for bcc metals by Emin et al.[78] (see also Section 10.2.1).

10.3.2. Localized Hydrogen Modes and the Hydrogen Potential

In the dilute α phases, and even in bcc-metal hydrides, collective hydrogen vibrations are of no importance and the hydrogen isotopes can be regarded as independent three-dimensional Einstein oscillators with frequencies determined by an effective single-particle potential. In this case, details on the shape of this potential can be unraveled from the isotopic shifts of the

10. HYDROGEN IN METALS

local mode frequencies as well as from the higher harmonics of the oscillator transitions. Precise data on the hydrogen potential help to facilitate a quantitative understanding of the metal–hydrogen interaction. For the bcc refractory metals, such knowledge is of particular importance in view of the inferred quantum-mechanical origin of the fast hydrogen-diffusion process. During the last five years, several groups reported new results on this subject that were partly possible only due to the availability of high-energy neutrons from hot and spallation sources. The data have been interpreted in terms of anharmonicity parameters. Besides semi-empirical approaches like that of Sugimoto and Fukai,[79–81] no first-principles calculations are available so far. In this section we discuss new results on bcc metals, and in Section 10.3.6 we shall come to results on palladium, titanium, and zirconium.

Hydrogen atoms dissolved in tantalum, niobium, or vanadium are independent oscillators with frequencies ω_1, ω_2, and ω_3. In pure niobium and tantalum, hydrogen occupies tetrahedral sites with tetragonal symmetry, and ω_2 and ω_3 are degenerate, forming a doublet. The double-differential neutron cross section $d^2\sigma/d\Omega\,d\omega$ for a harmonic three-dimensional oscillator can be given analytically and has the form[75]

$$\frac{d^2\sigma}{d\Omega\,d\omega} = \frac{\sigma_s}{4\pi}\frac{k_1}{k_0}\exp[-2W(Q)]\exp(\tfrac{1}{2}\beta\hbar\omega)$$

$$\times \sum_{n,m,l=-\infty}^{\infty} I_n(y_1)I_m(y_2)I_l(y_3)\delta(\hbar\omega - n\hbar\omega_1 - m\hbar\omega_2 - l\hbar\omega_3),$$

(10.33)

with

$$y_i = \frac{\hbar Q^2}{6M\omega_i}\operatorname{csch}(\tfrac{1}{2}\hbar\omega_i\beta);$$

σ_s is the total scattering cross section; M is the hydrogen (deuterium) mass; $\beta = 1/k_B T$; and $I_n(y)$ is the modified Bessel function of the first kind.

Now we consider small anharmonic disturbances of the hydrogen potential. The possible cubic and quartic anharmonic terms for a hydrogen on a site with tetragonal symmetry (point group D_d^2; $x = y \neq z$) can be found by group theory[82] and have the form

$$V_1 = ez(x^2 - y^2) + c_{4z}z^4 + c_{4x}(x^4 + y^4) + fx^2y^2 + gz^2(x^2 + y^2).$$

(10.34)

The vibrational corrections due to the third- and fourth-order terms are calculated in second- and first-order perturbation theory, respectively, and yield excitation energies $\hbar\omega_{lmn} \equiv e_{lmn} = E_{lmn} - E_{000}$, where E_{lmn} is the perturbed energy of the vibrational level corresponding to the l, m, n

harmonic state. The detailed results are lengthy expressions given by Eckert et al.[82] Here, we remark on the essential features.

(1) The anharmonicity of the z vibration c_{4z} and the corresponding harmonic frequency $\hbar\omega_z$ can be extracted directly from the data

$$\beta = \frac{3c_{4z}\hbar^2}{M^2\omega_z^2} = e_{200} - 2e_{100}, \tag{10.35}$$

$$\hbar\omega_z = 4e_{100} + e_{010} - e_{200} - e_{110}. \tag{10.36}$$

(2) The strength of the anharmonic coupling between z and x, y vibrations is measured by

$$\xi = e_{200} - \beta - \hbar\omega_z. \tag{10.37}$$

(3) Since the harmonic frequencies scale with the reciprocal square root of the isotope mass while the corrections are scaling with the reciprocal mass itself, the anharmonicity parameters also cause deviations from the harmonic isotope effect. For the z vibration, it is described by the expression

$$\frac{e_{100}^H}{e_{100}^{D(T)}} = \sqrt{2(3)} \left[\frac{1 + (1/\hbar\omega_z)(\xi^H + \beta^H)}{1 + (1/\sqrt{2(3)}\,\hbar\omega_z)(\xi^H + \beta^H)} \right]. \tag{10.38}$$

(4) Deviations from cubic symmetry due to hydride formation have to be considered carefully. In the case of the orthorhombic β phase, Eqs. (10.35) and (10.36) are still exact if e_{010} and e_{110} are replaced by average values of the then split x, y frequencies, while Eq. (10.37) is valid only approximately.

The observation of a well-defined second harmonic to the lower deuterium fundamental in niobium stimulated new interest in the investigation of local mode frequencies in bcc metals.[83] Most of the new results that are discussed here have been obtained with the berillium filter method, where neutrons with varying incident energy create phonons in the sample and are downscattered. Only neutrons with energies below 5.2 meV can pass the filter, which is placed in front of the detector, and are counted. The filter cutoff is well defined and the filter transmission below 5.2 meV is high. Therefore, at energies above 100 meV, the beryllium filter technique surpasses other neutron methods in intensity and resolution as long as the momentum transfer is not of importance. Apparently, the contributions of combined optic and acoustic multiphonon processes do not strongly contribute in these spectra.

Figure 20 presents the spectrum obtained from $NbD_{0.85}$, which for the first time showed the existence of a higher harmonic in a bcc hydride.[83] In addition to the two peaks corresponding to the two fundamentals under tetragonal symmetry, a third peak evolves at 170 meV, 7 meV below twice

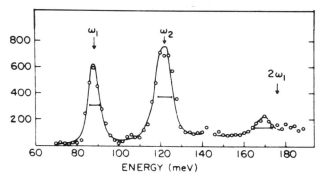

FIG. 20. Inelastic spectrum obtained on NbD$_{0.85}$ at 10 K. The bars give the resolution width for each peak, the arrow $2\omega_1$ shows the position of the second harmonic for an harmonic potential. (From Richter and Shapiro.[83])

the frequency of the lower fundamental $\hbar\omega_1$. Rush et al.[84] investigated the isotope shifts of the local hydrogen vibrations in niobium. Their results for hydrogen deuterium, and tritium in niobium (β phase) are shown in Fig. 21. Using neutrons from spallation sources allowed an observation of hydrogen excitations at much higher energies.[82, 85] Recent results on NbH$_{0.95}$ in the region above 200 meV are shown in Fig. 22. The peak at 231 (2) meV results from the 0–2 transition of the singlet, while the feature at 280 (4) meV stems from a simultaneous excitation of the singlet and the doublet. Its shift from the harmonic position allows the determination of the anharmonic coupling between these two modes. Above 300 meV, further structure appears that results from both the 0–3 singlet transition and the 0–2 excitation of the doublet. Its statistics, however, are not sufficient to extract quantitative information.

Table IV summarizes the accumulated data for niobium and tantalum. In addition the anharmonicity parameters β and ξ and the harmonic values for the singlet $\hbar\omega_z$ are given according to Eqs. (10.35)–(10.37). We note that the data from different groups agree very well, and that in all cases β and ξ are negative. Thus, the potential is widened trumpet-like compared to a parabolic shape.

With the help of the anharmonicity parameters obtained from β-NbH$_{0.92}$ at room temperature, Eq. (10.38) allows a prediction of the isotope effect on the frequencies. We arrive at values of 87 meV for β-NbD$_x$ and 72 meV for β-NbT$_x$, in excellent agreement with the observed frequencies. We note that a measurement of e_{100} and e_{200} alone is not sufficient to predict the isotope effect, if anharmonic coupling between the singlet and the doublet is important. The consistency of the anharmonicity parameters derived from hydrides and deuterides as well as the precise prediction of the isotope effects

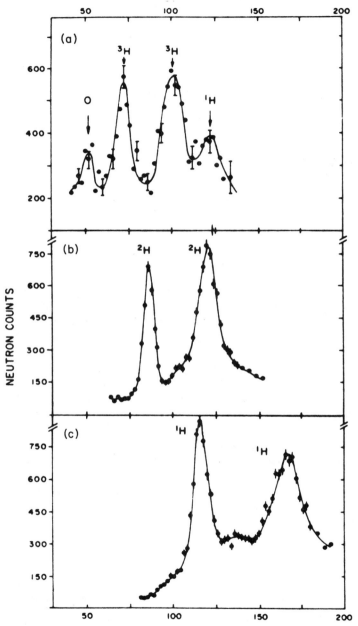

FIG. 21. Inelastic spectra for (a) $NbT_{0.2}$, (b) $NbD_{0.72}$, and (c) $NbH_{0.32}$ at 295 K. (From Rush et al.[84])

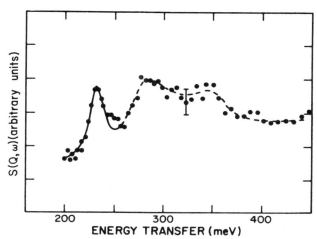

FIG. 22. Higher harmonics for NbH$_{0.95}$ at 15 K. Solid line represents a fit through e_{200}. Dashed line is a guide to the eye for the broad band consisting of e_{111}, e_{020}, and e_{300}. (From Eckert et al.[82])

demonstrate the success of the anharmonic oscillator model. They also show that within the accuracy of present experiments no isotope effect modifying the hydrogen potential itself is observed. In order to also allow an evaluation of the anharmonicity of the doublet as well as increase the accuracy, measurements at even higher energy transfers with better statistics are desirable. They can be expected when more powerful spallation sources come into operation.

The local vibrations of the hydrogen isotopes in vanadium are less well investigated. Two measurements on VH$_{0.5}$ have been reported,[86, 87] where below 172 C hydrogen precipitates into the β phase, occupying octahedral sites. The observed peak positions are displayed in Table V. As a consequence of the O site in bct β-VH$_{0.5}$, the local vibrations are split very widely and the anharmonicity of the lower mode is very strong. In contrast to niobium and tantalum, the parameter β is positive (close to square well potential) and amounts to about 30% of $\hbar\omega_1$. Therefore, we cannot expect the perturbation theory approach of Eqs. (10.34)–(10.38) to be still valid. For β-VD$_{0.5}$ $\hbar\omega_1$ was observed[87] at 47 meV, yielding an isotope shift of 1.19, while from the large positive β a shift toward values well above $\sqrt{2}$ is anticipated. This strong discrepancy may be either due to a potential with a shape completely different from a single well or to the closeness of the acoustic host vibrations, which may cause hybridization effects.

The observations of three distinctly different fundamentals in the ε phase excludes both O- or T-site occupation. Klauder et al.[86] propose an off-center

TABLE IV. Vibrational Excitations of Hydrogen Isotopes in Niobium and Tantalum (meV)

Substance and isotopes			e_{100}	e_{010}	e_{200}	e_{110}	β^a	ξ^a	$\hbar\omega_z^a$	Reference[b]
NbH$_{0.85}$	300 K	β	119.0 ± 0.2	116.8 ± 0.5						2
NbH$_{0.92}$	300 K	β	118.0 ± 1.0	162.0 ± 2.0	227.0 ± 3.0	273.0 ± 6	−9.0	−7.0	134.0	3
NbH$_{0.92}$	78 K	$\beta + \gamma$	120.0 ± 1.0	163.0 ± 2.0	227.0 ± 2.0	277.0 ± 5	−13.0	−6.0	139.0	3
NbH$_{0.95}$	15 K	$\beta + \gamma$	122.0 ± 1.0	166.0 ± 2.0	231.0 ± 2.0	280.0 ± 4	−13.0	−8.0	143.0	3
NbH$_{0.55}$	150 K	$\beta + \gamma$	121.0 ± 1.0	165.0 ± 2.0	230.0 ± 3.0	278.0 ± 3	−12.0	−8.0	141.0	3
NbH$_{0.32}$	300 K	β	116.0 ± 7.0	167.0 ± 1.5						4
NbD$_{0.85}$	300 K	β	87.2 ± 0.2	118.1 ± 0.3						2
NbD$_{0.85}$	10 K	$\xi + \gamma$	88.4 ± 0.3	121.6 ± 0.4	170.0		−6.8			2
NbD$_{0.8}$	78 K	$i + v$	86.7 ± 0.9	120.7 ± 0.9	166.0 ± 3.0	205.0 ± 5	−7.4	−2.4	96.5	3
NbD$_{0.7}$	300 K	β	86.0 ± 1.0	120.0 ± 1.5						4
NbT$_{0.2}$	300 K	β	72 1.0	101 ± 1.0						4
TaH$_{0.71}$	78 K	ζ	130.0 ± 1.0	170.0 ± 2.0	245.0 ± 3.0		−15.0			3
TaH$_{0.08}$	77 K	β	121.3 ± 0.2	163.4 ± 0.4	227.1 ± 2.0		−15.5			5
TaH$_{0.1}$	30 K	β	121.0	162.0	220.0		−22.0			6
TaD$_{0.14}$	77 K	β	88.4 ± 0.4	118.7 ± 0.6	167.8 ± 1.6		−9.0			5

[a] β and ζ are the anharmonicity parameters according to Eqs. (10.35) and (10.36), $\hbar\omega_z$ is the harmonic frequency in z direction. The hydride phases are marked according to Köbler and Welter.[1]

[b] References
1. U. Köbler and J. M. Welter, *J. Less-Common Met.* **84**, 225 (1982).
2. D. Richter and S. M. Shapiro, *Phys. Rev.* B **22**, 599 (1980).
3. J. Eckert, J. A. Goldstone, D. Tonks, and D. Richter, *Phys. Rev.* B **27**, 1980 (1983).
4. J. J. Rush, A. Magerl, J. M. Rowe, J. M. Harris, and J. L. Provo, *Phys. Rev.* B **24**, 4903 (1981).
5. R. Hempelmann, D. Richter, and A. Kollmar, *Z. Phys.* B **44**, 159 (1981).
6. S. Ikeda and N. Watanabe, *Natl. Lab. High Energy Phys.*, *KENS* **KENS-Rep. IV** (1983).

TABLE V. Vibrational Excitations in Vanadium Hydride (meV)

Substance			$\hbar\omega_1$	$\hbar\omega_2$	$\hbar\omega_3$	$\hbar\omega_{1/2}$	β	Reference
$VH_{0.51}$	295 K	β		56.0	221.0	125.0	13.0	1
$VH_{0.51}$	495 K	ε	52.5	112.5	210.0			1
$VH_{0.51}$	498 K	α		113.0	180.0			1
$VH_{0.32}$	30 K	β	51.0	57.0	220.0	130.0	22.0	2

1. D Klauder, V. Lottner, and H. Scheuer, *Solid State Commun.* **32**, 617 (1979).
2. S. Ikeda and N. Watanabe, *Natl. Lab. High Energy Phys.*, **KENS-Rep. IV** (1983).

position in between O and T sites, which could account for the smaller tetragonality of the ε phase compared to the β phase and correspondingly for the observed local mode frequencies. On the other hand, in the α phase, the observed frequencies are consistent with a T-site occupation.

Finally, we comment on implications of the observed deep potential wells on the fast diffusion mechanism of the hydrogen isotopes in bcc transition metals: the very high self-diffusion coefficients D for hydrogen in those metals are essentially related to a very small activation energy E^* (~ 100 meV and below) as found from the slope of $\ln D$ versus T^{-1}. It is noteworthy that E^* is smaller than the lowest vibrational energies in the localized modes and drastically below the energies of the higher harmonics, which range well above 300 meV. This shows that E^* is not the height of the potential well in which the hydrogen is bound; it is rather related to the activation of certain phonons (see also Section 10.2.2), which induce or enable the diffusive jump from one to another tetrahedral site.[78] Furthermore, at least for low temperatures, the high diffusivity is also caused by tunneling.[78]

Sugimoto and Fukai[79-81] proposed a semi-empirical potential for hydrogen in transition metals that reproduces the experimental observations reasonably well. In order to account for the (experimentally observed) double-force tensor, as well as for the local vibrations in bcc metals, they modeled the metal-hydrogen interaction by double Born-Mayer pair potentials:

$$V(r) = V_1 \exp[-(r - r_t)\alpha_1] + V_2 \exp[-(r - r_t)\alpha_2]. \qquad (10.39)$$

In this equation r_t is the nearest neighbor hydrogen-metal distance ($r_t = \sqrt{5}/4a$ for T sites in bcc metals). For niobium, tantalum, and vanadium, the respective parameters are $V_1 = 0.1\ (0.09)$ eV, $V_2 = 0.6\ (0.4)$ eV, $\alpha_1 = 40/a$, and $\alpha_2 = 3/a$. Very recently Akai *et al.*[88] gave a microscopic justification for such a potential. According to them the short-range part results from a bonding state between the 1s state of the hydrogen and the *d* states of the metal. The long-range part originates mainly from screening charges.

In order to find the configuration of minimal energy for hydrogen and host atoms, Sugimoto and Fukai[79] solved the Schrödinger equation for the hydrogen atom in the force field of the surrounding metal atoms. Utilizing the phonon Greens function, self-consistency with respect to the metal displacements was achieved. These results were obtained: first, for hydrogen and its heavier isotopes in niobium and tantalum the T site is the most stable configuration. The self-trapping energy for hydrogen in niobium of 476 meV is more than twice the classical continuum theory value of 200 meV.[89] This is to be understood as a consequence of the rather large breadth of the hydrogen wave function together with the very short range of the potential. Second, contour maps of excited-state wave functions were calculated and are shown in Fig. 23. We note that one of the two excitations of the doublet has a large amplitude near the neighboring pair of T sites. It has an excitation energy of 148 meV and a low-excitation probability for neutrons because of the small overlap of the wave functions. The proof of its existence would be an important test for the proposed potential. Third, a trumpet-like shape of the potential was calculated for T sites, while for O sites the potential is predicted to be deformed in the direction of a square well (niobium: $e^D_{200}/e^D_{100} = 1.9$; palladium: $e^H_{100}/e^D_{100} = 1.54$). In bcc niobium a second harmonic was found only for the deuterium singlet, while for hydrogen no such excitation should exist. This is in obvious disagreement with experiments on hydrides and a weak point of the theory. Fourth, as a consequence of the short-range potential, the local vibrations are very sensitive to displacements of the surrounding metal atoms. Qualitatively this explains the large broadening of the local modes in the α' phases and the strong narrowing at the α'-β ordering transition observed by Richter and Shapiro.[83] Fifth, using the potential parameters determined from the T-site occupation, calculations for hydrogen and deuterium on O sites in bct vanadium were carried out.[81]

The excitation energies of 50 meV for the doublet and 262 meV for the singlet are in fair agreement with experiment (cf. Table V). The calculations predict a potential deformed in the direction of a square well ($e^H_{100}/e^D_{100} = 1.8$), in qualitative agreement with what would be expected from the positive shift of the higher harmonic in $VH_{0.5}$, but in distinct disagreement with the actually observed isotope shift. Since both experiment and theory lead to apparently inconsistent results, new efforts are necessary to solve the problem. Sixth, Ross *et al.*[90] demonstrated the systematic variation of the hydrogen excitation energies with the metal-hydrogen distance in fluorite structures and proposed a $1/R^{3/2}$ dependence. Fukai and Sugimoto[91] reconsidered the problem and showed that within their potential a $1/R$

10. HYDROGEN IN METALS

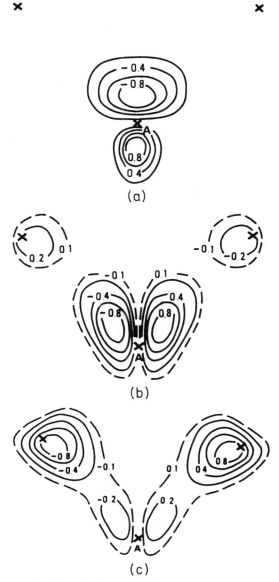

FIG. 23. Contour maps of the excited wave functions of a hydrogen atom in tantalum, which, in its ground state, occupies the T site labeled A (X's are neighboring T sites). (a) The excited-state wave function of the singlet; (b) and (c) excited states of the doublet. (From Sugimoto and Fukai.[80])

dependence follows naturally (it describes the experiments equally well), if the range parameter α varies in proportion to the metal–hydrogen distance, or if αr_t is constant.

10.3.3. Hydrogen Vibrations as a Local Probe

The sensitivity of the hydrogen vibrations to their local environment constitutes a powerful spectroscopic means of assigning hydrogen interstitial positions. The short-range character of the metal–hydrogen potentials makes them susceptible to distortions in the neighborhood. Important information on the local environment can be extracted from a measurement of hydrogen vibrations in two areas: (1) for impurity trapped hydrogen, the interstitial site and the binding energy are determined; (2) in amorphous metals, the use of hydrogen as a probe leads to information on the local geometry in the amorphous structure.

10.3.3.1. Local Vibrations in Hydrogen Trapping. As discussed in Section 10.2.3, interstitial impurities such as oxygen or nitrogen in niobium act as trapping centers for dissolved hydrogen. A large amount of experimental work has been performed in order to obtain microscopic information on the trapping process and the local dynamics of hydrogen in the trapped state. Despite these efforts, the information on the local site symmetry of a trapped proton has remained either contradictory or nonexistent. The effects of substitutional impurities on hydrogen in niobium have been investigated to a lesser extent and generally were found to be much weaker than those of oxygen or nitrogen.[92] Recent internal friction measurements, however, have indicated that the Ti–H interaction in niobium may be strong enough to prevent hydrogen from precipitating at low temperatures.

A microscopic interpretation of the trapping phenomena and the local dynamics at the trap requires knowledge of hydrogen site and hydrogen potential, both of which can be addressed by inelastic neutron scattering.

10.3.3.1.1. INTERSTITIAL IMPURITIES. Figure 24 presents spectra from hydrogen in $NbN_{0.004}H_{0.003}$ at various temperatures.[63] Though the fraction of trapped protons changes drastically between 295 and 10 K,[26] the observed peak positions remain practically unchanged and at low temperatures are distinctly different from what is expected from precipitated hydrogen (dashed line). Similar results have also been obtained on $NbO_{0.01}H_{0.01}$ and demonstrate that nitrogen and oxygen in niobium are efficient traps that suppress hydride phase separation.

Furthermore, the local vibration frequencies of hydrogen trapped at oxygen or nitrogen impurities in niobium are close to the frequencies of hydrogen in the dilute α phase, and the doublet degeneracy is not lifted. This proves that the protons occupy T sites in the trapped state as well. In view

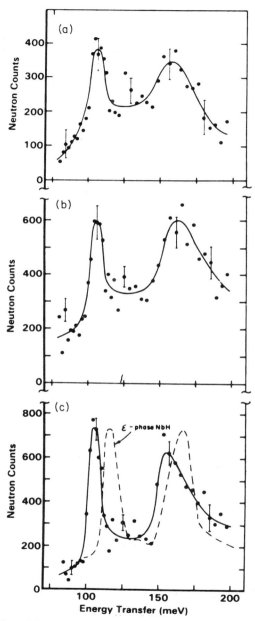

FIG. 24. Spectra from $NbN_{0.004}H_{0.003}$ at (a) 295 K, (b) 150 K, and (c) 10 K. Dashed curve represents ε-phase results. (From Magerl et al.[63])

of the short-range metal–hydrogen potential, we must assume that the nearest-neighbor niobium atoms can only be weakly displaced from their regular positions. Otherwise the frequencies would be strongly changed and the upper mode would split. In particular, there is no evidence for a direct (electronic) force between an impurity and hydrogen that would affect the tetragonal symmetry of the force field seen by the hydrogen. The observed T-site occupation disagrees with the suggestion of tunneling states involving tetrahedral and triangular sites proposed on the basis of internal friction measurements.[93] It also disproves the idea of a so-called 4T configuration suggested by Fukai and Sugimoto[94] on the basis of calculations. In such a 4T state the proton is spread over the tetrahedral sites on the face of the cube forming a common self-trapping distortion with the symmetry of an O site. Finally, channeling experiments that resulted in a deuterium position between O and T sites in TaN_xD_y,[95] are not valid for the system NbN_xH_y and thus may be questionable in general.

Magerl et al.[63] tried to assign the pair configuration taking into account (1) the very short-range Nb–H interaction potential and (2) the requirement of at least two crystallographically equivalent nearest-neighbor hydrogen sites in order to allow for a tunnel-split ground state.[70] Condition (1) excludes all hydrogen sites surrounded by niobium atoms that are nearest neighbors to the nitrogen atom; condition (2) suggests the fifth-nearest-neighbor dumbell configuration in the (111) direction (see Fig. 7, sites 6' and 5'). The resulting H–N distance of about one lattice spacing agrees with recent diffuse x-ray scattering results.[96]

At such large distances, it is reasonable that direct electronic forces are of no importance and trapping is governed by elastic interactions. Recently, on the basis of lattice elasticity theory. Shirley et al.[97] calculated binding energies for various O(N)–H pair configurations in niobium, tantalum, and vanadium. Assuming electronic repulsions, the most probable pair is a third-nearest-neighbor configuration in the (100) direction. The calculated binding energies of 120 meV (OH) and 100 meV (NH) are in good agreement with experimental values, and the position is compatible with the local-mode experiments. However, it does not allow for a tunnel-split ground state. On the other hand, the fifth-neighbor position leads to a binding energy of only 25 meV, which is much too small to account for the experiments.

10.3.3.1.2. SUBSTITUTIONAL IMPURITIES. Figure 25 presents first spectroscopic results[64] on hydrogen trapping at a substitutional titanium impurity in niobium. The peak positions and resolution-corrected widths are displayed in Table VI. The lower fundamental stiffens slightly with decreasing temperature. The higher fundamental, which is only vaguely defined at room temperature, sharpens considerably with decreasing temperature and is shifted by 10 meV below the corresponding α-phase

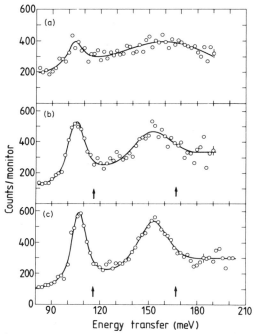

FIG. 25. Inelastic spectra obtained from NbTi$_{0.01}$H$_{0.009}$ at (a) 295 K, (b) 77 K, and (c) 10 K. The arrows indicate the peak positions in the hydride phase. (From Richter et al.[64])

TABLE VI. Vibrational Frequencies of Hydrogen in NbTi$_{0.01}$H$_{0.009}$[a]

Temperature (K)	e_{100} (meV)	FWHM (meV)	e_{010} (meV)	FWHM (meV)
297	104.0 ± 1.0	8 ± 3.0		
77	105.0 ± 0.5	9 ± 1.0	155.0 ± 2.0	29.0 ± 6.0
10	107.0 ± 0.5	7 ± 1.0	153.0 ± 1.0	20.0 ± 2.0

[a] D. Richter, J. J. Rush, and J. M. Rowe, *Phys. Rev. B* **27**, 6227 (1983).

value. The complete absence of any peaks at the positions of the niobium-hydride phase shows that titanium is a strong trap with a binding energy larger than the enthalpy of formation of the hydride phase (120 meV). Again, the vibrational frequencies immediately show that the protons trapped at titanium impurities occupy sites with an atomic arrangement very close to that of a T site in niobium. The large linewidth of the upper fundamental even at 10 K suggests an unresolved splitting of the doublet. This may indicate deviations from tetragonality as expected for a titanium nearest neighbor, which also could be the origin of the large frequency shift of the upper mode. A nearest-neighbor hydrogen position should be contrasted to the large hydrogen impurity distance inferred for interstitial impurities. While trapping at interstitial impurities is almost certainly strain related, for titanium the hydrogen association has presumably more the character of a chemical bond or is a consequence of electronic forces. In particular this follows from the observation that the elastic deformations induced by titanium are small ($\Delta a/\Delta c = -0.32 \times 10^{-3}$ Å/at. %) compared to those caused by oxygen or nitrogen ($\Delta a/\Delta c = 6.0 \times 10^{-3}$ Å/at. % for nitrogen).

If the vibrational frequencies in the α phase and in the trapped state are different, the transfer of protons from the free to the bound state can be observed directly by inelastic neutron scattering. Since the peak intensities are directly proportional to the fraction of protons in the respective states, a measurement of its temperature dependence reveals the binding energy at the trap. Figure 26 presents the outcome of such an experiment on NbCr$_{0.01}$H$_{0.009}$ and shows the gradual intensity transfer from a high-temperature peak around 106 meV to a higher energy peak at lower temperatures. A second peak develops near 123 meV and grows with decreasing temperature. At 210 K separate peaks are no longer distinguishable. Below 210 K a new center of vibrational intensity develops around 117 meV, which takes up all intensity at low temperature.

A careful analysis of the data showed that in NbCrH both trapping and precipitation of the hydride phase takes place, that is, the binding energy ΔE at the trap is smaller than the enthalpy of precipitation H. A simple

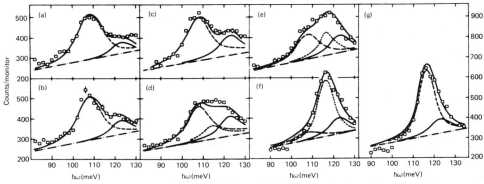

FIG. 26. Temperature dependence of the intensity distribution in the region of the lower fundamental vibration in NbCr$_{0.009}$H$_{0.009}$ at (a) 250 K, (b) 220 K, (c) 200 K, (d) 150 K, (e) 235 K, (f) 210 K, and (g) 170 K. The upper solid line represents the result of a fit with the combined trapping and precipitation model, the broken line represents the α-phase contribution, the dashed line the intensity from protons in the hydride phase, and the lower solid line displays the intensity originating from protons in the trap. Finally, the dot–dashed line shows the background level. (From Richter et al.[64])

thermodynamical approach was used to model this situation: each impurity was associated with n trapping sites, the potential energy of which is lowered by ΔE with respect to the undisturbed lattice. In thermal equilibrium the chemical potentials of trapped and dilute protons must be equal:

$$kT \ln\left(\frac{c_d}{6}\right) = -\Delta E + kT \ln\left[\frac{c_t}{n(c_i - c_t)}\right]. \qquad (10.40)$$

Here, c_d and c_t are the hydrogen concentrations in the dilute phase and in the trapped state, respectively; c_i is the impurity concentration. The factor $\frac{1}{6}$ enters because the number of T sites in niobium is six times higher than the number of niobium atoms. In addition it was assumed that one impurity atom is saturated already by one proton.[26] For the concentration of trapped protons, it follows that

$$c_t = \frac{1}{2}\left(c_H + c_i + \frac{6}{n}e^{-\Delta E/kT}\right) - \frac{1}{2}\left[\left(c_H + c_i + \frac{6}{n}e^{-\Delta E/kT}\right)^2 - 4c_H c_i\right]^{1/2}. \qquad (10.41)$$

If the solubility in the α phase, c_α, becomes smaller than c_d, then in Eq. (10.40) c_d has to be replaced by c_α, and we get

$$c_t = \frac{c_\alpha c_i}{c_\alpha + 6/n \exp(-\Delta E/kT)}. \qquad (10.42)$$

The result of a fit with this model is indicated by the various lines in Fig. 26. It is evident that the combined trapping and precipitation model allows a good description of the experimental data. In particular it accounts for the gradual growth of the trapping peak above 210 K as well as for the sudden intensity increase of the central β-phase peak below 210 K. For the binding energy, a value of $\Delta E = 105 \pm 10$ meV smaller than the enthalpy of the hydride phase $\Delta H = 120$ meV was determined. The vibrational energy of the trapping peak amounts to $\hbar\omega_t = 123$ (1) meV. Its shift from the α-phase value of 14% is the largest observed so far in the context of hydrogen trapping in niobium and indicates a severely stiffened potential. Though a shallow trap like chromium does not prevent precipitation at low temperatures, it shifts the phase boundary toward lower values. Such a shift of the phase boundary has also been observed[63] for $NbV_{0.008}H_{0.005}$ and most probably can be explained in the same way.

10.3.3.2. Hydrogen Vibrations in Amorphous Metals. The investigation of the structure and interatomic bonding of metallic glasses has attracted increasing scientific interest. Structural investigations are mostly performed with x rays and also by neutron scattering[98] and can be interpreted only in terms of pair correlation functions. However, many questions, in particular

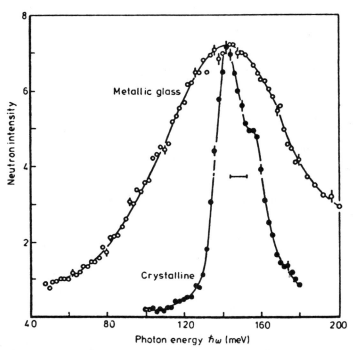

FIG. 27. Neutron spectra measured at 78 K for crystalline TiCuH$_{0.93}$ and amorphous TiCu$_{1.3}$H$_x$. The energy resolution (FWHM) near the peak is indicated by the horizontal bar. (From Rush et al.[65] Copyright of The Institute of Physics.)

about the local topology in the glassy state, are still open. Since the hydrogen vibrations are sensitive essentially to the immediate surrounding of the proton, they are well suited as local probes to address this problem.

First measurements were reported by Rush et al.[65] who compared the vibrational spectra of glassy and crystalline CuTiH. Their results are shown in Fig. 27. Large differences in the density-of-states distribution between both samples are evident. The crystalline sample shows a narrow frequency distribution with features at 142 and 157 meV and is very similar to that of γ-TiH$_2$. Since in both structures hydrogen occupies regular T sites with 4 Ti next-nearest neighbors, this similarity is not surprising. In the amorphous metal the peak occurs at the same frequency but with strongly increased width (75 meV). Thus, on the average, the hydrogen sites occupied in amorphous and crystalline structure are very similar. The large width indicates that the tetrahedron may be heavily distorted and that it fluctuates in its chemical composition. The wing toward frequencies below 100 meV may also show that O-site occupation occurs as well. The observation of a broad density of states disagrees with the microcrystalline or microcluster model of metglass, where the local environment would not differ from that in the crystal.

Later experiments on other glasses such as ZrNiH,[99-102] Ti$_2$NiH,[100] and Zr$_2$Pd[103] qualitatively all revealed the same result: the local hydrogen spectra in the amorphous substance center around similar frequencies to those in the crystalline state and the frequency distributions are generally much broader and washed out. Thus, on the basis of these measurements, we conclude that in amorphous metals the hydrogen atoms remain essentially on the same polyhedral sites as in the crystalline reference substance; changes in topology are rather restricted.

10.3.4. Hydrogen Vibrations and Superconductivity

One of the most exciting results in the field of hydrogen in metals was the observation in palladium of an increased superconducting transition temperature T_c upon hydrogenation[66] and an associated inverse isotope effect[67] on T_c. As an anomalous feature, superconductivity in PdH$_x$(D) is supposed to arise from strong electron–phonon coupling to the optic hydrogen vibrations, whereas normally it originates from coupling to acoustic phonons.

For a superconductor T_c is given by the McMillan formula, which, in a simplified form, can be written as[104]

$$T_c = \langle\omega\rangle \exp(-1/\lambda), \qquad (10.43)$$

where $\langle\omega\rangle$ is an average phonon frequency. The electron–phonon coupling parameter $\lambda = \eta/M\langle\omega^2\rangle$ is mainly influenced by an electronic contribution hidden in η and by the average force constant $M\langle\omega^2\rangle$. Since, in general, λ should be isotope independent, the normal isotope effect is that $T_c \sim \langle\omega\rangle \sim 1/\sqrt{M}$. The explanation of the anomalous isotope effect observed in palladium ($T_c^H = 9$ K, $T_c^D = 11$ K) remains controversial.

Essentially two different mechanisms have been suggested: (1) anharmonicity in the optic phonons was proposed as a cause of the anomalous isotope effect;[105,106] and (2) the existence of different electronic structures around the two hydrogen isotopes was postulated, which would result in isotope-dependent effective hydrogen potentials.[107] This could be understood by the fact that the effective potential at the hydrogen site may depend on the amplitude of the hydrogen isotope under consideration.

Phonon spectra from superconducting tunneling[108,109] and neutron scattering[110-112] revealed deviations from a harmonic isotope shift. Measurements of the optic-mode Grüneisen constant[113,114] revealed an anharmonic hydrogen potential, while NMR measurements[115] were interpreted in terms of different electronic structures for hydrogen and deuterium in palladium.

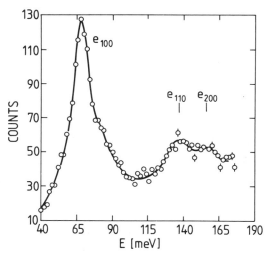

FIG. 28. Inelastic spectrum from $PdH_{0.014}$; e_{100}, e_{200}, and e_{110} assign the fundamental and the first two higher harmonics. (From Rush et al.[68])

In order to clarify the situation, Rush et al.[68] performed a measurement of the local hydrogen vibrations in α-PdH_x, studying the isotope shifts and higher harmonics. As discussed in Section 10.3.2, such a measurement directly yields the hydrogen potential and makes it possible to decide between purely anharmonic or isotope-dependent potentials.

For hydrogen on an O site in palladium, the anharmonic potential of Eq. (10.34) has only two independent anharmonicity parameters: $H_1 = c_4(x^4 + y^4 + z^4) + f(x^2y^2 + x^2z^2 + y^2z^2)$. There are two second-harmonic excitations and one first excited state: $e_{100} = \hbar\omega + \beta + \gamma$; $e_{110} = 2e_{100} + \gamma$; $e_{200} = 2e_{100} + \beta$ with β from Eq. (10.35), and $\gamma = f\hbar^2/M^2\omega^2$. According to Eq. (10.33), the intensities of e_{110} and e_{200} should scale as 2 : 1 in a beryllium filter experiment.

Figure 28 presents the spectrum observed on $PdH_{0.014}$. Besides the fundamental at 69 ± 0.5 meV, there is a double peak structure at 137 ± 2 meV and 156 ± 3 meV. The intensity of the lower peak is roughly twice that of the second peak. The anharmonicity parameters are obtained as $\gamma = 0 \pm 2$ meV and $\beta = 18 \pm 4$ meV. From these values an isotope ratio of the frequencies of 1.53 ± 0.04 is predicted. This is in good agreement with the value of 1.49 ± 0.02 obtained from a measurement of the deuterium excitation. Thus, the observed isotope effect for α-PdH_x can be explained satisfactorily by the anharmonicity of the hydrogen potential. Comparing the α-phase measurement with data obtained from the β phase, the authors prove that in both phases the same isotope ratio is observed. Furthermore, estimates on

TABLE VII. Local Vibrations in Zirconium and Titanium

Substance, Phase	e_{100}	e_{200}	$e_{200} - e_{100}$	ω^H/ω^D	Reference
α-Zr (492°C)	143.1 ± 6.0	277.7 ± 3.4	−8.5	1.36 ± 0.02	1
(600°C)	144.0 ± 1.0				2
δ-Zr (216°C)	141.4 ± 0.4	279.4 ± 1.2	−3.4	1.34 ± 0.01	1
δ-Zr (27°C)	140.3	273.6	−7.0		3
α-Ti (326°C)	105.5 ± 2.0[a]				1
	108.5 ± 2.0				
α-Ti (315°C)	141.0 ± 1.0[a]				2, 3
δ-Ti (25°C)	150.5 ± 1.0			1.39 ± 0.01	1
TiH$_2$ (27°C)	142.8	280.8	−4.8		3

[a] The reasons for the discrepancy between the different authors are not clear and a new experiment is advisable.
1. R. Hempelmann, D. Richter, and B. Stritzker, *J. Phys. F* **12**, 79 (1982).
2. R. Khoda-Bakhsh and D. K. Ross, *J. Phys. F* **12**, 15 (1982).
3. S. Ikeda and N. Watanabe, *Natl. Lab. High Energy Phys.*, KENS-Rep. IV (1983).

the basis of a Born–Mayer potential show that the metal–hydrogen potentials in the α and β phases are identical.

Thus a detailed study of the local vibrations in palladium has made a significant contribution to settling the old controversy about the origin of the T_c anomaly in PdH(D). It shows that within experimental accuracy there is no evidence for an isotope-dependent effective potential, or for different electronic structures in the vicinity of hydrogen or deuterium in palladium.

Recent observations of an enhanced T_c in α-TiH$_x$ and α-ZrH$_x$[116] stimulated neutron experiments on the hydrogen potential also in these substances.[69] In contrast to palladium, the local vibrations for titanium and zirconium occur at high frequencies and exhibit only small anharmonicities (see Table VII). There is no systematic relation between the position of the optic mode frequencies, their anharmonicity, and the superconducting behavior. This suggests that, except in PdH(D), the coupling to the acoustic modes dominates the electron–phonon coupling.

10.4. Hydrogen Tunneling

Some years ago, Sellers *et al.*[117] reported the observation of an isotope-dependent specific heat anomaly in NbH(D). Birnbaum and Flynn[118] ascribed this phenomenon to proton tunneling around the four tetrahedral and triangular interstices on the face of the cube. Morkel *et al.*[119] showed that the specific heat anomaly can only be observed if nitrogen or oxygen *and*

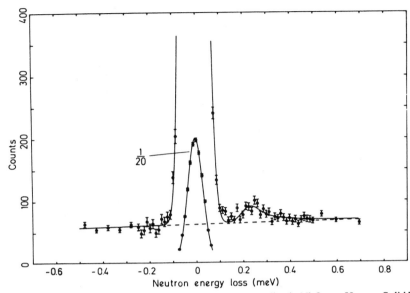

FIG. 29. Inelastic neutron scattering due to hydrogen tunneling in $NbO_{0.0019}H_{0.0018}$. Solid line shows result of a fit with Eq. (10.47) taking into account a Lorentzian distribution function for the disturbances. (From Wipf et al.[120])

hydrogen are present simultaneously in the niobium host and concluded that hydrogen tunneling must be associated with the traps provided by interstitial impurities. Recently, inelastic neutron scattering experiments allowed a first direct spectroscopic observation of hydrogen tunneling in NbO_xH_y.[70, 120]

Figure 29 presents the spectroscopic result obtained from $Nb_{0.002}H_{0.002}$ at 150 mK. Besides a very strong elastic peak, it shows the tunneling transition on the neutron-energy-loss side at about 200 µeV; since nearly all protons are in the ground state, neutron energy gain processes at this temperature are very unlikely.

Quantitatively, the data were interpreted in terms of a two-site tunneling model, where the hydrogen is assumed to tunnel between two neighboring tetrahedral sites whose potential wells are shifted randomly with respect to each other. The energy shifts ε thereby result from strain-induced interaction effects of different O–H pairs. For a random distribution of O–H pairs, ε is distributed according to a Lorentzian[121] $Z(\varepsilon) = (1/\pi)\varepsilon_0/(\varepsilon^2 + \varepsilon_0^2)$.

For a disturbed two-site tunneling system with a tunneling matrix element J, the energy splitting of the ground state is given by[122]

$$\Delta E = \sqrt{J^2 + \varepsilon^2}. \qquad (10.44)$$

The corresponding proton wave functions are

$$|1(2)\rangle = \alpha|e(u)\rangle \underset{(-)}{+} \beta|u(e)\rangle, \tag{10.45}$$

where $|e(u)\rangle$ are oscillator wave functions situated at the lower (e) or upper (u) pocket of the potential. The coefficients α and β are given by

$$\left.\begin{array}{c}\alpha^2\\ \beta^2\end{array}\right\} = \frac{J^2 \pm \varepsilon^2}{2\sqrt{J^2 + \varepsilon^2}}. \tag{10.46}$$

The orientationally averaged scattering function is given by

$$S_i(Q, \omega) = \frac{1}{2}\left\{\frac{J^2 + 2\varepsilon^2}{J^2 + \varepsilon^2} + \frac{J^2}{J^2 + \varepsilon^2}\frac{\sin Ql}{Ql}\right\}\delta(\omega)$$

$$+ \frac{1}{2}\frac{J^2}{J^2 + \varepsilon^2}\left\{1 - \frac{\sin Ql}{Ql}\left[\frac{\delta(\hbar\omega + \Delta E)}{1 + \exp(-\beta\,\Delta E)}\right.\right.$$

$$+ \left.\left.\frac{e^{-\beta\Delta E}}{1 + \exp(-\beta\,\Delta E)}\delta(\hbar\omega - \Delta E)\right]\right\}, \tag{10.47}$$

where l is the distance between the two sites. For a distribution of disturbances $Z(\varepsilon)$, the appropriate average has to be carried out. These features are important: (1) for $\varepsilon = 0$ the elastic part is the orientationally averaged Fourier transform of a dumbell and can be understood as an interference pattern from a proton distributed equally over both potential minima; (2) with increasing ε the elastic part increases since an increasing part of the proton wave function is forced into the deeper well; (3) at $T = 0$ only neutron energy loss processes are possible, with increasing T the energy gain part increases at the *expense* of the energy loss part. (In the case of phonon scattering both would grow [cf. Eq. (10.27)]).

The result of a fit of Eq. (10.47) to the measured data is shown by the solid line in Fig. 29. From the fit a tunneling matrix element $J = 0.21 \pm 0.03$ meV and a width of the strain energy distribution $\varepsilon_0 = 3.6 \pm 1.5$ meV were obtained. The value of J is in good agreement with the specific heat result of $J = 0.19 \pm 0.02$ meV.[119] The large width of $Z(\varepsilon)$ leads to a large majority of tunneling systems with $J \ll \varepsilon$. In this case the protons sit mainly in the deep well and the neutron cross section for tunneling transitions drops to zero. Therefore, the elastic peak is much stronger than the inelastic intensity.

The measurement is consistent with a tunneling distance of about 1 Å. Combining this with other information discussed in Section 10.3.3.1, it is highly probable that tunneling occurs between the sites denoted as 5' and 6' in Fig. 7.

10.5. Concluding Remarks

As regards the investigation of the diffusion process, quasi-elastic scattering has provided a good background for the qualitative understanding of the elementary diffusive step, at least for the simple bcc and fcc hydrides. On the other hand, there are no theoretical concepts for calculating, for instance, the characteristic residence times, or the host-lattice relaxation times connected with the diffusional jump on a first-principle basis, since the corresponding many-particle problem of the lattice is very difficult to solve.

Diffusion in metals with dilute impurities is approximately understood on a semi-empirical basis, and also a certain understanding exists of interesting phenomena, such as trapping modes and tunneling states of hydrogen around traps. Some theoretical and recent experimental work has been started to study disordered, glass-like, and recently nondilute hydrides. Coherent quasi-elastic scattering has not been studied so far. As a by-product of the research on diffusion, it has turned out that quasi-elastic scattering is a powerful tool for determining self-diffusion constants, in particular for metal alloy hydrides and storage materials, where other methods either are difficult to apply or fail.

A great wealth of information exists on phonon dispersion and localized modes in metal hydrides and, more recently, on the higher harmonics of these modes. This information was used to determine local force constants and potentials, including their anharmonicity. In particular they lead to a better understanding of the influence of dissolved hydrogen or deuterium on the superconductivity in palladium. Hydrogen vibrations were also used as a local probe in order to gain information on the hydrogen site symmetry in hydrogen impurity clusters and in amorphous metal hydrides. However, there exist no *ab initio* interpretations of these potentials in terms of the electronic properties of the hydrides. In general, there is a great variety of experimental data from neutron scattering on hydrides, but deeper theoretical interpretation is still required.

Acknowledgments

We thank Dr. R. Hempelmann and Dr. K. Kehr for valuable discussions.

References

1. K. Sköld, *in* "Hydrogen in Metals, I" (B. Alefeld and J. Völkl, eds), Topics in Applied Physics, Vol. 28, p. 267. Springer-Verlag, Berlin and New York, 1978; T. Springer, *in* "Hydrogen in Metals, I," p. 76.
2. T. Springer, *in* "Dynamics of Solids and Liquids" (S. W. Lovesey and T. Springer, eds.), Topics in Current Physics, Vol. 3, p. 255. Springer-Verlag, Berlin and New York, 1977.

3. D. Richter, in "Springer Tracts in Modern Physics" (C. Höhler, ed.), Vol. 101, p. 85. Springer-Verlag, Berlin and New York, 1983.
4. A. F. Anderson et al., in "Metal Hydrides 1982," Proceedings of the International Symposium on the Properties and Applications of Hydrides (W. E. Wallace, T. Schober, and S. Suda, eds.), Vol. 1. Elsevier Sequoia, Lausanne, 1983; V. A. Somenkov and S. S. Shil'stein, Prog. Mat. Sci. 24, 267 (1979).
5. V. Lottner, A. Heim, and T. Springer, Z. Phys. B 32, 157 (1979).
6. C. T. Chudley and R. J. Elliott, Proc. Phys. Soc., London 77, 353 (1961).
7. G. Blaesser and J. Peretti, Vacancies Interstitials Met., Proc. Int. Conf., Rep. Jul-Conf.-2 2, 886 (1968); J. M. Rowe, K. Sköld, H. E. Flotow, and J. J. Rush, J. Phys. Chem. Solids 32, 41 (1971).
8. W. Gissler and H. Rother, Physica (Amsterdam) 50, 380 (1970).
9. H. C. Torrey, Phys. Rev. 92, 262 (1953).
10. T. Springer, in "Quasielastic Neutron Scattering for the Investigation of Diffusive Motions" (G. Höhler, ed.), Springer Tracts in Modern Physics, Vol. 64, p. 54. Berlin and New York, 1972.
11. V. Lottner, U. Buchenau, and W. J. Fitzgerald, Z. Phys. B 35, 35 (1979).
12. V. Lottner, J. W. Haus, A. Heim, and K. W. Kehr, J. Phys. Chem. Solids 40, 557 (1979).
13. K. Sköld and G. Nelin, J. Phys. Chem. Solids 28, 2369 (1967).
14. J. M. Rowe, J. J. Rush, L. A. de Graaf, and C. A. Ferguson, Phys. Rev. Lett. 29, 1250 (1972).
15. I. S. Anderson, D. K. Ross, and C. J. Carlile, in "Neutron Inelastic Scattering," Vol. 2, p. 421. IAEA, Vienna, 1978.
16. J. M. Rowe, J. J. Rush, and H. E. Flotow, Phys. Rev. B 9, 5039 (1974).
17. R. Kutner and I. Sosnowska, J. Phys. Chem. Solids 38, 741, 747 (1977).
18. G. S. Bauer, E. Seitz, M. Horner, and W. Schmatz, Solid State Commun. 17, 161 (1975).
19. A. Magerl, W. D. Teuchert, and R. Scherm, J. Phys. C 11, 2175 (1978).
20. A. Magerl, N. Stump, W. D. Teuchert, V. Wagner, and G. Alefeld, J. Phys. C 10, 2783 (1977).
21. V. Lottner, H. R. Schober, and W. J. Fitzgerald, Phys. Rev. Lett. 42, 1162 (1979).
22. S. Emin, Phys. Rev. B 3, 1321 (1971).
23. H. R. Schober and V. Lottner, Z. Phys. Chem. (Wiesbaden) 114, 203 (1979).
24. W. Gissler, B. Jay, R. Rubin, and L. A. Vinhas, Phys. Lett. A 43, 279 (1973).
25. I. S. Anderson, J. E. Bonnet, A. Heidemann, D. K. Ross, and S. K. P. Wilson, and M. W. McKergow, J. Less-Common Met. 101, 405 (1984).
26. D. Richter and T. Springer, Phys. Rev. B 18, 126 (1978).
27. K. W. Kehr, D. Richter, and R. H. Swendsen, J. Phys. F 8, 433 (1978).
28. J. W. Haus, K. W. Kehr, and J. W. Lyklema, Phys. Rev. B 25, 2905 (1982).
29. D. Richter and H. Wipf, unpublished work, 1983.
30. Zh. Qi, J. Völkl, and H. Wipf, Scr. Metall. 16, 850 (1982).
31. D. Richter, ref. 3, p. 170.
32. D. Richter, B. Alefeld, A. Heidemann, and N. Wakabayashi, J. Phys. F. 7, 569 (1977).
33. U. Potzel, R. Raab, J. Völkl, H. Wipf, A. Magerl, D. Salomon, and G. Wortman, J. Less-Common Met. 101, 343 (1984).
34. I. S. Anderson, C. J. Carlile, D. K. Ross, and D. L. T. Wilson, Z. Phys. Chem. (Wiesbaden) 115, 165 (1979).
35. R. Hempelmann, D. Richter, R. Pugliesi, and L. A. Vinhas, J. Phys. F 13, 59 (1983).
36. M. W. Johnson, U.K. At. Energy Res. Establ., Rep. AERE-R7682 (1974).
37. D. Richter, R. Hempelmann, and L. A. Vinhas, J. Less-Common Met. 88, 353 (1982).
38. R. Hempelmann, D. Richter, and A. Heidemann, J. Less-Common Met. 88, 343 (1982).

39. D. Noréus and L. G. Olsson, *J. Chem. Phys.* **78**, 2419 (1983); see also J. Töpler, H. Bucher, A. Säuferer, K. Knorr, and Prandl, *J. Less-Common Met.* **88**, 397 (1982).
40. J. C. Achard, C. Lartigue, A. Percheron-Guegan, A. J. Dianoux, and F. Tasset, *J. Less-Common Met.* **88**, 89 (1982).
41. E. Lebsanft, D. Richter, and J. M. Töpler, *Z. Phys. Chem. (Wiesbaden)* **116**, 175 (1969).
42. E. Lebsanft, D. Richter, and J. M. Töpler, *J. Phys. F* **9**, 1057 (1979) (corrected version of ref. 41).
43. D. K. Ross and D. L. T. Wilson, *in* "Neutron Inelastic Scattering," Vol. 2, p. 383. IAEA, Vienna, 1978.
44. K. W. Kehr, R. Kutner, and K. Binder, *Phys. Rev. B* **23**, 4931 (1981).
45. R. A. Tahir-Kheli and R. J. Elliott, *Phys. Rev. B* **27**, 844 (1983).
46. R. A. Tahir-Kheli, *Phys. Rev. B* **27**, 7229 (1983).
47. A. D. Le Claire and A. B. Lidiard, *Philos. Mag.* **1**, 518 (1956).
48. J. Bardeen and C. Herring, *in* "Imperfections in Nearly Perfect Crystals" (W. Shockley, ed.), p. 261. Wiley, New York, 1952.
49. E. W. Montroll and G. H. Weiss, *J. Math. Phys.* **6**, 167 (1965).
50. R. Kutner, *Phys. Lett. A* **81A**, 239 (1981).
51. J. Völkl, *Ber. Bunsenges. Phys. Chem.* **76**, 797 (1972).
52. H. C. Bauer, J. Völkl, J. Tretkowski, and G. Alefeld, *Z. Phys. B* **29**, 17 (1978).
53. J. M. Rowe, N. Vagelatos, J. J. Rush, and H. E. Flotow, *Phys. Rev. B* **12**, 2959 (1975).
54. S. M. Shapiro, Y. Noda, T. O. Brun, J. Miller, H. Birnbaum, and T. Kajitani, *Phys. Rev. Lett.* **41**, 1051 (1978).
55. S. M. Shapiro, D. Richter, Y. Noda, and H. Birnbaum, *Phys. Rev. B* **23**, 1594 (1981).
56. N. Stump, G. Alefeld, and D. Tochetti, *Solid State Commun.* **19**, 805 (1976).
57. C. J. Glinka, J. M. Rowe, J. J. Rush, G. G. Libowitz, and A. Maeland, *Solid State Commun.* **22**, 541 (1977).
58. P. Vorderwisch, S. Hautecler, and J. B. Suck, *Phys. Status Solidi B* **94**, 569 (1979).
59. P. Vorderwisch, S. Hautecler, and W. Wegener, *J. Less-Common Met.* **74**, 117 (1980).
60. W. Wegner, P. Vorderwisch, and S. Hautecler, *Phys. Status Solidi B* **98**, K171 (1980).
61. P. P. Parshin, M. G. Zemlyanov, M. E. Kost, A. Yu. Rumyantsev, and N. A. Chernoplekov, *Sov. Phys.—Solid State (Engl. Transl.)* **22**, 275 (1980).
62. W. A. Kamitakahara and R. K. Crawford, *Solid State Commun.* **41**, 843 (1982).
63. A. Magerl, J. J. Rush, J. M. Rowe, D. Richter, and H. Wipf, *Phys. Rev. B* **27**, 927 (1983).
64. D. Richter, J. J. Rush, and J. M. Rowe, *Phys. Rev. B* **27**, 6227 (1983).
65. J. J. Rush, J. M. Rowe, and A. J. Maeland, *J. Phys. F* **10**, L283 (1980).
66. T. Soskiewisz, *Phys. Status Solidi A* **11**, K123 (1972).
67. B. Stritzker and W. Buckel, *Z. Phys.* **257**, 1 (1972).
68. J. J. Rush, J. M. Rowe, and D. Richter, *Z. Phys.* **B55**, 783 (1985).
69. R. Hempelmann, D. Richter, and B. Stritzker, *J. Phys. F* **12**, 79 (1982).
70. H. Wipf, A. Magerl, S. M. Shapiro, S. K. Satija, and W. Thomlinson, *Phys. Rev. Lett.* **46**, 947 (1981).
71. A. C. Switendick, ref. 1, p. 101.
72. E. Burkel, H. Behr, H. Metzger, and J. Peisl, *Phys. Rev. Lett.* **46**, 1078 (1981).
73. E. Burkel, H. Behr, H. Metzger, J. Peisl, and G. Eckold, *Z. Phys. B* **53**, 27 (1983).
74. A. Magerl, J. M. Rowe, and D. Richter, *Phys. Rev. B* **23**, 1605 (1981).
75. W. Marshall and S. W. Lovesey, "Theory of Thermal Neutron Scattering." Oxford Univ. Press, London and New York, 1971.
76. W. Bührer, A. Furrer, W. Hälg, and L. Schlapbach, *J. Phys. F* **9**, L141 (1979).
77. W. Bührer, *Recent Dev. Condens. Matter Phys.* [*Pap. Gen. Conf.*], *1st, Antwerp, 1980* **4**, 353 (1981).

78. D. Emin, M. I. Baskes, and W. D. Wilson, *Phys. Rev. Lett.* **42**, 791 (1979).
79. H. Sugimoto and Y. Fukai, *Phys. Rev.* B **22**, 670 (1980).
80. H. Sugimoto and Y. Fukai, *J. Phys. Soc. Jpn.* **50**, 3709 (1981).
81. H. Sugimoto and Y. Fukai, *J. Phys. Soc. Jpn.* **51**, 2554 (1982).
82. J. Eckert, J. A. Goldstone, D. Tonks, and D. Richter, *Phys. Rev.* B **27**, 1980 (1983).
83. D. Richter and S. M. Shapiro, *Phys. Rev.* B **22**, 599 (1980).
84. J. J. Rush, A. Magerl, J. M. Rowe, J. M. Harris, and J. L. Provo, *Phys. Rev.* B **24**, 4903 (1981).
85. S. Ikeda and N. Watanabe, *Nat. Lab. High Energy Phys. KENS (Jpn)* **KENS-IV**, 177 (1983).
86. D. Klauder, V. Lottner, and H. Scheuer, *Solid State Commun.* **32**, 617 (1979).
87. J. M. Rowe, *Solid State Commun.* **11**, 1299 (1972).
88. M. Akai, H. Akai, and J. Kanamori, *J. Magn. Magn. Mater.*, in press.
89. K. W. Kehr, ref. 1, p. 197.
90. D. K. Ross, P. F. Martin, W. A. Oates, and R. Khoda Bakhsh, *Z. Phys. Chem. (Wiesbaden)* **114**, 341 (1979).
91. Y. Fukai and H. Sugimoto, *J. Phys. F* **11**, L137 (1981).
92. D. Pine and R. M. Cotts, *Phys. Rev.* B **28**, 641 (1983).
93. P. E. Zapp and H. K. Birnbaum, *Acta Metall.* **28**, 1275, 1523 (1980).
94. Y. Fukai and H. Sugimoto, *Proc. JIMIS—Z Hydrogen Met., Tokyo* **21**, 41 (1980).
95. H. D. Carstanjen, *Phys. Status Solidi A* **59**, 11 (1980).
96. H. Metzger, private communication.
97. A. T. Shirley, C. K. Hall, and N. J. Prince, *Acta Metall.* **31**, 985 (1983).
98. K. Suzuki, *J. Less-Common Met.* **89**, 183 (1983).
99. H. Kaneko, T. Kajitani, M. Hirabayashi, M. Ueno, and K. Suzuki, *Proc. Int. Conf. Rapidly Quenched Met., 4th, Sendai*, p. 1605 (1981).
100. K. Kai, N. Hayashi, Y. Tomizaka, S. Ikeda, N. Watanable, and K. Suzuki, *Natl. Lab. High Energy Phys.*, **KENS-Rep. IV** (1983).
101. H. Hirabayashi, H. Kaneko, T. Kajitani, H. Suzuki, and M. Ueno, *AIP Conf. Proc. No.* 89, p. 87 (1982).
102. T. Kajitani, H. Kaneko, and M. Hirabayashi, *Sci. Rep. Tohoku Univ., Ser. 8* **29**, 210 (1981).
103. A. Williams, J. Eckert, X. L. Yeh, M. Atzmon, and K. Samwer, Los Alamos Meson Phys. Fac. (prepr.).
104. W. L. McMillan, *Phys. Rev.* **167**, 331 (1968).
105. B. N. Ganguly, *Z. Phys.* **265**, 433 (1973).
106. B. N. Ganguly, *Phys. Rev.* B **14**, 3848 (1976).
107. R. J. Miller and C. B. Satterthwaite, *Phys. Rev. Lett.* **34**, 144 (1975).
108. R. C. Dienes and J. P. Garno, *Bull. Am. Phys. Soc.* **20**, 422 (1973).
109. A. Eichler, H. Wühl, and B. Stritzker, *Solid State Commun.* **17**, 213 (1975).
110. J. M. Rowe, J. J. Rush, H. G. Smith, M. Mostoller, and H. E. Flotow, *Phys. Rev. Lett.* **33**, 1297 (1974).
111. A. Rahman, K. Sköld, C. Pelizzari, S. K. Sinha, and H. E. Flotow, *Phys. Rev.* B **14**, 3630 (1976).
112. C. J. Glinka, J. M. Rowe, J. J. Rush, A. Rahman, S. K. Sinha, and H. E. Flotow, *Phys. Rev.* B **17**, 448 (1978).
113. R. Abbenseth and H. Wipf, *J. Phys. F* **10**, 353 (1980).
114. O. Blaschko, J. P. Burger, R. Klemencic, and O. Pepy, *J. Phys. F* **11**, 2015 (1981).
115. P. Jena, C. L. Wiley, and F. Y. Fradin, *Phys. Rev. Lett.* **40**, 578 (1978).
116. J. D. Meyer and B. Stritzker, *Conf. Ser.—Inst. Phys.* No. 55, p. 591 (1981).

117. G. J. Sellers, A. C. Anderson, and H. K. Birnbaum, *Phys. Rev. B* **10**, 2771 (1974).
118. H. K. Birnbaum and C. P. Flynn, *Phys. Rev. Lett.* **37**, , 25 (1976).
119. C. Morkel, H. Wipf, and K. Neumeyer, *Phys. Rev. Lett.* **40**, 947 (1978).
120. H. Wipf, K. Neumaier, A. Magerl, A. Heidemann, and W. Stirling, *J. Less-Common Met.* **101**, 317–326 (1984).
121. A. M. Stoneham, *Rev. Mod. Phys.* **41**, 82 (1969).
122. Y. Imry *in* "Tunneling Phenomena in Solids" (E. Burstein and S. Lundqvist, eds.), p. 563. Plenum, New York, 1969.

Note Added in Proof

After finishing the review a number of publications have appeared in this field. In the following we give a selection of some of these articles, in order to indicate the scope of new developments, naturally without claiming completeness.

Hydrogen Diffusion

1. H. Dosch, J. Peisl, and B. Dorner, Coherent quasielastic neutron scattering from NbD_x, *Phys. Rev. B* **35** (1987).
2. H. Dosch and J. Peisl, Local defect structure of highly mobile deuterium in niobium, *Phys. Rev. Lett.* **56**, 1385 (1986).
3. R. Hempelmann and J. J. Rush, Neutron vibrational spectroscopy of disordered metal–hydrogen systems, *in* "Hydrogen in Disordered and Amorphous Solids" (G. Bambakidis and R. C. Bowman, eds.), p. 283. Plenum, New York, 1986.
4. R. Hempelmann, Diffusion of hydrogen in metals, *J. Less-Common Met.* **101**, 69 (1984).
5. D. Richter, G. Driesen, R. Hempelmann, and I. S. Anderson, Hydrogen-diffusion mechanism in amorphous $Pd_{85}Si_{15}H_{7.5}$: a neutron-scattering study, *Phys. Rev. Lett.* **57**, 731 (1986).

Hydrogen Vibrations

6. I. S. Anderson, J. J. Rush, T. Udovic, and J. M. Rowe, Hydrogen pairing and anisotropic potential for hydrogen isotopes in yttrium, *Phys. Rev. Lett.* **57**, 2822 (1986).
7. J. M. Rowe, J. J. Rush, J. E. Schirber, and J. M. Mintz, *Phys. Rev. Lett.* **57**, 2955 (1986).
8. R. Hempelmann, D. Richter, and D. L. Price, High energy neutron vibrational spectroscopy on β-V_2H, *Phys. Rev. Lett.*, **58**, 1016 (1987).
9. K. M. Ho, H. J. Tao, and W. J. Zhu, First principle investigation of metal hydrogen interactions in NbH, *Phys. Rev. Lett.* **53**, 1586 (1984).

Hydrogen Tunneling

10. A. Magerl, A. J. Dianoux, H. Wipf, K. Neumaier, and I. S. Anderson, Concentration dependence and temperature dependence of hydrogen tunneling in $Nb(OH)_x$, *Phys. Rev. Lett.* **56**, 159 (1986).
11. H. Wipf, D. Steinbinder, K. Neumayer, P. Gutsmiedl, A. Magerl, and A. J. Dianoux, Influence of electrons on the tunneling state of a hydrogen atom in a metal, "Proceedings of Workshop on Quantum Motion in Solids," Prog. in Phys., Springer-Verlag, Heidelberg and New York, 1987.

11. FAST ION CONDUCTORS

N. H. Andersen, K. N. Clausen, and J. K. Kjems

Physics Department
Risø National Laboratory
DK-4000 Roskilde, Denmark

11.1. Introduction

11.1.1. Short History

In the nineteenth century it was realized that ions may conduct electric currents in a solid with mobilities comparable to those of liquid electrolytes ($\sigma \gtrsim 10$ S/m). The first observations date from 1839 when Faraday[1] reported on ion conduction in PbF_2, and later, Nernst[2] (1899) found high oxygen conductivity in Y_2O_3 stabilized ZrO_2 (YSZ) at elevated temperatures. Despite their potential for a number of applications in electrochemical devices, only a few solid electrolytes were discovered up to 1960. One of the few important results from this period is the observation[3] of high silver mobility in AgI above the first-order structural phase transition at 146°C, indicating that high ionic mobilities may exist down to ambient temperatures. Another classic example[4] is the high-temperature, lithium-conducting α phase of Li_2SO_4.

The development from the 1960s onward was spurred by the work on the β aluminas and has revealed many new materials with high values of conductivity, mainly with H^+, Li^+, Na^+, K^+, Cu^+, Ag^+, Cl^-, F^-, or O^{2-} as the mobile ions. Except for O^{2-} only monovalent ions with relatively small ionic radii give rise to rapid diffusion.

11.1.2. Classical Model

For a classical transport process in a crystalline solid it is expected that ionic motion is most favorable for monovalent ions with small ionic radii. The fixed ions establish a periodic potential where the barriers between the potential wells increase with the charge of the moving ions and clearly go to infinity when the bottleneck in the diffusion pathway is smaller than the ions.

The presence of potential barriers in the diffusion pathway is observed in the temperature dependence of the ionic conductivity that follows an Arrhenius type of behavior, corresponding to an activated process:

$$\sigma = \rho \frac{q^2 v d^2}{k_B T} \exp(-\Delta/k_B T). \tag{11.1}$$

Equation (11.1) is the simple classical expression for a jump-diffusion type of ionic conductivity for either vacancies or interstitial ions of concentration ρ and charge q, oscillating in a potential well of height $\Delta (\gg k_B T)$ with an attempt frequency v and an average jump distance d along the field. This model assumes an excess of available sites compared to mobile ions and thus that the crystal structure is disordered to some degree. All solids have crystalline disorder either as a result of impurities or by thermal generation of Schottky or Frenkel defects. The ionic motion may take place either by an interstitialcy mechanism where ions are moving between interstitial sites in the lattice, and by vacancy diffusion where the motion is between regular sites. In most ionic crystals the concentrations of mobile ions or vacancies created by thermal formation or dissociation of impurity complexes are usually low because of the large values of the respective activation energies. Similarly the potential barrier Δ for migration is usually high, and the model result given by Eq. (11.1) is expected to be valid. However, the properties of solid ionic conducting materials, which have conductivities comparable to those of liquid electrolytes, do not match these assumptions because the activation energies are low (0.1–0.3 eV), and the number of ions involved in the transport process is so high that their interactions play an essential role. It is generally accepted that there is a connection between the large number of ions participating in the conduction process and the low activation energy of motion. Only for correlated motion is it realistic to obtain potential barriers in a solid as low as 0.1–0.3 eV, and it may be shown theoretically that an array of strongly correlated ions may move in a periodic potential with reduced activation energy of motion if the average distance between the ions is different from the period of the potential.[5] The physical explanation for this behavior is that some ions are moving up the potential barriers while others are moving down. In the limit the effective potential barrier may be zero.

In the majority of the reported studies of fast ion conductors, the ionic conductivity is analyzed in terms of simple expressions like Eq. (11.1) or with small modifications thereof, even though it is obvious that the concept of simple interstitialcy or vacancy motion is no longer valid. To understand the conduction process, one must establish how large numbers of correlated interstitial ions and/or vacancies are created and how they move in the lattice. For this purpose it is vital to obtain structural information about possible sites in the lattice, partial occupancies, correlations between ions and vacancies, and ion dynamics.

11.1.3. Neutron Scattering Methods

Several of the neutron scattering techniques discussed extensively in Chapter 1 (Part A) are applicable, and some of them offer unique possibilities for studying the structural properties of fast ion conductors. Neutron scattering is often complementary to the analogous x-ray techniques, and the combination can be especially valuable. Technically, the methods differ in that many of the fast conducting ions are light and therefore weak x-ray scatterers, whereas neutron scattering lengths vary unsystematically over the periodic system, and by isotope exchange it is possible to increase the scattering contrast. Accordingly, all the fast conducting ions may be seen by neutrons in almost any reference lattice. The ability of neutrons to penetrate materials is also useful, since most fast ion conducting phases are found at elevated temperatures. The application of furnaces is therefore needed, and the neutron penetration facilitates the design.

Diffraction techniques may be used in the usual way to derive basic crystallographic properties such as crystal symmetry, unit cell parameters, ionic sites and their average occupancies, and harmonic temperature factors. More special analysis based on probability-density-function methods and on expansion of thermal parameters to higher order than the harmonic approximation are also appropriate, due to the very shallow and anharmonic single-ion potentials of the mobile ions. For the same reasons combinations of the different types of Fourier synthesis and least-squares refinements are useful in the data analysis.

Correlations between defects in the lattice and ionic diffusion may be studied by quasi-elastic diffuse neutron scattering. Integration over the energy transfer around zero energy of the coherent scattering gives an instantaneous picture of the defect correlations. If the correlated defect motion is dynamic on a time scale shorter than about 10^{-10} s, it may be studied from the **Q**-dependent energy width. Similarly, the incoherent scattering may be used to study the self-diffusion of ions.

Inelastic neutron scattering is useful for elucidating the vibrational modes of the ion conductors. Low-lying modes are of special interest because they may be related directly to the vibrations of mobile ions, and the acoustic contributions to the thermal diffuse scattering (TDS) are important for a proper interpretation of both the Bragg peak intensities and the quasi-elastic diffuse data. Higher lying vibrational modes in fast ion conductors are often strongly damped because the single-ion potentials of the mobile ions are strongly anharmonic and the force constant network is defective due to disorder in site occupancies.

11.1.4. Terminology

There is still an ongoing discussion about the most appropriate name for this group of ion conductors, perhaps reflecting that a clear and unique definition of their characteristics has not yet been given. At present the most widely used and least controversial name is fast ion conductors, which is adopted in the present presentation. The more restrictive name solid electrolytes is used for fast ion conductors with negligible electronic conductivity. In order to avoid misleading associations with the electronic superconducting phase, we do not use the commonly adopted adjective "superionic" in connection with these materials and their behavior, although the meaning, when appearing in the literature, is the same. In this connection it should be mentioned that what we call a fast ion phase is not necessarily separated from a normal ion conducting phase by a distinct phase transition. This is the case for a large group of fast ion conductors, e.g., many of those with the cubic fluorite structure (CaF_2 structure), where a diffuse type of anomaly is observed in the specific heat without any change in the crystallographic symmetry. In any case, it is well established that a fast ion phase is closely related to a disordering process in which the mobile ions gain entropy corresponding to fusion. In combination with the observation of liquid-like conductivity this has introduced another controversial term "molten sublattice" for the structure of the mobile ions. Although it has been shown in several systems that the potential variation along the conduction pathway in the fast ion phase is very shallow, and thereby that the mobile ions are performing large amplitude anharmonic vibrations around the equilibrium sites in combination with the translatory jumps, the mobile ions do not occupy all available sites with equal probability. One way of characterising fast ion motion is in terms of the mean residence time τ_r on a site relative to the mean jump time τ_j that it takes for the ion to move to a neighboring equilibrium site. For fast ion motion, τ_j can be comparable to τ_r.

11.1.5. Classification

As mentioned, the definition of a fast ion conducting phase is rather unclear; the same holds true for the classification of ion conductors. O'Keeffe[6,7] has divided solid ion conductors into three classes. Class I contains normal ion conductors that attain high ionic conductivity only on melting; these materials are not fast ion conductors. The alkali halides are examples of class I materials. Class II contains systems which acquire fast ion conduction via a first-order phase transition with change in the crystallographic symmetry. Actually, O'Keeffe divides class II into two subclasses: class IIa contains materials where the crystallographic space group of the fast ion phase is not a supergroup of the normal conducting one (major changes of crystal symmetry for all ions), while class IIb contains those where this is the case (minor changes of crystal symmetry for the immobile ions). The best known examples of class II materials are silver iodide, sulfide, and selenide. Class III contains materials where ionic conduction is obtained via a diffuse transition with no change in the overall crystallographic symmetry, often called the Faraday transition because Faraday[1] first observed the diffuse transition in PbF_2. Examples from this group of materials are the fluorides PbF_2, CaF_2, and BaF_2. Pardee and Mahan[8] have chosen a slightly different classification scheme that relates directly to the nature of the phase transitions. They do not distinguish between class IIa and IIb but add a new class of materials that enter the fast ion conducting phase via a second-order phase transition. An important example of this class is $RbAg_4I_5$, which shows fast ion conduction at room temperature.

11.1.6. Phase Transitions

As noted above, the basic origin of fast ion conduction is the disorder and the low activation energy of motion associated with correlated motions of the ions. Therefore, the fast ion conducting phases are found in systems that may transform crystallographically or disorder thermally without too much gain in enthalpy. The phase stability is ensured by a simultaneous gain in entropy on disordering. In order to establish many equivalent sites for disordering, it is clear that fast ion conducting phases should have a high crystallographic symmetry. This general trend is assured for systems obeying the first Landau rule[9] of phase transitions, which in this context requires that the space group of the fast ion conducting phase is a supergroup of the ordered one. Second-order phase transitions and those of first order corresponding to O'Keeffe's subclass IIb are Landau transitions. However, also for the fast ion conductors belonging to subclass IIa, it is observed that the fast ion conducting phases almost always belong to the highly symmetric cubic or hexagonal space groups. For a more detailed discussion of phase transitions in fast ion conductors, see Salamon.[10]

11.1.7. Stoichiometry

Significant enhancements of the ionic conductivities may be obtained by doping. It is a characteristic of many crystallographic systems that may ultimately transform into a fast ion conducting phase by themselves that they are highly susceptible to doping, because the enthalpies involved in lattice distortions are small. Aliovalent ions substituted into a system create disorder that directly influences the conduction process. In the classical picture of individual point defect jumps, this involves a dissociation energy to be overcome for motion. In fast ion conducting systems the situation gets more complicated because more complex defect structures are formed as a result of interactions between the doped ions and the thermally generated defects. In general, these interactions lower the activation energy of formation and/or motion of mobile ions. In O'Keeffe's class III systems, where fast ion conduction is created without changing the overall crystallographic symmetry, high ionic conductivity is observed at much lower temperatures. BaF_2 doped with LaF_3 is an example of such behavior. With 50 mole % LaF_3 substituted into BaF_2, the fluorine conductivity at room temperature is enhanced by more than eight orders of magnitude.[11] Alternatively, a fast ion conducting phase may be stabilized by doping or introducing nonstoichiometry into the system. One of the famous examples is the oxygen conductor ZrO_2, which has a fast ion conducting cubic α phase above 2643 K. On doping (e.g., with Y_2O_3, CaO, or MgO), the α phase may be stabilized down to room temperature,[12] although the conductivity behavior at low temperatures cannot be characterized as fast ion conduction. Another important group of materials are the β aluminas, e.g., Na β aluminas that have hexagonal structures composed of spinel blocks ($Al_{11}O_{16}$) separated by Na conducting basal planes of Na and O ions. The nominal composition is $Na_2O \cdot 11Al_2O_3$, but single phase Na β alumina is stable only with an excess of about 20–33% Na. More details, also about other stable compositions in these systems, are given in the reviews by Kennedy[13] and Collongues et al.[14]

11.1.8. Related Physics

The increasing activity in the field of fast ion conductors, starting in the 1960s but accelerating significantly in the 1970s, is clearly related to their potential for applications. At the same time, in solid-state physics and chemistry, a more general interest developed in properties relating to deviations from the simple idealized periodicity of three-dimensional crystalline structures. Examples are surface properties, incommensurate structures, chaotic behavior, phase transitions, spin glasses, amorphous materials, one-dimensional conductors, liquid crystals, and fractal dimensions. Defect properties, as observed in fast ion conducting systems, are in line with this development. Apart from the interesting studies of transport-structure relations, these systems pose the fundamental question: to what extent are basic solid-state concepts such as Brillouin zones, phonons, and energy bands valid in the presence of massive disorder in the crystalline lattice.

11.1.9. Applications

The major field of application for fast ion conducting materials is that of electrochemical cells. Among these, batteries are the most familiar and their potentials for load leveling and traction purposes are well known. Fuel cells have similar technological potential, and their function is conceptually similar to that of batteries but the chemical reactions always involve oxidation ("combustion") of reducing gases like hydrogen, carbon monoxide, or methane. Other applications comprise various gas monitoring cells, such as oxygen sensors for traditional combustion control, electrochemical cells for direct determination of the thermodynamic functions involved in chemical reactions, and electrolytes in electrical condensers. A more exotic example is the application in electrochromatic displays of certain ion conductors that allow incorporation of ions with an associated change in color (e.g., Na^+ in WO_3). For a more extensive overview of applications the reader should consult the reviews by van Gool,[15] Geller,[16] Hagenmuller and van Gool,[17] Subbarao,[18] and Chandra.[19]

In order to elucidate the materials requirement for successful application, it is appropriate to mention some basic properties of electrochemical cells. Their function is based on the utilization of chemical reactions for generating voltage differences or vice versa. The process is controlled by an electrolytic separation of the constituents in the chemical reaction such that only a specific ion is able to diffuse until the difference in chemical potential is balanced by an electrostatic potential. Electrodes are necessary for the function of the cell. They supply or drain the electrons needed in the reaction and connect to external circuits when the cell is operating.

In most electrochemical cells the materials requirements are similar. Electrodes and electrolytes should be chemically and mechanically compatible, and electrolytes should have high ionic and negligible electronic conductivities to avoid ohmic losses and self-discharge. Ideally, electrode materials should be both electron and ion conducting with high conductivities, and in some cases (e.g., fuel cells and oxygen sensors) a good catalytic function is also required. For technological applications, especially in relation to energy storage and conversion, low-cost materials and high-power densities are essential and require that the materials in the cells are light and that the chemical reaction is between strongly electropositive and electronegative reactants.

11.1.10. Outline

In Sections 11.2–11.6 the use of neutron scattering techniques for studies of fast ion conductors is exemplified. The presentation does not aim at a thorough review of all the neutron scattering studies performed on fast ion conductors. Emphasis is put on the specific information obtained from the different techniques. As model systems we mainly use the bcc structure Ag ion conductors and the fluorite structure halide conductors, which are among the most extensively studied systems. For reference and when an elucidation of the neutron scattering methods make it necessary, studies on other systems will also be mentioned. Further information can be found in Refs. 15–20 and concerning neutron scattering especially in the contribution by Shapiro and Reidinger in Ref. 20. The remaining part of this chapter is organized in this way: Section 11.2 contains a summary of the scattering formalism appropriate for analysis of neutron scattering data on fast ion conductors, and Section 11.3 discusses the basic properties of the model systems. Selected studies on fast ion conductors by the different neutron scattering techniques, their interpretation, and the physical properties that may be derived are presented in Sections 11.4, 11.5, and 11.6. Section 11.4 deals with diffraction studies, Section 11.5 with quasi-elastic diffuse scattering, and Section 11.6 with inelastic neutron scattering. Finally, Section 11.7 contains general conclusions about the structural and dynamic properties of fast ion conductors found from neutron scattering studies and how this information may be used to explain the transport properties.

11.2. Scattering Formalism

11.2.1. General Formulation

Almost all the features of nuclear scattering of neutrons can be utilized in the study of ion conductors. In this section we start from the general formulation introduced in Part A, Chapter 1, and give a brief outline of the nuclear scattering formalism suitable for the present purpose.

A general master equation for nuclear scattering may be derived from Eq. (1.40) with minor changes that emphasize the effects of the disorder in the system. A notation similar to the one introduced in Section 1.5.3. is used:

$$\frac{d^2\sigma}{d\Omega\, dE} = \frac{k_1}{k_0} \sum_{s,s'} \sqrt{N_s N_{s'}}\, \bar{b}_s^* \bar{b}_{s'}\, S_c^{ss'}(\mathbf{Q}, E) + \frac{k_1}{k_0} \sum_s N_s \frac{\sigma_i^s}{4\pi} S_i^s(\mathbf{Q}, E), \quad (11.2)$$

$$S_c^{ss'}(\mathbf{Q}, E) = \frac{1}{2\pi\hbar} \frac{1}{\sqrt{N_s N_{s'}}} \int_{-\infty}^{\infty} \sum_{\substack{j \in s \\ j' \in s'}} \langle e^{-i\mathbf{Q}\cdot\mathbf{R}_j(0)} e^{i\mathbf{Q}\cdot\mathbf{R}_{j'}(t)}\rangle e^{-iEt/\hbar}\, dt, \quad (11.3)$$

and

$$S_i^s(\mathbf{Q}, E) = \frac{1}{2\pi\hbar} \frac{1}{N_s} \int_{-\infty}^{\infty} \sum_{j \in s} \langle e^{i\mathbf{Q}\cdot[\mathbf{R}_j(t) - \mathbf{R}_j(0)]}\rangle e^{-iEt/\hbar}\, dt. \quad (11.4)$$

Equation (11.2) is a sum of the coherent and the incoherent parts of the differential cross sections expressed in terms of the Fourier transforms of the pair correlation and selfcorrelation functions for a system of particles j of different species s, with numbers N_s, coherent scattering amplitudes $\bar{b}_s$, and incoherent scattering cross sections σ_i^s. The j summations in Eqs. (11.3) and (11.4) are restricted to the specific species s, s'. It should be emphasized that our differential scattering cross sections are defined for the whole scattering system and *not* as in Chapter 1 (Part A) normalized to the scattering per particle. The coherent scattering function $S_c^{ss'}(\mathbf{Q}, E)$ in Eq. (11.3) is defined to be independent of the scattering lengths. This is most convenient for simple systems with only one kind of atom because the results may be compared directly with those of other scattering techniques. However, for more complex physical systems, this definition is rather inconvenient since in general the response from the system depends on correlations between all pairs. A complete coherent scattering function may be defined as

$$S_c(\mathbf{Q}, E) = \frac{1}{2\pi\hbar} \frac{1}{N_c} \sum_{s,s'} \bar{b}_s^* \bar{b}_{s'} \int_{-\infty}^{\infty} \sum_{\substack{j \in s \\ j' \in s'}} \langle e^{-i\mathbf{Q}\cdot\mathbf{R}_j(0)} e^{i\mathbf{Q}\cdot\mathbf{R}_{j'}(t)} \rangle e^{-iEt/\hbar} \, dt$$

$$= \frac{1}{2\pi\hbar} \int_{-\infty}^{\infty} I_c(\mathbf{Q}, t) e^{-iEt/\hbar} \, dt, \quad (11.5)$$

in terms of which the coherent part of Eq. (11.2) alone is given as

$$\left(\frac{d^2\sigma}{d\Omega \, dE}\right)_c = \frac{k_1}{k_0} N_c S_c(\mathbf{Q}, E), \quad (11.6)$$

where N_c is the number of unit cells. Equation (11.5) defines the complete coherent intermediate scattering function $I_c(\mathbf{Q}, t)$. In the first line of the equation, the summation is over the particles j at positions $\mathbf{R}_j$. In defective crystalline solids, the diffuse scattering results from correlations between occupied and empty sites in the lattice. The change from particle to site representation may be performed by the substitution

$$\mathbf{R}_j(t) = \sum_{l,d} P_{ldj}(t)[\mathbf{l} + \mathbf{d} + \mathbf{u}_{ldj}(t)], \quad (11.7)$$

where $P_{ldj}(t) = 1$ or 0, depending on whether or not particle j is at site d in unit cell l at time t, and $\mathbf{u}_{ldj}(t)$ is the positional deviation from the equilibrium site $\mathbf{l} + \mathbf{d}$ of particle j at time t. By use of Eq. (11.7), the double j summation in Eq. (11.5) may be converted into a summation over sites (l, d). It is recalled that for each particle j only one site (l, d) gives $P_{ldj}(t) = 1$, and that the coherent cross section depends on the ion species s on the site and not on the specific particle j. One gets

$$S_c(\mathbf{Q}, E) = \frac{1}{2\pi\hbar} \frac{1}{N_c} \sum_{s,s'} \bar{b}_s^* \bar{b}_{s'} \int_{-\infty}^{\infty} \left\langle \sum_{\substack{l,d \\ l',d'}} P_{lds}(0) P_{l'd's'}(t) \right.$$

$$\left. \times e^{-i\mathbf{Q}\cdot[\mathbf{l}+\mathbf{d}+\mathbf{u}_{lds}(0)]} e^{i\mathbf{Q}\cdot[\mathbf{l}'+\mathbf{d}'+\mathbf{u}_{l'd's'}(t)]} \right\rangle e^{-iEt/\hbar} \, dt, \quad (11.8)$$

where $P_{lds}(t)$ and $\mathbf{u}_{lds}(t)$ are defined for species s in a similar way as $P_{ldj}(t)$ and $\mathbf{u}_{ldj}(t)$. Defining the time-average occupation of species s on sites d in the lattice,

$$c_{ds} = \left\langle \frac{1}{N_c} \sum_l P_{lds}(t) \right\rangle_t, \quad (11.9)$$

and the deviations from average occupation,

$$\Delta P_{lds}(t) = P_{lds}(t) - c_{ds}, \quad (11.10)$$

Eq. (11.8) may be expressed as

$$S_c(\mathbf{Q}, E) = \frac{1}{2\pi\hbar} \frac{1}{N_c} \sum_{s,s'} \bar{b}_s^* \bar{b}_{s'} \sum_{\substack{l,d \\ l',d'}} e^{i\mathbf{Q}\cdot(\mathbf{l}'-\mathbf{l})} e^{i\mathbf{Q}\cdot(\mathbf{d}'-\mathbf{d})}$$

$$\times \left\{ c_{ds} c_{d's'} \int_{-\infty}^{\infty} \langle e^{i\mathbf{Q}\cdot[\mathbf{u}_{l'd's'}(t)-\mathbf{u}_{lds}(0)]} \rangle e^{-iEt/\hbar} \, dt \right.$$

$$+ \int_{-\infty}^{\infty} \langle \Delta P_{lds}(0) \Delta P_{l'd's'}(t) e^{i\mathbf{Q}\cdot[\mathbf{u}_{l'd's'}(t)-\mathbf{u}_{lds}(0)]} \rangle e^{-iEt/\hbar} \, dt$$

$$\left. + \text{cross terms} \right\}. \quad (11.11)$$

The first ensemble average in the parenthesis gives rise to the Debye–Waller factor and the phonons (see Part A, Chapter 1, or standard textbooks like Squires[21]). The second term contains the average in the combined time evolution of the correlations in the defect occupations $\Delta P_{lds}(t)$ and the phonon coordinates $\mathbf{u}_{lds}(t)$. Since the time scales for the configurational rearrangements and the lattice vibrations are usually significantly different, we may decouple their correlations:

$$\langle \Delta P_{lds}(0) \Delta P_{l'd's'}(t) e^{i\mathbf{Q}\cdot[\mathbf{u}_{l'd's'}(t)-\mathbf{u}_{lds}(0)]} \rangle$$

$$= \langle \Delta P_{lds}(0) \Delta P_{l'd's'}(t) \rangle \langle e^{i\mathbf{Q}\cdot[\mathbf{u}_{l'd's'}(t)-\mathbf{u}_{lds}(0)]} \rangle. \quad (11.12)$$

In the same approximation the cross terms [between c_{ds} and $\Delta P_{l'd's'}(t)$] in Eq. (11.11), vanish and one gets by traditional methods[21]

$$S_c(\mathbf{Q}, E) = S_c^B(\mathbf{Q}, 0) + S_{c,ph}(\mathbf{Q}, E) + S_c^{diff}(\mathbf{Q}, E), \quad (11.13)$$

where

$$S_c^B(\mathbf{Q}, 0) = \frac{(2\pi)^3}{v_0} \left| \sum_{d,s} \bar{b}_s c_{ds} e^{-W_{ds}(\mathbf{Q})} e^{i\mathbf{Q}\cdot\mathbf{d}} \right|^2 \sum_\tau \delta(\mathbf{Q} - \tau) \quad (11.14)$$

is the Bragg-scattering contribution,

$$S_{c,ph}(\mathbf{Q}, E) = \sum_n S_{c,ph}^{\pm n}(\mathbf{Q}, E) \quad (11.15)$$

is an expansion in n-phonon emission ($+$) and absorption ($-$) processes, and

$$S_c^{diff}(\mathbf{Q}, E) = \frac{1}{2\pi\hbar} \frac{1}{N_c} \sum_{s,s'} \bar{b}_s^* \bar{b}_{s'} \sum_{\substack{l,d \\ l',d'}} e^{i\mathbf{Q}\cdot(l'-l)} e^{i\mathbf{Q}\cdot(\mathbf{d}'-\mathbf{d})} e^{-W_{ds}(\mathbf{Q})} e^{-W_{d's'}(\mathbf{Q})}$$

$$\times \int_{-\infty}^{\infty} \langle \Delta P_{lds}(0) \Delta P_{l'd's'}(t) \rangle e^{-iEt/\hbar} \, dt \quad (11.16)$$

is the diffuse scattering contribution. In the equations v_0 is the unit cell volume, τ a reciprocal lattice vector, and $\exp[-2W_{ds}(\mathbf{Q})]$ the Debye–Waller factor for ion species s on sites d in the lattice. In Eq. (11.16) we have retained only the Debye–Waller term from the evaluation of the second factor on the right-hand side of Eq. (11.12).

11.2.1.1. *Coherent Scattering.* Most of the ion conductors that will be discussed have long-range crystalline order, albeit sometimes heavily perturbed by the ionic disorder. Hence, the dominant coherent elastic scattering is the Bragg scattering, which from Eqs. (11.6) and (11.14) has the cross sections

$$\left[\frac{d\sigma}{d\Omega}\right]_{c,el} = \frac{N_c(2\pi)^3}{v_0} \sum_\tau |F(\mathbf{Q})|^2 \delta(\mathbf{Q} - \tau), \quad (11.17)$$

where the structure factor is

$$F(\mathbf{Q}) = \sum_{d,s} \bar{b}_s c_{ds} e^{-W_{ds}(\mathbf{Q})} e^{i\mathbf{Q}\cdot\mathbf{d}}. \quad (11.18)$$

It is important to note that the Bragg scattering only measures the average occupation of the atomic species that may reside on a given site weighted with their coherent scattering lengths. Therefore, one cannot deduce any information about correlations in the occupancy of neighboring sites based on the observed Bragg intensities. Such information is contained in the diffuse coherent scattering given by Eqs. (11.6) and (11.16). In general the diffuse scattering is inelastic or at least quasi-elastic but, as already mentioned, on a different energy scale than the phonons. The phonon modes give rise to inelastic features typically in the range 2–50 meV, whereas the disorder scattering from diffusive configurational rearrangement of defects generally has quasi-elastic character with an energy width inversely proportional to a typical relaxation time, that is in the range 0–2 meV.

Important information may be obtained from an energy integral over the quasi-elastic diffuse scattering,

$$S_c^{\text{diff}}(\mathbf{Q}) = \int_{\substack{\text{quasi}\\\text{elastic}}} S_c^{\text{diff}}(\mathbf{Q}, E) dE, \tag{11.19}$$

which corresponds to a determination of the intermediate coherent scattering function $I_c(\mathbf{Q}, t)$ at $t = 0$ [cf. Eqs. (11.5) and (11.16)] and hence gives an instantaneous picture of the disorder configurations. From Eqs. (11.16) and (11.19) one may obtain

$$S_c^{\text{diff}}(\mathbf{Q}) = \frac{1}{N_c} \left| \sum_{lds} \Delta P_{lds} \bar{b}_s e^{-W_{ds}(\mathbf{Q})} e^{i\mathbf{Q}\cdot(\mathbf{l}+\mathbf{d})} \right|^2. \tag{11.20}$$

The difference between the information obtained from the diffuse scattering and the Bragg scattering can be made more transparent if the definition of ΔP_{lds} [Eq. (11.10)] is considered:

$$S_c^{\text{diff}}(\mathbf{Q}) = \frac{1}{N_c} \left| \sum_{lds} (P_{lds}\bar{b}_s - c_{ds}\bar{b}_s) e^{-W_{ds}(\mathbf{Q})} e^{i\mathbf{Q}\cdot(\mathbf{l}+\mathbf{d})} \right|^2. \tag{11.21}$$

The second term in the parenthesis of Eq. (11.21) is representative for the average crystallographic structure, and when it is considered alone it gives exactly the Bragg scattering. If defects are defined as deviations from the average structure, it is clear that the correlations between defects are contained in the diffuse scattering. One can therefore analyze the observed integrated quasi-elastic scattering by static models for different defect configurations as has been done for the fluorites.[22]

The phonon cross sections are also of interest for studies of ion conductors. From a practical point of view, they provide a convenient internal reference for the other types of scattering and hence can be used for intensity calibrations. From standard procedures[21] one may obtain the one-phonon scattering functions

$$S_{c,\text{ph}}^{\pm 1}(\mathbf{Q}, E) = \frac{(2\pi)^3}{v_0} \frac{\hbar^2}{N_c} \sum_m \frac{\langle n_m + \tfrac{1}{2} \pm \tfrac{1}{2} \rangle}{2E_{\text{ph}}^m} \left| \sum_{d,s} \frac{\bar{b}_s c_{ds}}{\sqrt{M_s}} e^{-W_{ds}(\mathbf{Q})} e^{i\mathbf{Q}\cdot\mathbf{d}} (\mathbf{Q}\cdot\mathbf{e}_m^d) \right|^2$$

$$\times \sum_\tau \delta(\mathbf{Q} \mp \mathbf{q} + \boldsymbol{\tau}) \delta[E \mp E_{\text{ph}}^m(\mathbf{q})], \tag{11.22}$$

where M_s is the ionic mass of species s, $\langle n_m \rangle$ is the Bose factor, $\mathbf{e}_m^d$ is a unit polarization vector such that $\sum_d |e_m^d|^2 = 1$, $E_{\text{ph}}^m(\mathbf{q})$ is the phonon energy, and m is a mode index.

To establish the calibration formula, we consider a standard neutron scattering experiment where the observed count rate for a given setting of the spectrometer $(\mathbf{Q}, E)$, normalized to monitor counts, is given as

$$\mathcal{F}(\mathbf{Q}, E) = CN_c \int S(\mathbf{Q}', E') R(\mathbf{Q}, \mathbf{Q}', E, E') \, d\mathbf{Q}' \, dE', \qquad (11.23)$$

where C is a spectrometer constant including effects like the sample absorption, and $R(\mathbf{Q}, \mathbf{Q}', E, E')$ is the resolution function for the instrument. The integrated intensity for either a constant $\mathbf{Q}$ scan of a phonon group, or an energy scan of the quasi-elastic diffuse scattering, is

$$\mathcal{F}(\mathbf{Q}) = \int \mathcal{F}(\mathbf{Q}, E) \, dE. \qquad (11.24)$$

For acoustic one-phonon emission in the long wavelength limit where the mean square amplitudes of the ions are equal, this can be approximated by

$$\mathcal{F}_{\text{ph}}^{+1}(\mathbf{Q}) = CV_0 V_1 N_c \frac{\hbar^2 (\mathbf{Q} \cdot \hat{\mathbf{e}}_m)^2}{E_{\text{ph}}^m 2M_c} \langle n_m + 1 \rangle$$

$$\times \left| \sum_{d,s} b_s c_{ds} e^{-W_{ds}(\mathbf{Q})} e^{i\mathbf{Q} \cdot \mathbf{d}} \right|^2, \qquad (11.25)$$

where V_i are resolution volume factors depending on the incident and outgoing wave vectors k_i and on the spectrometer configuration, $\hat{\mathbf{e}}_m$ is the unit polarization vector ($|\hat{\mathbf{e}}_m|^2 = 1$) for the phonon mode m, and M_c is the total ionic mass in the unit cell.

Similarly, we have for the diffuse scattering

$$\mathcal{F}^{\text{diff}}(\mathbf{Q}) = CV_0 V_1 N_c S_c^{\text{diff}}(\mathbf{Q}). \qquad (11.26)$$

Thus, a comparison between $\mathcal{F}_{\text{ph}}^{+1}(\mathbf{Q})$ and $\mathcal{F}^{\text{diff}}(\mathbf{Q})$ allows a determination of the defect structure on an absolute scale. The phonon cross section is also of interest in its own right, and it can reveal the effects of lattice disorder through the linewidth and through the renormalization of the phonon energies. Phonon line shapes are most often analyzed by means of standard Green's function theory, which yields a damped harmonic oscillator form for the response.[23] For a given acoustic mode m the last delta function in Eq. (11.22) is then replaced by a Lorentzian

$$\frac{1}{\pi} \frac{\Gamma}{(E \mp E_{\text{ph}}^m)^2 + \Gamma^2}, \qquad (11.27)$$

where

$$E_{\text{ph}}^m = \sqrt{(E_{\text{ph}}^{m0})^2 + a\hbar q^2},$$

the renormalized phonon energy, and

$$\Gamma = \hbar[(1/2)(D + \eta) + b]q^2/2, \tag{11.28}$$

the phonon width resulting from disorder, are given by the parameters a and b, ionic diffusion (diffusion constant D), and anharmonicity (viscosity coefficient η) (see reference 24 for details).

11.2.1.2. Incoherent Scattering. The coherent scattering reflects the interparticle correlations and yields information on both the average structure and defect configurations and their temporal evolution. The incoherent scattering reflects the single-particle motion, information about which can also be derived from tracer diffusion experiments. Generally the incoherent scattering is given by [see Eq. (11.2)]

$$\left(\frac{d^2\sigma}{d\Omega\, dE}\right)_i = \frac{k_1}{k_0} \sum_s N_s \frac{\sigma_i^s}{4\pi} S_i^s(\mathbf{Q}, E) \tag{11.29}$$

where $S_i^s(\mathbf{Q}, E)$ is given by Eq. (11.4)

The self-intermediate scattering function $I_{\text{self}}^s(\mathbf{Q}, t)$ and the self-correlation function $G_{\text{self}}^s(\mathbf{R}, t)$ are given by

$$I_{\text{self}}^s(\mathbf{Q}, t) = \frac{1}{N_s} \sum_{j \in s} \langle e^{-i\mathbf{Q} \cdot \mathbf{R}_j(0)} e^{i\mathbf{Q} \cdot \mathbf{R}_j(t)} \rangle$$

$$= \int G_{\text{self}}^s(\mathbf{R}, t) e^{i\mathbf{Q} \cdot \mathbf{R}}\, d\mathbf{R}. \tag{11.30}$$

The incoherent scattering spectra also reflects the different kinds of ionic motion. The lattice vibrations give inelastic scattering with a spectrum given by the amplitude weighted density of states

$$\left(\frac{d^2\sigma}{d\Omega\, dE}\right)_{i,\text{ph}}^{\pm 1} = \frac{k_1}{k_0} \sum_{s,d} \frac{c_{ds}\hbar^2}{2M_s} \frac{\sigma_i^s}{4\pi} e^{-2W_{ds}(\mathbf{Q})}$$

$$\times \sum_m \frac{|\mathbf{Q} \cdot \mathbf{e}_m^d|^2}{E_{\text{ph}}^m} \left\langle n_m + \frac{1}{2} \pm \frac{1}{2} \right\rangle \delta(E \mp E_{\text{ph}}^m). \tag{11.31}$$

The diffusional motion gives rise to a quasi-elastic contribution, which often is of Lorentzian form

$$S_i^{\text{diff}}(\mathbf{Q}, E) = \frac{1}{\pi} \sum_s \frac{\hbar f_s(\mathbf{Q})}{E^2 + \hbar^2 f_s^2(\mathbf{Q})} e^{-2W_s(\mathbf{Q})}, \tag{11.32}$$

where the width of the scattering depends on the diffusion jump geometry and has the low-Q limit $f_s(\mathbf{Q}) = D_s Q^2$, where D_s is the macroscopic diffusion constant. A more detailed discussion of the incoherent scattering from diffusional modes is given in Chapter 10 with reference to hydrogen motion in metals.

11.2.2. Simple Model Cases

In order to illustrate the difference between coherent and incoherent quasi-elastic scattering, it is instructive to go through a simple example[22] of the random motion of vacancies in a Bravais lattice with N_c sites occupied by N_p identical particles with unit scattering length $\bar{b}_s = 1$ and N_v vacancies ($N_v + N_p = N_c$). This is a generalized version of the Chudley–Elliot model.[25] It is assumed that the transit time is short compared to the residence time, and that jumps are uncorrelated. We consider the two limits of $N_v \ll N_p$ and $N_p \ll N_v$. In the first case, the diffuse part of the intermediate scattering function [Eqs. (11.5) and (11.16)] can be written as a sum over the vacant sites R_v in the lattice:

$$I_c^{\text{diff}}(\mathbf{Q}, t) = \frac{1}{N_c} \sum_{R_v, R_{v'}} \langle e^{-i\mathbf{Q} \cdot \mathbf{R}_v(0)} e^{i\mathbf{Q} \cdot \mathbf{R}_{v'}(t)} \rangle e^{-2W(\mathbf{Q})}, \quad (11.33)$$

where $e^{-2W(\mathbf{Q})}$ is the Debye–Waller factor for the particles. The random vacancy motion can be described through the probability that a given vacancy has made n jumps by time t,

$$P_v(n, t) = \frac{1}{n!} \left(\frac{|t|}{\tau_v}\right)^n e^{-|t|/\tau_v}, \quad (11.34)$$

and by the jump geometry factor

$$F_v(\mathbf{Q}) = \sum_{k=1}^{N_j} P_k e^{i\mathbf{Q} \cdot \mathbf{R}_k}, \quad (11.35)$$

where $\mathbf{R}_k$ spans the N_j possible sites to which the vacancy may jump with probability P_k. The characteristic time τ_v is the average vacancy residence time. In the following scattering functions, S and I, the superscripts "v" and "p" denote, respectively, diffuse scattering in the limits of few vacancies and few particles. We can now write

$$e^{i\mathbf{Q} \cdot \mathbf{R}_v(t)} = e^{i\mathbf{Q} \cdot \mathbf{R}_v(0)} \sum_{n=0}^{\infty} P_v(n, t) F_v^n(\mathbf{Q})$$

$$= e^{i\mathbf{Q} \cdot \mathbf{R}_v(0)} e^{-[1 - F_v(\mathbf{Q})]|t|/\tau_v}, \quad (11.36)$$

so that

$$I_c^v(\mathbf{Q}, t) = I_c^v(\mathbf{Q}, 0)e^{[1 - F_v(\mathbf{Q})]|t|/\tau_v}, \quad (11.37)$$

and hence

$$S_c^v(\mathbf{Q}, E) = S_c^v(\mathbf{Q}) \frac{\hbar}{2\pi} \frac{2\tau_v[1 - F_v(\mathbf{Q})]}{\hbar^2[1 - F_v(\mathbf{Q})]^2 + (E\tau_v)^2}. \quad (11.38)$$

In the dilute particle case $N_p \ll N_c$, we find similarly

$$S_c^p(\mathbf{Q}, E) = S_c^p(\mathbf{Q}) \frac{\hbar}{2\pi} \frac{2\tau_p[1 - F_p(\mathbf{Q})]}{\hbar^2[1 - F_p(\mathbf{Q})]^2 + (E\tau_p)^2}. \quad (11.39)$$

The incoherent scattering follows from

$$\begin{aligned}I_{\text{self}}(\mathbf{Q}, t) &= \frac{1}{N_p} \sum_j \langle e^{i\mathbf{Q} \cdot \mathbf{R}_j(0)} e^{-i\mathbf{Q} \cdot \mathbf{R}_j(t)} \rangle \\ &= e^{-2W(\mathbf{Q})} e^{-[1 - F_p(\mathbf{Q})]|t|/\tau_p}, \quad (11.40)\end{aligned}$$

and we get

$$S_i(\mathbf{Q}, E) = e^{-2W(\mathbf{Q})} \frac{\hbar}{2\pi} \frac{2\tau_p[1 - F_p(\mathbf{Q})]}{\hbar^2[1 - F_p(\mathbf{Q})]^2 + (E\tau_p)^2}. \quad (11.41)$$

Thus, in the limit $N_p \ll N_c$, one finds the Vineyard approximation,[26]

$$S_c^p(\mathbf{Q}, E) = S_c^p(\mathbf{Q})S_i(\mathbf{Q}, E)e^{2W(\mathbf{Q})}. \quad (11.42)$$

Returning to the other limit $N_v \ll N$, we first note that $F_v(\mathbf{Q}) = F_p(\mathbf{Q})$, and since $N_p \tau_p^{-1} = N_v \tau_v^{-1}$ we get

$$S_c^v(\mathbf{Q}, E) = S_c^v(\mathbf{Q}) \frac{N_v}{N_p} S_i\left(\mathbf{Q}, \frac{N_v}{N_p} E\right) e^{2W(\mathbf{Q})}, \quad (11.43)$$

that is, the coherent scattering has the same spectral shape as the incoherent scattering, but the energy scale is increased by the factor N_p/N_v. Hence, in such cases we expect the coherent quasi-elastic scattering to be considerably broader in energy than the incoherent scattering.

11.3. Model Systems

In the discussion of the structural methods, we mainly consider the Ag ion conductors α-AgI, α-Ag$_3$SI, and β-Ag$_2$S (also an electronic conductor) and examples from the large group of ion conductors with the fluorite structure (CaF$_2$ structure).

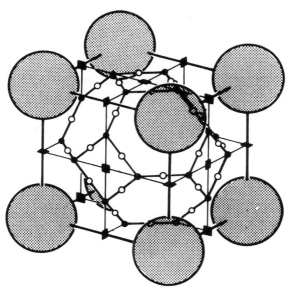

FIG. 1. The unit cell of the bcc crystal structure of α-AgI, α-Ag$_3$SI, and β-Ag$_2$S. The large shaded circles are the iodine or sulfur atoms. The small symbols refer to the 42 possible silver ion sites first suggested by Strock:[27] the octahedrally coordinated 6 b sites (■); the tetrahedrally coordinated 12 d sites (●) and the triangularly coordinated 24 h sites (○). (After Shapiro and Reidinger.[20])

Figure 1 shows the bcc crystal structure (space group $Im3m$) of α-AgI, α-Ag$_3$SI, and β-Ag$_2$S as first suggested from x-ray powder diffraction studies by Strock,[27] Reuter and Hardel,[28] and Rahlfs,[29] respectively. The first-order structural phase transitions to the fast ion conducting phases occur at 147°C for α-AgI, 242°C for Ag$_3$SI, and 177°C for β-Ag$_2$S. As indicated, there are 42 available sites in the unit cell, comprising octahedrally coordinated 6 b sites at $0\frac{1}{2}\frac{1}{2}$, tetrahedrally coordinated 12 d sites at $0\frac{1}{4}\frac{1}{2}$, and triangularly coordinated 24 h sites at $0xx$ ($x \approx \frac{3}{8}$), among which the Ag ions are distributed. Since each unit cell contains two, three, and four Ag ions in α-AgI, α-Ag$_3$SI, and β-Ag$_2$S, respectively, there is considerable disorder in the structure of these systems, although steric considerations preclude some of the statistically possible configurations. The x-ray powder diffraction studies suggested an even distribution among the 42 available sites in the unit cell in all three systems. Also the observed activation energies of Ag-ion migration are qualitatively similar and in the range of 0.05–0.1 eV (see Ref. 30 for an overview). These very low values of the activation energy are on the order of thermal energies ($k_B T = 0.04$ eV at 200°C), indicating that the conduction process may be almost liquid-like at higher temperatures.

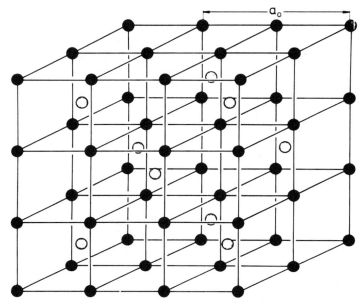

FIG. 2. The regular fluorite (CaF_2) type of fcc crystal structure: the cation sites (○) and the regular anion sites (●). (After Hutchings et al.[22])

The ordered fluorite structure (space group $Fm3m$) is shown in Fig. 2. The cations (or in antifluorite systems, which also exist, the anions) form a simple fcc lattice, whereas the anions occupy the sites corresponding to a simple cubic lattice obtained by displacements $\pm(\frac{1}{4}, \frac{1}{4}, \frac{1}{4})$ relative to the cations. This arrangement leaves every second simple cube empty and open for displacements and diffusion of the anions.

In many fluorite systems fast ion conduction results from a thermal disordering process within the fluorite lattice itself (e.g., the fluorine ion conductors CaF_2, BaF_2, and β-PbF_2), while in other cases (e.g., in the oxygen conductor ZrO_2), the disordered fluorite phase is established via a structural phase transition. An interesting property of the fluorite lattice is the accessibility to doping (in fluorine ion conductors, for example, with LaF_3 and UF_4), which may stabilize the fast ion conducting phases down to considerably lower temperatures. The thermal properties of the disordering process of pure and doped systems have been subject of several studies (see, e.g., Schröter and Nölting,[31] Oberschmidt,[32] Ouwerkerk et al.,[33] and Andersen et al.[34]). These indicate that the diffuse Faraday transition should be regarded as a cooperatively enhanced activated process and not as a regular phase transition.

11.4. Diffraction Studies

11.4.1. General Consideration

The formal neutron scattering cross section for coherent elastic scattering at the Bragg positions in reciprocal space was given in Eqs. (11.17) and (11.18). For simplicity, the thermal factors were given in the harmonic approximation as the Debye-Waller factors $\exp[-2W_{ds}(\mathbf{Q})]$. Since the mobile ions perform large amplitude motion, it is more appropriate to replace the Debye-Waller factors by more general thermal factors, $T_{ds}(\mathbf{Q})$, for ion species s on site d. Then the Bragg scattering expression becomes

$$\left(\frac{d\sigma}{d\Omega}\right)_{c,\,el} = \frac{N_c(2\pi)^3}{v_0}\left|\sum_{d,s} c_{ds}\bar{b}_s T_{ds}(\mathbf{Q})e^{i\mathbf{Q}\cdot\mathbf{d}}\right|^2 \sum_\tau \delta(\tau - \mathbf{Q}). \quad (11.44)$$

The structural information about the scattering length densities and thermal motion parameters is contained in the structure factors

$$F(\mathbf{Q}) = \sum_{d,s} c_{ds}\bar{b}_s T_{ds}(\mathbf{Q})e^{i\mathbf{Q}\cdot\mathbf{d}} \quad (11.45)$$

at the Bragg positions $\mathbf{Q} = \tau$. In the harmonic approximation, the thermal parameter (on a cubic site for simplicity) is given in terms of the mean square displacement amplitude $\langle u_{ds}^2 \rangle$ as

$$T_{ds}(\mathbf{Q}) = \exp[-W_{ds}(\mathbf{Q})], \quad (11.46)$$

where

$$W_{ds}(\mathbf{Q}) = \langle u_{ds}^2 \rangle Q^2/6. \quad (11.47)$$

For a classical oscillator

$$\langle u_{ds}^2 \rangle = 3k_B T/M_s \omega_{ds}^2, \quad (11.48)$$

where M_s is the mass and ω_{ds} the Einstein frequency for an ion of type s on site d. In a shallow potential ω_{ds} is small, and in addition to the possible site disorder ($c_{ds} < 1$), this causes a strong reduction in the effective coherent scattering amplitudes $c_{ds}\bar{b}_s \exp[-W_{ds}(\mathbf{Q})]$ for the mobile ions. Accordingly, the immobile ions in the lattice are dominant in the structure factors. Further, since the typical barrier for fast ion diffusion is comparable to the thermal energy, it is clear that the effective single-ion potential for the moving ions is inadequately accounted for in the harmonic approximation. Strong anharmonic terms or even higher order contributions to the potential functions are required for a proper description of the thermal motion. As discussed in the following, models that account for thermal motion in such potentials can be constructed but require a large number of parameters. It is obvious that the low signal-to-background ratio of powder Bragg intensities (as a rule of thumb approximately 500–1000 times smaller than for single crystals) prevents the use of powder data for this purpose and in most cases even for a determination of anisotropic harmonic parameters (the Debye–Waller factors). Another problem in powder diffration studies is the sintering and recrystallization processes that are especially pronounced in fast ion conductors due to the relatively high temperatures and their plastic properties.

11.4.2. Data Correction

In the determination of structure factors from measured Bragg intensities, the usual problems arising from absorption and extinction of the neutron beam are present. Absorption is easily taken care of for well-characterized geometric shapes, provided the sample does not contain isotopes with very high absorption cross sections. In single crystals extinction is a more serious problem. Although methods for corrections are available for well–characterized sample geometries,[35,36] annealing may reduce the mosaic spread of the crystal and thereby increase the extinction (see, e.g., Dickens et al.[37]). Checks by recurrence to selected standard Bragg peaks during data collection are therefore very important.

Thermal diffuse scattering gives rise to another crucial problem to which too little attention has been paid in studies of fast ion conductors. TDS is the part of the background intensity caused by the inelastic phonon scattering processes. Contributions from optical phonons and multiphonon processes may be corrected for because they are slowly varying as a function of wave vector transfer $\mathbf{Q}$, in contrast to the scattering contributions from acoustic phonons, which peak at the Bragg positions. Qualitatively this may be seen from the scattering cross sections for one-phonon scattering, which in the appropriate classical limit for fast ion conductors are proportional to

$$k_B T/[E_{ph}^m(\mathbf{q})]^2, \qquad (11.49)$$

where $E_{ph}^m(\mathbf{q})$ is the phonon energy of the mth phonon branch at wave vector $\mathbf{q} = \mathbf{Q} - \boldsymbol{\tau}$. For the low lying acoustic modes

$$E_{ph}^m(\mathbf{q}) = \hbar q v_m, \qquad (11.50)$$

which results in a divergence in the cross sections for $q \to 0$ (v_m is the sound velocity). With a finite instrumental resolution, these modes may give significant contributions to the scattering intensity around the Bragg points at high temperatures.

Accurate calculations of the TDS contributions to the Bragg intensities may be accomplished only on the basis of detailed knowledge about the resolution properties and the phonon spectrum, and in general only by numerical calculations. Waller[38] first considered the contributions from

TDS. Nilsson[39] performed calculation of the TDS corrections to the measured integrated Bragg intensities of cubic crystals from ω scans (transverse scans) assuming an infinite vertical aperture. Cooper and Rouse[40] performed similar calculations for standard $\omega - 2\theta$ scans (longitudinal scans) and suggested an approximate analytical method of evaluation in which the volume in reciprocal space covered in the scan is assumed to be a sphere. Further approximations include almost isotropic elastic constants (although this requirement may be partially relaxed[41]) and negligible resolution effects from the divergence of the incident beam and the mosaic spread of the crystal. It is also important for the validity of the results given in the following that the neutron velocity is considerably higher than the sound velocity.[42]

The results obtained by Cooper and Rouse are instructive. They give the total scattering intensity as

$$I = I_0(1 + \alpha) \tag{11.51}$$

in terms of the true Bragg intensity I_0 and the correction coefficient α given by

$$\alpha = \frac{4\pi k_B T \kappa \sigma}{3\lambda^3} \sin 2\theta \sin^2 \theta, \tag{11.52}$$

where 2θ is the scattering angle, λ the neutron wavelength, and κ a function of the elastic constants. The factor σ contains the integral over the appropriate volume in reciprocal space, which by approximation to a sphere of radius q_{max} around the Bragg point τ equals

$$\sigma = \frac{2\lambda}{\pi \sin 2\theta} q_{max}. \tag{11.53}$$

Accordingly, Eq. (11.52) may be converted to the simple form

$$\alpha = \frac{k_B T \kappa q_{max}}{6\pi^2} \tau^2. \tag{11.54}$$

If α is small, the correction amounts to

$$I = I_0 e^\alpha, \tag{11.55}$$

and the main influence from TDS is an effective reduction in the Debye–Waller factor [compare Eqs. (11.45)–(11.48)]. Accordingly, a refinement of diffraction data without considering TDS may be successful and give correct positional parameters but unrealistic mean square amplitudes. As an example, Cooper[42] discusses TDS corrections for diffraction data obtained from the fluorine conductor BaF_2 at 400°C.

As a test of proper data corrections, a comparison of symmetry equivalent reflections is important. Proper sample quality and alignment, as well as extinction and absorption corrections, may be examined in this way. The instrumental scale factor may also be used. When the influence of the lattice expansion (see Bacon,[43] p. 67 and Dickens et al.[37]) has been considered, it should be temperature independent. For the reliability of data refinements based on more elaborate structural models of the high-temperature, fast ion conducting phases, it is important to compare these with scale factors obtained at lower temperatures where the structure is more regular. Cooper et al.[44] have demonstrated that a proper treatment of extinction and TDS leads to temperature-independent scale factors for BaF_2 from room temperature up to 600°C. It should be mentioned, however, that the onset of fast ion conduction in BaF_2 occurs at about 975°C.

11.4.3. Fourier Synthesis

The method of structure determination by Fourier synthesis is based on the structure factors $F(\tau)$ being simply the Fourier transform of the periodic part of the nuclear scattering amplitudes in the crystal, that is, the average distribution of coherent scattering amplitudes within the unit cell, also called the probability density function (PDF) $\rho(\mathbf{r})$. In Eq. (11.45) the structure factor is expressed formally in terms of a distribution of ions around the equilibrium sites d, but it is generally true that the structure factor is the Fourier transform of $\rho(\mathbf{r})$. The PDF may therefore be derived from an inverse Fourier transform of the structure factors

$$\rho(\mathbf{r}) = \frac{1}{v_0} \sum_\tau F(\tau) \exp(-i\tau \cdot \mathbf{r}), \tag{11.56}$$

where v_0 is the unit cell volume. From the experimental data the norm-squared structure factor $|F_0(\tau)|^2$ may be determined for a finite number of reciprocal lattice vectors τ. In order to derive $\rho(\mathbf{r})$, it is necessary that the phase factors $\exp[i\phi(\tau)]$ are known. However, for centrosymmetric systems $\exp(i\phi(\tau)) = \pm 1$, and since the immobile ions usually are dominant in the scattering, $\phi(\tau)$ may often be determined from the calculated structure factors based on these ions alone. Alternatively, the phases may be determined from a proposed model, the consistency of which should be tested by comparison with the derived densities.

The finite range of measurable Bragg peaks restricts the details of information that may be obtained. From reciprocal lattice vectors smaller than τ_{max}, the corresponding positional accuracy by which the scattering densities may be derived has been given by James[45] in the approximate form

$$\Delta r = 0.715 \times 2\pi/\tau_{max}. \tag{11.57}$$

Despite their restrictions, the different kinds of Fourier syntheses are very useful as a guide to the ion densities in the unit cell. Direct Fourier synthesis will give a density distribution that is strongly dominated by the immobile ions. If not known in advance, the immobile ion positions may be determined this way, but the details of the mobile ion distributions may be more or less disguised. From their Fourier synthesis on the fast Ag ion conductor β-Ag$_2$S at 186°C, Cava et al.[46] quote the maximum Ag ion scattering density to be only 14.6% of the corresponding one for the S ions. As convincingly demonstrated by Cava et al.[46,47] in two papers based on single-crystal neutron diffraction data on the fast Ag ion conductors α-AgI and β-Ag$_2$S, this problem may be overcome by use of partial Fourier synthesis. In this method the structure factors calculated for the ions, which have a known structure, are subtracted from the observed structure factors, $F_0(\tau)$. If, for example, a proper model for the I ions in α AgI is chosen, the difference (or "partial") density

$$\Delta\rho(\mathbf{r}) = \frac{1}{v_0} \sum_\tau [F_0(\tau) - F_c(\tau)] \exp(i\phi(\tau)) \exp(-i\tau \cdot \mathbf{r}) \qquad (11.58)$$

should be representative for the Ag ions alone. In Eq. (11.58), $F_c(\tau)$ are calculated structure factors for the immobile I ions alone, and $\phi(\tau)$ are the phases of the structure factors including all ions.

The results of two extremes of the partial Fourier analysis performed by Cava et al. are shown in Fig. 3 as contour maps of the Ag ion scattering densities in the $(0, y, z)$ plane of α-AgI at 160°C and β-Ag$_2$S at 325°C, obtained from Eq. (11.58) by subtraction of suitable models for the I and S ions. The structural difference is obvious: α-AgI has density maxima at the 12 d sites, and in contrast to the x-ray powder diffraction results,[27] significant minima at the 6 b sites. Qualitatively these structural features are maintained up to the highest temperatures studied (300°C). The Ag ion distribution in β-Ag$_2$S at 325°C is far more delocalized with very high densities in (100) "bands" of direct space passing through the 6 b and 12 d sites. At lower temperatures the distribution gets a little more localized with weak maxima at both the 6 b and 12 d sites.

Neutron diffraction studies have not been performed on α-Ag$_3$SI, but the results obtained by Perenthaler et al.[48] from single-crystal, x-ray diffraction data confirm the relation to α-AgI and β-Ag$_2$S, and show that the structure of α-Ag$_3$SI is intermediate to these systems. The results of their direct Fourier synthesis [based on the x-ray analog of Eq. (11.56)] of data obtained at 300°C is shown in Fig. 4. The strong dominance of the scattering from the immobile S and I ions at the origin, also mentioned to be present in the direct Fourier synthesis on β-Ag$_2$S, is clearly seen, but the density distribution of the

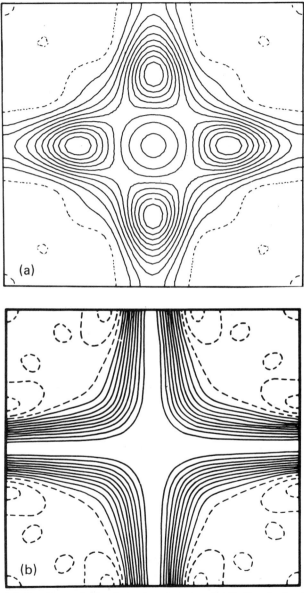

FIG. 3. Partial Fourier synthesis of the Ag scattering distribution in the (100) plane of (a) α-AgI at $T = 160°C$ and (b) β-Ag$_2$S at $T = 325°C$. The scattering contributions from I and S ions have been subtracted. Contour intervals are (a) 3.8×10^{-15} cm/Å^3 and (b) 4.8×10^{-15} cm/Å^3. Dashed lines indicate negative scattering densities. (After Cava et al.[46,47])

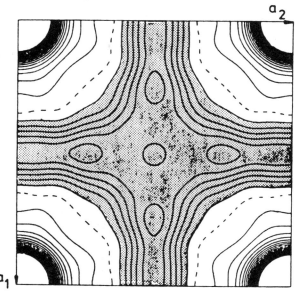

FIG.4. Direct Fourier synthesis of the x-ray scattering density in the (100) plane of α-Ag$_3$SI at $T = 300°C$. (After Perenthaler et al.[48])

Ag ions may be resolved to the same degree as in Fig. 3. From the density maps shown in Figs. 3 and 4, it is obvious that there are significant Ag densities around the 24 h sites. Whether this density results from actual equilibrium sites in the structure or from large amplitude anharmonic thermal vibrations on the 6 b and 12 d sites cannot be judged, due to the finite resolution in the density maps given by Eq. (11.57). Figure 5 shows the partial density of Ag ions in α-AgI at 160°C obtained by subtraction of the scattering contributions from the I ions and the harmonic part of the Ag ion distributions. The result is indicative of additional equilibrium sites at the bridging points [24 h symmetry] between the tetrahedral 12 d sites. However, repeated least-squares refinements showed that the scattering is better accounted for by inclusion of third- and fourth-order thermal vibrational parameters for the Ag ions at the 12 d tetrahedral sites.

Significance tests are important in particular for model refinements including several higher order terms in a series expansion of the thermal parameters. This may lead to strong correlations between the parameters, and thereby to ambiguity and doubt as to the physical relevance of the results. A specific model may be tested by difference Fourier synthesis, which, following Cava et al.,[46,47] is defined as a calculation by Eq. (11.58) of the final model structure Factors $F_c(\tau)$ relative to the observed ones $F_0(\tau)$, and

FIG. 5. The partial scattering density of Ag ions in the (100) plane of α-AgI at $T = 160°C$ obtained by subtraction of the I ion and the harmonic part of the Ag ion scattering contributions. The contour intervals are 9.1×10^{-16} cm/Å (cf. Fig. 3a). (After Cava et al.[47])

may therefore be used to give an error map of the model. In Fig. 6 an example is shown of a difference or error map resulting from a model refinement to data from β-Ag$_2$S at 186°C. It should be mentioned that the terminology used in the Fourier methods is by no means standard. Koto et al.[49] consider what they call difference maps between difference Fourier maps in their x-ray studies on the defect structure of fast ion conducting β-PbF$_2$ in order to reduce systematic errors. In this terminology they determine the defective part of the structure from the difference between the partial PDFs obtained at higher temperatures and the corresponding ones at room temperature, where the structure is ordered (we recall that fast ion conduction is established in β-PbF$_2$ without change of crystallographic symmetry). The partial PDF is calculated from the experimental data by subtraction of a model for the regular β-PbF$_2$ structure.

Despite the obvious advantages of Fourier syntheses, relatively few structural studies have been performed by these methods. Clearly the need for single crystals for this purpose restricts the possibilities, because for many systems it is difficult to grow single crystals that survive the structural phase transitions on cooling to room temperature. Exceptions are the fluorite systems, where the crystalline phase with fast ion conduction is often stable

Fig. 6. Difference Fourier synthesis in the (100) plane of the observed scattering densities relative to the final structural model for β-Ag_2S at $T = 186°C$. The contour intervals are 3.1×10^{-16} cm/Å^3. (After Cava et al.[46])

between room temperature and the melting point. Catlow et al.[50] have used partial Fourier synthesis (they use the terminology difference Fourier sections for their maps) to establish the sites for the defective ions in LaF_3- and ErF_3-doped CaF_2. On the other hand, Dickens et al.[37] note that extinction problems in β-PbF_2 reduce the applicability of Fourier methods to establish the thermally generated defect structure in the fast ion conducting phase ($T \geq 438°C$).

Another example is the study by Horiuchi et al.[51] on single-crystal 0.84 $ZrO_2 \cdot 0.16$ Y_2O_3. Their time-of-flight neutron diffraction data were analyzed by partial (the authors use the word "difference") Fourier synthesis with the purpose of elucidating the change in defect structure as a result of increased temperature and an applied field.

Powder diffraction data may be analyzed by Fourier synthesis, but the significantly lower signal-to-background ratio compared with single-crystal data further reduces the Q range of investigation and thereby the resolution in the Fourier maps. In addition, degeneracies in the Bragg-peak positions of powder patterns require models to distinguish the intensity contributions.

The information level is therefore quite limited as the results obtained by Axe et al.[52] from Fourier analysis of powder data on BaF_2 and β-PbF_2 exemplify. They deduced that at elevated temperatures up to 20% of the F ions leave their regular sites and accumulate near the octahedral interstitial sites, but they were unable to resolve the details of the defect structure. Improvements were obtained if combinations of Fourier synthesis and least-squares refinements were used (see Shapiro and Reidinger[20]).

For completeness we will also mention the work by French groups (see, e.g., Soubeyroux *et al.*[53,54]) on several fluoride and oxide–fluoride systems with fluorite structure where Fourier analysis of neutron powder diffraction data was used to determine defect structure sites and possible diffusion pathways.

For evaluations based on partial and difference Fourier synthesis, it is a prerequisite that the parameters used in or refined from the models are of physical relevance, and large negative scattering densities extending over regions larger than the spatial resolution given by Eq. (11.57) should not occur. It should be recalled that the reliability of both the model parameters and the data correction procedures, like extinction, absorption, and TDS, may be tested by Fourier synthesis.

11.4.4. Model Refinements

In traditional crystallographic studies, the ions are assumed to occupy well-defined equilibrium sites, and they perform thermal vibrations in a harmonic potential, which, depending on the site symmetry, may be anisotropic. Assuming the harmonic approximation for the vibration of the ions around their equilibrium sites, the validity of Eqs. (11.17) and (11.18) may be derived exactly from the formal scattering cross sections by an expansion of the deviations from the average ion positions **d** in phonon coordinates (see Part A, Chapter 1 or standard textbooks such as Squires[21]). The general expression for the Debye–Waller factor for any site symmetry may be given in matrix notation as $\exp[-2W_{ds}(\mathbf{Q})]$ with

$$W_{ds}(\mathbf{Q}) = \tfrac{1}{2}\mathbf{Q}^t \sigma_{ds} \mathbf{Q}, \qquad (11.59)$$

where σ_{ds} is the dispersion matrix for an ion of type s on site d. By a suitable coordinate transformation, it may be diagonalized and the diagonal components of σ_{ds} are then simply the mean square displacements of the ion along the principal axes,

$$\sigma_{ds}^{ii} = \langle u_{ds}^{ii\,2} \rangle, \qquad i = 1, 2, 3. \qquad (11.60)$$

The Debye–Waller factor enters as a reduction of the effective coherent elastic scattering intensities, because the ion distributions around the equilibrium positions are uncorrelated on a long-time scale. As already mentioned the general form of the structure factors [Eq. (11.45)] is the Fourier transform of the distribution of coherent scattering densities $\rho(\mathbf{r})$ in the unit cell

$$F(\mathbf{Q}) = \int_{\text{unit cell}} \rho(\mathbf{r}) e^{i\mathbf{Q}\cdot\mathbf{r}}\, d\mathbf{r}. \qquad (11.61)$$

It is easy to show that Gaussian distributions of coherent scattering densities from ions of type s around the sites d in the unit cell with occupation probabilities c_{ds} lead to the structure factors in Eqs. (11.45) and (11.46) with $W_{ds}(\mathbf{Q})$ given by Eq. (11.59). To prove this we choose for simplicity a cartesian coordinate system with axes along the principal directions of the thermal motions around side d. The Fourier transform of

$$\rho_{ds}(\mathbf{r}) = \frac{c_{ds}\bar{b}_s}{(2\pi)^{3/2}} \prod_{i=1}^{3} \frac{1}{\sqrt{\sigma_{ds}^{ii}}} \exp\left[-\frac{(r_i - d_i)^2}{2\sigma_{ds}^{ii2}}\right] \quad (11.62)$$

gives the structure factor

$$F_{ds}(\mathbf{Q}) = c_{ds}\bar{b}_s \exp[-W_{ds}(\mathbf{Q})] \exp(i\mathbf{Q}\cdot\mathbf{d}), \quad (11.63)$$

where $W_{ds}(\mathbf{Q})$ has the diagonalized matrix form of Eq. (11.59). For practical use Eq. (11.62) should be expressed in terms of the general trivariate Gaussian distribution functions

$$\rho_{ds}(\mathbf{r}) = \frac{c_{ds}\bar{b}_s}{(2\pi)^{3/2}} [\det \mathbf{p}_{ds}]^{1/2} \exp\left[-\frac{1}{2}(\mathbf{r}-\mathbf{d})^t \mathbf{p}_{ds}(\mathbf{r}-\mathbf{d})\right], \quad (11.64)$$

where $\mathbf{p}_{ds}$ is the inverse of the dispersion matrix σ_{ds}.[55] Without loss of generality we may then write the unit cell probability density function as

$$\rho(\mathbf{r}) = \sum_{d,s} \rho_{ds}(\mathbf{r}) \quad (11.65)$$

and obtain the desired result.

It is instructive to note that an ion moving in an effective single-ion harmonic potential has a Gaussian probability density function. In this case, the Hamiltonian is given by

$$H(\mathbf{p}, \mathbf{r}) = p^2/2M + V(\mathbf{r}). \quad (11.66)$$

Being concerned only about the spatial variations of the ion distribution, the ionic probability density function for one ion in the unit cell is given from classical statistical mechanics as

$$f(\mathbf{r}) = \frac{\exp[-V(\mathbf{r})/k_B T]}{\int_{\text{unit cell}} \exp[-V(\mathbf{r})/k_B T]\, d\mathbf{r}}. \quad (11.67)$$

If the potential function for an ion of type s is an anisotropic harmonic oscillator around site d, then

$$V_{ds}(\mathbf{r}) = \tfrac{1}{2}(\mathbf{r}-\mathbf{d})^t \alpha_{ds}(\mathbf{r}-\mathbf{d}), \quad (11.68)$$

where α_{ds} is a force constant matrix. The coherent scattering density functions are given by

$$\rho_{ds}(\mathbf{r}) = c_{ds}\bar{b}_s f_{ds}(\mathbf{r}), \quad (11.69)$$

and the result of Eq. (11.64) is obtained by substituting

$$\mathbf{p}_{ds} = \mathbf{a}_{ds}/k_B T \qquad (11.70)$$

and performing the integral in the denominator of Eq. (11.67). Similar results may be found from a quantum statistical treatment.[21]

Early crystallographic studies and refinements of powder diffraction data were performed with models based on the harmonic approximations (see, e.g., Thomas[56]). From such studies and those to be discussed, where small anharmonic corrections to the thermal vibrations are included, it is clear that the vibrational amplitudes are very large in fast ion conducting phases.

Dawson et al.,[57] Cooper et al.,[44] and Willis[58] realized the significance of higher order terms in the potentials for anion motion in fluorite crystals like UO_2, ThO_2, CeO_2, CaF_2, and BaF_2, even at temperatures where fast ion conduction is not observed. The cation sites in the fluorite lattice (see Fig. 2) have cubic symmetry and therefore isotropic harmonic temperature factors. However, it is obvious that the preferred direction of motion on the noncentrosymmetric anion sites (site symmetry $\bar{4}3m$) is along the (111) directions, corresponding to the body diagonals away from the cation sites. The simplest way to account for this motion is to introduce a cubic term in the potential function:[57,58]†

$$V_{ds}(\mathbf{u}) = \tfrac{1}{2}\alpha_{ds}(u_1^2 + u_2^2 + u_3^2) + \beta_{ds} u_1 u_2 u_3, \qquad (11.71)$$

where $\mathbf{u} = \mathbf{r} - \mathbf{d}$. If the cubic term in the potential expansion is small, the probability density function based on Eqs. (11.67) and (11.71) may be expanded:

$$\rho_{ds}(\mathbf{u}) = \frac{c_{ds}\bar{b}_s \alpha_{ds}^{3/2}}{(2\pi)^{3/2}(k_B T)^{3/2}} \exp\left[-\frac{1}{2}\frac{\alpha_{ds}}{k_B T}(u_1^2 + u_2^2 + u_3^2)\right]\left(1 - \frac{\beta_{ds}}{k_B T} u_1 u_2 u_3\right), \qquad (11.72)$$

and the corresponding contribution to the structure factor is

$$F_{ds}(\mathbf{Q}) = c_{ds}\bar{b}_s \exp\left(-\frac{1}{2}\frac{k_B T}{\alpha_{ds}^2} Q^2\right)\left[1 - i\frac{\beta_{ds}}{\alpha_{ds}^3}(k_B T)^2 Q_1 Q_2 Q_3\right]. \qquad (11.73)$$

Dickens et al.[37] have used the refinement method in an extensive study of the crystallographic properties of the fluorite β-PbF_2 at several temperatures both below and above the transition temperature ($T_c = 711$ K). They recognized the conclusion drawn by Cooper et al.[44] (see also Ref. 58) from fluorites below T_c, that a cubic term in the temperature factors for the

---

† Although in the fluorites and the other examples mentioned only one type of ion can reside on a given site d, we maintain the general formulation.

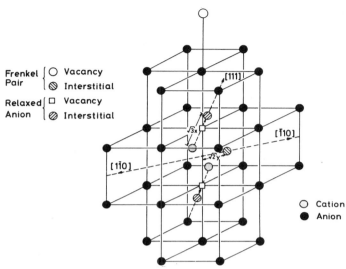

FIG. 7. Interstitial (I) and relaxed (R) types of equilibrium sites in the fluorite lattice being populated in the temperature range around and above the Faraday transition. The defect structure shown, consisting of one Frenkel pair and two relaxed ions, corresponds to the so-called 3 : 1 : 2 defect clusters determined from quasi-elastic diffuse neutron scattering (Section 11.5). (After Hutchings et al.[22])

fluorine ions on the regular sites suffices for a proper refinement. However, at temperatures approaching T_c and above they found evidence for significant site disorder due to population of the empty cube interstitial sites shown in Fig. 7. Dickens et al. distinguish between I-site interstitials, which are true interstitials due to formation of Frenkel pairs, and R-site interstitials, which are fluorine ions relaxed away from their regular sites in the (111) direction (see Fig. 7). Altogether they considered eight models that were tested against their experimental data obtained at 11 temperatures. The principal result of their analyses indicates that the relative occupation on I and R sites is between 0.5 and 1.0, corresponding to a mixture of the so-called 3 : 1 : 2 and 4 : 2 : 2 cluster defects, which emerge more directly from the quasi-elastic diffuse scattering studies discussed in Section 11.5. Figure 7 shows the 3 : 1 : 2 cluster configuration; a 4 : 2 : 2 cluster contains two of the Frenkel pairs together with two relaxed ions. The two models contain 10 adjustable parameters among which the most significant is the fractional number of vacancies D in the regular fluorine lattice. A correction for TDS was calculated at each temperature by use of the method developed by Cooper and Rouse.[40] The temperature variation of D is shown in Fig. 8, where the extension of the symbols indicates the combined uncertainties resulting from the fit to each

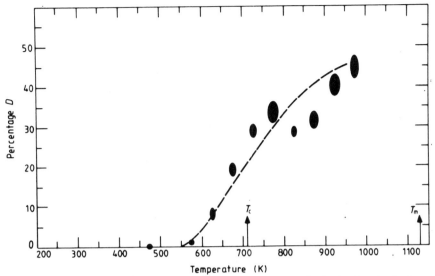

FIG. 8. The temperature variation of the density D of vacancies in the fluorine lattice of β-PbF$_2$ determined from diffraction studies; D results from formation of both I and R interstitials. (After Dickens et al.[37])

of the two models and their mutual disagreement. Consistent values of very large vacancy concentrations are observed. However, such large vacancy concentrations are also found from thermodynamic model calculations based on specific heat data (Andersen et al.[34]). Another result of interest is the temperature variation of the Debye-Waller factors for the regular Pb^{2+} and F$^-$ sites shown in Fig. 9. Although significant site disorder is allowed for in the models, there is still an anomalous increase around T_c. The Pb^{2+} ions, which are assumed to be immobile, are also influenced by the onset of fast ion conduction, probably because they perform large amplitude vibrations and hence allow for the ion diffusion process.

Bachmann and Schulz[59] analyzed the data of Dickens et al.[37] from a different point of view. They considered to what extent the experimental data could be interpreted by use of only the regular sites in the fluorine lattice but with the inclusion of higher order expansions in the temperature factors. The potential expansion approach by Dawson et al.[57] and Willis[58] has been established to higher order in the potential functions $V(\mathbf{r})$ than the cubic one in Eq. (11.71). However, Bachmann and Schulz used the Gram-Charlier series expansion method, which focusses on the PDF rather than the potentials.[55] In this approach the anharmonic PDF on site d, $\rho_{ds}^a(\mathbf{u})$, is considered as a series expansion of the harmonic (Gaussian) distribution

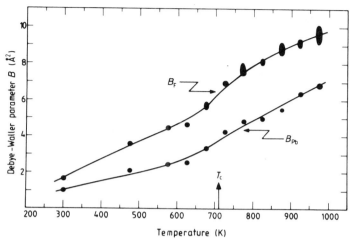

FIG. 9. The temperature variation of the isotropic Debye–Waller parameters $B = 8\pi^2 \langle u^2 \rangle / 3$ of Pb and F ions in β-PbF$_2$. (After Dickens et al.[37])

function $\rho_{ds}(\mathbf{u})$ [Eq. (11.64) with $\mathbf{u} = \mathbf{r} - \mathbf{d}$]:

$$\rho_{ds}^a(\mathbf{u}) \approx [1 - c_s^j D_j + (1/2!) c_s^{jk} D_j D_k - (1/3!) c_s^{jkl} D_j D_k D_l + (1/4!) c_s^{jklm} D_j D_k D_l D_m - \cdots] \rho_{ds}(\mathbf{u}), \quad (11.74)$$

where $D_j = \partial/\partial u_j$ is a partial differential operator, and a repeating index summation is assumed. By use of general multidimensional Hermite polynomial tensors

$$H_{\alpha_1 \cdots \alpha_n}(\mathbf{u}) = (-1)^n \exp(\tfrac{1}{2}\mathbf{u}'\mathbf{p}\mathbf{u}) D_{\alpha_1 \cdots \alpha_n} \exp(-\tfrac{1}{2}\mathbf{u}'\mathbf{p}\mathbf{u}), \quad (11.75)$$

Eq. (11.74) may be expressed as

$$\rho_{ds}^a(\mathbf{u}) \approx \hat{\rho}_{ds}(\mathbf{u})[1 + (1/3!) c_s^{jkl} H_{jkl}(\mathbf{u}) + (1/4!) c_s^{jklm} H_{jklm}(\mathbf{u}) + \cdots, \quad (11.76)$$

where $\hat{\rho}_{ds}(\mathbf{u})$ is a Gaussian distribution function chosen relative to $\rho_{ds}(\mathbf{u})$ such that the linear and quadratic terms vanish. The structure factor obtained from a Fourier transform of Eq. (11.76) is

$$F_{ds}^a(\mathbf{Q}) = F_{ds}(\mathbf{Q})[1 + (i^3/3!) c_s^{jkl} Q_j Q_k Q_l + (i^4/4!) c_s^{jklm} Q_j Q_k Q_l Q_m + \cdots], \quad (11.77)$$

where $F_{ds}(\mathbf{Q})$ is the harmonic expression given by Eq. (11.63). Neglecting terms of order higher than cubic, the formal equivalence with Eq. (11.73) may easily be established.

Bachmann and Schulz[59] used thermal parameters up to sixth order in their refinements (cf. the additional information given by Bachmann and Schulz[60]) on the data by Dickens et al.[37] They concluded that up to 673 K the experimental data are well accounted for with only the regular sites in the lattice being occupied. At higher temperatures population of sites corresponding to the I sites deduced by Dickens et al. improved the fit significantly. Attempts to include the R sites as interstitials did not lead to improvements. They concluded that the refinements alone were insufficient for a discrimination between anharmonic motion and population on the interstitial sites and suggested that the population on the I sites may be a purely mathematical aid in the refinements. Further, they suggested that a proper way to test this possibility is to consider the joint, single-particle density function for species s (fluorine ions)

$$f_s(\mathbf{r}) = \sum_d f_{ds}^a(\mathbf{r}) c_{ds}, \qquad (11.78)$$

which is calculated from the single-site particle density functions

$$f_{ds}^a(\mathbf{r}) = \rho_{ds}^a(\mathbf{r})/(c_{ds}\bar{b}_s). \qquad (11.79)$$

Since the results obtained for $f_s(\mathbf{r})$ turned out to be almost independent of the model, they concluded that the effective one-particle potential defined by inverting a generalized version of Eq. (11.67)

$$V_s(\mathbf{r}) = V_0 - k_B T \ln\{f_s(r)\} \qquad (11.80)$$

is a more unique function from which a proper particle density in the unit cell may be derived. The fluorine particle probability density function they obtain in the (110) plane of direct space is given in Fig. 10. Apart from the regular fluorine sites the fluorine density distribution shows no local maxima, indicating that the disorder is mainly large-amplitude, anharmonic thermal vibrations around these sites, with preference for motion along the (111) directions away from the Pb^{2+} ions, with some density in the regions connecting the regular sites and the I sites obtained by Dickens et al. From the estimated errors in the density functions a maximum concentration of 5% of defects, interpreted as vacancies in the regular fluorine lattice, is deduced. From Eq. (11.80) an effective potential barrier of motion along the pathway suggested in Fig. 10 is deduced. The good agreement between the value thus obtained (0.26 eV) and the activation energy derived from the ionic conductivity measurements about T_c is adduced as support for the validity of the effective one-particle potential approach. Further, they observe that the effective one-particle potential functions are essentially temperature independent. From their arguments this is a prerequisite for the absence of site disorder and thereby for the agreement between the effective potential barriers obtained from the two types of experiments.

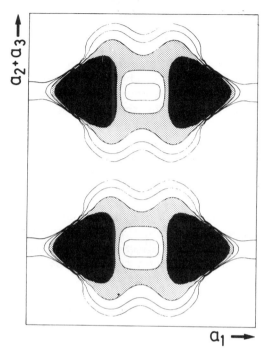

Fig. 10. The fluorine probability density function in the (110) plane of β-PbF$_2$ at $T = 773$ K determined by Bachmann and Schulz[60] from the data obtained by Dickens et al.[37]

At this point it is appropriate to include some comments on the effective one-particle potential approach, especially since there is still some disagreement with respect to the results derived from different methods of analyses of the diffraction data. Indeed, it is very interesting that agreement between the potentials derived from diffraction and from ionic conductivity studies may be obtained, not only for β-PbF$_2$ but also in several other systems.[59-62] It should be recalled, however, that the effective one-particle potential actually derives from an ensemble average over all possible configurations in phase space of the ions in the lattice (see Willis[58] for a discussion and references to such approaches). With reference to the results of the quasi-elastic diffuse neutron scattering studies discussed in Section 11.5, which give a snapshot picture of all possible configurations, it is well established that the fluorine ions in β-PbF$_2$ are located in defect configurations involving both the I and R sites on a time scale longer than a characteristic phonon vibration period. Accordingly, the effective one-particle potential function, on both regular sites and the I and R sites, must contain an average over the potentials, which may be derived from such defect configurations. It may be argued that if only small concentrations of these defect clusters are present in the fast ion conducting phase, then the effective one-particle potential functions are determined mainly by the regular site configuration. However, if the conduction process is mediated by cooperative motion, for example as a result of reorientation type of processes in the defect cluster system, as suggested by Andersen et al.,[34] the configurations decisive for the effective potential barrier for conduction are those involving the defects and not those of the regular undisturbed lattice.

Another point of concern is the role of TDS in the derivation of the effective one-particle potential functions. As already discussed, TDS gives significant contributions to the Bragg peak scattering at elevated temperatures.[44,58] When not corrected for, TDS gives an effective reduction of the Debye–Waller factors that result from the leading term of the series expansion of the PDF. As a consequence, the dominant harmonic term in the effective one-particle potential function may be significantly increased. TDS results from long wavelength phonon scattering and thereby from correlated motion due to long-range potential energy interactions. Since these interactions are difficult to break, it might be argued that corrections for TDS should not be performed if the aim of the crystallographic study is to derive effective potential barriers for ion motion. Clearly, the role of TDS needs more consideration before the connection between diffraction data and effective one-particle potential functions is satisfactorily established.

It has been mentioned by several authors that it is difficult to obtain unique results from model refinements involving both higher order thermal parameters and site disorder because of correlations between the parameters. Qualitatively, this ambiguity may be seen from inspection of Eqs. (11.78) and (11.79). If c_{ds} is small, then $\rho^a_{ds}(\mathbf{r})$ is a function giving rise to small scattering densities and quite large uncertainties in $f^a_{ds}(\mathbf{r})$ result. Therefore, putting small fractions c_{ds} of ions on site d with large-amplitude anharmonic ionic distribution functions $f^a_{ds}(\mathbf{r})$ does not influence the structure factor very much, and the reduction on the regular site may be compensated for by introducing anharmonic temperature factors. In the same way, close lying sites may be imitated by large-amplitude anharmonic vibrations. Formally, this ambiguity may be derived from a simple one-dimensional model involving two closely lying ions at sites $d = 0$ and $d = x_0$ with unit scattering length and harmonic distribution functions $\exp[-\frac{1}{2}a(x - d)^2]$. The PDF is

$$\rho(x) \propto \exp(-\tfrac{1}{2}ax^2) + \exp[-\tfrac{1}{2}a(x - x_0)^2]$$
$$= \exp(-\tfrac{1}{2}ax^2)[1 + \exp(-\tfrac{1}{2}ax_0^2)\exp(ax_0 x)]. \qquad (11.81)$$

If the last exponential is expanded and a straightforward renaming of constants is performed, Eq. (11.81) becomes

$$\rho(x) \propto A \exp(-\tfrac{1}{2}ax^2)(1 + \alpha x + \beta x^2 + \gamma x^3 + \cdots). \quad (11.82)$$

Apart from the lack of incorporated linear and quadratic terms in the harmonic distribution function, the formal equivalence with Eq. (11.76) is established. Equation (11.82) shows that occupation on two closely lying sites with only harmonic distribution functions may be expressed in terms of one site with an anharmonic distribution function. Allowing for both site disorder and anharmonic thermal parameters clearly increases the possibilities for correlations. More generally it should be noted that any probability density function in the one-dimensional unit cell may be expressed in the form of Eq. (11.82) with a fixed value of the harmonic parameter a, because the polynomial in the bracket may be expressed in terms of a complete set of orthogonal polynomials with respect to the weighting factor $\exp(-ax^2)$ in a unique way. Accordingly, allowing a to vary causes an ambiguity in the determination of the parameters.

Cava et al.[46,47] also use higher order terms in their thermal parameters in order to refine the structural data on α-AgI and β-Ag$_2$S. They use a refinement program developed by C. K. Johnson (unpublished), which allows thermal parameters up to fourth order to be incorporated in the models, but do not include TDS in the data corrections. In order to reduce the correlations between the thermal parameters of order p and $p + 2$, a variable α is introduced, and an expansion in terms of orthogonalized Hermite polynomials $H_{jkl}(\alpha, \mathbf{Q})$ and $H_{jklm}(\alpha, \mathbf{Q})$ is carried out in place of the simple polynomial in Eq. (11.77). It is mentioned that this method is an effective improvement relative to both Eq. (11.77) and the early methods based on higher order cumulant expansions, which result from the Edgeworth series expansion of the PDF.[55,63] By careful combinations of least-squares refinement methods and the use of the different types of Fourier synthesis, Cava et al. obtained the structural results already discussed. It appears highly recommended that the crystallographic data are collected with such accuracy that both the different types of Fourier synthesis and the least-squares refinement methods may be used in order to obtain the best model for the defect structure in fast ion conductors.

11.5. Quasi-Elastic Diffuse Scattering

In discussing quasi-elastic neutron scattering, it is very important to make the distinction between incoherent and coherent scattering. This is conceptually easy to do, especially for those who have familiarized themselves with the nature of the neutron scattering probe. However, in practice the separation is far from trivial since the type of scattering one observes from a given specimen cannot always be controlled but depends on the isotopic composition. Fortunately, nature has enough variety so that there are important examples of fast ion conductors in which either the incoherent or the coherent quasi-elastic scattering dominates. In general one has to consider both types of scattering and ensure that they are taken into account in the data analysis. Here one is often helped by their different characteristics, which give the contributions different weight depending on momentum transfer and temperature. Isotopic substitution is also an important means of separation.

11.5.1. Coherent Quasi-Elastic Scattering

The coherent scattering from an ideal, harmonic crystalline solid consists of the strictly elastic Bragg scattering and the inelastic phonon scattering with well-defined resonances at the energies of the lattice vibrational normal modes. Any defect in such a system will give rise to an additional diffuse coherent scattering which in turn can be utilized to determine both the structure of the defect configurations and their temporal evolution. To illustrate these points we consider one of the simplest possible defects—vacancies in a simple cubic system with lattice constant a containing only one type of atom with coherent scattering length $\bar{b}$. Given a vacancy at $\mathbf{R}_v(t)$ with the six nearest neighbor atoms relaxed toward the vacancy, then the resulting defect cluster can be described by vacancies in seven positions $\mathbf{R}_v^i(t) = \mathbf{R}_v(t) + \delta_i$ where $\delta_i = (0, 0, 0)$, $(\pm a, 0, 0)$, $(0, \pm a, 0)$, and $(0, 0, \pm a)$ and by new equilibrium sites for the six relaxed ions at $\mathbf{R}_R^j(t) = \mathbf{R}_v(t) + \varepsilon_j$ with $\varepsilon_j = \pm((1 - \delta)a, 0, 0)$, $\pm(0, (1 - \delta)a, 0)$, and $\pm(0, 0, (1 - \delta)a)$. If we assume that we are in the low-density limit, and that the relaxation field follows the jumps of the vacancy, the diffuse scattering results from distinct defect clusters and can be added incoherently $S_c^{\text{diff}}(\mathbf{Q}, E)$ is then given by Eq. (11.38) with

$$S_c^{\text{diff}}(\mathbf{Q}) \approx N_D \cdot \frac{\bar{b}^2}{N_c} \left| \sum_{j=1}^{6} e^{i\mathbf{Q} \cdot \varepsilon_j} - \sum_{i=1}^{7} e^{i\mathbf{Q} \cdot \delta_i} \right|^2 e^{-2W(\mathbf{Q})}$$

$$= \frac{N_D}{N_c} |F_{\text{clust}}(\mathbf{Q})|^2, \qquad (11.83)$$

where N_D is the number of clusters, and $F_{\text{clust}}(\mathbf{Q})$ is the cluster structure factor. From Eq. (11.83) it can be seen that, in the limit where cluster–cluster correlations are neglected, the integrated quasi-elastic scattering is proportional to the number of clusters and is uniform in $\mathbf{Q}$ if the relaxations δ are negligible. The width in energy reflects the jump geometry and the average residence time of the cluster at a given site. This is illustrated by the example discussed in Section 11.2.2 and quantified in Eq. (11.38).

In real systems different types of generally noncentrosymmetric clusters are found, and $S_c^{\text{diff}}(\mathbf{Q})$ can be written as

$$S_c^{\text{diff}}(\mathbf{Q}) = \sum_i \sum_j \frac{N_D^{ij}}{N_c} |F_{\text{clust}}^{ij}(\mathbf{Q})|^2, \tag{11.84}$$

where N_D^{ij} and $F_{\text{clust}}^{ij}(\mathbf{Q})$ are the number and structure factor for a cluster of type i with orientation j. The integrated coherent diffuse scattering then reflects an instantaneous picture of the defects, properly averaged over all the possible defect cluster configurations and orientations. Figure 11 shows an example of $S_c^{\text{diff}}(\mathbf{Q})$ observed in CaF_2 in the fast ion phase at high temperatures.[22] The data are presented in the form of a contour map that clearly shows the anisotropic nature of the scattering. Maps of this kind have been analyzed in detail by Hutchings et al.,[22] who have also discussed the formalism needed for the interpretation in terms of specific models of the defect clusters. This formalism neglects the interference between different clusters, and this has proved to be justified for the intrinsic fast ion conductors, where the disorder is thermally induced.[22] In doped or nonstoichiometric systems the cluster–cluster interference must be included.[64,65] It should also be pointed out that $S_c^{\text{diff}}(\mathbf{Q})$ observed by neutron scattering in many cases are similar to what can be observed in diffuse x-ray scattering. The only difference is the possible contributions from phonon scattering, which can be discriminated against in the neutron experiments.

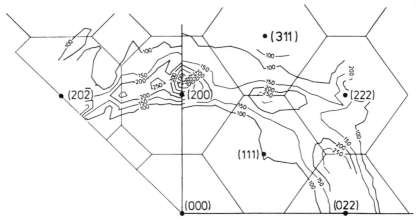

FIG. 11. Contour map of the integrated diffuse coherent quasi-elastic scattering observed in CaF_2 in the fast ion conducting phase at $T = 1473$ K. (After Hutchings et al.[22])

11.5.1.1. Fluorites. There has been a considerable number of diffuse neutron scattering experiments on fast ion conductors with the fluorite structure. In many of these systems (e.g., CaF_2 and PbF_2), the observed scattering is predominantly coherent. The ion motion in these compounds is promoted either by thermal activation or by doping with valence defect metal ions. In all cases the defect structures appear to be related to variants of the much discussed 4 : 2 : 2 defect cluster, first invoked from diffraction studies on nonstoichiometric compounds.[37, 66, 67] The nomenclature refers to the number of vacancies, true interstitials, and relaxed ions, respectively, and Fig. 7 gives an example of a 3 : 1 : 2 cluster. The diffuse scattering for the 3 : 1 : 2 cluster with additional relaxation of the next nearest neighbors[22] (the so-called 9 : 1 : 8 cluster) is shown in Fig. 12 and correlates well with the $S_c^{\text{diff}}(Q)$ observed for CaF_2. Having established this correspondence in the static approximation, one can then proceed to analyze the intensities in order to quantify the extent of the disorder. To this end it is important to have an internal reference for the scattered intensity, and in the work on the fluorites,[22] the scattering from selected acoustic phonons with well-known structure factors were used following the procedure outlined in Section 11.2. In this manner a number for the fraction of anions n_d that leave their regular sites either as true interstitials or as relaxed ions could be derived. The results for PbF_2, $SrCl_2$, and CaF_2 are shown in Fig. 13. As discussed in the previous section, n_d can also be obtained from diffraction data through an analysis that is consistent with the cluster models,[37] and fair agreement between these

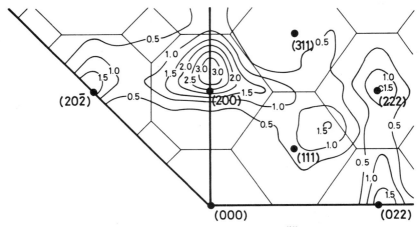

FIG. 12. Model calculation of the diffuse coherent scattering $S_c^{\text{diff}}(Q)$ with parameters relevant for CaF_2 at $T = 1473$ K. The 9 : 1 : 8 cluster, which is derived from the 3 : 1 : 2 cluster (Fig. 7) by additional next-nearest-neighbor relaxations, has been used in this case. (After Hutchings et al.[22]).

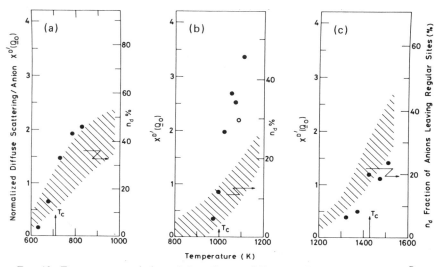

FIG. 13. Temperature variation of the coherent diffuse quasi-elastic scattering, $X^D(\mathbf{Q}_0)$, normalized per anion, for (a) β-PbF$_2$, (b) SrCl$_2$, and (c) CaF$_2$ at a scattering vector near the maximum intensity, $\mathbf{Q}_0$ ($X^D(\mathbf{Q}_0) = S_c^{\text{diff}}(\mathbf{Q}_0) \times e^{2W(Q_0)}/\bar{b}_a^2$ [cf. Eq. (11.84)]). The right-hand ordinate gives the value, n_d, of displaced fluorine ions determined by comparison with cluster models. The shaded areas represent the range of values obtained by neutron diffraction (given as D in Fig. 8 for β-PbF$_2$). (After Hutchings et al.[22])

different methods has been obtained for the intrinsic fluorite structure compounds, as indicated in the figure.

In all the cases discussed here, the energy widths of the coherent quasi-elastic scattering could easily be resolved (an example of the actual data for CaF$_2$ was shown in Fig. 14 of Part A, Chapter 1). The observed energy widths show some wave vector dependence and for CaF$_2$ and SrCl$_2$, it was found that they tend toward zero near the Bragg points. This was interpreted as a sign of dynamic correlations between similar clusters appearing at different times separated by the nearest-neighbor distance, and it is consistent with the expectations based on the approximation in the example given in Section 11.2.2 [Eq. (11.43)]. The temperature dependence of the energy widths were found to be characteristic of an activated process, and the same holds for the integrated intensities. These data show no evidence for a phase transition associated with the onset of fast ion conduction, but rather a gradual increase of the defect population, which in the beginning is aided by the defect interactions. Ultimately, the same interactions limit the amount of disorder that shows signs of saturation before the melting point is reached.

Diffuse scattering studies have also been carried out on fluorite structure compounds, where the defects have been induced by the solid solution of metal ions with different valence. Examples are $Ba_{1-x}La_xF_{2+x}$,[64,65] $Pb_{1-x}U_xF_{2+2x}$,[34] and $ZrO_2(Y_2O_3)_x$.[68,69] In these systems strong diffuse scattering was observed at low temperatures with unresolvable energy widths. The intensity distribution in reciprocal space shows similarities with that observed for the pure compounds at elevated temperatures but often with more pronounced variations indicating cluster–cluster correlations over considerable range. In some cases[64,65,68,69] clear trends toward superlattice formation are observed.

11.5.1.2. *Silver Conductors.* The classic fast ion conductor α-AgI and other highly Ag-conducting compounds like $RbAg_4I_5$, Ag_2S, Ag_2Se, and Ag_3SI have been the subject of several quasi-elastic diffuse neutron scattering investigations.[70-74] Silver has about equal coherent and incoherent cross sections, and in most of these studies it has not been possible to clearly separate the two components of the quasi-elastic scattering. An exception is β-Ag_2S, where the Ag–Ag correlations give considerable enhancement of the coherent scattering in certain regions of reciprocal space and hence make the identification straightforward.[72] In this case the $S_c^{\text{diff}}(\mathbf{Q})$ obtained by neutron scattering was in good agreement with the diffuse x-ray scattering data[75] that were explained in terms of a specific microdomain model for the short range correlations.

The data and the model are illustrated in Fig. 14. The subsequent neutron data showed that the scattering is quasi-elastic, and the energy widths have both significant wave vector and temperature dependencies. As was the case for the fluorites, an exhaustive theoretical model for dynamics of the fast ion state with considerable amounts of short-range correlations has not yet been worked out, but progress has been made in relating the coherent structure factor to molecular dynamics simulations.[76]

11.5.1.3. *Lithium Conductors.* Lithium conductors are of considerable interest from both a scientific and technical point of view. However, owing to the fact that the lithium scattering length is small for both available isotopes (^{6}Li and ^{7}Li), that the incoherent cross section is small, and that ^{6}Li has significant absorption, only a few studies have so far been carried out in which the diffuse coherent scattering could be analyzed in detail. The best example is the pseudo-one-dimensional lithium conductor β-eucryptite ($LiAlSiO_4$), where the ion transport occurs in channels along the hexagonal c axis in the oxide framework. The lithium ions are disordered in the β phase above 490 K, and strong diffuse sheets of scattering are observed indicating one-dimensional correlations over a considerable distance.[77] In a subsequent high-resolution experiment, the energy width of this scattering was resolved, and the scattering showed an anisotropy consistent with the picture of the channel directions as the preferred direction for the lithium ion motion.

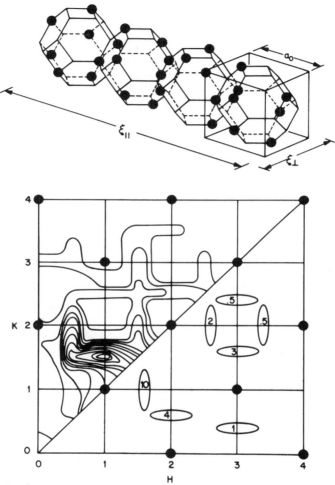

Fig. 14. Idealized model for the microdomains of ordered Ag ions on the cubo-octahedra of tetrahedral sites in β-Ag$_2$S. The lower figure shows the calculated intensity of the diffuse scattering for this model compared with a schematic representation of the observed diffuse scattering in the ($hk0$) plane. The calculated intensity is above the $h = k$ mirror plane and the observed intensity below. The observed intensities are height above background, normalized to the intensity of the disk at (1.6, 1, 0). (After Cava and McWhan.[75])

11.5.2. Incoherent Quasi-Elastic Scattering

The incoherent scattering is determined by the self-correlation function $G_s(\mathbf{R}, t)$, which traces the correlation between a given particle at time 0 and the *same* particle at a later time t. This function is often easier to relate to a physical picture than the more complex pair-correlation function, which includes correlations with all atoms at time t and which governs the coherent scattering from a given system. The incoherence in the scattering may originate in the spin states for a given isotope or stem from the isotopic mixture, but the interpretation is the same in both cases. Typical cross sections are, with the notable exception of the proton, of the same order of magnitude as the typical coherent counterparts, and hence one must take care when identifying these components and also remember that more than one atomic species may contribute. In the identification one is aided by the fact that the integrated incoherent scattering is isotropic, apart from the well-known Debye–Waller factor, and it is often, but not always, the dominating component in the low to medium Q region, 0.5–1.0 Å^{-1}. In the present context the analysis of the energy widths of the quasi-elastic scattering, $S_i^s(\mathbf{Q}, E)$ is almost always carried out in the spirit of the Chudley–Elliot (CE) model for jump diffusion on a lattice.[25] This model has been extensively discussed in Chapter 10, which was devoted to hydrogen diffusion in metals. The relevant formulas for the neutron scattering cross section were quoted in Section 11.2.

11.5.2.1. Fluorites. The best studied compound with fluorite structure in this context is $SrCl_2$, where the incoherent scattering cross section of the chlorine atoms, 5.5 ± 0.5 b, is sizeable.[78] The separation from the coherent scattering can be done reliably because (1) the incoherent scattering is dominant in the low Q region, $Q < 1.5$ Å^{-1}, and (2) the coherent scattering was known[79] to have a relatively large width in energy so that the separation can be performed even in Q regions with significant coherent scattering by a judicious choice of spectrometer energy resolution. This requirement could in this case be fulfilled by the use of a cold source, high-resolution, triple-axis spectrometer. The structural information about the conduction process is contained in the wave vector dependence along the different directions in the reciprocal unit cell of the energy width of the quasi-elastic scattering [Eq. (11.32)]. It is therefore essential that the experiments are carried out on single-crystal specimens. The results for $SrCl_2$ in the fast ion phase above 1000 K are shown in Fig. 15. They show a very satisfactory agreement with the simple CE model and have enough detail to discriminate between different refinements of this model.[78] It is important to note that the process of ion motion can be described by a single temperature-dependent characteristic time, namely, the average residence time of the migrating species. It is assumed that transit times are small relative to this. Intuitively, one can associate the time scale reflected in the coherent scattering with the typical transit time, since it is related to the lifetimes of the defect clusters that may be formed during the transit. For $SrCl_2$ at 1000 K the typical $\tau_c \approx 10^{-11}$ s, whereas $\tau_i \approx 10^{-10}$ s is an order of magnitude larger,[78] in good agreement

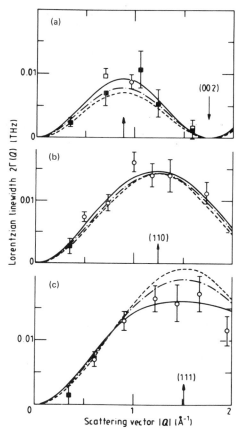

FIG. 15. Variation of the FWHM of the incoherent quasi-elastic scattering from $SrCl_2$ at 1053 K with scattering vector in three principal crystal directions: (a) [001], (b) [110], and (c) [111]. The symbols □ and ○ denote data taken with $k_i = 1.047$ Å^{-1} and 1.160 Å^{-1}, respectively; full and open symbols denote data from two different samples. The broken (NN hops) and full (NN and NNN hops) curves are the best fit of the Chudley-Elliott expression, and the chain curve that of Wolf's encounter model.[83] (After Dickens et al.[78])

with the assumptions of the CE model (cf. Section 11.2.2) and the considerations leading to Eq. (11.43). It can be concluded that in the conduction process a vacancy diffusion mechanism is dominant, with anion hops mainly to nearest-neighbor (NN) and next-nearest-neighbor (NNN) regular anion sites. This picture of the diffusion seen over an intermediate time scale is consistent with the interpretation of the coherent scattering[22] in terms of short-lived defect clusters with lifetimes in the range 10^{-11}–10^{-12} s and that can be associated with the process of anion transit between the regular sites.

11.5.2.2. Silver Conductors. As already mentioned, there have been several studies of the classic fast ion conductor AgI in its cubic α phase above the first-order structural transition at 147°C. The most comprehensive analysis of the incoherent scattering from a single-crystal specimen has been carried out by Funke et al.[80] They found that the energy width of the quasi-elastic scattering could be accounted for by the CE model involving jumps between the tetrahedral sites on the faces of the cubic unit cell in accordance with the most probable paths that were derived from the structural studies. Similar experiments have been carried out on single crystals of α-Ag_2Se with some difficulty in clearly separating the coherent and incoherent scattering.[73] However, it was possible to conclude that the diffusion appeared more isotropic than predicted by the CE model, possibly indicating that the assumption of negligible transit times is not justified in this case.

Diffusional broadening of the quasi-elastic incoherent scattering in the fast ion phases of $RbAgI_4$, Ag_3SI, and Ag_3SBr has been looked for but so far with insufficient energy resolution to be observed.

11.5.2.3. Lithium Conductors. Neutron scattering experiments that aim to observe lithium diffusion are very difficult to make because of the small scattering cross sections of both the lithium isotopes and the high absorption of ^6Li. They can only be made for compounds in which the other constituents have a negligible incoherent scattering. Fortunately, two of the most interesting lithium fast ion conductors meet this criterion, namely, LiAl (which can be used as an electrode metal)[81] and $LiAlSiO_4$ (pseudo-one-dimensional conductor).[82] In both cases the contributions due to the coherent scattering

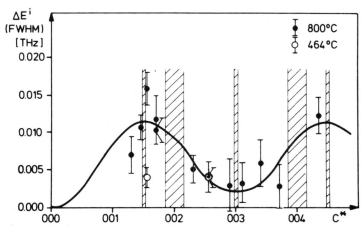

FIG. 16. The Q_z dependence of the quasi-elastic linewidth in $LiAlSiO_4$. The pronounced maximum for $Q_z = \frac{3}{2}c^*$ at 800°C is invisible at 464°C. The solid line represents a calculation based on a jump diffusions model. (After Renker et al.[82])

could be separated out because of their much larger energy widths and, as a consistency check, it could be established that the integrated intensity followed the expected Debye–Waller factor. Again the CE model modified in an encounter version[83] for LiAl gave an excellent description of the data. As an example the results for LiAlSiO$_4$ are shown in Fig. 16. It is clearly seen that the period of the variation of the energy width corresponds to real space jumps of $c/3$ rather than the full c axis unit.

For LiAl it was shown that the Q dependence of the energy width in the low Q region followed $\Gamma = DQ^2$ with $D = (6 \pm 1) \times 10^{-6}$ cm^2/s at 800 K. The value of D was found to be in agreement with both NMR and chemical diffusion measurements, and it was concluded that the lithium diffusion occurs via rapid diffusion of vacancies residing on the lithium sublattice.

11.5.3. Summary of Quasi-Elastic Scattering

Both the coherent and the incoherent quasi-elastic neutron scattering contain important information about ion diffusion in solids, and in both cases the use of single-crystal specimens in the experiments is essential for making the analysis. In most cases the incoherent scattering can be interpreted in terms of the Chudley–Elliot model, or variants thereof, and through such an analysis the most important diffusion jump mechanisms can be identified. The incoherent scattering relates directly to the macroscopic diffusion.

The coherent scattering has so far mainly been used to obtain structural information about the short-lived, cluster-like defect configurations that appear during the ion motion and, in the static approximation, several successful and very specific models have been derived. However, comprehensive dynamical models for a full description of the observed scattering are still lacking, although some progress has been made through comparisons with the results of molecular dynamics simulation of fast ion conductors.

11.6. Inelastic Scattering

11.6.1. Low-Temperature Phonons

Most of the ion conductors are well-behaved ionic crystals at low temperatures and show the normal spectrum of phonon modes that can be determined most efficiently by the classical constant Q method using triple-axis spectrometers. Such experiments have been performed for a range of fast ion compounds.[84–91] A thorough study of β-PbF$_2$ has been carried out by Dickens and Hutchings,[84] from which the low-temperature ($T = 10$ K) dispersion relations are shown in Fig. 17, together with two different model

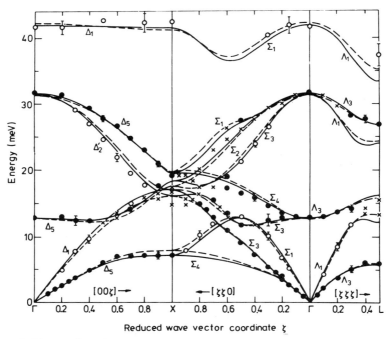

FIG. 17. Phonon dispersion relations in the [001], [110], and [111] directions of β-PbF$_2$ at 10 K. Closed circles are predominantly transverse mode and open circles are longitudinal modes. The crosses represent observed points that were not included in the fitting procedure. The broken curves are calculated from the best fit to an eight-parameter shell model and the full curves from the best fit to a 13-parameter shell model. (After Dickens and Hutchings.[84])

fits. Both results are from shell model calculations based on 8 and 13 adjustable parameters, respectively. These include the nearest- and second-nearest-neighbor repulsive core interactions and the electrical and mechanical polarizabilities of the Pb and F ions (alternatively and equivalently, the effective shell charges and the shell core coupling constants) for the eight-parameter models (dashed lines). The long-range electrostatic interactions between the ions in the crystal are calculated by the method developed by Kellermann[92] with the assumption of an effective ionic charge of $2|e|$ and e on Pb and F ions, respectively. In the 13-parameter fit the effective charges were varied (with the necessary constraint for charge neutrality), thus allowing for some covalency in the Pb–F bonding. In addition, the short-range repulsive potentials for third nearest neighbors were included, which resulted in the slight improvements of the fit represented by the full curve. It should be emphasized that a rigid ion model may be fitted convincingly to the acoustic branches of the phonon dispersion relations, but fails decisively for the optical modes. When it is recalled that acoustic and optical modes are essentially in-phase and antiphase motions of nearest-neighbor ions, respectively, this is not surprising. Only the antiphase motions are greatly facilitated by polarization effects. This means that the effective ionic charges and the short-range repulsive core potentials may be determined from fits to the acoustic modes. These parameters may be used with or without the polarizability contributions in computer-based molecular dynamics simulations. Shell models have also been applied in the analysis of the low-temperature phonon dispersion of copper halides,[88] and also in this case an excellent phenomenological description of the experimental data was achieved. In most cases, however, the parameters could not be given a direct physical meaning. They have to be regarded primarily as compact descriptions of the data with qualitative information on the relative importance of the different types of interactions.[90,91]

The phonon dispersions alone contain little information that can be related directly to the ion mobility. In some cases one finds low-lying optical modes that signal an incipient instability for certain patterns of distortions. This is the case for β-AgI, where the low-frequency modes of $\Gamma_6(E_2)$ symmetry are believed to promote the onset of ionic disorder.[89] Similarly, flat optical modes have been observed in the β-aluminas[87] and in $RbAg_4I_5$.[70] These modes have been associated with the mobile metal ions and, regarded as localized oscillations, their frequencies give reasonable estimates of the attempt frequencies that enter in the classical Nernst–Einstein expression [Eq. (11.1)] for the ionic conductivity.

11.6.2. Temperature Effects

Quite dramatic anharmonic effects are expected in the phonon modes as the temperature is raised, since they are clearly visible already in the diffraction data. The fluorites are again good model systems for these effects, particularly for studying the influence of ionic disorder, since the transition to high ionic conductivity occurs without change in crystal symmetry. Dickens et al.[24] have studied the temperature evolution of the acoustic phonons in β-PbF_2 up to 900 K, and in their analysis they were able to estimate the relative importance of the individual contributions to the changes in frequencies and linewidths from effects like anharmonicity, defects, and ion hopping. Examples of their results are shown in Fig. 18. They find that the anharmonicity gives the dominant contribution at low temperatures, where it leads to a linear decrease of the elastic constant and a linear increase in the observed width of the form $\Gamma = ATq^2$ [cf. Eq. (11.28)]. At the onset of high ion mobility near $T_c = 700$ K, they observe a large additional contribution, primarily from the defects that cause breakdown of

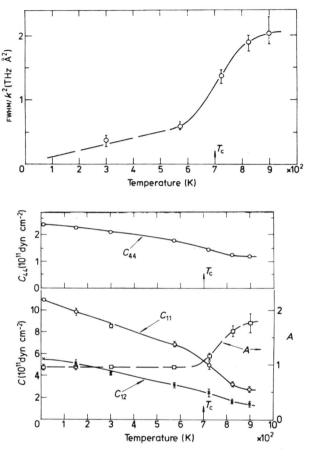

FIG. 18. Temperature variation of (a) the phonon width parameter $2\Gamma/q^2$ (the parameter is isotropic) and (b) the elastic constants C_{44}, C_{11}, and C_{12} and the anisotropy parameter $A = 2C_{44}/(C_{11} - C_{12})$ in β-PbF$_2$. (After Dickens et al.[24])

the phonon coupling selection rules and hence give new channels for phonon decay. It is noteworthy that the simple phenomenological form of the response [Eq. (11.27)], based on hydrodynamic theory,[23] gives a good description of the acoustic phonon profiles at all temperatures. This is in contrast to the optical modes that generally broaden very quickly and rarely can be resolved in the fast ion phases. This phenomenon is not surprising as the optic mode potentials are determined mainly by the nearest-neighbor interactions and, therefore, are very sensitive to vacancies and interstitials in the lattice.

In summary, one can say that phonons, like Bragg scattering, reflect the long-range crystalline order and hence give only indirect information about the ionic disorder and mobility. In some cases particular low-lying phonon modes may give hints about vibrational patterns that can lead to anharmonic motion of large amplitude and hence provide a mechanism for the ion jumps. However, most often the phonon frequencies are much larger than the inverse configurational relaxation times and hence the two phenomena can be treated separately. This separation can be seen directly in the experimental data, for example in the fluorites where the quasi-elastic coherent and incoherent scattering occur in a frequency range 10–100 times smaller than the phonon band width. It is also striking that there appears to be little correlation between phonon structure factors and the observed intensity variation of the coherent quasi-elastic scattering.

11.7. Conclusions

Neutron scattering techniques have been applied to study a variety of phenomena in relation to the structure and dynamics of fast ion conductors. The structural studies give descriptions of the average crystallographic order with the distribution of atoms in the unit cell presented either in terms of discrete sites or by more general density distributions, which, in many cases, can lead to the identification of the most probable paths for the ion migration. The diffuse scattering studies focus more directly on the disorder that is inherent in these systems. Such studies reveal that the ion motion is generally strongly correlated and may involve the formation of short-lived defect configurations or fluctuations that affect tens, sometimes hundreds, of atoms. This emphasizes the cooperative nature of the dynamics, which is also seen indirectly through the thermal properties with their characteristically low activation energies for fast ion conductors. However, when one focuses on the motion of a particular ion, as is done through the incoherent scattering cross section, a simpler picture emerges that corresponds to the jump diffusion models of the type first suggested by Chudley and Elliot. In this description, the migrating ion resides most of the time at regular lattice sites and occasionally makes a transition to another site in the immediate neighborhood. The apparent conflict between the picture of highly correlated collective motion and the very simple picture of the motion of the individual ions can be reconciled when the single-ion transit time is short compared to the residence time. The strong correlations occur during the transit in which the migrating particle may become a member of an extended defect-induced configurational fluctuation. This correlated motion may leave the particle at a new site. This example illustrates the kind of information about the correlation between structure and ion transport that can be obtained from neutron scattering studies. For specific systems the analysis can be made quantitative and can be related to studies of other physical properties, such as macroscopic diffusion, electrical resistivity, and specific heat, as well as their dependence on temperature and doping.

An important parallel development to the activity in neutron scattering has been the molecular dynamics computer simulation of fast ion conductors. This method allows a very direct analysis of the complex pattern of ion motion and has led to detailed understanding of several transport mechanisms, especially in the fluorites. The studies rely on the availability of parameters for the interaction potentials that most often are derived from fits to structural as well as spectroscopic data, with phonon dispersion obtained by neutron scattering as an important component. So far, there has been a very fruitful interplay with considerable mutual inspiration between the molecular dynamics and neutron scattering studies of fast ion conductors, which has helped the interpretation of the experiments.

To conclude this chapter, we reiterate some of the specific points already made concerning the use of neutron scattering in the study of fast ion conductors. The neutron probe is particularly well suited to such studies because of the bulk penetration and the sensitivity to light elements and isotopic composition. From the structural point of view, neutron methods are complementary to those of x rays, but they offer distinct advantages for dynamical studies because of the possible distinction between different dynamical processes and the ability to probe both collective modes through the coherent scattering and single-particle motion through the incoherent scattering.

The structure of crystalline ion conductors combines the translational symmetry with different degrees of local disorder, and the analysis of scattering data for such systems requires considerable care when standard methods are applied. Large anisotropic temperature factors are encountered and large thermal diffuse scattering often must be corrected for. The best results have so far been obtained on the basis of very extensive and accurate data treated in an iterative combination of Fourier methods and fits to discrete models. Similarly, the diffuse scattering can be analyzed in terms of discrete models for the defect configurations. A full dynamical description of a fast ion conductor including correlation effects is still lacking. Although considerable insight has been gained by combining quasi-elastic neutron scattering data, simple model calculations, and molecular dynamics simulations, there is still need for substantial progress in our understanding of these systems.

Acknowledgments

It is a pleasure to thank M. H. Dickens, W. Hayes, M. T. Hutchings, P. G. Schnabel and C. Smith for fruitful collaboration on several parts of the work described in this chapter. This work is partly supported by the Danish Ministry of Energy (EPF-contract 1443/85-7).

References

1. M. Faraday, "Experimental Researches in Electricity," Artic. 1340. Taylor & Francis, London, 1839.
2. W. Nernst, *Z. Elektrochem.* **6**, 41 (1899).
3. C. Tubandt and E. Lorenz, *Z. Phys. Chem.* **87**, 513 (1914).
4. A. Benrath and K. Drekopf, *Z. Phys. Chem.* **99**, 57 (1921).
5. H. U. Beyeler, L. Pietronero, and S. Strässler, *Phys. Rev. B* **22**, 2988 (1980).
6. M. O'Keeffe and B. G. Hyde, *Philos. Mag.* **33**, 219 (1976).
7. M. O'Keeffe, in "Superionic Conductors" (G. D. Mahan and W. L. Roth, eds.), p. 101. Plenum, New York, 1976.
8. W. J. Pardee and G. D. Mahan, *J. Solid State Chem.* **15**, 310 (1975).
9. L. D. Landau and E. M. Lifschitz, "Statistical Physics," 2nd Ed. Pergamon, Oxford, 1970.
10. M. B. Salamon, in ref. 20, p. 175.
11. K. E. D. Wapenaar, *J. Phys. Colloq. (Orsay, Fr.)* **41**(C6), 220 (1980).
12. E. C. Subbarao, *Adv. Ceram.* **3**, 1 (1981).
13. J. H. Kennedy, in ref. 16, p. 105.
14. R. Collongues, J. Théry, and J. P. Boilot, in ref. 17, p. 253.
15. "Fast Ion Transport in Solids" (W. van Gool, ed.). North-Holland Publ., Amsterdam, 1973.
16. "Solid Electrolytes" (S. Geller, ed.), Topics in Applied Physics. Springer-Verlag, Berlin and New York, 1977.
17. "Solid Electrolytes: General Principles, Characterization, Materials, Applications" (P. Hagenmuller and W. van Gool, eds.). Academic Press, New York, 1978.
18. "Solid Electrolytes and their Applications" (E. C. Subbarao, ed.). Plenum, New York, 1980.
19. S. Chandra, "Superionic Solids: Principles and Applications." North-Holland Publ., Amsterdam, 1981.
20. "Physics of Superionic Conductors" (M. B. Salamon, ed.), Topics in Current Physics. Springer-Verlag, Berlin and New York, 1979.
21. G. L. Squires, "Introduction to The Theory of Thermal Neutron Scattering." Cambridge Univ. Press, London and New York, 1978.
22. M. T. Hutchings, K. Clausen, M. H. Dickens, W. Hayes, J. K. Kjems, P. G. Schnabel, and C. Smith, *J. Phys. C* **17**, 3903 (1984); also AERE-R11127-MPD-NBS-244. At. Energy Res. Establ., Harwell, UK (1983).
23. R. Zeyher, *Z. Phys. B* **31**, 127 (1978).
24. M. H. Dickens, W. Hayes, M. T. Hutchings, and W. G. Kleppman, *J. Phys. C* **12**, 17 (1979).
25. C. T. Chudley and R. J. Elliot, *Proc. R. Soc. London* **77**, 353 (1961).
26. G. H. Vineyard, *Phys. Rev.* **110**, 999 (1958).
27. L. W. Strock, *Z. Phys. Chem., Abt. B* **25**, 411 (1934); **31**, 132 (1936).
28. B. Reuter and K. Hardel, *Z. Anorg. Allgem. Chem.* **340**, 168 (1965).
29. P. Rahlfs, *Z. Phys. Chem., Abt. B* **31**, 157 (1936).
30. K. Funke, *Prog. Solid State Chem.* **11**, 345 (1976).
31. W. Schröter and J. Nöltig, *J. Phys. Colloq. (Orsay, Fr.)* **41**(C6), C6-20 (1980).
32. J. Oberschmidt, *Phys. Rev. B* **23**, 5038 (1981).
33. M. Ouwerkerk, E. M. Kelder, J. Schoonman, and J. C. Miltenburg, *Solid State Ionics* **9/10**, 531 (1983).
34. N. H. Andersen, K. Clausen, and J. K. Kjems, *Solid State Ionics* **9/10**, 543 (1983); also in "Transport-Structure Relations in Fast Ion and Mixed Conductors" (F. W. Poulsen, N. H. Andersen, K. Clausen, S. Skaarup, and O. Toft Sørensen, eds.), p. 171. Risø Nat. Lab., Roskilde, Denmark, 1985.

35. W. H. Zachariassen, *Acta Crystallogr.* **23**, 558 (1967).
36. P. J. Becher and P. Coppens, *Acta Crystallogr., Sect. A* **A30**, 129 (1974); **A30**, 148 (1974); **A31**, 417 (1975).
37. M. H. Dickens, W. Hayes, M. T. Hutchings, and C. Smith, *J. Phys. C* **15**, 4043 (1982).
38. I. Waller, *Ann. Phys. (Leipzig)* **83**, 153 (1927).
39. N. Nilsson, *Ark. Fys.* **12**, 247 (1957).
40. M. J. Cooper and K. D. Rouse, *Acta Crystallogr., Sect. A* **A24**, 405 (1968).
41. C. B. Walker and D. R. Chipman, *Acta Crystallogr., Sect. A* **A25**, 395 (1969).
42. M. J. Cooper, in "Thermal Neutron Diffraction" (B. T. M. Willis, ed.), p. 51. Oxford Univ. Press, London and New York, 1970.
43. G. E. Bacon, "Neutron Diffraction." Oxford Univ. Press (Clarendon), London and New York, 1975.
44. M. J. Cooper, K. D. Rouse, and B. T. M. Willis, *Acta Crystallogr., Sect. A* **A24**, 484 (1968).
45. R. W. James, *Acta Crystallogr.* **1**, 132 (1948).
46. R. J. Cava, F. Reidinger, and B. J. Wunsch, *J. Solid State Chem.* **31**, 69 (1980).
47. R. J. Cava, F. Reidinger, and B. J. Wunsch, *Solid State Commun.* **24**, 411 (1977).
48. E. Perenthaler, H. Schultz, and H. U. Beyeler, *Solid State Ionics* **5**, 493 (1981).
49. K. Koto, H. Schulz, and R. A. Huggins, *Solid State Ionics* **3/4**, 381 (1981).
50. C. R. A. Catlow, A. V. Chadwick, and J. Corish, *Radiat. Eff.* **75**, 61 (1983); also *J. Solid State Chem.* **48**, 65 (1983).
51. H. Horiuchi, A. J. Schultz, P. C. W. Leung, and J. M. Williams, *Acta Crystallogr., Sect. B* **B40**, 367 (1984).
52. J. D. Axe, S. M. Shapiro, and N. Wakabayashi, unpublished; see also Shapiro and Reidinger, in ref. 20, p. 45.
53. J. L. Soubeyroux, J. M. Reau, S. F. Matar, P. Hagenmuller, and C. Lucat, *Solid State Ionics* **2**, 215 (1981).
54. J. L. Soubeyroux, S. F. Matar, J. M. Reau, and P. Hagenmuller, *Solid State Ionics* **14**, 337 (1984).
55. C. K. Johnson and H. A. Levy, *Int. Tables X-Ray Crystallogr.* **4**, 311 (1974).
56. M. W. Thomas, *Chem. Phys. Lett.* **40**, 111 (1976).
57. B. Dawson, A. C. Hurley, and V. W. Maslen, *Proc. R. Soc. London, Ser. A* **298**, 289 (1967).
58. B. T. M. Willis, *Acta Crystallogr., Sect. A* **A25**, 277 (1969).
59. R. Bachmann and H. Schulz, *Solid State Ionics* **9/10**, 521 (1983).
60. R. Bachmann and H. Schulz, *Acta Crystallogr., Sect. A* **A40**, 668 (1984).
61. H. Schulz and U. H. Zucker, *Solid State Ionics* **5**, 41 (1981).
62. R. J. Cava, F. Reidinger, and B. J. Wuensch, in "Fast Ion Transport in Solids" (P. Vashishta, J. N. Mundy, and G. K. Shenoy, eds.), p. 217. Elsevier/North-Holland, New York, 1979.
63. C. K. Johnson, in "Thermal Neutron Diffraction" (B. T. M. Willis, ed.), p. 132. Oxford Univ. Press, London and New York, 1970.
64. J. K. Kjems, N. H. Andersen, J. Schoonman, and K. Clausen, *Physica B (Amsterdam)* **120B**, 357 (1983).
65. N. H. Andersen, K. Clausen, J. K. Kjems, and J. Schoonman, *J. Phys. C* **19**, 2377 (1986).
66. A. K. Cheetham, in "Chemical Applications of Thermal Neutron Scattering" (B. T. M. Willis, ed.), p. 225. Oxford Univ. Press, London and New York, 1973.
67. B. T. M. Willis and R. G. Hazell, *Acta Crystallogr., Sect. A* **A36**, 582 (1980).
68. N. H. Andersen, K. Clausen, M. A. Hackett, W. Hayes, M. T. Hutchings, J. E. Macdonald, and R. Osborn, in "Transport-Structure Relations in Fast Ion and Mixed Conductors" (F. W. Poulsen, N. H. Andersen, K. Clausen, S. Skaarup, and O. Toft Sørensen, eds.), p. 279. Risø Nat. Lab., Roskilde, Denmark, 1985.

69. R. Osborn, N. H. Andersen, K. Clausen, M. A. Hackett, W. Hayes, M. T. Hutchings, and J. E. Macdonald, *Mater. Sci. Forum* **7**, 55 (1986).
70. S. M. Shapiro and M. B. Salamon, in "Fast Ion Transport in Solids" (P. Vashishta, J. N. Mundy, and G. K. Shenoy, eds.), p. 237. Elsevier/North-Holland, New York, 1979.
71. G. Eckold, K. Funke, J. Kalus, and R. E. Lechner, *Phys. Lett. A* **55A**, 125 (1975).
72. D. H. Grier, S. M. Shapiro, and R. J. Cava, *Phys. Rev. B* **29**, 3810 (1984).
73. K. Funke, *Solid State Ionics* **6**, 93 (1982).
74. S. Hoshino, H. Fujishita, M. Jakashige, and T. Sakuma, *Solid State Ionics* **3/4**, 35 (1981).
75. R. J. Cava and D. B. McWhan, *Phys. Rev. Lett.* **45**, 2046 (1980).
76. P. Vashishta, I. Ebbsjö, R. Dejus, and K. Sköld, *J. Phys. C* **18**, L291 (1985).
77. W. Press, B. Renker, H. Schultz, and H. Böhm, *Phys. Rev. B* **21**, 1250 (1980).
78. M. H. Dickens, W. Hayes, P. Schnabel, M. T. Hutchings, R. E. Lechner, and B. Renker, *J. Phys. C* **16**, L1 (1983).
79. M. H. Dickens, M. T. Hutchings, J. K. Kjems, and R. E. Lechner, *J. Phys. C* **11**, L583 (1978).
80. K. Funke, A. Höch, and R. E. Lechner, *J. Phys. (Orsay, Fr.)* **41**(C6) (1980).
81. T. O. Brun, S. Susman, J. M. Rowe, and J. J. Rush, *Solid State Ionics* **5**, 417 (1981).
82. B. Renker, H. Bernotat, G. Heger, N. Lehner, and W. Press, *Solid State Ionics* **9/10**, 1331 (1983).
83. D. Wolf, *Solid State Commun.* **23**, 853 (1977).
84. M. H. Dickens and M. T. Hutchings, *J. Phys. C* **11**, 461 (1978).
85. M. M. Elcombe and A. W. Pryor, *J. Phys. C* **3**, 492 (1970).
86. A Sadoc, F. Moussa, and G. Pepy, *J. Phys. Chem. Solids* **37**, 197 (1976).
87. D. B. McWhan, S. Shapiro, J. P. Remeika, and G. Shirane, *J. Phys. C* **8**, L487 (1975).
88. S. Hoshino, Y. Fujii, J. Harada, and J. D. Axe, *J. Phys. Soc. Jpn.* **41**, 965 (1976).
89. W. Bührer and P. Brüesch, *Solid State Commun.* **16**, 155 (1975).
90. W. Kress, H. Grumm, W. Press, and J. Lefebvre, *Phys. Rev. B* **10**, 4620 (1980).
91. T. O. Brun, J. E. Robinson, S. Susman, D. F. R. Mildner, R. Dejus, and K. Sköld, *Solid State Ionics* **9/10**, 485 (1983).
92. E. W. Kellermann, *Philos. Trans. R. Soc. London, Ser. A* **238**, 513 (1940).

12. GLASSES

Kenji Suzuki

The Research Institute for Iron, Steel, and Other Metals
Tohoku University
Katahira 2-1-1, Sendai, Japan

12.1. Introduction

The term "glass" implies an amorphous solid having a disordered atomic-scale structure and glass transition, regardless of the chemical constitution and method of preparation. Metallic and ceramic glasses have attracted strong attention for the past ten years because of two reasons. One is the fundamental point of view that the glass provides a promising approach to materials physics and chemistry as opposed to the perfect crystal. The other reason comes from the application of glasses as engineering materials in high-tech fields. The properties of glasses are particularly structure sensitive, because the glass usually stays in the nonequilibrium state and has a great degree of freedom in atomic motion and configuration.

Among the various experimental techniques, radiation scattering is one of the most direct methods for studying the spatial arrangement and motion of constituent atoms in glasses. In this method, we measure the momentum and energy transferred between radiation and target sample during the scattering process. The scattering experiments so far performed on condensed materials using conventional radiation are summarized in Table I.

The atomic spacing is several Ångström units and the interaction energy between atoms is of the order of 100 meV in glasses. The ideal choice of the wavelength and energy of the radiation used must be of this order. When electrons and x rays have a wavelength comparable with the average atomic spacing in glasses, their energy is too high to examine the dynamics of glasses. The wavelength of light is too large compared with the atomic distance in glasses, but its energy resolution is in the correct range for finding the atomic vibrations. Thermal neutrons are a unique and powerful probe for investigating simultaneously the structure and dynamics of glasses, because their wavelength and energy satisfy the condition of ideal choice. In particular, pulsed neutron sources based on electron and/or proton accelerators are quite favorable for the structure characterization of glasses, because the

TABLE I. Radiation Scattering Experiments Used for Structure Characterization of Condensed Materials

Radiation	Range	Wavelength	Energy	Experiment[a]
Photon	Microwave	1 ~ 100 cm	10^{-4} ~ 10^{-6} eV	NMR, ESR
	Infrared	> 7700 Å	< 1.6 eV	IR, Raman
	Visible	3800 ~ 7700 Å	1.6 ~ 3.3 eV	V, Raman
	Ultraviolet	< 3970 Å	> 3.1 eV	UV, Raman
	X-Ray	0.01 ~ 100 Å	1240 keV ~ 124 eV	Diffraction XPS, Emission, EXAFS
	γ-Ray	< 1 Å	> 12.4 keV	Mössbauer Compton, γ-γ
Electron		1 ~ 0.037 Å	150 eV ~ 100 keV	Diffraction, EELS
Neutron	Cold	> 4 Å	< 5 meV	SAS, INS Diffraction
	Thermal	0.5 ~ 4 Å	330 ~ 5 meV	Diffraction INS
	Epithermal	< 0.5 Å	< 330 meV	Diffraction INS

[a] NMR: nuclear magnetic resonance; ESR: electron spin resonance; IR: infrared spectroscopy; V: visible spectroscopy; UV: ultraviolet spectroscopy; XPS: x-ray photoelectron spectroscopy; EXAFS: extended x-ray absorption fine structure; γ-γ: time differential gamma-gamma perturbed angular correlation; SAS: small-angle scattering; INS: inelastic neutron scattering; and EELS: electron energy loss spectroscopy.

neutron flux in the epithermal energy region is relatively high and the time-of-flight technique can be appropriately used.

Thermal neutrons are scattered by the nucleus and unpaired electrons in the atom. The nuclear scattering is usually used to investigate the atomic arrangement and dynamics in condensed materials. Since the neutron scattering length of an individual nucleus varies with different isotopes and their spin states, coherent and incoherent scattering takes place according to the existence and absence of interference occurring between neutron waves scattered by different nuclei. Coherent neutron scattering provides the information of collective structure and motion of atoms such as atomic pair correlation functions, phonon dispersion, etc. On the other hand, localized vibrations and atomic diffusion in terms of individual motions of atoms are observed in incoherent scattering experiments. Several reviews[1-5] on neutron scattering studies of amorphous solids have been published.

This chapter describes high-resolution observations of the short-range structure of metallic and ceramic glasses by coherent total scattering, the hydrogen atom environment in metallic glasses by coherent total and incoherent inelastic scattering, and the atomic vibrations in metallic and oxide glasses by coherent inelastic scattering using short-wavelength pulsed neutrons generated by accelerators.

12.2. Structure Factors and Pair Correlation Functions

12.2.1. Total Scattering Experiment

A total neutron scattering measurement is performed in the experimental arrangement as shown schematically in Fig. 1. Incident neutrons with energy E_0 and wave vector $\mathbf{k}_0$ impinge on a sample, from which neutrons of energy E_1 and wave vector $\mathbf{k}_1$ come out, scattered at an angle 2θ. Each scattering event follows the law of energy and momentum conservation:

$$E_0 - E_1 = E, \quad \mathbf{k}_0 - \mathbf{k}_1 = \mathbf{Q}, \quad (12.1)$$

where E and $\mathbf{Q}$ are the energy and momentum transferred from the neutron to the sample during the scattering event. The scattering process is therefore characterized in terms of the energy- and momentum-dependent double differential cross section.

The double differential cross section for the coherent neutron scattering $d^2\sigma_c/d\Omega\, dE$ is proportional to the Fourier transform of the Van Hove[6] space–time correlation function $G(\mathbf{r}, t)$:

$$\frac{d^2\sigma_c}{d\Omega\, dE} = N \frac{k_1}{k_0} \frac{b^2}{2\pi\hbar} \iint dr\, dt\, e^{i(\mathbf{Q}\cdot\mathbf{r} - \omega t)} G(\mathbf{r}, t), \quad (12.2)$$

where N is the number of nuclei in the target sample and b is the coherent scattering length per nucleus (Part A, Chapter 1).

The Van Hove space–time correlation function $G(\mathbf{r}, t)$ is expressed as the sum of two terms in the classical limit

$$G(\mathbf{r}, t) = G_s(\mathbf{r}, t) + G_d(\mathbf{r}, t). \quad (12.3)$$

Here, $G_s(\mathbf{r}, t)$ is the "self" space–time correlation function representing the probability density that if an atom is located at the origin at time zero, the

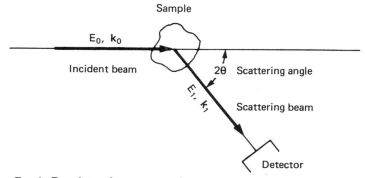

FIG. 1. Experimental arrangement for a neutron scattering measurement.

same atom is displaced to position **r** at time t. On the other hand $G_d(\mathbf{r}, t)$ is the "distinct" space-time correlation function that gives the probability density that there is a different atom at position **r** at time t when an atom is found at the origin at time zero. Therefore the physical meaning of the pair density function $\rho(\mathbf{r})$, which represents the probability density of finding a pair of atoms with a spacing of **r** at the same instant of time, corresponds to the instantaneous distinct space-time correlation function

$$G_d(\mathbf{r}, 0) = \rho(\mathbf{r}), \tag{12.4}$$

while the instantaneous self space-time correlation function can be expressed by the delta function

$$G_s(\mathbf{r}, 0) = \delta(\mathbf{r}). \tag{12.5}$$

The angular dependent differential cross section for coherent total scattering $(d\sigma_c/d\Omega)_{\text{total}}$ is directly related to the total structure factor $S(Q)$:

$$(d\sigma_c/d\Omega)_{\text{total}} = N\langle b\rangle^2 S(Q), \tag{12.6}$$

and $S(Q)$ is defined as

$$S(Q) = \frac{1}{2\pi} \iiint d\mathbf{r}\, dt\, e^{i(\mathbf{Q}\cdot\mathbf{r}-\omega t)} \{G(\mathbf{r}, t) - \rho_0\}\bigg|_{\text{constant } Q} d\omega, \tag{12.7}$$

$$= \iint d\mathbf{r}\, dt\, e^{i\mathbf{Q}\cdot\mathbf{r}} \delta(t)\{G(\mathbf{r}, t) - \rho_0\},$$

$$= \int \{G(\mathbf{r}, 0) - \rho_0\} e^{i\mathbf{Q}\cdot\mathbf{r}}\, d\mathbf{r},$$

$$= 1 + \int \{\rho(\mathbf{r}) - \rho_0\} e^{i\mathbf{Q}\cdot\mathbf{r}}\, d\mathbf{r}, \tag{12.8}$$

where ρ_0 is the average number density of the sample and Eqs. (12.3)–(12.5) are used to move from Eq. (12.7) to Eq. (12.8). Thus, integrating $d^2\sigma_c/d\Omega\, dE$ over all energy transfers at a constant value of the momentum transfer gives the Fourier transform of the pair density function $\rho(r)$, if the static approximation ($k_0 \sim k_1$) is valid.

In the practice of measuring the angular dependent differential cross section $(d\sigma_c/d\Omega)_{\text{total}}$ by coherent total neutron scattering, the energy integration is automatically performed by a detector with a detection efficiency $f(E_1)$ at a constant scattering angle 2θ, using incident neutrons of a finite energy E_0:

$$\left(\frac{d\sigma_c}{d\Omega}\right)_{\text{total}} = \int_{-\infty}^{E_0} \frac{d^2\sigma_0}{d\Omega\, dE} f(E_1)\bigg|_{\text{constant } 2\theta}\, dE. \tag{12.9}$$

The total structure factor $S(Q)$ defined by Eq. (12.7) is asymptotically accessible from Eq. (12.9), if we use incident neutrons with energy high enough ($E_0 \to \infty$) to realize the static approximation ($k_0 \sim k_1$), if constant $2\theta \to$ constant Q and if the detector efficiency $f(E_1) \sim$ constant. This means that epithermal neutrons of incident energy higher than several electron volts are needed in usual total scattering measurements. Use of high-flux epithermal neutrons generated by the electron Linac (γ, n) reaction and/or proton accelerator spallation reaction is recommended to minimize errors in measuring the instantaneous short-range structure of liquids and glasses.

Figure 2 shows schematic views of total neutron scattering spectrometers installed at the Tohoku University 300-MeV electron Linac[7,8] and at the National Laboratory for High-Energy Physics 500-MeV proton booster synchrotron.[9] The essential features of these spectrometers include shielding located inside the beam tube collimators and around the samples, the small-angle scattering system, and the geometrical and electronic focusing arrangements of detectors. High Q measurements at a low scattering angle of $2\theta < 30°$ are particularly useful for approaching an ideal $S(Q)$ of a glass.

12.2.2. High-Resolution Observation of Short-Range Structure

Even if $S(Q)$ is ideally observed by the total scattering experiment using neutrons of high incident energy, we still need to obtain $\rho(r)$ through the Fourier transformation defined as

$$\rho(r) = \rho_0 + \frac{1}{2\pi^2 r} \int_0^\infty Q\{S(Q) - 1\} \sin Qr \, dQ. \tag{12.10}$$

In the measurement of $S(Q)$, the maximum momentum transfer is limited to a finite value Q_{max} corresponding to the incident neutron energy and to the highest angle used. Therefore, the Fourier transformation of Eq. (12.10) has to be truncated at Q_{max}. This truncation has a great effect on the resolution of $\rho(r)$ in the case of amorphous solids.

In contrast to simple liquids, the atomic arrangement in amorphous solids tends to preserve polyhedral structure units with less fluctuation in bond length and angle. Here, we demonstrate how the fine short-range structure in amorphous solids is resolved by measuring the high-momentum transfer structure factor with short-wavelength neutrons, using a model structure factor:

$$S_M(Q) = 1 + \sum_{i=1}^{N} n_i \frac{\sin(Qr_i)}{Qr_i} \exp\left(-\frac{\Delta_i^2}{2} Q^2\right), \tag{12.11}$$

where n_i is the number of atoms at a position r_i from the origin, Q is the scattering vector, and $\Delta_i^2 = \langle (r_i - \langle r_i \rangle)^2 \rangle$ is the mean square fluctuation of the atomic spacing of the atom at r_i.

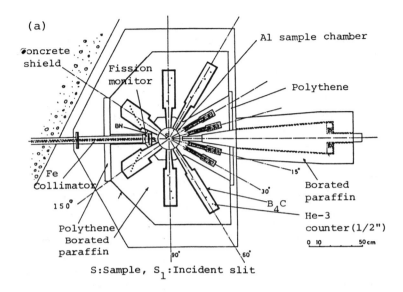

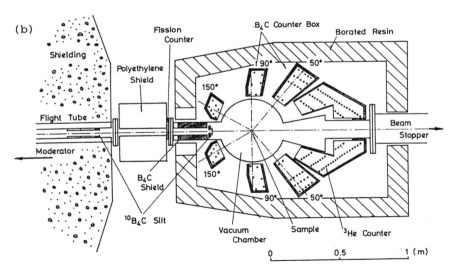

FIG. 2. Schematic views of neutron total scattering spectrometers. (a) MARK-II installed at Tohoku University 300-MeV electron Linac and (b) HIT at the National Laboratory for High Energy Physics, Tsukuba.

The Fourier transform of Eq. (12.11) truncated at Q_{max} gives the model pair correlation function (PCF) as

$$\text{PCF} = \frac{2}{\pi} \int_0^{Q_{max}} Qr\{S_M(Q) - 1\} \sin Qr \, dQ, \quad (12.12)$$

which approaches the Gaussian function

$$\lim_{Q_{max} \to \infty} \text{PCF} = \sum_{i=1}^{N} \frac{n_i}{\sqrt{2\pi\Delta_i^2}} \exp\left\{-\frac{(r - r_i)^2}{2\Delta_i^2}\right\}, \quad (12.13)$$

when Q_{max} is extended to infinity.

Misawa[10] and co-workers have discussed the effect of truncating the scattering vector on the position, height, width, and area of a single-peak PCF for the case of $i = 1$. The results are summarized in Fig. 3, where the fluctation in the atomic spacing is used as a parameter. The behavior of PCF with a small value of Δ_i^2 is sensitive to the truncation of the scattering vector in the Fourier transformation $S_M(Q) \to \text{PCF}$. On the other hand, Fukunaga[11] and co-workers have examined how the mutual separation between two peaks in the PCF is modified by truncating the Fourier transformation $S_M(Q) \to \text{PCF}$, when two Gaussian functions come near each other for the case of $i = 2$. Figure 4 shows a comparison between two different models of $S_M(Q)$:

Model 1 $i = 1$ $r = 2.9$ Å $2\sqrt{2\ln 2}\,\Delta_1 = 0.08$ Å $n_1 = 2$;

Model 2 $i = 2$ $\begin{cases} r = 2.8 \text{ Å} & 2\sqrt{2\ln 2}\,\Delta_1 = 0.08 \text{ Å} & n_1 = 1 \\ r = 3.0 \text{ Å} & 2\sqrt{2\ln 2}\,\Delta_2 = 0.08 \text{ Å} & n_2 = 1. \end{cases}$

Only small differences between amplitudes of both of the oscillatory $S_M(Q)$'s are found over the low Q region below 10 Å^{-1}. The phases in both the oscillations, however, are clearly reversed in the high-Q region above 18 Å^{-1}.

The PCFs obtained by truncating the Fourier transformation $S_M(Q) \to$ PCF at several different values of Q_{max} are compared in Fig. 5. The PCFs gradually approach the original Gaussian functions with increasing Q_{max}. In particular, the reappearance of the complete separation between the two Gaussian functions needs an extraordinarily high value of Q_{max} of more than 40 Å^{-1}. If the value of Q_{max} is truncated at less than 20 Å^{-1}, we cannot determine whether the PCF has a single peak or double peak. The measurement of the high-momentum transfer structure factor is necessary to find the fine short-range structure as well as the precise position and coordination number among nearest-neighbor atoms in amorphous solids. Therefore, short-wavelength neutrons in the epithermal energy region generated from accelerator sources are powerful probes for obtaining a highly resolved short-range structure.[12]

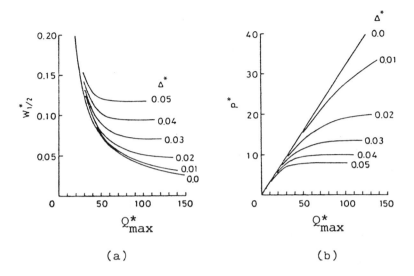
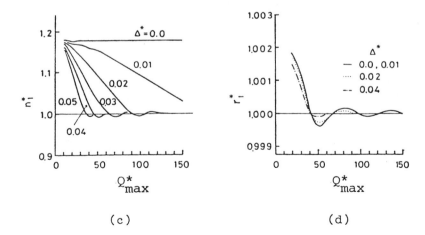

FIG. 3. Variation in the profile of single-peak PCF in case of $i = 1$ for Eq. (12.11) as a function Q^*_{max} and Δ^*. (a) FWHM $w^*_{1/2}$; (b) peak height p^*; (c) area n^*; and (d) position r^*_1. Asterisk superscript means the dimensionless reduced quantity, $Q^* = Qr_1$, $\Delta^* = \Delta/r_1$; Q is the scattering vector, r_1 the atomic distance, and $\Delta^2 = \langle (r_1 - \langle r_1 \rangle)^2 \rangle$. (After Misawa.[10])

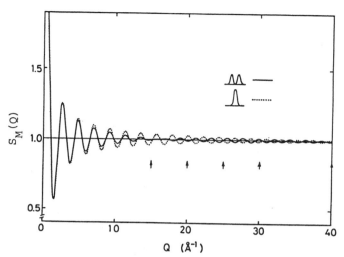

FIG. 4. Model structure factors $S_M(Q)$ in case of $i = 1$ (single peak) and 2 (double peak) for Eq. (12.11). Vertical arrows in the figure mean the Q_{max} position where the $S_M(Q) \to$ PCF Fourier transformation, Eq. (12.12), is truncated.

12.2.3. Partial Pair Density Function

Except for a few covalent amorphous solids such as selenium, arsenic, germanium, etc., the majority of amorphous solids are stabilized by containing several chemical species. The chemical and geometrical short-range structures of multicomponent amorphous solids have been discussed so far in terms of the Faber–Ziman[13] formalism and the Bhatia–Thornton[14] formalism.

The angular differential cross section $(d\sigma_c/d\Omega)_{total}$ for an amorphous solid measured by total neutron scattering is related to the Faber–Ziman total structure factor $S^{FZ}(Q)$:

$$\left(\frac{d\sigma_c}{d\Omega}\right)_{total} = N\langle b\rangle^2 S^{FZ}(Q) + N(\langle b^2\rangle - \langle b\rangle^2), \qquad (12.14)$$

where $\langle b \rangle$ is the coherent scattering length averaged over a chemical formula, $\langle b^2 \rangle - \langle b \rangle^2$ is the mean-square deviation of the scattering lengths from their average value, N is the number of chemical formula units involved in the amorphous solid sample, and $Q = 4\pi \sin\theta/\lambda$.

The Faber–Ziman total structure factor $S^{FZ}(Q)$ for an A–B binary system is expressed as the weighted sum of the three partial structure factors $S_{AA}(Q)$, $S_{AB}(Q)$, and $S_{BB}(Q)$ associated with A–A, A–B, and B–B pair correlations, respectively, in the amorphous solid:

$$S^{FZ}(Q) = w_{AA} S_{AA}(Q) + 2w_{AB} S_{AB}(Q) + w_{BB} S_{BB}(Q), \qquad (12.15)$$

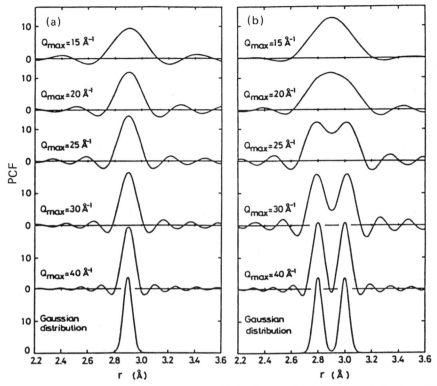

FIG. 5. Pair correlation functions (PCF) with (a) single and (b) double peak ($i = 1$ and 2, respectively), and effect of Q_{max} truncation in the $S_M(Q) \to$ PCF Fourier transformation [Eq. (12.12)] on their peak profiles. Truncated values of Q_{max} are shown on each curve.

where the weighting factor w_{ij} for the i–j pair correlation is written as

$$w_{ij} = \frac{c_i c_j \bar{b}_i \bar{b}_j}{\langle b \rangle^2}, \qquad (i, j = A, B); \tag{12.16}$$

c_i and $\bar{b}_j$ are the concentration fraction and coherent neutron scattering length of the ith atom, and $\langle b \rangle = c_A \bar{b}_A + c_B \bar{b}_B$.

The partial atom–atom pair density function $\rho_{ij}(r)$ representing the i–j pair correlation in real space is given as the Fourier transform of $S_{ij}(Q)$:

$$\rho_{ij}(r) = \rho_0 + \frac{1}{2\pi^2 r} \int_0^\infty Q\{S_{ij}(Q) - 1\} \sin Qr\, dQ. \tag{12.17}$$

The total pair density function $\rho(r)$, defined as the Fourier transform of $S^{FZ}(Q)$, is written exactly as the weighted sum of the three partial pair density functions $\rho_{AA}(r)$, $\rho_{AB}(r)$, and $\rho_{BB}(r)$ in the case of neutron scattering:

$$\rho(r) = \rho_0 + \frac{1}{2\pi^2 r}\int_0^\infty Q\{S^{FZ}(Q) - 1\}\sin Qr\, dQ \quad (12.18)$$

$$= w_{AA}\rho_{AA}(r) + 2w_{AB}\rho_{AB}(r) + w_{BB}\rho_{BB}(r), \quad (12.19)$$

because the nuclear scattering length for neutrons is independent of scattering vector Q, in contrast to the x-ray scattering amplitude.

The three partial structure factors $S_{ij}(Q)$ or partial pair density functions $\rho_{ij}(r)$ can be resolved from Eq. (12.15) or (12.19), if three independent measurements of $S^{FZ}(Q)$ or $\rho(r)$ are carried out by using three different values for each weighting factor w_{ij}. We know from Eq. (12.16) that the variation in w_{ij} values can be achieved by changing the value of the scattering length $\bar{b}_A$ and $\bar{b}_B$ through the utilization of isotope substitution in a neutron scattering experiment. However, if the first-neighbor distances in each partial pair density function are reasonably isolated, as expected in metal-metalloid amorphous alloys, we can separate peaks assigned to A–A, A–B, and B–B first-neighbor correlations in the total pair density function $\rho(r)$ by improving the experimental resolution in real space, without the difficulties often involved in the partial structure separation procedure. The extension of the magnitude of scattering vector Q up to an extraordinarily high value, such as $Q \geq 50\,\text{Å}^{-1}$, in $S(Q)$ measurements using short-wavelength neutrons generated by accelerators is quite effective for realizing high resolution in the measurement of $\rho(r)$.

12.2.4. Number Density–Concentration Partial Correlation Function

According to the proposition of Bhatia and Thornton,[14] the total structure factor $S^{BT}(Q)$ is divided into the three partial structure factors representing the number density fluctuation $S_{NN}(Q)$, the concentration fluctuation $S_{CC}(Q)$, and the cross correlation between density and concentration $S_{NC}(Q)$:

$$S^{BT}(Q) = \frac{1}{\langle b \rangle^2}\{\langle b\rangle^2 S_{NN}(Q) + 2\Delta b\langle b\rangle S_{NC} + (\Delta b)^2 S_{CC}(Q)\},$$
$$(12.20)$$

where the experimental differential cross section for coherent total neutron scattering $(d\sigma_c/d\Omega)_{\text{total}}$ is related to $S^{BT}(Q)$ as

$$(d\sigma_c/d\Omega)_{\text{total}} = N\langle b^2\rangle S^{BT}(Q), \quad (12.21)$$

$$\langle b^2\rangle = c_A \bar{b}_A^2 + c_B \bar{b}_B^2, \qquad \langle b\rangle = c_A \bar{b}_A + c_B \bar{b}_B,$$

and

$$\Delta b = |\bar{b}_A - \bar{b}_B|.$$

The physical meaning of $S_{NN}(Q)$ corresponds exactly to the geometrical short-range order in terms of the spatial correlations between number density fluctuations, while $S_{CC}(Q)$ describes only the chemical short-range order defined as the atomic-scale spatial correlation between concentration fluctuations. In contrast to x-ray diffraction, neutron diffraction makes it possible to measure $S_{CC}(Q)$ selectively, because there are some nuclei with negative neutron scattering lengths.

If the average coherent neutron scattering length of an amorphous solid is adjusted to zero by isotope substitution (i.e., $\langle b \rangle = 0$), the experimental $S^{BT}(Q)$ presents $S_{CC}(Q)/c_A c_B$ exclusively, as shown by Eq. (12.20):

$$S^{BT}(Q)|_{\langle b \rangle = 0} = \frac{(\Delta b)^2}{\langle b^2 \rangle} S_{CC}(Q) = \frac{S_{CC}(Q)}{c_A c_B}. \qquad (12.22)$$

Then, we can find the chemical short-range structure of the amorphous solid in real space by calculating the concentration partial correlation function $G_{CC}(r)$ defined as the Fourier transform of $S_{CC}(Q)$:

$$G_{CC}(r) = 4\pi r \rho_{CC}(r) = \frac{2}{\pi} \int_0^\infty Q \left[\frac{S_{CC}(Q)}{c_A c_B} - 1 \right] \sin Qr \, dQ. \qquad (12.23)$$

The generalized Warren[15] chemical short-range order parameter α, indicating the preference for unlike- or like-atom neighbors, is calculated from Eq. (12.23) as

$$\alpha = \frac{1}{Z} \int_{\Delta r} r G_{CC}(r) \, dr = 1 - \frac{Z_{AB}}{c_B Z} \qquad (12.24)$$

$0 < \alpha < 1$: $Z_{AB} < c_B Z$, preference for like-atom neighbors,

$\alpha = 0$: $Z_{AB} = c_B Z$, statistically random distribution,

$-c_A/c_B < \alpha < 0$: $Z_{AB} > c_B Z$, preference for unlike-atom neighbors,

where Z is the coordination number between nearest-neighbor atoms defined from the area under the first peak of the radial distribution function $4\pi r^2 \rho(r)$, Z_{AB} is the coordination number of B atoms surrounding an A atom, and Δr is the range of the first peak of the radial distribution function.

Relations between the atom–atom pair partial structure factors in the Faber–Ziman formalism and the number density–concentration partial structure factors in the Bhatia–Thornton formalism are established as[16]

$$S_{AA}(Q) = [c_A^2 S_{NN}(Q) + 2c_A S_{NC}(Q) + S_{CC}(Q) - c_A c_B]/c_A^2,$$

$$S_{AB}(Q) = [c_A c_B S_{NN}(Q) + (c_B - c_A) S_{NC}(Q) - S_{CC}(Q) + c_A c_B]/c_A c_B,$$

$$S_{BB}(Q) = [c_B^2 S_{NN}(Q) - 2c_B S_{NC}(Q) + S_{CC}(Q) - c_A c_B]/c_B^2, \qquad (12.25)$$

$$S_{NN}(Q) = c_A^2 S_{AA}(Q) + 2c_A c_B S_{AB}(Q) + c_B^2 S_{BB}(Q),$$
$$S_{NC}(Q) = c_A c_B [c_A S_{AA}(Q) + (c_B - c_A) S_{AB}(Q) - c_B S_{BB}(Q)], \quad (12.26)$$
$$S_{CC}(Q) = c_A c_B [c_A c_B S_{AA}(Q) - 2 S_{AB}(Q + c_A c_B S_{BB}(Q) + 1].$$

12.3. Geometrical and Chemical Short-Range Structure

12.3.1. Metal–Metalloid Amorphous Alloys

The Pd–Si alloy system is an excellent candidate in which to examine the chemical short-range structure of metal–metalloid alloys in both the liquid and glass state by total neutron scattering, because palladium and silicon atoms consist of nuclei with adequate magnitudes of the coherent scattering length and, in addition, have no contribution from magnetic scattering.

Figure 6 shows the experimental $S^{FZ}(Q)$ of $Pd_{80}Si_{20}$ glass[17] and liquid[17] measured by pulsed total neutron scattering using the Tohoku University electron Linac (γ, n) source. The oscillation in $S^{FZ}(Q)$ of $Pd_{80}Si_{20}$ glass definitely persists up to values of $Q \geq 25$ Å^{-1}. The shoulder on the high Q side of the second peak, a well-known feature common in the structure factor of amorphous metals, disappears in the liquid state, but the second peak itself preserves an asymmetrically distorted profile.

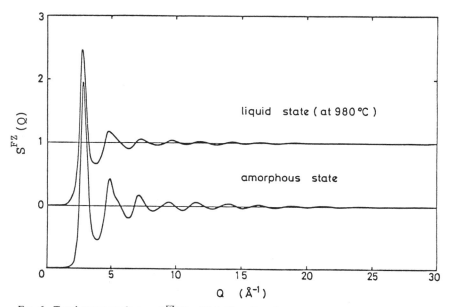

FIG. 6. Total structure factors $S^{FZ}(Q)$ of $Pd_{80}Si_{20}$ alloy glass at room temperature and liquid at 980°C.

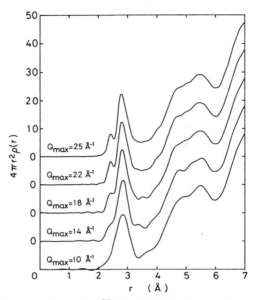

FIG. 7. Effect of Q_{max} truncation in the $S^{FZ}(Q) \to$ RDF Fourier transformation on the profile of the RDF of $Pd_{80}Si_{20}$ alloy glass.

The effect of truncating the Fourier transformation $S^{FZ}(Q) \to \rho(r)$ on the resolution of the first and second peak profiles of the radial distribution function (RDF) $4\pi r^2 \rho(r)$ of $Pd_{80}Si_{20}$ glass is illustrated in Fig. 7.[18] The RDF of $Pd_{80}Si_{20}$ glass clearly indicates the first peak split into two subpeaks and the second peak having three small humps. The subpeak on the low r side ($r \approx 2.4$ Å) of the first peak corresponds to the Si-Pd pair correlation subjected to the strong chemical bond, while the large subpeak at $r \approx 2.8$ Å is almost entirely contributed by the Pd-Pd pair correlation. This assignment is confirmed with the comparison of RDFs between the $Pd_{80}Si_{20}$ glass and Pd_3Si crystalline compound, as shown in Fig. 8. Such a first peak splitting is not observed in the $\rho(r)$ of $Pd_{80}Si_{20}$ liquid, although Fig. 9 shows that a slight hump appears at the position corresponding to the Si-Pd pair correlation of $Pd_{80}Si_{20}$ glass. It is noteworthy that the chemical short-range structure of $Pd_{80}Si_{20}$ alloy is essentially identical among the crystal, glass, and liquid states.

The peak widths and coordination numbers corresponding to the Si-Pd and Pd-Pd correlations in the first peak of the RDF of Pd-Si glasses are shown as a function of silicon concentration in Figs. 10 and 11.[17] The peak width of the Si-Pd correlation is much narrower than that of the Pd-Pd correlation, and both widths are independent of silicon concentration. The

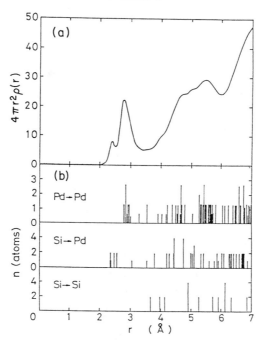

FIG. 8. Neutron RDF of (a) $Pd_{80}Si_{20}$ alloy glass $[S^{FZ}(Q) \to \rho(r)$ Fourier transform truncated at $Q_{max} = 25$ Å^{-1}] and (b) crystal structure of Pd_3Si compound (orthorhombic type).

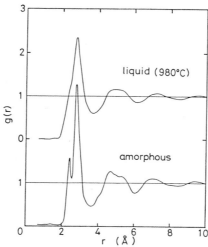

FIG. 9. Total pair distribution functions $g(r)$ of $Pd_{80}Si_{20}$ alloy glass at room temperature and liquid at 980°C.

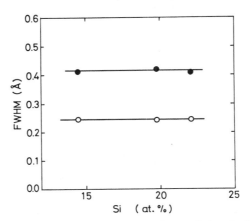

FIG. 10. First peak widths of Pd-Si (○) and Pd-Pd (●) correlations in the neutron RDF of Pd-Si alloy glasses ($Q_{max} = 25$ Å^{-1}).

coordination number of palladium atoms around a palladium atom (n_{Pd-Pd}) is about 10.6 palladium atoms below 20 at. % silicon and then approaches the value found in the Pd$_3$Si crystalline compound with increasing silicon content. The coordination number of palladium atoms surrounding a silicon atom (n_{Si-Pd}) decreases linearly and approaches again the value in the Pd$_3$Si crystalline compound with increasing silicon concentration.

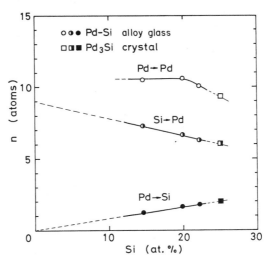

FIG. 11. Coordination numbers obtained from the neutron RDF of Pd-Si alloy glasses, where the $S^{FZ}(Q) \to \rho(r)$ Fourier transform is truncated at $Q_{max} = 25$ Å^{-1}. In the figure, A → B means the number of B atoms around an A atom.

We can recognize from Fig. 8 that the shortest spacing between silicon atoms in $Pd_{80}Si_{20}$ glass is never located at the contact position, because there is no peak at the position corresponding to the silicon atom diameter of about 2 Å in the RDF. Fugunaga and Suzuki[19] distinguished the three partial pair distribution functions of Pd-Pd, Pd-Si, and Si-Si correlation in $Pd_{80}Si_{20}$ glass by using a combination of x-ray, electron, and neutron diffraction and found that the first peak in $\rho_{SiSi}(r)$ is located at the position of about 3.2 Å, rather close to the Si-Si separation existing in the Pd_3Si crystalline compound.

The neutron diffraction result supports Gaskell's[20] proposition that the structure of metal-metalloid amorphous solid alloys can be interpreted in terms of the nonperiodic connection of stereochemically defined polyhedra, which have a chemical short-range structure almost similar to that of corresponding crystalline compounds. This fact has been confirmed by Hayashi et al.[21] in a high-resolution measurement of the RDF for Pd-Ge glasses, where the local environment surrounding a Ge atom consists of about six palladium atoms forming prismatic packing over the entire glass-forming composition range.

The Ni-B alloys are quenched into the glass state from their melts over a wide composition range from 18 to 40 at. % B. This composition range includes three different kinds of crystalline compounds, Ni_3B, Ni_2B, and Ni_4B_3, in the equilibrium phase diagram, as shown in Fig. 12. The local environments around a boron atom in both crystalline Ni_3B and Ni_4B_3 are constructed from trigonal prismatic packing of six nickel atoms surrounding

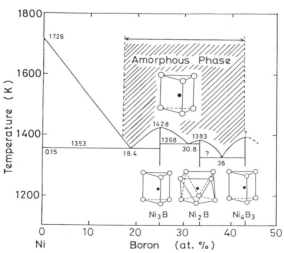

FIG. 12. Nickel-boron binary diagram and local B atom environments in crystalline compounds and alloy glasses.

a boron atom, while crystalline Ni_2B has a boron atom situated in the central hole of an Archimedian antiprism consisting of eight nickel atoms. Therefore, the Ni–B alloy system is one of the interesting materials for examining how the short-range structure is conserved in going between glass and crystal states.

The total scattering measurement using the KENS spallation neutron source has shown that the coordination number of nickel atoms surrounding a boron atom lies in the range Z_{BNi} = 5 to 6 neighbors over the whole boron content.[22] This implies that the Ni–B glasses prefer to have the trigonal prismatic packing of 6 nickel atoms surrounding a boron atom, which corresponds to the boron-atom site environments found in crystalline Ni_3B and Ni_4B_3. The Archimedian antiprism of eight nickel atoms around a boron atom in crystalline Ni_3B is not preserved but modified into the trigonal prism of six nickel atoms surrounding a boron atom in the glass. The neutron results agree with the NMR observations[23] on the electric field gradient around a boron atom in Ni–B glasses.

The Ni–B glasses described above were prepared from natural nickel and boron. This indicates that a reliable $S^{FZ}(Q)$ can be obtained even for an alloy glass with constituents such as natural boron with a high absorption cross section for thermal neutrons, if short-wavelength pulsed neutrons generated from a spallation source are utilized.

In order to specify the individual structure of metal–metalloid alloy glasses, we must find the mutual connectivity among the polyhedral structure units having the specified chemical short-range order. Therefore, it is most important to characterize the metalloid–metalloid correlations existing inherently in metal–metalloid alloy glasses, because these describe the center-center correlations between the polyhedra. High-quality separation of three partial structures in $Fe_{80}B_{20}$ and $Ni_{81}B_{19}$ alloy glasses has been successfully achieved using neutron diffraction with isotope substitution by Lamparter et al.[24] Their results show a drastic variation in the B–B pair correlation between the two alloy glasses, suggesting obviously different ways of connecting the polyhedral structure units, in contrast to the similarity of metal–metal and metal–boron pair correlations between the two glasses. The precise determination of the metalloid–metalloid pair correlations in various kinds of metal–metalloid amorphous alloys is still needed to elucidate the intermediate-range structure.

12.3.2. Metal–Metal Amorphous Alloys

It is interesting to examine what kinds of chemical short-range order still exist in metal–metal amorphous alloys cohesively stabilized by metallic bonds. Figure 13 shows $S^{BT}(Q) = S_{CC}(Q/c_{Ni}c_{Ti})$ of Ni–Ti neutron zero-scattering

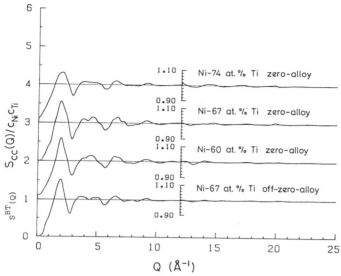

FIG. 13. Concentration–concentration structure factors $S_{CC}(Q)/c_{Ni}c_{Ti}$ of Ni–Ti neutron zero-scattering alloy glasses and Bhatia–Thornton total structure factor $S^{BT}(Q)$ of $Ni_{33}Ti_{67}$ neutron off-zero-scattering alloy glass.

glasses with three different alloy compositions of 74, 67, and 60 at. % Ti.[25] These glasses were prepared by melt-quenching from natural nickel metal ($b_{Ni} = 10.3 \times 10^{-15}$ m), ^{60}Ni-isotope (enriched to 99.08%) metal ($b_{Ni-60} = 2.82 \times 10^{-15}$ m), and natural titanium metal ($b_{Ti} = -3.37 \times 10^{-15}$ m). Characteristic oscillations in $S_{CC}(Q)/c_{Ni}c_{Ti}$ obviously persist up to very high scattering vectors in the region $Q \geq 20$ Å^{-1}. The main peak located at $Q \approx 1.95$ Å^{-1} in $S_{CC}(Q)/c_{Ni}c_{Ti}$ becomes sharp and high with decreasing titanium content.

The $G_{CC}(r)$ functions for the Ni–Ti neutron zero-scattering alloy glasses defined by Eq. (12.23) are shown in Fig. 14. The negative first peak located at $r \approx 2.54$ Å, whose position shifts slightly to larger distance with increasing titanium content, indicates the preference for Ni–Ti unlike-atom pairs in Ni–Ti alloy glasses. The Ni–Ti unlike-atom pairs appear at the first-neighbor position in the $NiTi_2$ crystalline compound, too. Figure 15 shows a comparison between the $G_{CC}(r)$ of the $Ni_{33}Ti_{67}$ neutron zero-scattering alloy glass and the crystal structure of the stoichiometric $NiTi_2$ crystalline compound. The negative peak at $r \approx 2.54$ Å in $G_{CC}(r)$ corresponds exactly to the Ni–Ti unlike-atom pairs in the crystalline compound.

The two positive peaks at $r \approx 3.0$ and 4.1 Å in $G_{CC}(r)$ are found to be predominantly contributed by Ti–Ti and Ni–Ni like-atom pairs in the $NiTi_2$

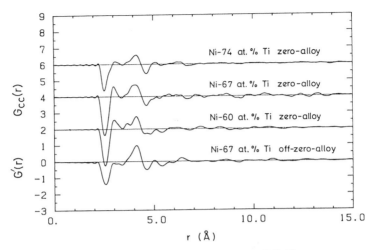

FIG. 14. Concentration partial correlation functions $G_{CC}(r)$ of Ni–Ti neutron zero-scattering alloy glasses and Bhatia–Thornton total pair correlation function $G'(r)$ of $Ni_{33}Ti_{67}$ neutron off-zero-scattering alloy glass.

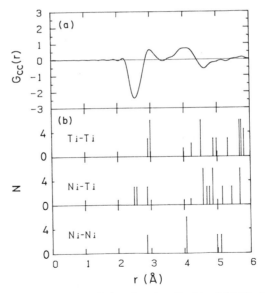

FIG. 15. Concentration partial correlation function $G_{CC}(r)$ of (a) $Ni_{33}Ti_{67}$ neutron zero-scattering alloy glass and (b) atom–atom pair correlations in $NiTi_2$ crystalline compound, where A–B in the figure means the number of B atoms around an A atom.

TABLE II. Generalized Warren Chemical Short-Range Order Parameter α for Ni-Ti Alloy Glasses[a]

	α	α_{COA}	α/α_{COA}	Z^{b}	Z_{NiTi}	Z_{NiTi}^{R}
Ni–74 at. % Ti	−0.102	−0.351	0.290	11.85	9.66	8.77
Ni–67 at. % Ti	−0.141	−0.493	0.287	12.07	9.23	8.09
Ni–60 at. % Ti	−0.116	−0.667	0.175	12.77	8.55	7.66

[a] $\alpha_{COA} = -c_{Ni}/c_{Ti}$ for a completely ordered alloy, ratio α/α_{COA}, total coordination number Z obtained from the first peak area of RDF, experimental coordination number of titanium atoms around a nickel atom in Ni-Ti alloy glasses, Z_{NiTi}, and coordination number of titanium atoms around a nickel atom in a random distribution Ni-Ti alloy, Z_{NiTi}^{R}.
[b] X-ray diffraction.

crystal structure. Therefore, the chemical short-range structure in Ni–Ti alloy glasses is quite analogous to that in the crystalline compound.

The values of the Warren chemical short-range order parameter α calculated from Eq. (12.24) for Ni–Ti neutron zero-scattering alloy glasses are shown in Table II, together with the values of $\alpha_{COA} = -c_{Ni}/c_{Ti}$ for the completely ordered alloy ($Z = Z_{AB}$) and of the ratio α/α_{COA}. Negative values of α for Ni–Ti neutron zero-scattering alloy glasses certainly imply the preferential existence of Ni–Ti unlike-atom pairs at the nearest-neighbor position. The value of α for the $Ni_{33}Ti_{67}$ alloy glass corresponding to the stoichiometric $NiTi_2$ crystalline compound is larger than those for the two off-stoichiometric Ni–Ti alloy glasses. However, the α/α_{COA} ratio has an almost similar value in high-titanium-content alloys of $Ni_{26}Ti_{74}$ and $Ni_{33}Ti_{67}$ alloy glasses. In contrast, a drastic decrease in the α/α_{COA} value is found in the lower titanium content alloy of $Ni_{40}Ti_{60}$ alloy glass.

Table II also shows the values of the total coordination number Z obtained by x-ray diffraction and the coordination numbers of titanium atoms around a nickel atom in Ni–Ti neutron zero-scattering alloy glasses (Z_{NiTi}) and in statistically random distribution alloys (Z_{NiTi}^{R}) calculated from the values of α and Eq. (12.24). The value of Z increases with decreasing titanium content in the alloys. It is noteworthy that near-stoichiometric $Ni_{33}Ti_{67}$ alloy glass has about nine titanium atoms around a nickel atom. Figure 15 shows that the stoichiometric $NiTi_2$ crystalline compound has six titanium atoms at $r \approx 2.5$ Å and three titanium atoms at $r \approx 2.9$ Å around a nickel atom. Therefore, the chemical short-range order in the $NiTi_2$ crystalline and amorphous solid alloys must be very similar, if it is supposed that three titanium atoms at $r \approx 2.9$ Å shift toward the position of six titanium atoms located at $r \approx 2.5$ Å due to the strong chemical bond in $NiTi_2$ alloy glasses. In fact, a dotted line connecting experimental coordination numbers of titanium atoms around a nickel atom in Ni–Ti alloy glasses passes by those in both the $NiTi_2$ and NiTi crystalline compounds, as shown in Fig. 16.

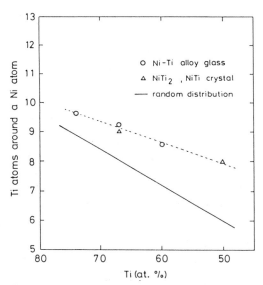

FIG. 16. Coordination numbers of Ti atoms around a Ni atom in Ni–Ti alloy glasses (○) and NiTi$_2$ and NiTi crystalline compounds (△). The solid line gives the coordination number of Ti atoms around a Ni atom in the case of statistically random distribution of both atoms under the total coordination number of 12 atoms.

It is concluded that the chemical short-range structure even in metal–metal amorphous alloys is quite analogous to that in corresponding crystalline compounds over the entire glass-forming composition range. Such a conclusion has been confirmed in the chemical short-range structure of Cu–Ti alloy glasses,[26] where the Cu–Cu pair correlation appears predominantly at the nearest-neighbor position as in the CuTi crystalline compound. There are, however, some differences in the chemical short-range order between Ni–Ti and Cu–Ti alloy glasses, which may originate from the different nature of the chemical bonds between Ni–Ti and Cu–Ti pairs.

Quite recently, reliable partial structures have been successfully resolved for Ni$_{40}$Ti$_{60}$[27] and Ni$_{50}$Zr$_{50}$[28] alloy glasses by using clever choices of isotope substitution to avoid the ill-conditioned matrix for weighting factors w_{ij} defined as Eq. (12.16). In case of Ni$_{50}$Zr$_{50}$ alloy glass, the weighting factors for Ni–Zr unlike-atom pairs w_{NiZr} in three total structure factors $S^{\text{FZ}}(Q)$ were adjusted to have the same value, so that the calculation of the determinant for w_{ij} was simplified.

The atom-atom partial structure factors $S_{ij}(Q)$ and reduced partial pair correlation functions $G_{ij}(r) = 4\pi r\{\rho(r) - \rho_0\}/\rho_0$ for Ni$_{50}$Zr$_{50}$ alloy glass are shown in Figs. 17 and 18, respectively. It is noteworthy that the Ni–Ni pair correlation is weak at the first-neighbor position and rather emphasized in

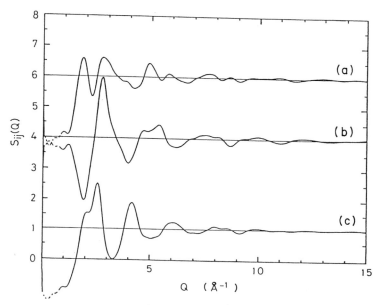

FIG. 17. Atom–atom partial structure factors (a) $S_{NiNi}(Q)$, (b) $S_{NiZr}(Q)$, and (c) $S_{ZrZr}(Q)$ of $Ni_{50}Zr_{50}$ alloy glass.

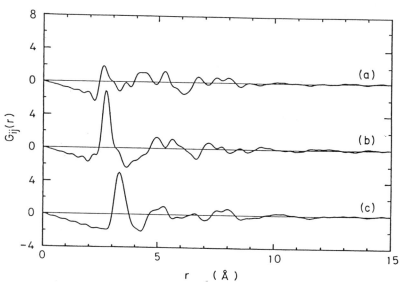

FIG. 18. Atom–atom partial distribution function (a) $G_{NiNi}(r)$, (b) $G_{NiZr}(r)$, and (c) $G_{ZrZr}(r)$ of $Ni_{50}Zr_{50}$ alloy glass.

TABLE III. Interatomic Distances (r) and Coordination Numbers (n) of the Nearest-Neighbor Ni-Ni, Ni-Zr, and Zr-Zr Correlations in ZrNi Alloy Glass and Crystalline Compound[a]

Atomic pair	NiZr glass		NiZr crystal		
	r (Å)	n (atoms)	r (Å)	n (atoms)	
Ni—Ni	2.63	3.3	2.49	2	(4)
			3.26	2	
Ni—Zr	2.73	6.7	2.66	2	
			2.70	4	(7)
			2.87	1	
Zr—Zr	3.32	7.8	3.26	2	
			3.41	4	(8)
			3.46	2	

[a] The coordination number for A-B pair means the number of B atoms surrounding an A atom.

highest order neighbor positions. A similar behavior of the Ni-Ni pair correlation is also found in $Ni_{40}Ti_{60}$ alloy glass.

The coordination numbers and interatomic distances for each partial pair correlation in $Ni_{50}Zr_{50}$ alloy glass are listed in Table III, together with those of the NiZr crystalline compound. From this table it is concluded that the Ni-Ni and Zr-Zr interatomic distances in the alloy glass are shorter than the corresponding ones in the crystalline compound, but longer than the nickel and zirconium atomic diameters ($D_{Ni} = 2.48$ Å and $D_{Zr} = 3.20$ Å). The Ni-Zr interatomic distance in the alloy glass is close to that in the crystalline

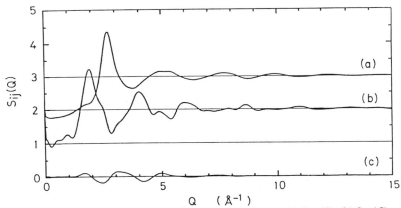

FIG. 19. Number density-concentration partial structure factors (a) $S_{NN}(Q)$, (b) $S_{NC}(Q)$, and (c) $S_{CC}(Q)$ of $Ni_{50}Zr_{50}$ alloy glass.

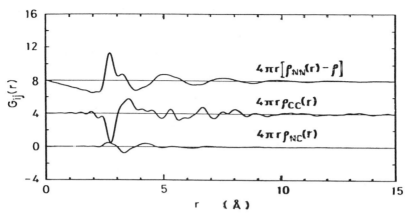

FIG. 20. Number density–concentration partial correlation functions $G_{NN}(r)$, $G_{NC}(r)$, and $G_{CC}(r)$ of $Ni_{50}Zr_{50}$ alloy glass.

compound, but smaller than the mean value of the nickel and zirconium atomic diameters. The coordination numbers of Ni–Ni, Ni–Zr, and Zr–Zr pair correlations in $Ni_{50}Zr_{50}$ alloy glass are slightly less than those in the crystalline compound. The chemical short-range structures are very similar.

The number density–concentration partial structure factors $S_{NN}(Q)$, $S_{NC}(Q)$, and $S_{CC}(Q)$ of $Ni_{50}Zr_{50}$ alloy glass can be calculated directly from the atom–atom partial structure factors $S_{NiNi}(Q)$, $S_{NiTi}(Q)$, and $S_{TiTi}(Q)$ given in Fig. 17 by using Eq. (12.25). The results are shown in Fig. 19. Figure 20 is the number density–concentration partial pair correlation functions $4\pi r[\rho_{NN}(r) - \rho_0]$, $4\pi r \rho_{NC}$, and $4\pi r \rho_{CC}(r)$ for $Ni_{50}Zr_{50}$ alloy glass, defined as the Fourier transforms of $S_{NN}(Q)$, $S_{NC}(Q)$, and $S_{CC}(Q)$ respectively.

From a comparison between Figs. 17 and 19, it is concluded that the large positive peak at $Q \approx 1.9 \text{ Å}^{-1}$ in $S_{CC}(Q)$, which corresponds to the prepeak appearing in the total structure factor $S^{FZ}(Q)$ of $Ni_{50}Zr_{50}$ alloy glass, is mainly contributed by the Ni–Ni pair correlation in the alloy glass. Furthermore, the oscillations in $4\pi r \rho_{CC}(r)$ and $G_{NiZr}(r)$ are found to be in an antiphase relation, with peaks in one corresponding exactly to valleys in the other, as indicated in Figs. 18 and 20.

12.3.3. Bridging and Nonbridging Oxygen Atoms in Oxide Glasses

The discrimination between the bridging oxygen atom (O^0) and nonbridging oxygen atom (O^-) bound to a central glass-forming cation in polyhedral structure units making up the three-dimensional network structure is one of the most important keys for characterizing the chemical short-range structure of oxide glasses. So far, x-ray photoelectron spectroscopy measurements

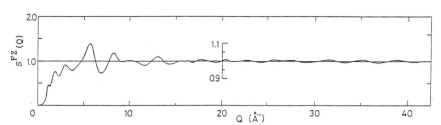

FIG. 21. Neutron total structure factor $S^{FZ}(Q)$ of $P_2O_5 \cdot Na_2O$ glass.

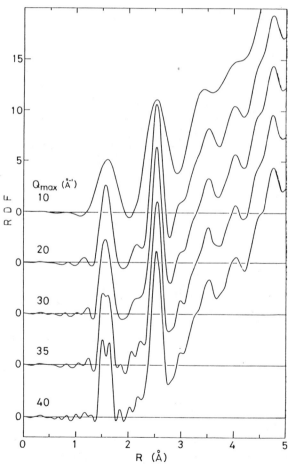

FIG. 22. Radial distribution functions RDF of $P_2O_5 \cdot Na_2O$ glass, where the $S^{FZ}(Q) \rightarrow \rho(r)$ Fourier transform is truncated at various values of Q_{max}.

have been carried out to identify the bridging (O^0), nonbridging (O^-), and free (O^{2-}) oxygen atoms existing in silicate and phosphate glasses.

From the high-resolution measurement of the neutron radial distribution function, Misawa et al.[29] have concluded that the Si—O^- bond length is significantly smaller than the Si—O^0 bond length in $Na_2O \cdot 2SiO_2$ glass as well as in the $Na_2O \cdot 2SiO_2$ crystalline compound. Unfortunately, the spatial resolution in the Misawa et al. experiment is not enough to definitively resolve the Si—O^0 and Si—O^- bond lengths which are too close to each other in the silicate glass.

It is well known that there is a significant difference in the bond lengths between the bridging (P—O^0) and the nonbridging (P—O^- and P=O^0) oxygen—phosphorus bond in the $NaPO_3$ crystalline compound. Recently, Suzuki and Ueno[30] have succeeded in directly discriminating between P—O^0 and P—O^- (or P=O^0) bonds in the total radial distribution function of $P_2O_5 \cdot Na_2O$ glass by obtaining the neutron total structure factor $S^{FZ}(Q)$ up to a very high value of the magnitude of scattering vector Q, about 40 Å^{-1}, using an electron Linac pulsed neutron source.

The experimental $S^{FZ}(Q)$ of $P_2O_5 \cdot Na_2O$ glass still has an oscillatory

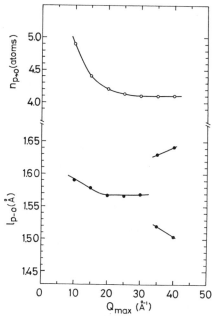

FIG. 23. The Q_{max} dependence of the position and area of the first peak in the neutron RDF of $P_2O_5 \cdot Na_2O$ glass, which correspond to the P—O bond length and coordination number of O atoms surrounding a P atom in the glass.

behavior at $Q = 40$ Å^{-1} as shown in Fig. 21. Figure 22 shows the RDF defined as the Fourier transform of the $S^{FZ}(Q)$ truncated at various values of Q_{max}. The first peak in the RDF is split into two subpeaks when the Q_{max} value is extended beyond 35 Å^{-1}. The first subpeak is located at $r = 1.50$ Å, while the position of the second subpeak is $r = 1.64$ Å. It is not certain that these values have finally converged, because Fig. 23 shows that the Q_{max} dependence of the split peak positions is not saturated even at $Q_{max} = 40$ Å^{-1}. If the short-range structure of $P_2O_5 \cdot Na_2O$ glass is assumed to resemble that of the corresponding $NaPO_3$ crystalline compound, the first and second subpeak can be surely assigned to the $P-O^-$ and $P-O^0$ bonds existing in the glass.

Based on least-squares fitting using Gaussian functions, it is concluded that the coordination numbers of O^- and O^0 atoms bound to a phosphorus atom are 2.1 and 2.0 ± 0.1 in $P_2O_5 \cdot Na_2O$ glass. Therefore the basic structural unit making up $P_2O_5 \cdot Na_2O$ glass is a distorted tetrahedron consisting of four oxygen atoms whose central hole is occupied by a phosphorus atom. The central phosphorus atom is bound to two O^- atoms with the $P-O^-$ bond length of 1.50 Å and two O^0 atoms with the $P-O^0$ bond length of 1.64 Å. Both bond lengths in the glass are longer than those in the crystalline compounds. A similar behavior has been observed in the case of silicate glasses. A small peak located around $r = 3.0$ Å in the RDF corresponds to the $P-P$ correlation in $P_2O_5 \cdot Na_2O$ glass. This means that the average $\angle P-O^0-P$ bond angle is about $130°$ in $P_2O_5 \cdot Na_2O$ glass.

12.4. Structural Anisotropy and Fluctuations in the Intermediate Range

12.4.1. CVD Amorphous Si_3N_4

Bulk samples of amorphous Si_3Ni_4 can be prepared by chemical vapor deposition (CVD). The neutron total structure factor $S^{FZ}(Q)$ observed by using a (γ, n) pulsed neutron source installed on an electron Linac is shown in Fig. 24.[31] The oscillations in the $S^{FZ}(Q)$ persist up to a very high scattering vector $Q \geq 30$ Å^{-1} and so-called small-angle scattering intensity is found in the low Q region of $Q \leq 0.1$ Å^{-1}.

Figure 25 shows the radial distribution functions that are Fourier transformed by truncating $S^{FZ}(Q)$ at various values of the magnitude of scattering vector Q_{max}.[31] The first peak appearing around $r = 1.7$ Å in the RDF corresponds to the Si–N bond in CVD amorphous Si_3N_4. The Q_{max} dependence of the position and area of the first peak is demonstrated in Fig. 26. The most likely values of the Si—N bond length and the coordination number of nitrogen atoms binding to a silicon atom should be deduced from the asymptotic values of both curves at the limit of high Q_{max}. The numerical

12. GLASSES

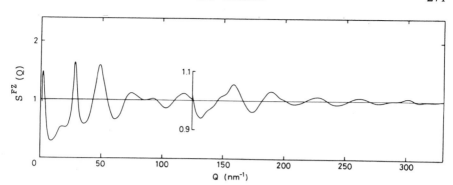

Fig. 24. Neutron total structure factor $S^{FZ}(Q)$ of CVD amorphous Si_3N_4.

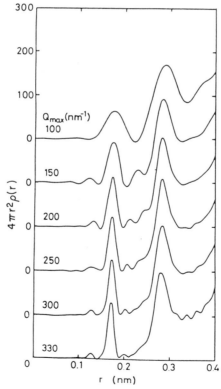

Fig. 25. Radial distribution functions of CVD amorphous Si_3N_4, where the $S^{FZ}(Q) \to \rho(r)$ Fourier transform is truncated at various values of Q_{max}.

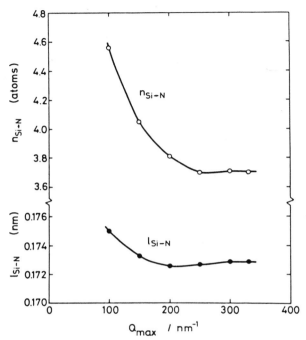

FIG. 26. The Q_{max} dependence of the Si—N bond length and coordination number of N atoms bound to a Si atom in CVD amorphous Si_3N_4.

results are listed in Table IV, together with the root-mean-square displacement of the Si—N bond length obtained by Gaussian fitting of the first peak in the RDF corresponding to $Q_{max} = 33$ Å^{-1}.

Misawa et al.[31] have also distinguished N-N and Si-N pair correlations in CVD amorphous Si_3N_4 from the second peak of the RDF by combining

TABLE IV. Coordination Number (n), Interatomic Distance (l), Root-Mean Square Displacement (Δl), and Bond Angle (ϕ) for Si—N, N—N, and Si—Si Pair Correlations in CVD Amorphous Si_3N_4[a]

	Si–N	N–Si	N–N	Si–Si
n (atoms)				
experimental	3.7	2.8	7.7	6.5
ideal	4.0	3.0	9.0	8.0
l (Å)	1.729	1.729	2.83	3.01
Δl (Å)	0.007	0.007	0.016	0.016
ϕ (deg)			109.8	121.0
			(N-Si-N)	(Si-N-Si)

[a] The coordination number of A–B pair means the number of B atoms surrounding an A atom.

x-ray[32] and neutron total scattering measurements. The atomic spacings and coordination numbers for both like-atom pairs are summarized in Table IV. The bond angle at the silicon atom is 109.8°, which is very close to the tetrahedral bond angle (109.47°). The value of 121.0° for the bond angle around a nitrogen atom is in good agreement with that of a perfect triangle (120°).

The $S^{FZ}(Q)$ observed at a scattering angle $2\theta = 5°$ in a TOF neutron total scattering experiment has two contributions: one is the small-angle scattering term $S_{SA}(Q)$ and the other is the regular structure factor extrapolated smoothly toward $Q = 0$ from the high Q region, as shown in Fig. 27. The $S_{SA}(Q)$ is related to the characteristic function $\gamma(r)$, which specifies the fluctuation in a sample[33]:

$$S_{SA}(Q) = A\rho_0 \int_0^\infty 4\pi r^2 \gamma(r) \frac{\sin Qr}{Qr} dr, \qquad (12.27)$$

$$A = (1 - w_f)w_f \frac{1}{\rho_0^2 \langle b \rangle^2} (\rho_m b_m - \rho_f b_f)^2, \qquad (12.28)$$

where w_f is the volume fraction of the fluctuating region, ρ_m, ρ_f and ρ_0 are the atomic number densities of the matrix, flucturating region, and entire sample, and b_m, b_f, and $\langle b \rangle$ are the average scattering lengths of nuclei in

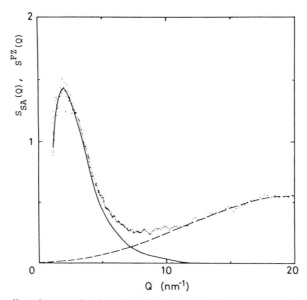

FIG. 27. Small-angle scattering intensity $S_{SA}(Q)$ from CVD amorphous Si_3N_4 (solid line). Dots in the figure represent the experimental neutron total structure factor $S^{FZ}(Q)$ observed at the scattering angle $2\theta = 5°$, and the broken line is an extrapolation of $S^{FZ}(Q)$ toward $Q = 0$.

the matrix, fluctuating region, and entire sample, respectively. The area of interface per unit volume (S/V) is calculated from

$$S/V = \pi(1 - w_f)w_f \lim_{Q \to \infty} [Q^4 S_{SA}(Q)]/2\pi^2 A\rho_0. \qquad (12.29)$$

The mean volume of each fluctuation region v_f and the length of heterogeneity l_c are

$$v_f = \int_0^\infty 4\pi r^2 \gamma(r)\, dr \qquad (12.30)$$

and

$$l_c = 2\int_0^\infty \gamma(r)\, dr. \qquad (12.31)$$

The nature of the fluctuating region with physically reasonable properties can be assigned by using Eq. (12.28). It is concluded that it must consist of voids about 10 Å in average diameter. The possibility of other kinds of fluctuating regions, such as SiO_2 and Si- or N-rich Si_3N_4, are ruled out because unreasonably large volume fractions are estimated from the experimental small-angle scattering intensity. The values of the parameters characterizing the defect structure in CVD amorphous Si_3N_4 are summarized as w_f = 4.2 vol. %, $S/V = 630\, m^2/cm^3$, $l_c = 7$ Å, and $v_f = 440$ Å^3. Misawa et al.[31] have reasonably interpreted the coordination number deficiencies from the ideal values, as listed in Table IV, on the assumption that there are dangling bonds only on the surface of the void in CVD amorphous Si_3N_4. This means that the distortion of covalent bonds in amorphous Si_3N_4 is relaxed by introducing many small voids into the bulk during the CVD processing.

12.4.2. Sputter-Deposited Amorphous SiO₂ and Ni—V Alloy

SiO_2 is one of the materials used extensively in various engineering applications. In particular, great attention has been paid to characterizing and controlling the structure defect included inherently in the SiO_2 glass film prepared by deposition from the gaseous phase. Misawa et al.[34] have investigated with pulsed neutron total scattering what structural differences exist between sputter-deposited SiO_2 amorphous film (hereafter called a-SiO_2) and ordinary melt-quenched SiO_2 glass (hereafter called g-SiO_2).

The Faber–Ziman total structure factors $S^{FZ}(Q)$ observed for a-SiO_2 and g-SiO_2 are compared in Fig. 28. There are remarkable differences between the two structure factors. Intense small-angle scattering is clearly observed in the $S^{FZ}(Q)$ of a-SiO_2, indicating the existence of inhomogeneity in the sample. The first peak in $S^{FZ}(Q)$ of a-SiO_2 is drastically attenuated compared with that of g-SiO_2.

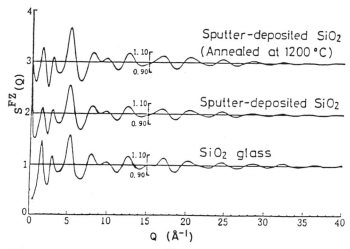

FIG. 28. Neutron total structure factors $S^{FZ}(Q)$ of sputter-deposited SiO₂ amorphous film and SiO₂ melt-quenched glass.

The RDFs obtained as the Fourier transform of $S^{FZ}(Q)$ truncated at $Q_{max} = 40$ Å are shown in Fig. 29. From the first peak of the RDF, the structural unit in a-SiO₂ is confirmed to be the [SiO₄] tetrahedron as in g-SiO₂. However, the fluctuations of Si—O bond length and O–O atomic distance are more pronounced in a-SiO₂.

In order to evaluate the dedgree of disorder in a-SiO₂, one may introduce a broadening function $\alpha(r)$ with which the pair distribution function $g(r)$ of

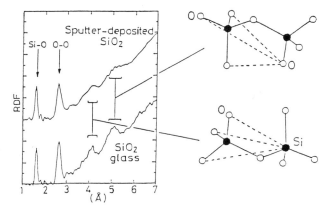

FIG. 29. Radial distribution functions RDF of sputter-deposited SiO₂ amorphous film and SiO₂ melt-quenched glass.

a-SiO$_2$ is given as

$$g(r) - 1 = \alpha(r)[g_0(r) - 1], \qquad (12.32)$$

where $g_0(r)$ is the pair distribution function of g-SiO$_2$. One of the simplest approximations for $\alpha(r)$ is presented by the equation

$$\alpha(r) = 1 - \frac{3}{2}\left(\frac{r}{2R_s}\right) + \frac{1}{2}\left(\frac{r}{2R_s}\right)^2, \qquad (12.33)$$

where $2R_s$ denotes the characteristic length beyond which atomic correlations disappear. The $S(Q)$'s calculated as the Fourier transform of Eq. (12.33) as a function of R_s are shown in Fig. 30. The $S(Q)$ with $R_s = 7$ Å reproduces well the experimental $S^{FZ}(Q)$. Therefore, it is concluded that the atomic correlation in a-SiO$_2$ disappears rapidly beyond about $2R_s = 14$ Å, while g-SiO$_2$ retains significant atomic correlations up to $2R_s \geq 30$ Å. Such a drastic loss of the correlation length in a-SiO$_2$ can be also verified from the fact that the Si-O and O-O correlations between neighboring [SiO$_4$] tetrahedra obviously tend to blur in the RDF of a-SiO$_2$ as shown in Fig. 29.

Analysis of the small-angle scattering intensity in the low Q region with $Q \leq 0.1$ Å^{-1}, shown in Fig. 28 by using Eqs. (12.28)–(12.31), suggests that a-SiO$_2$ includes voids of about 10 Å in diameter with a volume fraction of about 4%. After annealing at 1200°C, these voids remain in the glass, while the height of the first peak in $S^{FZ}(Q)$ of a-SiO$_2$ is almost restored to that of g-SiO$_2$, indicating that the geometrical distortion and connection of [SiO$_4$] tetrahedra in a-SiO$_2$ approaches the state found in g-SiO$_2$.

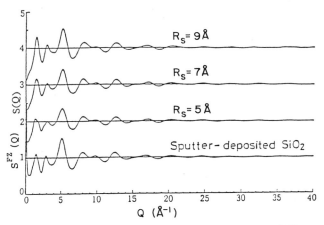

FIG. 30. The model $S(Q)$ defined as the Fourier transform of Eq. (12.32) as a function of correlation length R_s and the experimental $S^{FZ}(Q)$ of sputter-deposited SiO$_2$ amorphous film.

It is well known that amorphous alloys often have anisotropy in their atomic-scale structure, which depends sensitively on the method of preparation. The control of structural anisotropy is one of the most important aspects of engineering applications of amorphous alloys as well as oxide glasses. The neutron total scattering experiment combining time-of-flight measurement and an accelerator-pulsed neutron source is a powerful technique for detecting the small differences between direction-dependent structure factors. The first experiment was carried out by Windsor et al.[35] on $Pd_{80}Si_{20}$ alloy glass ribbon prepared by a single-roll melt-quenching method. They found that the short-range structure parallel to the long ribbon direction is less ordered than that in the other directions.

Fukunaga et al.[36] have simultaneously measured the neutron total structure factors $S^{FZ}(Q)$ of an $Ni_{58}V_{42}$ amorphous alloy plate prepared by argon-plasma DC sputter-deposition in two directions with the scattering vector Q parallel and perpendicular to the direction of deposition. Figures 31 and 32 indicate that there is significant anisotropy in the Ni–Ni partial structure of the sputter-deposited $Ni_{58}V_{42}$ amorphous alloy.

It is noteworthy that the long-range order of the Ni—Ni correlation is rather isotropic because of an isotropic prepeak appearing at $Q \approx 2 \text{ Å}^{-1}$, while the density of Ni—Ni pairs situated at the first-neighbor position is higher parallel to the direction of deposition than perpendicular to it. This identification may also be helpful in interpreting the magnetic anisotropy existing in ferromagnetic amorphous alloy thin plates including iron atoms.

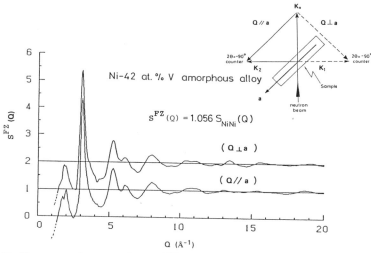

FIG. 31. Neutron total structure factors $S^{FZ}(Q)$ of sputter-deposited $Ni_{58}V_{42}$ amorphous alloy plate, when the scattering vector **Q** is parallel and perpendicular to the plate surface (**a**).

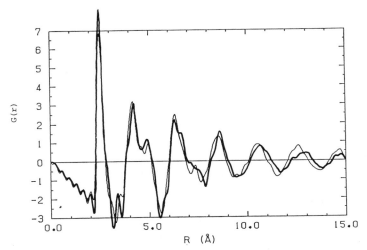

FIG. 32. Pair correlation functions $G(r)$ of sputter-deposited $Ni_{58}V_{42}$ amorphous plate, when the scattering vector **Q** of the structure factor $S^{FZ}(\mathbf{Q})$ is parallel (thin solid line) and perpendicular (thick solid line) to the plate surface (**a**).

12.4.3. GeO_2—Na_2O and TeO_2—BaO Glasses

The structure of GeO_2 glass is described in terms of the continuous random network (CRN) of $[GeO_4]$-tetrahedral structural units. When a modifier oxide such as Na_2O is mixed into GeO_2 glass, a transformation of the structural unit from $[GeO_4]$ tetrahedron to $[GeO_6]$ octahedron may occur. This phenomenon has been termed the "germanate anomaly" in the nonmonotonic dependence of physical properties on glass composition. Ueno and co-workers[37] have studied the structural origin of the germanate anomaly by pulsed neutron total scattering using an electron Linac neutron source.

The experimental neutron total structure factors $S^{FZ}(Q)$ for GeO_2–Na_2O glasses are shown as a function of Na_2O content in Fig. 33. The radial distribution functions (RDF) shown in Fig. 34 are the Fourier transforms of the $S^{FZ}(Q)$ truncated at $Q = 30$ Å^{-1}. The first peaks in the RDF are assigned to the Ge–O correlation within a $[GeO_4]$-tetrahedral structural unit. The average number of oxygen atoms surrounding a germanium atom and the average number of germanium atoms bound to an oxygen atom are shown as a function of Na_2O content in Figs. 35 and 36, respectively. In GeO_2 glass a germanium atom is exactly coordinated with four oxygen atoms. When Na_2O is added into the glass, the average coordination number of oxygen atoms surrounding a germanium atom apparently increases and reaches the maximum of 4.5 at the composition of 20 mole % Na_2O. Further increase in Na_2O content results in a symmetric decrease in the coordination number of

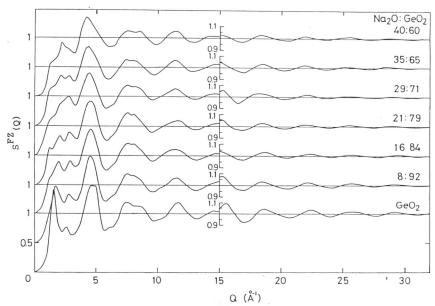

FIG. 33. Neutron total structure factors $S^{FZ}(Q)$ of GeO_2 and GeO_2–Na_2O glasses.

oxygen atoms around a germanium atom, returning to four oxygen atoms at and beyond the composition of 33 mole % Na_2O.

The RDF for GeO_2–Na_2O glasses containing 8 to 21 mole % Na_2O clearly indicate additional shoulders appearing on the high r side of the first peak and on the low r side of the second peak. The shoulders correspond to the Ge–O and O–O correlations within a [GeO_6]-octahedral structural unit contained in the glass. From least-squares fitting using two Gaussian curves, the geometries and fractions of the [GeO_4]-tetrahedral structural units and [GeO_6]-octahedral units were determined.

1. The Ge—O bond lengths are 1.74 Å in a [GeO_4]-tetrahedral structural unit and 1.95 Å in a [GeO_6]-octahedral structural unit.
2. By adding one unit of Na_2O, one unit of [GeO_4] tetrahedron is converted into one unit of [GeO_6] octahedron without formation on nonbridging oxygen atoms. In fact, Fig. 36 shows that the number of germanium atoms bound to an oxygen atom remains 2.0 up to the composition of 20 mole % Na_2O.
3. The maximum fraction of [GeO_6] octahedra is 0.25 at the composition of 20 mole % Na_2O.
4. Over the composition range of 20 to 33 mole % Na_2O, one unit of [GeO_6] octahedron returns to one unit of [GeO_4] tetrahedron to form four nonbridging oxygen atoms by adding one unit of Na_2O.

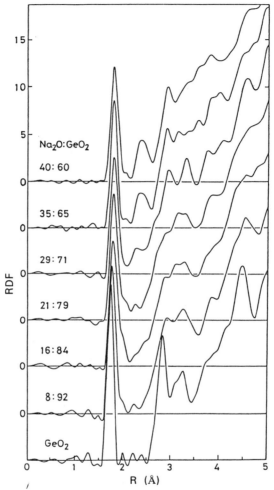

FIG. 34. Radial distribution functions RDF of GeO_2 and GeO_2–Na_2O glasses, where the $S^{FZ}(Q) \to \rho(r)$ Fourier transform is truncated at $Q_{max} = 30$ Å^{-1}.

5. There are only [GeO$_4$] tetrahedra in the glass, and addition of one unit of Na$_2$O produces two nonbridging oxygen atoms, at the composition beyond 33 mole % Na$_2$O.

Based on the information about the elemental structural units described, Ueno[38] and co-workers have proposed a model for the mode of connection between [GeO$_4$]-tetrahedral structural units and [GeO$_6$]-octahedral ones,

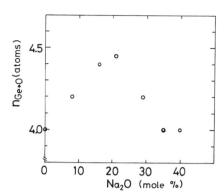

FIG. 35. Average coordination numbers of O atoms bound to a Ge atom in GeO$_2$–Na$_2$O glasses.

and for the spatial configuration of nonbridging oxygen atoms and sodium ions in GeO$_2$–Na$_2$O glasses. However, we are still very far from the full description of the intermediate-range structure of glasses.

In the case of TeO$_2$–BaO glasses, the average coordination number of oxygen atoms surrounding a tellurium atom has been confirmed to decrease monotonically from four to three oxygen atoms by Ueno and Suzuki.[39] Their high-resolution observation of the RDF by high Q measurements of $S^{FZ}(Q)$ has thrown light on the detailed mechanism of the transformation of the chemical short-range structure involved in tellurite glasses.

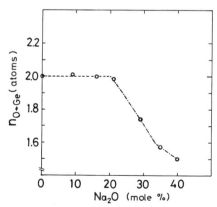

FIG. 36. Average coordination numbers of Ge atoms bound to an O atom in GeO$_2$–Na$_2$O glasses.

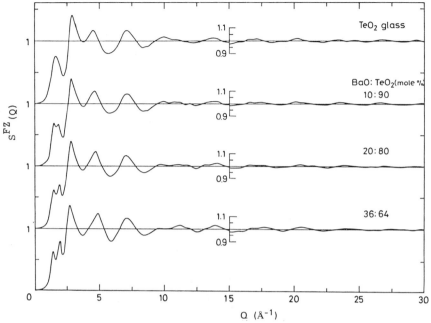

FIG. 37. Neutron total structure factors $S^{FZ}(Q)$ of TeO_2 and TeO_2–BaO glasses.

Figures 37 and 38 show the $S^{FZ}(Q)$ of TeO_2–BaO glasses obtained by total neutron scattering experiments using an electron Linac pulsed neutron source and the RDF defined as the Fourier transform of the $S^{FZ}(Q)$ truncated at $Q_{max} = 30 \text{ Å}^{-1}$, respectively. The first peak in the RDF of TeO_2 glass is divided into two subpeaks over the range of $r = 1.9$ to 2.2 Å, which are assigned to two kinds of Te—O bond with different bond lengths. The coordination numbers of oxygen atoms bound to a tellurium atom and the bond lengths in the two Te—O bonds can be calculated from the area under the each subpeak, based on least squares fitting using two Gaussian functions.

The structural unit making up TeO_2 glass is found to be a $[TeO_{3+1}]$-distorted tetrahedron, in which three oxygen atoms are bound to a central tellurium atom with the bond length of $r = 1.95$ Å, and the remaining oxygen atom is located at the position of $r = 2.18$ Å, far from the tellurium atom. When a modifier oxide such as BaO is added into TeO_2 glass, the small subpeak located at $r = 2.18$ Å gradually decreases as shown in Fig. 38. Figure 39 shows how the coordination number of oxygen atoms surrounding a tellurium atom depends on the BaO content in TeO_2–BaO glasses. A broken

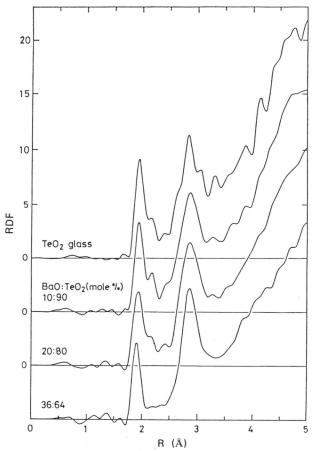

FIG. 38. Radial distribution functions RDF of TeO$_2$ and TeO$_2$–BaO glasses, where the $S^{FZ}(Q) \to \rho(r)$ Fourier transform is truncated at $Q_{max} = 30$ Å^{-1}.

line in the figure means that one unit of [TeO$_{3+1}$] tetrahedron is modified into one unit of [TeO$_3$]-trigonal pyramid with the formation of one non-bridging oxygen atom, as one unit of BaO is added into the glass.

It is noteworthy that the Te–O bond length in a [TeO$_3$]-trigonal pyramidal structural unit decreases with increasing BaO content and reaches a value of 1.86 Å at the composition of TeO$_2$·BaO glass, which exactly corresponds to the Te—O bond length found in the TeO$_2$·BaO crystalline compound. The [TeO$_{3+1}$] → [TeO$_3$] transition in TeO$_2$–BaO glasses deduced from the neutron total scattering experiment is satisfactorily supported by the infrared absorption spectra observed by Uneo and co-workers.[38]

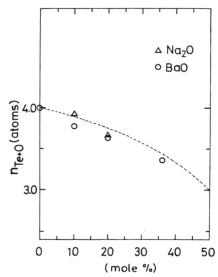

FIG. 39. Coordination numbers of O atoms bound to a Te atom in TeO_2 and TeO_2–BaO glasses.

12.5. Hydrogen Atoms as a Probe for Structure Characterization of Metallic Alloy Glasses

12.5.1. Static Environment around Hydrogen Atoms in Metallic Glasses

Hydrogen and deuterium atoms are easily absorbed into the interstitial hole in metallic atom polyhedra existing in metal–metal amorphous alloys consisting of early and late transition metals or rare earth metals.[40] Therefore, amorphous alloys are expected to be promising candidates for hydrogen energy conversion and storage materials, while hydrogen and deuterium atoms can be used as a sensitive probe for investigating the atomic-scale structure of the amorphous alloys.[40]

Neutron scattering experiments have the unique advantage of effectively detecting hydrogen and deuterium atoms in amorphous alloys, since the nucleus of the hydrogen atom has a large incoherent neutron scattering cross section (79.7 b) and the deuterium atom contains a nucleus with a reasonable coherent neutron scattering cross section (5.6 b).

The atomic spacing between deuterium atoms and metallic atoms is fairly short and the vibrational energy of hydrogen atoms is considerably higher in amorphous alloys, because the size and mass of hydrogen and deuterium atoms are extremely small. Therefore, short-wavelength neutrons with epithermal energy are necessary to obtain well-resolved experimental observations of the local environment around a deuterium atom and the localized vibration of hydrogen atoms in amorphous alloys. Concerning this point, the use of pulsed neutrons generated from accelerators with a high flux above the thermal energy region is a more prudent choice than the use of steady-state reactor neutrons.

Zirconium–nickel alloy glasses can store two H(D) atoms per zirconium atom over the whole composition range from 30 to 75 at. % Zr.[41] Furthermore, hydrogen absorption in Zr–Ni alloy glasses obeys Sievert's law in various hydrogen concentration ranges.[41] These facts mean that some of the hydrogen atom sites are selectively occupied depending on the hydrogen content in the alloy glasses.

Figure 40 shows the neutron total structure factors $S^{FZ}(Q)$ of ZrNiD$_x$($x = 0 \sim 1.72$) alloy glasses observed by using a spallation neutron source.[42] As deuterium atoms are introduced, the first and split second peak over $Q = 2.5 \sim 6 \text{ Å}^{-1}$ decrease drastically, while a prepeak around $Q = 1.7 \text{ Å}^{-1}$ originating from the Ni–Ni pair correlation does not change significantly. Oscillations in $S^{FZ}(Q)$ become enhanced in the high scattering vector region of $Q \gtrsim 10 \text{ Å}^{-1}$ with increasing deuterium content.

The total pair distribution functions $g(r)$ of ZrNiD$_x$ alloy glasses defined as the Fourier transform of $S^{FZ}(Q)$ truncated at $Q_{max} = 25 \text{ Å}^{-1}$ are shown in Fig. 41.[42] When deuterium atoms are absorbed into ZrNi alloy glass, a small peak first appears at the position of $r = 2.1 \text{ Å}$, corresponding to the

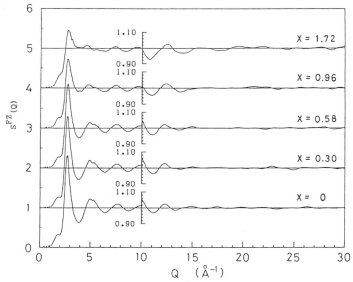

FIG. 40. Neutron total structure factors $S^{FZ}(Q)$ of glassy ZrNiD$_x$ ($x = 0 - 1.72$).

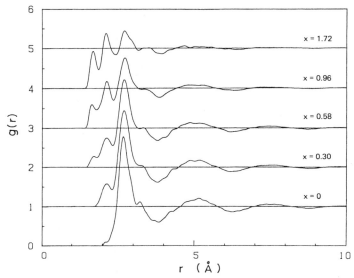

FIG. 41. Neutron total pair distribution functions $g(r)$ of glassy $ZrNiD_x$ ($x = 0 - 1.72$).

D–Zr atomic distance, and then a second peak appears around the position of $r = 1.7$ Å where a D–Ni atomic distance is located. These atomic distances correspond well to the values found in ZrNiD and $ZrNiD_3$ crystalline compounds.

The coordination numbers of zirconium and nickel atoms surrounding a deuterium atom in $ZrNiD_x$ alloy glasses can be calculated from the area under the two peaks representing the D–Zr and D–Ni pair correlations in the RDF. As shown in Fig. 42, deuterium atoms in $ZrNiD_x$ alloy glasses with low deuterium content prefer to occupy the interstitial hole surrounded by four zirconium atoms. With increasing deuterium content, deuterium atoms asymptotically occupy the four-coordinated site consisting on the average of three zirconium and one nickel atom.

Hydrogen atoms in the ZrNiH crystalline compound selectively occupy the interstitial hole tetrahedrally surrounded by four zirconium atoms (*a* site) to induce the lattice distortion from an orthorhombic to a triclinic structure, while hydrogen atoms in $ZrNiH_3$ crystalline compound sit in two different kinds of sites, namely, the tetrahedral hole surrounded by three zirconium and one nickel atom (*c* site) and the hexahedral hole surrounded by three zirconium and two nickel atoms (*b* site), emptying the *a* site and reverting the lattice to the orthorhombic structure.[43] Figure 43 shows the geometrical dimensions of the *a*, *b*, and *c* sites in ZrNiH and $ZrNiH_3$ crystalline compounds.

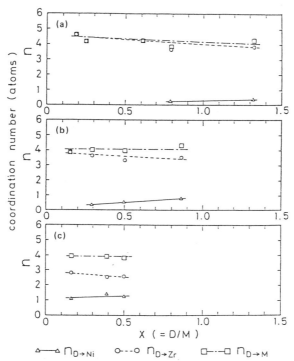

FIG. 42. Coordination numbers of Zr and Ni atoms surrounding a deuterium atom in (a) glassy $Zr_{0.67}Ni_{0.33}D_x$, (b) $Zr_{0.50}Ni_{0.50}D_x$, and (c) $Zr_{0.35}Ni_{0.65}D_x$.

The D–Zr atomic distance in the four-zirconium atom tetrahedron occupied by a deuterium atom in $ZrNiD_{0.3}$ alloy glass is about 2.10 Å, which is obviously shorter than the D–Zr atomic distance (2.20 Å) in the a site in the ZrNi crystalline compound and rather close to the D–Zr atomic distance (2.08 Å) in the four-zirconium tetrahedron in the ZrD_2 crystalline compound.

Since $NiTi_2$ alloy glass has a composition near the neutron zero-scattering alloy, it is interesting to observe the total structure factor $S^{BT}(Q)$ of $NiTi_2D_x$ alloy glasses as a function of deuterium content.[44] Figure 44 shows the total pair correlation function $G'(r)$ defined as the Fourier transform of $S^{BT}(Q)$ truncated at $Q_{max} = 25$ Å^{-1} as[44]

$$G'(r) = \frac{2}{\pi} \int_0^{Q_{max}} Q[S^{BT}(Q) - 1] \sin Qr \, dQ. \qquad (12.34)$$

A negative peak at $r = 2.56$ Å in the $G'(r)$ of $NiTi_2$ alloy glass comes mainly from the preference for Ni–Ti unlike atom pairs. This peak vanishes and a new negative peak appears at the position around $r = 1.95$ Å with

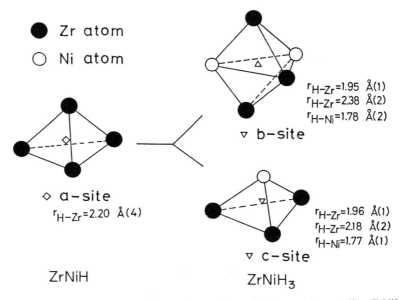

FIG. 43. Polyhedral sites occupied by hydrogen (deuterium) atoms in crystalline ZrNiH(D) and ZrNiH(D)$_3$ compound proposed by Westlake et al.[43]

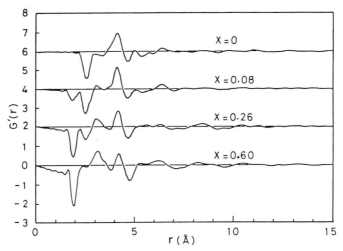

FIG. 44. Total pair correlation functions $G'(r)$ defined as in Eq. (12.34) for glassy $Ni_{0.33}Ti_{0.67}D_x$ ($x = 0 - 0.60$).

increasing deuterium content. The new negative peak comes mainly from D-Ti unlike atom pairs with a short atomic distance in $NiTi_2$ alloy glasses.

According to x-ray and neutron diffraction experiments of $NiTi_2D_{1.5}$ crystalline hydride,[45] deuterium atoms are known to occupy the octahedral site surrounded by six titanium atoms as well as the tetrahedral site consisting of four titanium atoms. The average atomic distances for D-Ni and D-Ti pairs in $NiTi_2D_{1.5}$ crystalline hydride are $r_{DNi} = 1.8$ Å and $r_{DTi} = 2.0$ Å, respectively. Both atomic distances are too close to each other to be resolved in the first-peak profile of $G'(r)$. Therefore, the first dips appearing at $r = 1.95$ Å in $G'(r)$ are expected to result from the overlapping of the D-Ti negative peak and the D-Ni positive peak. From the area of the first dips in $G'(r)$, the ratio of nickel atoms to titanium atoms can be obtained as a function of deuterium content, if it is assumed that deuterium atoms occupy the tetrahedral ($n = 4$) or octahedral ($n = 6$) sites in the alloy glass. As described in Section 12.5.2, a larger fraction of hydrogen atoms prefers to sit in the tetrahedral site in $NiTi_2$ alloy glass.

The local hydrogen atom environment obtained by total neutron scattering experiments suggests that the tetrahedral structural units are quite favorable in the short-range structure of metallic glasses, as shown in the computer simulation of the dense random packing of soft spheres by Finney and Wallace.[46]

12.5.2. Local Vibrations of Hydrogen Atoms in Metallic Glasses

The energy spectrum of local vibrations of hydrogen atoms absorbed in metallic glasses provides useful knowledge of the geometrical distortion and chemical fluctuation of the local environment around a hydrogen atom in a metallic glass. As the hydrogen atom has a large incoherent neutron scattering cross section [$\overline{b^2} - (\overline{b})^2$] as well as a high vibrational energy ($\hbar\omega = E = E_0 - E_1$), the frequency distribution (or density of states) $Z(\omega)$ for vibrations of hydrogen atoms in metallic glasses can be obtained directly from the measurement of the double-differential cross section for the incoherent single-phonon scattering process[47] with the use of epithermal pulsed neutrons:

$$\frac{d^2\sigma_i}{d\Omega\, dE} = \frac{k_1}{k_0}\{\overline{b^2} - (\overline{b})^2\}\frac{Q^2\langle u^2 \rangle}{2M}\left\{\frac{1}{\exp(\hbar\omega/k_B T) - 1} + \left(\frac{1}{2} \pm \frac{1}{2}\right)\right\}$$
$$\times e^{-2W} N \frac{Z(\omega)}{\omega}, \qquad (12.35)$$

where $Z(\omega)$ is the density of states of hydrogen atom vibrations, $\exp(-2W)$ is the Debye-Waller factor, and $\langle u^2 \rangle$ is the mean square vibrational amplitude of hydrogen atoms [see Eq. (1.70) in Part A, Chapter 1, Section 1.3.4].

Figure 45 shows the double-differential cross sections for incoherent inelastic scattering from hydrogen atoms[44] in crystalline TiH_2, crystalline

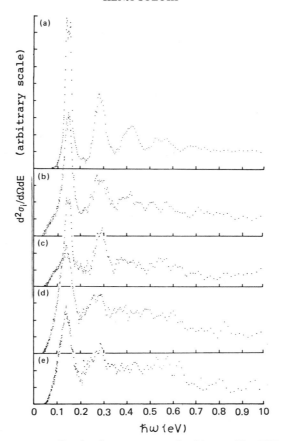

FIG. 45. Hydrogen atom vibrational energy spectra for (a) crystalline TiH_2, (b) $NiTi_2H_{2.7}$, (c) $NiTi_2H_{1.5}$, (d) glassy $NiTi_2H_{1.5}$ observed at room temperture and for (e) glassy $NiTi_2H_{1.5}$ at 40 K.

$NiTi_2H_{2.7}$, crystalline $NiTi_2H_{1.5}$, and glassy $NiTi_2H_{1.5}$ observed at room temperature, and that for glassy $NiTi_2H_{1.5}$ at a low temperature of 40 K, by using a crystal analyzer time-of-flight spectrometer (CAT)[48] installed at the spallation neutron source at the KEK 500-MeV proton booster synchrotron. The spectra of the hydrogenated metallic glasses have very broad peaks even at low temperature compared with those of the hydrogenated crystals. However, peak positions are nearly identical in both glass and crystal. Such observations have been previously reported for glassy $TiCuH_{1.3}$,[49] $NiZrH_{1.8}$,[50] and $NiZr_2H_{4.4}$.[51]

A striking result shown in Fig. 45 is the fact that higher harmonics of the hydrogen atom vibrational frequency are clearly observed even in the glasses.

TABLE V. Central Position and Width of Peaks in Hydrogen Atom Vibrational Energy Spectra for Crystalline TiH$_2$, Crystalline (c-) NiTi$_2$H$_{1.5}$, NiTi$_2$H$_{2.7}$, and Glassy (a-) NiTi$_2$H$_{1.5}$ (meV)

	First peak		Second peak		Third peak	
	$\hbar\omega$	FWHM	$\hbar\omega$	FWHM	$\hbar\omega$	FWHM
a-NiTi$_2$H$_{1.5}$						
R.T.	139 ± 2	74 ± 5	271 ± 3	106 ± 10		
L.T.	138 ± 2	66 ± 5	281 ± 3	93 ± 10	415 ± 10	140 ± 20
c-NiTi$_2$H$_{1.5}$	146 ± 2	36 ± 5	282 ± 3	67 ± 10	406 ± 8	90 ± 20
c-NiTi$_2$H$_{2.7}$	$\genfrac{}{}{0pt}{}{140}{151} \pm 2$	47 ± 5	282 ± 3	90 ± 10	408 ± 8	130 ± 20
TiH$_2$	$\genfrac{}{}{0pt}{}{138}{149} \pm 2$	31 ± 2	$\genfrac{}{}{0pt}{}{260}{281} \pm 3$	55 ± 5	$\genfrac{}{}{0pt}{}{390}{424} \pm 8$	96 ± 10

In particular, the fourth- and fifth-order harmonics can be found for glassy NiTi$_2$H$_{1.5}$ at 40 K as well as for crystalline NiTi$_2$H$_{1.5}$. The peak positions and widths in the vibrational frequency spectra shown in Fig. 45 are summarized in Table V, together with those for crystalline TiH$_2$(CaF$_2$-type structure). The second peak position shifts by about 10 meV at room temperature from the exact harmonic position $2\hbar\omega_1$ in glassy NiTi$_2$H$_{1.5}$ as well as in the corresponding crystalline NiTi$_2$H$_{1.5}$. At 40 K, however, the positions of the second and third peak for glassy NiTi$_2$H$_{1.5}$ are very close to its exact harmonic positions $n\hbar\omega_1$ ($n = 2$ and 3). This implies that the local configuration of metal atoms around a hydrogen atom in NiTi$_2$ glass is sensitively modified as a function of temperature.

As shown in Fig. 45 and Table V, the first peak in the hydrogen atom vibrational spectra of NiTi$_2$H$_{1.5}$ is located around $\hbar\omega = 140$ meV in both the crystalline and glassy states, which is also very close to the energy position of the hydrogen atom vibrational spectrum of crystalline TiH$_2$. This means that the local environment around a hydrogen atom is rather similar in the three different phases, glassy NiTi$_2$H$_{1.5}$, crystalline NiTi$_2$H$_{1.5}$, and crystalline TiH$_2$.

Since hydrogen atoms occupy the tetrahedral site consisting of four titanium atoms in crystalline TiH$_2$ and NiTi$_2$(D)$_{1.5}$, it may be assumed that a large fraction of hydrogen atoms also sits in the tetrahedral site surrounded by four titanium atoms in the glasses. According to x-ray and neutron diffraction experiments of crystalline NiTi$_2$D$_{1.5}$, deuterium atoms are known to occupy the octahedral site surrounded by six titanium atoms as well as the tetrahedral site consisting of four titanium atoms.

Figure 45 clearly shows that there is extra intensity over the low-energy region below $\hbar\omega = 100$ meV in the hydrogen atom vibrational spectra of crystalline NiTi$_2$H$_{1.5}$ and NiTi$_2$H$_{2.7}$. The spectrum of crystalline TiH$_2$ has

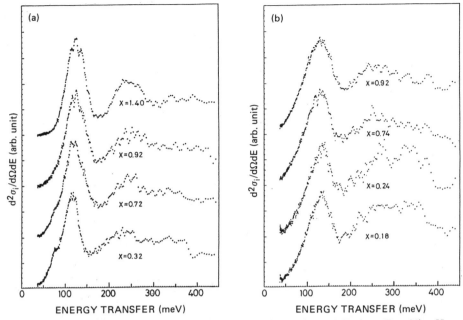

FIG. 46. Vibrational energy spectra of H atoms in (a) crystalline and (b) glassy $Zr_{0.50}Ni_{0.50}H_x$ observed at room temperature.

no such shoulders on the lower energy side of the first peak. Therefore, this shoulder is assigned to the vibrational mode of hydrogen atoms sitting in the octahedral site surrounded by six titanium atoms. Although considerable broadening of peaks is found for glassy $NiTi_2H_{1.5}$ even at the low temperature of 40 K compared with the crystalline counterpart of $NiTi_2H_{1.5}$, both the tetrahedral and octahedral sites are certainly occupied by hydrogen atoms in the glass.

The double differential cross section for incoherent inelastic neutron scattering from glassy and crystalline $ZrNiH_x$ is shown as a function of energy transfer $\hbar\omega$ in Fig. 46.[42] In the local vibrational spectrum of hydrogen atoms in crystalline $ZrNiH_{0.64}$, the main peak is divided into three peaks spread over the energy region $\hbar\omega = 110 \sim 120$ meV and a shoulder appears around $\hbar\omega = 80$ meV. With increasing hydrogen content, the shoulder gradually disappears, while the main peak shifts toward higher energy. The main peak in the spectrum for crystalline $ZrNiH_{2.8}$ is still separated into three peaks located at $\hbar\omega = 112, 123,$ and 134 meV, respectively, and another shoulder takes place on the high-energy side of the main peak around $\hbar\omega = 150$ meV.

The local vibration energy spectra of hydrogen atoms in glassy $ZrNiH_x$ do not indicate the disappearance and energy shift of peaks and shoulders as shown in Fig. 46. The peaks are generally broader even in glassy $ZrHiH_x$ with low hydrogen content. Higher order vibrations of hydrogen atoms are still observed in glassy $ZrNiH_x$ as well as in the crystalline material. It is noteworthy that the main peak position is almost fixed around $\hbar\omega = 130 \sim 135$ meV over the whole hydrogen content of the glass. In particular, the position of the main peak in the spectrum of low hydrogen content glasses, such as glassy $ZrNiH_{0.3}$ and $ZrNiH_{0.48}$, is about 10 to 15 meV higher than that of crystalline $ZrNiH_{0.64}$. This supports the conclusion that the geometrical size of the four-zirconium-tetrahedral site occupied by an hydrogen atom in glassy $ZrNiH_x$ is rather close to that in crystalline ZrH_2.

12.6. Atomic Vibrations by Inelastic Pulsed Neutron Scattering

12.6.1. Vibrational Density of States in Metallic Glasses

Since powerful pulsed neutron sources based on high-energy proton spallation accelerators have now become available, we are gaining deeper insight into the vibrational dynamics of amorphous solids by means of inelastic neutron scattering techniques. Because of the lack of long-range lattice periodicity, the E-Q dispersion relation conventionally used to describe phonon modes in crystals is no longer an appropriate way of describing the atomic vibrations in amorphous solids, except for the limit of long-wavelength acoustic modes. Therefore, the vibrational density of states $Z(E)$ is the physical quantity usually employed to describe the vibrational dynamics in amorphous solids.

If the vibrating atoms in an amorphous solid are incoherent scatterers for neutrons, for example, hydrogen atoms absorbed in metallic glasses, the vibrational density of states can be directly obtained from incoherent inelastic neutron scattering measurements as described in Section 12.5.2. However, amorphous solids often include coherently scattering nuclei. To obtain the equivalent of the vibrational density of states from coherent inelastic scattering measurements, the dynamic structure factor $S(Q, E)$ is averaged over an extended range of the high momentum transfer Q to cancel the correlations between distinct atom motions. This procedure corresponds to the incoherent approximation.

The total neutron scattering experiments described in Section 12.3 show that the chemical short-range structures preserved in amorphous alloy glasses are essentially similar to those found in their crystalline counterparts. A pioneering investigation to distinguish the chemical interactions between metal–metal and metal–metalloid atoms in $Pd_{80}Si_{20}$ alloy glass was carried out by Windsor and co-workers[52] using a beryllium filter spectrometer installed at a steady-state reactor source.

Recently, Lustig and co-workers[53,54] have performed high-resolution observations of the vibrational interaction between transition metal–transition metal and transition metal–metalloid pairs in $Fe_{78}P_{22}$ and $Ni_{82}B_{18}$ alloy glasses, using the combination of a chopper spectrometer and relatively high-energy incident neutrons supplied by the pulsed spallation neutron source IPNS at Argonne National Laboratory. This experiment provides the neutron-weighted vibrational density of states $G(E)$ defined by averaging $G(Q, E)$ over the Q range of 5–9 Å^{-1} (incoherent approximation):

$$G(E) = \langle G(Q, E) \rangle_{\text{average over } Q = 5-9 \text{Å}} \quad (12.36)$$

$$= \sum_i c_i \frac{\bar{b}_i^2}{\langle \bar{b}^2 \rangle} \exp(-2W_i) \frac{(\mathbf{Q} \cdot \mathbf{e}_i^\lambda)^2}{M_i} Z_i(E) \quad (12.37)$$

where

$$G(Q, E) = \frac{4\pi}{\sigma_{\text{coh}}} \frac{k_0}{k_1} \frac{E}{n(E)} \left. \frac{d^2\sigma_{\text{coh}}}{d\Omega \, dE} \right|_{\text{one phonon}}. \quad (12.38)$$

The term $d^2\sigma_{\text{coh}}/d\Omega\, dE|_{\text{one phonon}}$ is the one-phonon term of the double differential cross section for coherent neutron scattering per unit chemical formula, $n(E) = [\exp(E/kT) - 1]^{-1}$ the population factor for mode energy E, $\exp(-2W_i)$ the Debye–Waller factor, $\mathbf{e}_i^\lambda$ the displacement vector, M_i the mass, and $\bar{b}_i$ the coherent scattering length of atoms of type i.

The experimental $G(E)$ of $Fe_{68}P_{22}$ and $Ni_{82}B_{18}$ alloy glasses are shown in Fig. 47. A comparison of $G(E)$ between $Fe_{78}P_{22}$ alloy glass and Fe_3P crystalline compound is also shown in Fig. 48.[54] The $G(E)$ spectra in both figures are normalized to the same area to make comparison easier. Each $G(E)$ spectrum consists of a more intense low-energy band peaked around 20 meV and a weaker high-energy band located above 40 meV. The low-energy band is primarily associated with metal–metal (Fe–Fe, Ni–Ni) interactions having a larger neutron weighting factor, while the high-energy band originates from metal–metalloid (Fe–P, Ni–B) interactions with a smaller neutron weighting factor in the alloy glasses.

The Fe–Fe interaction peak is located about 1.4 meV below the Ni–Ni interaction peak. This means that the Ni–Ni interaction in $Ni_{82}B_{18}$ alloy glass is about 20% stronger than the Fe–Fe interaction in $Fe_{78}P_{22}$ alloy glass, based on a simple effective force constant model $k_{\text{eff}} = mE^2$. The higher energy peaks separated at about 56 and 68 meV in $Ni_{82}B_{18}$ alloy glass correspond to those at about 44 and 52 meV in $N_{78}P_{22}$ alloy glass. Lustig et al.[53] have concluded that the ratio of effective force constants for the metal–metalloid interactions yields $k_{\text{Fe-P}}/k_{\text{Ni-B}} = 1.2$, resulting from the stronger covalent

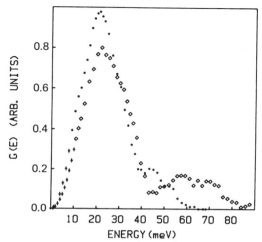

FIG. 47. Neutron-weighted vibrational density of states $G(E)$ of $Fe_{78}P_{22}$ (O) and $Ni_{82}B_{18}$ (◇) alloy glasses, where the Debye model is assumed for extrapolating the spectra to the lower energy region below 10 meV (+).[53]

bonds of Fe-P pairs due to greater p-d hybridization. These high-energy bands may arise from the optic or cage-like modes of phosphorus (or boron) atoms interacting with iron (or nickel) atoms.

Figure 48 indicates that the overall behavior of the two $G(E)$ spectra shows considerable similarity between $Fe_{78}P_{22}$ alloy glass and the Fe_3P crystalline compound. This is strongly suggestive that a similar average short-range structure exists in the alloy glass and the corresponding crystalline compound. However, there is a marked difference in that the $G(E)$ spectrum of $Fe_{78}P_{22}$ alloy glass is significantly enhanced on the low-energy side of the Fe-Fe interaction peak, compared with that of the Fe_3P crystalline compound.

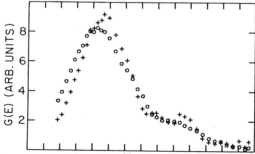

FIG. 48. Comparison of the neutron-weighted vibrational densities of states $G(E)$ of $Fe_{78}P_{22}$ (O) alloy glass and Fe_3P (+) crystalline compound.[54]

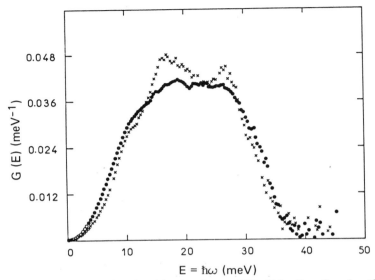

FIG. 49. Neutron-weighted vibrational density of states $G(E)$ of $Mg_{70}Zn_{30}$ alloy glass (●) and the same sample after crystallization (×) measured at 296 K, where the low-energy region below 3.5 meV is fitted to the Debye model (○ and +).[56]

A similar low-energy enhancement in the vibrational density of states has been extensively observed in metal–metal amorphous alloys such as $Cu_{46}Zr_{54}$[55] and $Mg_{70}Zn_{30}$[56] alloy glasses by Suck and co-workers using time-of-flight inelastic neutron scattering facilities at the Institut Laue-Langevin.

In contrast to metal–metalloid amorphous alloys, the vibrational density of states in metal–metal amorphous alloys has been found to contain only a single band, comparatively broad compared with the crystalline counterparts. As an example, the $G(E)$ spectrum of $Mg_{70}Zn_{30}$ alloy glass measured by Suck et al.[56] is shown in Fig. 49 together with that of the crystallized alloy. The two maximum peaks in $G(E)$ for the crystallized alloy can be assigned to the vibration of magnesium atoms. Additional intensities found below 10 meV and above 30 meV in $G(E)$ of the alloy glass may indicate that amorphous alloys contain not only expanded regions but also compressed regions due to the introduction of free volume.

12.6.2. Correlated Motions in SiO_2 and $Mg_{70}Zn_{30}$ Glasses

Inelastic neutron scattering measurements of amorphous solids directly provide detailed information not only about the vibrational density of states but also on the correlations between motions of different atoms. This is a unique advantage of neutron spectroscopy that cannot be found in conventional optical spectroscopy such as Raman scattering, infrared absorption, etc. In 1975 Carpenter and Pelizzari[57] proposed a procedure for extracting explicitly the real-space displacement–displacement correlations in the atomic motions of a specified energy in an amorphous solid from coherent inelastic neutron scattering data. The first experimental demonstration for the proposal mentioned above has been shown quite recently in the study of SiO_2 glass by Carpenter and Price.[58]

Systematic measurements of coherent inelastic neutron scattering from a series of SiO_2, GeO_2, BeF_2, and $SiSe_2$ glasses have been performed with the pulsed spallation neutron source IPNS at Argonne National Laboratory. As an example, the two-dimensional plot of the experimental $S(Q, E)$ for SiO_2 glass is shown in Fig. 50, where a very broad range of energy and momentum transfer is covered by using monochromatized incident neutrons with the relatively high energy of 218 meV.[58] This figure clearly indicates pronounced oscillations in $S(Q, E)$ as a function of Q, which represent the characteristic correlations of the motions and positions between different atoms, as well as peaks in the energy coordinate corresponding to certain modes with energy $E = \hbar\omega$.

The experimentally measured neutron scattering function $S(Q, E)$ can be developed in the conventional harmonic phonon expansion as

$$S(Q, E) = S^{(0)}(Q)\delta(E) + S^{(1)}(Q, E) + S^{(2)}(Q, E) + \cdots, \quad (12.39)$$

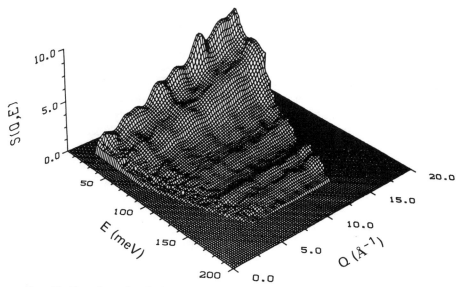

FIG. 50. Two-dimensional plot of neutron scattering function $S(Q, E)$ for SiO_2 glass measured at 33 K.[58]

where $S^{(0)}(Q)\delta(E)$ is the elastic scattering contribution. The second term $S^{(1)}(Q, E)$ in Eq. (12.39) represents the contribution from the one-phonon scattering and is formulated as[58]

$$S^{(1)}(Q, E) = \frac{1}{2N} \sum_{ij} \frac{\bar{b}_i \bar{b}_j}{\langle \bar{b}^2 \rangle} \exp[-(W_i + W_j)] \exp[i\mathbf{Q} \cdot (\mathbf{R}_i - \mathbf{R}_j)]$$

$$\times \sum_\lambda \hbar \frac{(\mathbf{Q} \cdot \mathbf{e}_i^\lambda)(\mathbf{Q} \cdot \mathbf{e}_j^\lambda)}{(M_i M_j)^{1/2} \omega_\lambda} \langle n_\lambda + 1 \rangle \delta(E - \hbar\omega_\lambda). \quad (12.40)$$

Therefore, $S^{(1)}(Q,E)$ describes the correlated motions with mode energy $E = \hbar\omega_\lambda$ between distinct atoms through the factor $(\mathbf{Q} \cdot \mathbf{e}_i^\lambda)(\mathbf{Q} \cdot \mathbf{e}_j^\lambda)$ in Eq. (12.40). The third and higher terms in Eq. (12.39) are contributed by double and higher multiple-phonon scattering, resulting in a smooth background under the one-phonon scattering. In the incoherent approximation as described in Section 12.6.1, $(\mathbf{Q} \cdot \mathbf{e}_i^\lambda)(\mathbf{Q} \cdot \mathbf{e}_j^\lambda)$ is replaced with $\frac{1}{3}Q^2(e_i^\lambda)^2\delta_{ij}$ by averaging $S^{(1)}(Q, E)$ over an extended range of Q.

Figure 51 shows the one-phonon density of states $G(E)$ for SiO_2 glass, which is calculated on the basis of the incoherent approximation by averaging $S^{(1)}(Q, E)$ over the Q range from 6 to 13 Å^{-1}. The peaks appearing at the energy E around 50, 100, and 130–150 meV in $G(E)$ are assigned to the rocking, bending, and stretching motions of Si—O—Si bonds, respectively. The detailed behavior of this $G(E)$ is very close to that obtained previously from beryllium filter spectroscopy by Galeener et al.,[59] except for the reversal in the peak height around 130–150 meV.

The Q dependence of $S(Q, E)$ for specific energy modes corresponding to the rocking, bending, and stretching motions of Si—O—Si bonds is plotted in Fig. 52. The oscillations in the higher energy vibrational modes for the bending and stretching motions are more pronounced than those for the low-energy rocking motion. The periods of the oscillations ranging from 1.6 to 1.8 Å^{-1} in the experimental $S(Q, E)$ in Fig. 52 are obviously shorter than that of $2\pi/\{\mathbf{R}_{Si(1)} - \mathbf{R}_{Si(2)}\} = 2.12$ Å^{-1}, which corresponds to the interference

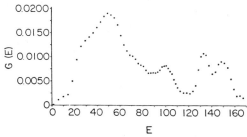

FIG. 51. Vibrational density of states $G(E)$ of SiO_2 glass obtained for the one-phonon contribution to the experimental $S(Q, E)$.[58]

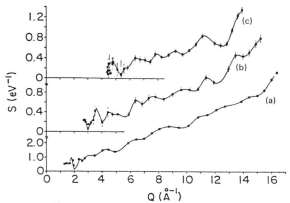

FIG. 52. Q dependence of $S(Q, E)$ for SiO_2 glass at (a) $E = 49.2$, (b) 98.5, and (c) 147.5 meV, corresponding to rocking, bending, and stretching motion of Si—O—Si bonds, respectively.[58]

between motions of second-nearest-neighbor silicon atoms. This implies that the correlated motions in SiO_2 glass definitely extend to an intermediate range beyond the second-neighbor interaction.

Correlated vibrational motions between different atoms have also been observed for metallic glasses by Suck et al.[56] using coherent inelastic neutron scattering measurements. Figure 53 shows $S(Q, E)$ at a low energy transfer value of $E = \hbar\omega = 5.3$ meV as a function of Q for $Mg_{70}Zn_{30}$ alloy glass and its crystallized alloy. Additional intensity in the $S(Q, E)$ of the alloy glass appears in the large Q range compared with the crystallized alloy. The origin of this type of correlated motion associated with such a low excitation energy in amorphous alloys is not yet clear, although there are some speculations

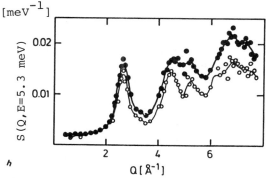

FIG. 53. Neutron scattering functions $S(Q, E)$ at $E = 5.3$ meV as a function of Q for $Mg_{70}Zn_{30}$ alloy glass (●) and its crystallized alloy (○) measured at 296 K.[56]

about the atomic-scale defect structure, density fluctuations, two-level system, etc.

Another topical problem in collective atomic motions in amorphous alloys is the extent of propagating short-wavelength collective excitations in the topologically disordered structure. Suck *et al.*[60,61] observed short-wavelength transverse collective modes in $Mg_{70}Zn_{30}$ alloy glass by time-of-flight inelastic neutron scattering measurements. They concluded that the dispersive modes decrease in frequency as the wave vector Q approaches the value of the first peak Q_p in the static structure factor $S(Q)$ and then reaches a clear minimum at Q_p. This observation was consistent with model calculations previously performed by Hafner.[62]

References

1. A. J. Leadbetter, in "Chemical Applications of Thermal Neutron Scattering" (B. T. M. Willis, ed.), p. 146. Oxford Univ. Press, London and New York, 1973.
2. J. F. Sadoc and C. N. J. Wagner, in "Glassy Metals II" (H. Beck and H.-J. Güntherodt, eds.), p. 51. Springer-Verlag, Berlin and New York, 1983.
3. J.-B. Suck and H. Rudin, in "Glassy Metals II" (H. Beck and H.-J. Güntherodt, eds.), p. 217. Springer-Verlag, Berlin and New York, 1983.
4. K Suzuki, in "Amorphous Metallic Alloys" (F. E. Luborsky, ed.), p. 75. Butterworth, London, 1983.
5. R. N. Sinclair, *U.K. At. Energy Res. Establ., Rep.* **AERE-MPD/NBS 275** (1985).
6. L. Van Hove, *Phys. Rev.* **95**, 249 (1954).
7. K. Suzuki, M. Misawa, K. Kai, and N. Watanabe, *J. Nucl. Instrum. Methods* **147**, 519 (1977).
8. M. Ueno, M. Misawa, and K. Suzuki, *Kakuriken Kenkyu Hokoku (Tohoku Daigaku)* **13**, 254 (1980).
9. N. Watanable, T. Fukunaga, T. Shinohe, K. Yamada, and T. Mizoguchi, *Proc. 4th Meet. Int. Collab. Adv. Neutron Sources* p. 539. Nat. Lab. High Energy Phys., Tsukuba, Jpn., 1981.
10. M. Misawa, Ph.D. Thesis, Tohoku Univ., 1976.
11. T. Fukunaga, Ph.D. Thesis, Tohoku Univ., 1978.
12. K. Suzuki, *Proc. 4th Int. Conf. Rapidly Quenched Met.* **1**, 309. Jpn. Inst. Met., Sendai, 1982.
13. T. E. Faber and J. M. Ziman, *Philos. Mag.* **11**, 153 (1965).
14. A. B. Bhatia and D. E. Thornton, *Phys. Rev. B* **2**, 3004 (1970).
15. B. E. Warren, B. L. Averbach, and B. W. Roberts, *J. Appl. Phys.* **22**, 1493 (1951).
16. Y. Waseda, "The Structure of Non-Crystalline Materials," p. 16. McGraw-Hill, New York, 1980.
17. T. Fukunaga and K. Suzuki, *Sci. Rep. Tohoku Univ. Ser. A* **29**, 153 (1981).
18. K. Suzuki, T. Fukunaga, M. Misawa, and T. Masumoto, *Mater. Sci. Eng.* **23**, 215 (1976).
19. T. Fukunaga and K. Suzuki, *Sci. Rep. Tohoku Univ. Ser. A* **28**, 208 (1980).
20. P. H. Gaskell, *J. Non-Cryst. Solids* **32**, 207 (1979).
21. N. Hayashi, T. Fukunaga, M. Ueno, and K. Suzuki, *Proc. 4th Int. Conf. Rapidly Quenched Met.* **1**, 355. Jpn. Inst. Met., Sendai, 1982.
22. K. Suzuki, T. Fukunaga, F. Itoh, and N. Watanabe, *Proc. 5th Int. Conf. Rapidly Quenched Met.* p. 479. Elsevier, Amsterdam, 1985.

23. I. Bakonyi, P. Panissod, J. Durand, and R. Hasegawa, *J. Non-Cryst. Solids* **61/62**, 1189 (1984).
24. P. Lamparter, W. Sperl, E. Nold, G. Rainer-Harbach, and S. Steeb, *Proc. 4th Int. Conf. Rapidly Quenched Met.* **1**, 343. Jpn. Inst. Met., Sendai, 1982.
25. T. Fukunaga, N. Hayashi, K. Kai, N. Watanabe, and K. Suzuki, *Physica (Amsterdam) B + C* **120B + C**, 352 (1983).
26. T. Fukunaga, K. Kai, M. Naka, N. Watanable, and K. Suzuki, *Proc. 4th Int. Conf. Rapidly Quenched Met.* **1**, 347. Jpn. Inst. Met., Sendai, 1982.
27. T. Fukunaga, N. Watanabe, and K. Suzuki, *J. Non-Cryst. Solids* **61/62**, 343 (1984).
28. T. Fukunaga, N. Hayashi, N. Watanabe, and K. Suzuki, *Proc. 5th Int. Conf. Rapidly Quenched Met.*, p. 475. Elsevier, Amsterdam, 1985.
29. M. Misawa, D. L. Price, and K. Suzuki, *J. Non-Cryst. Solids* **37**, 85 (1980).
30. K. Suzuki and M. Ueno, *J. Phys. (Orsay, Fr.)* **46**, C8-261 (1985). (*Proc. 3rd Int. Conf. Struct. Non-Cryst. Mater., ILL, Grenoble, 1985*).
31. M. Misawa, T. Fukunaga, K. Niihara, T. Hirai, and K. Suzuki, *J. Non-Cryst. Solids* **34**, 313 (1979).
32. T. Aiyama, T. Fukunaga, K. Niihara, T. Hirai, and K. Suzuki, *J. Non-Cryst. Solids* **33**, 131 (1979).
33. A. Guinier and G. Fournet, "Small-Angle Scattering of X-Ray." Wiley, New York, 1955.
34. M. Misawa, Y. Kobayashi, and K. Suzuki, *Proc. Int. Ion Eng. Congr.—ISIAT '83 IPAT '83*, p. 957. Inst. Elecr. Eng. Jpn., Tokyo, 1983.
35. C. G. Windsor, D. S. Boudreaux, and M. C. Narasimhan, *Phys. Lett. A* **67A**, 282 (1978).
36. T. Fukunaga, S. Urai, N. Hayashi, N. Watanabe, and K. Suzuki, *Nat. Lab. High Energy Phys. KENS (Jpn.)* **KENS-V**, 40 (1984).
37. M. Ueno, M. Misawa, and K. Suzuki, *Physica (Amsterdam) B + C* **120B + C**, 347 (1983).
38. M. Ueno, Ph.D. Thesis, Tohoku Univ., 1985.
39. M. Ueno and K. Suzuki, *Kakuriken Kenkyu Hokoku (Tohoku Daigaku)* **16**, 49 (1983).
40. K. Suzuki, *J. Less-Common Met.* **89**, 183 (1983).
41. K. Aoki, A. Horata, and T. Masumoto, *Proc. 4th Int. Conf. Rapidly Quenched Met.* **2**, 1649. Jpn. Inst. Met., Sendai, 1982.
42. K. Suzuki, N. Hayashi, Y. Tomizuka, T. Fukunaga, K. Kai, and N. Watanabe, *J. Non-Cryst. Solids* **61/62**, 637 (1984).
43. D. G. Westlake, H. Shaked, P. R. Mason, B. R. McCart, M. H. Mueller, T. Masumoto, and M. Amano, *J. Less-Common Met.* **88**, 18 (1982).
44. K. Kai, S. Ikeda, T. Fukunaga, N. Watanabe, and K. Suzuki, *Physica (Amsterdam) B + C* **120B + C**, 342 (1983).
45. H. Bucher, M. A. Gutijar, K. D. Beccu, and M. Saufferer, *Z. Metallkd.* **63**, 497 (1972).
46. J. L. Finney and J. Wallace, *J. Non-Cryst. Solids* **43**, 165 (1981).
47. C. G. Windsor, *in* "Chemical Applications of Thermal Neutron Scattering" (B. T. M. Willis, ed.), p. 1. Oxford Univ. Press, London and New York, 1973.
48. S. Ikeda, K. Kai, and N. Watanabe, *Physica (Amsterdam) B + C* **120B + C**, 131 (1983).
49. J. J. Rush, J. M. Rowe, and A. J. Maeland, *J. Phys. F* **10**, L283 (1980).
50. H. Kaneko, T. Kajitani, M. Hirabayashi, M. Ueno, and K. Suzuki, *Proc. 4th Int. Conf. Rapidly Quenched Met.* **2**, 1605. Jpn. Inst. Met., Sendai, 1982.
51. H. Kaneko, T. Kajitani, M. Hirabayashi, M. Ueno, and K. Suzuki, *J. Less-Common Met.* **89**, 237 (1983).
52. C. G. Windsor, H. Kheyrandish, and M. C. Narasimhan, *Phys. Lett. A* **70A**, 485 (1979).
53. N. Lustig, J. S. Lannin, D. L. Price, and R. Hasegawa, *J. Non-Cryst. Solids* **75**, 277 (1985).
54. N. Lustig, J. S. Lannin, J. M. Carpenter, and M. Arai, *Proc. 5th Int. Conf. Rapidly Quenched Met.* p. 501. Elsevier, Amsterdam, 1985.

55. J. B. Suck, H. Rudin, H.-J. Güntherodt, H. Beck, J. Daubert, and W. Gläser, *J. Phys. C* **13**, L167 (1980).
56. J. B. Suck, H. Rudin, H.-J. Güntherodt, and H. Beck, *J. Phys. C* **14**, 2305 (1981).
57. J. M. Carpenter and C. A. Pelizzari, *Phys. Rev. B* **12**, 2391, 2397 (1975).
58. J. M. Carpenter and D. L. Price, *Phys. Rev. Lett.* **54**, 441 (1985).
59. F. L. Galeener, A. J. Leadbetter, and M. W. Stringfellow, *Phys. Rev. B* **27**, 1052 (1983).
60. J. B. Suck, H. Rudin, H.-J. Güntherodt, and H. Beck, *J. Phys. C* **13**, L1045 (1980).
61. J. B. Suck, H. Rudin, H.-J. Güntherodt, and H. Beck, *Phys. Rev. Lett.* **50**, 49 (1983).
62. J. Hafner, *J. Phys. C* **14**, L287 (1981).

13. SOLID AND LIQUID HELIUM

Henry R. Glyde

Physics Department
University of Delaware
Newark, Delaware 19716

Eric C. Svensson

Atomic Energy of Canada Limited
Chalk River Nuclear Laboratories
Chalk River, Ontario
Canada KOJ 1JO

13.1. Introduction

13.1.1. Condensed Helium

The condensed phases of the stable isotopes of helium, ^{3}He (Fermi particles of spin $\frac{1}{2}$), and ^{4}He (Bose particles of spin 0), continue to be of high interest to scientists in general and to neutron scatterers in particular. This interest stems from the fact that these substances form the prototype examples in nature of quantum solids and liquids of interacting bosons and fermions. Theories of Bose and Fermi liquids play an extremely important role throughout theoretical physics, and liquid ^{4}He and ^{3}He provide the best opportunities for extensive experimental tests of the predictions of these theories. A very meaningful comparison of the results of first-principles theoretical calculations with experiment is in fact possible for these materials because the interatomic potential is reasonably simple and well known. Different regions of this potential can readily be probed in the experiments by varying the pressure and temperature so as to vary the interatomic distance, an extremely important advantage.

The phase diagrams for ^{3}He and ^{4}He at low temperatures are shown in Fig. 1. Note that, under low pressures, both remain liquid right down to absolute zero, a consequence of the very weak interatomic interactions and the very large zero-point motion resulting from the small mass. The different zero-point motion for ^{3}He and ^{4}He gives rise to large differences in the solid phases; note in particular that approximately twice as much pressure is

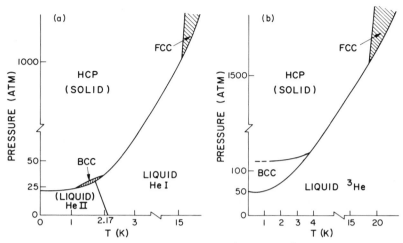

FIG. 1. Phase diagrams of (a) ^{4}He and (b) ^{3}He.

required to solidify ^{3}He at $T = 0$. The condensed phases of helium have extremely large molar volumes. Even in the solids, at least at modest pressures, the separation of the atoms is well beyond the minimum in the pair potential and the atoms exhibit very large amplitudes of vibration. The low-density solids are thus extremely anharmonic materials and provide the most severe tests of theories of anharmonic lattice dynamics. By raising the pressure, the anharmonic effects can be made to essentially disappear. Note also that the solids exhibit only simple crystallographic structures, which greatly facilitates a comparison of theory and experiment.

Kamerlingh Onnes first liquified ^{4}He in 1908, and a remarkably high interest in this fascinating fluid has been maintained ever since by a continuing stream of surprises and puzzling phenomena. Note that the liquid phase of ^{4}He is divided into He(I) and He(II) by a line known as the λ line. The discoveries during the 1930s of the "super properties"[1] of He(II) led to widespread speculation about the nature of the phase transition that took place on crossing this line. In 1938, Fritz London proposed[2] that this phase transition was a "Bose–Einstein Condensation" of a macroscopic fraction of the ^{4}He atoms into the zero-momentum state. This intriguing proposal, which started the "Condensate Saga," did not receive conclusive experimental verification until 44 years later (see Section 13.3). Superfluid ^{4}He [He(II)] is still the only example of a real system that exhibits a Bose condensate, although there has recently been great effort to achieve Bose condensation in atomic hydrogen, a presumed "Bose gas."

The first neutron measurements on liquid ^{4}He were reported[3] in 1951. At low temperatures ($T \approx 1$ K) in He(II) one observes, by neutron inelastic scattering, a branch of extremely sharp, essentially δ-function, excitations—the phonon–roton branch proposed by Landau[4,5] in the 1940s. Such sharp excitations are, among fluids, unique to He(II), and this suggests that they are a consequence of the Bose condensate, but exactly how is still not clear. In fact there is still no all-encompassing general theory for superfluid ^{4}He such as exists for superconductors or liquid ^{3}He. The existence of this single branch of excitations in an isotropic system makes He(II) an extremely favorable system for the study of interactions between excitations. Superfluid ^{4}He can also be made extremely pure (≈ 1 part in 10^{15}), which is of great importance for the production and storage of ultra-cold neutrons ($E \approx 10^{-7}$ eV) in "bottles" filled with He(II).

In 1949 ^{3}He was first liquified by Sydoriak et al.[6] In 1956, Landau[7,8] presented his celebrated phenomenological theory of normal Fermi liquids, which was later given a firm microscopic foundation.[9,10] In 1972, superfluidity (due to a Cooper pairing of ^{3}He atoms rather than to Bose condensation) was discovered[11] in liquid ^{3}He at temperatures below 3 mK. The first neutron measurements on normal liquid ^{3}He were reported[12] in 1974; there are still no neutron measurements on superfluid ^{3}He. Neutron studies on ^{3}He are hampered by the enormously large absorption cross section (see the appendix of Part A of this book), but in spite of this we have already learned a great deal about the zero sound and spin-fluctuation excitations in this Fermi fluid (see Section 13.4).

In 1926 ^{4}He was first solidified by Keesom. The first neutron diffraction measurement on solid ^{4}He were reported[13] in 1958, and the first inelastic scattering measurements[14] in 1967. Since then there have been measurements in all three solid phases shown in Fig. 1. In 1951 ^{3}He was first solidified by Osborne et al.[15] The large amplitudes of vibration of the atoms in the helium solids lead to rapid particle exchange, which is especially important for the fermion ^{3}He system where it leads to nuclear magnetic ordering at $T \approx 1$ mK in the bcc phase. The ordered magnetic spin structure is believed to be up-up-down-down (uudd), and the first neutron measurements on solid ^{3}He reported[16] in 1985 give a result that is consistent with this structure.

In this Chapter we will attempt to highlight some of the most important features of the structure and excitations of the condensed phases of ^{3}He and ^{4}He, concentrating on the results from neutron scattering measurements and on those theoretical developments that have direct consequences for neutron studies, even though in several cases the appropriate studies may not yet have been attempted. We will also largely restrict our attention to work of the last decade and will emphasize areas where major progress has been made since the previous extensive reviews by Woods and Cowley[17] and Price.[18] The two volumes edited by Bennemann and Ketterson[19] contain many excellent articles on the condensed phases of helium, and the books by Wilks[20] and Keller[21] are other excellent sources of general information.

For a more detailed discussion of ^{3}He, and especially of its superfluid phases, see the reviews by Wheatley,[22] Leggett,[23] and Vollhardt.[24] The review by Glyde[25] and the proceedings of the 75th Jubilee Conference on Helium-4, edited by Armitage,[26] are other very recent sources of information.

13.1.2. The Dynamic Form Factors

13.1.2.1. Condensed ^{4}He. Since the ^{4}He nucleus is spinless, the scattering of neutrons from condensed ^{4}He is coherent. Thus one always observes the coherent dynamic form factor

$$S_c(\mathbf{Q}, \omega) = \frac{1}{2\pi} \int_{-\infty}^{\infty} dt\, e^{i\omega t} \frac{1}{N} \langle \rho(\mathbf{Q}, t)\rho(-\mathbf{Q}, 0) \rangle, \qquad (13.1)$$

where $\hbar\mathbf{Q}$ and $\hbar\omega$ are the momentum and energy transferred from the neutron to the sample. Equation (13.1) may be obtained from Eq. (1.39a) of Chapter 1 (Part A) by reversing the sign of t. The operator

$$\rho(\mathbf{Q}, t) = \sum_{i}^{N} e^{-i\mathbf{Q}\cdot\mathbf{r}_i(t)} = \sum_{\mathbf{k}} a^\dagger_{\mathbf{k}-\mathbf{Q}}(t) a_{\mathbf{k}}(t) \qquad (13.2)$$

is the $\mathbf{Q}$th Fourier component of the particle number $\rho(\mathbf{r}, t)$ at time t. The second part of Eq. (13.2) is the second quantized representation appropriate for a fluid. The $\langle \rho(\mathbf{Q}, t)\rho(-\mathbf{Q}, 0)\rangle/N$ is a dimensionless, intensive function and at $Q = 0$, $\langle \rho(0, 0)\rangle = N$. We denote the static number density as $n = N/V$. Since ^{4}He is entirely a coherent scatterer, we will drop the subscript c when discussing pure ^{4}He.

13.1.2.2. Condensed ^{3}He. The ^{3}He nucleus has a large absorption cross section for neutrons (see the Appendix of Part A). To extract $S(\mathbf{Q}, \omega)$ from experiments, one must carefully allow for the effects of absorption.[27,28] We assume this is done and consider only the scattering component.

Since ^{3}He has nuclear spin $\frac{1}{2}$, the scattering length (b_i) of a specific nucleus (i) depends upon its spin $(\mathbf{I}_i)$ relative to that of the incoming neutron $(\mathbf{S})$. This dependence may be expressed as [27,29] (see Eq. (1.270), Chapter 1, Part A)

$$b_i = \bar{b} + \frac{2\bar{b}_i}{[I(I+1)]^{1/2}} \mathbf{S} \cdot \mathbf{I}_i \qquad (13.3)$$

where $\bar{b}$ and $\bar{b}_i$ are the coherent and incoherent scattering lengths defined in Eq. (1.271) of Chapter 1, Part A of ^{3}He nuclei. The inelastic scattering cross section of the collection is (see Eq. (1.32), Chapter 1, Part A)

$$\frac{d^2\sigma}{d\Omega\, dE} = \frac{N\bar{b}^2}{\hbar}\frac{k_1}{k_0} S(\mathbf{Q}, \omega),$$

where

$$S(\mathbf{Q}, \omega) = \frac{1}{\bar{b}^2}\frac{1}{2\pi}\int dt\, e^{i\omega t}\frac{1}{N}\langle \rho_b(\mathbf{Q}, t)\rho_b(-\mathbf{Q}, 0)\rangle \qquad (13.4)$$

is a total dynamic form factor and

$$\rho_b(\mathbf{Q}, t) = \sum_i b_i e^{-i\mathbf{Q}\cdot\mathbf{r}_i(t)} \qquad (13.5)$$

is the scattering length density. The total $S(\mathbf{Q}, \omega)$ describes the correlations in the scattering length density, which depend, via Eq. (13.3), on the total density and the spin density. If there are correlations among the spins (I_i), $S(\mathbf{Q}, \omega)$ does not separate into the usual coherent and incoherent parts.

For an unpolarized neutron beam ($\langle S \rangle = 0$), we obtain, on substituting Eq. (13.3) into Eq. (13.4),

$$S(\mathbf{Q}, \omega) = S_c(\mathbf{Q}, \omega) + (\sigma_i/\sigma_c)S_I(\mathbf{Q}, \omega). \qquad (13.6)$$

Here S_c is given by Eq. (13.1) and the spin-dependent component

$$S_I(\mathbf{Q}, \omega) = \frac{1}{2\pi}\int dt\, e^{i\omega t}\frac{1}{NI(I+1)}\langle \mathbf{I}(\mathbf{Q}, t)\cdot\mathbf{I}(-\mathbf{Q}, 0)\rangle, \qquad (13.7)$$

depends upon Fourier components of the spin density

$$\mathbf{I}(\mathbf{Q}, t) = \sum_i \mathbf{I}_i\, e^{-i\mathbf{Q}\cdot\mathbf{r}_i(t)} = \frac{1}{2}\sum_{\mathbf{k}\lambda} a^\dagger_{(\mathbf{k}-\mathbf{Q})\lambda}\hat{\sigma} a_{\mathbf{k}\lambda}. \qquad (13.8)$$

In the second term, $\hat{\sigma}$ is the Pauli spin matrix and $\lambda = \uparrow$ or $\downarrow$ is the spin projection.

For an isotropic liquid, $\mathbf{I}\cdot\mathbf{I} = 3\langle I_z I_z\rangle$, where I_z is the z component of I. Using σ_z with $I = \frac{1}{2}$,

$$S_I(\mathbf{Q}, \omega) = \frac{1}{2\pi}\int dt\, e^{i\omega t}\frac{1}{N}\langle \rho_z(\mathbf{Q}, t)\rho_z(-\mathbf{Q}, 0)\rangle, \qquad (13.9)$$

where $\rho_z(\mathbf{Q}) \equiv \rho_\uparrow(\mathbf{Q}) - \rho_\downarrow(\mathbf{Q})$ and $\rho_\uparrow(\mathbf{Q}) = \sum_\mathbf{k} a^\dagger_{(\mathbf{k}-\mathbf{Q})\uparrow} a_{\mathbf{k}\uparrow}$. In Eq. (13.1) $S_c(\mathbf{Q}, \omega)$ is similarly related to the total density $\rho = \rho_\uparrow + \rho_\downarrow$. If the spins are independent, $\mathbf{I}_i\cdot\mathbf{I}_{i'} = I(I+1)\delta_{ii'}$, and Eq. (13.9) reduces to the incoherent $S_i(\mathbf{Q}, \omega)$ [Eq. (1.39), Chapter 1, Part A]. In liquid ^{3}He, $S_I(\mathbf{Q}, \omega)$ appears to approach $S_i(Q, \omega)$ for $Q \gtrsim 1\,\text{Å}^{-1}$, since $S_i(Q) \equiv 1$ and the observed[30] $S_I(Q)$ appears to approach 1 at $Q \approx 1\,\text{Å}^{-1}$.

13.1.2.3. ^{3}He-^{4}He Mixtures. In ^{3}He-^{4}He mixtures, it is convenient to separate the scattering length density into a component from the ^{3}He and one from the ^{4}He,

$$\rho_b(\mathbf{Q}) = \sum_i^{N_3} b_{i3} e^{-i\mathbf{Q}\cdot\mathbf{R}_i} + \sum_i^{N_4} b_{i4} e^{-i\mathbf{Q}\cdot\mathbf{R}_i}. \tag{13.10}$$

Here b_{i3} is given by Eq. (13.3) and $b_{i4} = \bar{b}_4$, the same coherent scattering length for all ^{4}He atoms. Again, we ignore the issue of absorption[31] (^{3}He) and assume b_{i3} is purely real. Then, when Eq. (13.10) is substituted into Eq. (13.7), $S(\mathbf{Q}, \omega)$ has four terms that we discuss in Section 13.5.

13.1.2.4. Dynamic Susceptibilities and Sum Rules. To evaluate S_c and S_I, it is convenient to introduce the corresponding retarded dynamic susceptibility

$$\chi_c(\mathbf{Q}, t) = (-i/V)\theta(t)\langle[\rho(\mathbf{Q}, t), \rho(-\mathbf{Q}, 0)]\rangle. \tag{13.11a}$$

Its Fourier transform

$$\chi_c(\mathbf{Q}, \omega) = \int_{-\infty}^{\infty} dt\, e^{i\omega t} \chi_c(\mathbf{Q}, t) \tag{13.11b}$$

is related to S_c by

$$S_c(\mathbf{Q}, \omega) = -(1/n\pi)[n_B(\omega) + 1]\chi_c''(\mathbf{Q}, \omega), \tag{13.12}$$

where $n_B(\omega)$ is the Bose function and χ_c'' is the imaginary part. At $T = 0$ K, it is convenient to use the time-ordered (T) susceptibility $\chi_c(\mathbf{Q}, t) = -i\langle T\rho(\mathbf{Q}, t)\rho(-\mathbf{Q}, 0)\rangle/V$, which is also related to S_c by Eq. (13.12) at $T = 0$ K. At $T = 0$ K, the two susceptibilities are equal for $\omega > 0$. Similarly, we may introduce a $\chi_I(\mathbf{Q}, t)$ as in Eq. (13.11) using ρ_z, with $\chi_I(\mathbf{Q}, \omega)$ related to $S_I(\mathbf{Q}, \omega)$ as in Eq. (13.12).

We will make frequent use of the moments of S_c and S_I ($\alpha = c, I$),

$$\langle \omega^n \rangle_\alpha = \int d\omega\, \omega^n S_\alpha(\mathbf{Q}, \omega). \tag{13.13}$$

In particular, the static structure factor is

$$S_\alpha(\mathbf{Q}) = \int d\omega\, S_\alpha(\mathbf{Q}, \omega) \tag{13.14}$$

and the f-sum rule for S_c is

$$\int d\omega\, \omega S_c(\mathbf{Q}, \omega) = \omega_R = \hbar Q^2/2M, \tag{13.15}$$

where ω_R is the recoil frequency and M is the mass.

13.2. Solid Helium

13.2.1. Introduction

Helium was first solidified at the famous Kamerlingh Onnes Laboratories in Leiden by W. H. Keesom on 25 June 1926. The early work on solid helium is discussed in Keesom's book[32] and is elegantly reviewed by Domb and Dugdale.[33] Neutron inelastic scattering studies of solid ^{4}He opened [14, 34, 35] in 1967 and 1968. Bitter et al.[14] explored hcp polycrystals at $V = 20.9$ cm^3/mol, while Lipschultz et al.[14] and Minkiewicz et al.[34] investigated single-crystal hcp ^{4}He at $V = 21.1$ cm^3/mol. The dynamics of hcp ^{4}He at $V = 16.0$ cm^3/mol was examined by Brun et al.[35] and Reese et al.[35] The purpose of these experiments was to explore the phonon dynamics of these highly anharmonic, quantum crystals at low wave vector Q. In 1972 bcc ^{4}He at $V = 21$ cm^3/mol was thoroughly investigated,[36] including the dynamics at momentum transfers up to $Q = 5$ Å^{-1}. Traylor et al.[37] and Stassis et al.[38] measured phonon energies and lifetimes in higher density helium in the fcc phase at $V = 11.7$ cm^3/mol. Finally, fcc and hcp crystals under high pressure (V as low as 9.03 cm^3/mol) were investigated in 1977 and 1978 by Eckert et al.[39] and Thomlinson et al.[40] More recently, Hilleke et al.,[41] Sokol et al.,[41] and Simmons and Sokol[42] have observed scattering from hcp and bcc ^{4}He at high momentum transfer ($Q \sim 20$ Å^{-1}) to determine single-atom properties such as the atomic kinetic energy. In a *tour de force*, Benoit et al.[16] have observed magnetic ordering in solid ^{3}He below 1 mK using neutrons, but as yet solid ^{3}He/^{4}He mixtures have not been investigated.

The aim of early neutron experiments[34–40] on solid ^{4}He was to explore and test our basic ideas of solid helium. For example, because the interatomic potential is so weak (well depth ~ 10 K), harmonic models of helium using a Lennard–Jones potential predict an unstable solid.[33, 43, 44] Thus the harmonic theory suggested that phonons may not exist at all in solid helium at low density. It was therefore of great interest to determine the nature of the atomic dynamics. In low-density bcc ^{4}He phonons are indeed observed,[36] having low energies of ~ 2 meV $= 20$ K. In compressed[39, 40] fcc ^{4}He the phonon energies reach as high as 150 K. This shows that the strongly repulsive hard core of the interatomic potential plays a key role in the dynamics. The challenge was then to develop a theory[43, 44] that predicts a stable solid and reasonable phonon frequencies and lifetimes. This was accomplished in the late 1960s and early 1970s using the self-consistent phonon (SCP) theory,[45] with different treatments of the short-range correlations (SRC) between atoms.[46–50]

The neutron studies[36] soon showed that it was possible to determine single-phonon frequencies and lifetimes reliably in low-density helium using low scattering wave vectors only ($Q \lesssim 1$ Å^{-1}). At higher Q the incoming neutrons excite pairs, triplets, and higher multiples of phonons and these components make large contributions to $S(Q, \omega)$. The single and multiphonon components of $S(Q, \omega)$ also "interfere" due to the coupling of phonons via anharmonic terms.[36,51-55] Similar effects take place [39,56] in fcc ^{4}He but at higher $Q(Q \gtrsim 3$ Å$^{-1})$. The interference and multiphonon scattering contain very interesting physics but make it impossible to isolate single-phonon character except at $Q \lesssim 1$ Å^{-1}. At $Q \gtrsim 5$ Å^{-1} the scattering is characteristic[36] of scattering from nearly free single particles, as in liquid ^{4}He. It is therefore possible to determine the momentum distribution and kinetic energy (KE) of single atoms in solid He.

Our aim here is to display the understanding of the dynamics of solid helium revealed by neutron scattering. The dynamics has been extensively reviewed by Guyer,[46] Horner,[47] Koehler,[48] Glyde,[49] Varma and Werthamer,[50] and Price.[18] These reviews discuss neutron scattering results up to 1975. There are also recent reviews by Andreev[57] and Roger et al.[58] on the defect and magnetic properties of solid helium. Since these reviews, high-density helium [38-40] and high momentum transfers[41,42] have been investigated. We therefore emphasize the density dependence of excitations in solid helium revealed by recent experiments and the information this provides on the SCP theory and on treatments of SRC's, paying less attention to low-density helium results reviewed previously.[46-51] We also review the single-particle excitations observed at large Q. The properties of anharmonic interference between the one and multiphonon parts of $S(Q, \omega)$ have been thoroughly discussed by Glyde,[54] Price,[18] and Collins and Glyde.[56] Also, the similarity in the scattering intensity from liquid and solid ^{4}He has recently[25] been emphasized. This suggests that similar interference between the single- and multiphonon parts of $S(Q, \omega)$ takes place in liquid ^{4}He as well as in solid ^{4}He. Since anharmonic interference and its possible implications in liquid ^{4}He have been reviewed,[25] we devote little space to this interesting topic.

13.2.2. Physics of Solid Helium

13.2.2.1. Basic Dynamics.
The basic nature of solid helium may be exposed by examining the Debye–Waller factor

$$e^{-2W} \simeq e^{-\langle(\mathbf{Q}\cdot\mathbf{u})^2\rangle} = e^{-(1/3)Q^2\langle u^2\rangle}. \qquad (13.16)$$

Here we have approximated $2W$ by its first cumulant ($2W \approx \langle(\mathbf{Q}\cdot\mathbf{u})^2\rangle$). There is as yet no evidence that the second cumulant $\frac{1}{12}[\langle(\mathbf{Q}\cdot\mathbf{u})^4\rangle - 3\langle(\mathbf{Q}\cdot\mathbf{u})^2\rangle]$, which vanishes in the harmonic approximation, is significant in solid helium. The expression on the right of Eq. (13.16) is valid for cubic crystals. The

Debye expression for $\langle u^2 \rangle$ at $T = 0$ K is[49]

$$\langle u^2 \rangle = \frac{9}{4} \frac{\hbar^2}{M k_B \Theta_{DW}}, \qquad (13.17)$$

where Θ_{DW} is the Debye–Waller temperature. Fits of Eqs. (13.16) and (13.17) to the observed values of $2W$ in low-density ^{4}He at $V = 21$ cm^3/mol give[34] $\Theta_{DW} = 25$ K (hcp) and[36] $\Theta_{DW} = 22.5$ K (bcc). Since the crystals are at $T \approx 1 - 1.5$ K and the zero-point energy (in a Debye model) is $\approx \frac{9}{8}\Theta_{DW}$, it is a good approximation to ignore the thermal energy compared with the zero-point energy. Also, since Eq. (13.17) is fitted to the observed $2W$ in Eq. (13.16), the experimental values of Θ_{DW} are really just a way of representing the observed values of $\langle u^2 \rangle$.

In bcc ^{4}He the observed $\Theta_{DW} = 22.5$ K means, from Eq. (13.17), that $\langle u^2 \rangle^{1/2} \simeq 1.1$ Å. Since the interatomic spacing R is 3.56 Å, the Lindemann[59] ratio in bcc ^{4}He is $\delta = \langle u^2 \rangle^{1/2}/R \approx 0.3$. Thus, under low pressure, solid helium is a weakly bound crystal of light atoms executing large-amplitude vibrations. These amplitudes are even larger[49,55] in bcc ^{3}He.

Since δ is large, solid helium is highly anharmonic. This is illustrated in Fig. 2, which shows the potential well seen by a single atom due to two neighboring atoms on opposite sides. The total potential well set up by the two neighbors is clearly not well approximated by a harmonic well. Thus we do not expect a harmonic approximation to serve well. To date, the dynamics of solid helium is best described by the SCP theory[45] in which anharmonic terms are incorporated consistently at the outset rather than added as perturbations later on. Also, with $\langle u^2 \rangle^{1/2} \approx 1.1$ Å and $R = 3.56$ Å, the vibrations carry the atoms well into the hard-core region of the potential

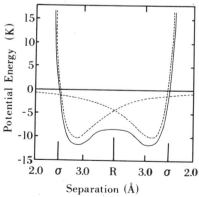

FIG. 2. Potential well (solid curve) seen by a helium atom in a chain due to its nearest neighbors. The dashed curves show the Lennard–Jones pair potentials.

(i.e., $R - \langle u^2 \rangle^{1/2} \approx 2.5$ Å), which is comparable with the hard-core radius, $\sigma = 2.556$ Å, for the Lennard–Jones potential. Thus an accurate treatment of the dynamics must include a description of the SRC's[47–50] in the atomic motion induced by the hard core of the potential. To date, this has been done by introducing[44] a Jastrow-type pair correlation function as in the correlated basis function (CBF)[60, 61] method for liquids or by using a Brueckner-type T-matrix as in nuclear matter[62] and liquids.[63] Neutron scattering experiments provide a critical test of these descriptions of solid helium.

By contrast, compressed solid helium is much less anharmonic. Neutron studies[39] of fcc ^{4}He under 4.93 kbar, for which $V = 9.03$ cm^3/mol, find $\Theta_{DW} = 154$ K. This Θ_{DW} corresponds to $\langle u^2 \rangle^{1/2} \approx 0.4$ Å, giving a Lindemann ratio of $\delta = 0.12$. Comparison of calculated[56] and observed[39] phonon dynamics suggests that an explicit treatment of SRC's is not required at these volumes. The term *quantum crystal* refers to solid helium in which the quantum zero-point energy dominates the thermal energy. The term also implies that an explicit description of SRC's is required. For this reason solid helium under very high pressure is not always referred to as a quantum crystal. (Under extreme pressure helium can be solidified at room temperature and treated as a classical crystal using molecular dynamics.[64]) We outline below the SCP theory as a basis for discussing the observed results.

13.2.2.2. *Self-Consistent Phonon Theory.* The lowest-order approximation to the SCP theory is the self-consistent harmonic (SCH) approximation.[45] The basic idea of the SCH theory is to assume a harmonic model and to find the best harmonic force constants that represent the real anharmonic crystal. These "best" harmonic force constants are most readily found using a variational principle.[45, 47–50] This leads to phonon frequencies $\omega_\lambda(\mathbf{q})$ given by the usual harmonic result,

$$\omega_\lambda(\mathbf{q})^2 = \sum_{\alpha\beta} e^\lambda_\alpha(\mathbf{q}) \left[\frac{1}{M} \sum_l (e^{i\mathbf{q}\cdot\mathbf{R}_l} - 1) \Phi^{0,l}_{\alpha\beta} \right] e^\lambda_\beta(\mathbf{q}), \tag{13.18}$$

with SCH force constants

$$\Phi_{\alpha\beta}(0l) = \left\langle \frac{\partial^2 v(|\mathbf{r}_l - \mathbf{r}_0|)}{\partial r_{0\alpha} \partial r_{l\beta}} \right\rangle, \tag{13.19}$$

where $R_l = l$ is the distance from the origin to lattice vector l. The SCH force constants are the second derivative of the potential $v(\mathbf{l} + \mathbf{u})$, averaged over the relative vibrational amplitude $\mathbf{u} = \mathbf{u}^l - \mathbf{u}^0$ of the atoms (in well such as shown in Fig. 2). For example, the averaged $v(\mathbf{l} + \mathbf{u})$ is

$$\langle v(\mathbf{l} + \mathbf{u}) \rangle = [(2\pi)^3 |\Lambda|]^{-1/2} \int d^3\mathbf{u}\, e^{-1/2 \mathbf{u}\cdot\Lambda^{-1}\cdot\mathbf{u}} v(\mathbf{l} + \mathbf{u}). \tag{13.20}$$

Since an atom, when vibrating, samples wall regions of the well where the second derivative is large, the $\Phi_{\alpha\beta}$ in Eq. (13.19) is much greater than the second derivative evaluated at the center of the well. Thus the SCH force constants become larger and larger as $\langle u^2 \rangle$ increases for interatomic potentials having a hard core such as depicted in Fig. 2. The width Λ of the Gaussian vibrational distribution is given by

$$\Lambda_{\alpha\beta}(0l) = \frac{1}{N} \sum_{\mathbf{q}\lambda} \left(\frac{\hbar}{M\omega_\lambda(\mathbf{q})}\right)(1 - e^{-i\mathbf{q}\cdot\mathbf{l}})e_\alpha^\lambda(\mathbf{q})e_\beta^\lambda(\mathbf{q}) \coth[\tfrac{1}{2}\beta\hbar\omega_\lambda(\mathbf{q})]. \quad (13.21)$$

Thus Eqs. (13.18), (13.19), (13.20), and (13.21) must be evaluated iteratively until consistent.

The SCH theory may also be derived by summing all the even anharmonic terms (quartic, sixth order, ...) that appear in the first-order correction to the harmonic frequencies.[45,65-67] This is depicted in Fig. 3(a). Since these even anharmonic terms are generally positive, we again expect the SCH frequencies to exceed the harmonic ones.

The next anharmonic term beyond the SCH approximation, in order of size, is the cubic anharmonic term, which appears first in second order[45] (all odd derivative terms vanish to first order). This term is either added as a perturbation[48-50,68-70] to the SCH theory or included in an approximate iteration scheme [47,71] with the SCH theory. With the cubic term added, the one-phonon part of the dynamic form factor, Eq. (13.1), is[47-50,67]

$$S_1(\mathbf{Q}, \omega) = \frac{1}{2\pi}[n(\omega) + 1]|F_1(\mathbf{Q}, \mathbf{q}\lambda)|^2 A(\mathbf{q}\lambda, \omega), \quad (13.22)$$

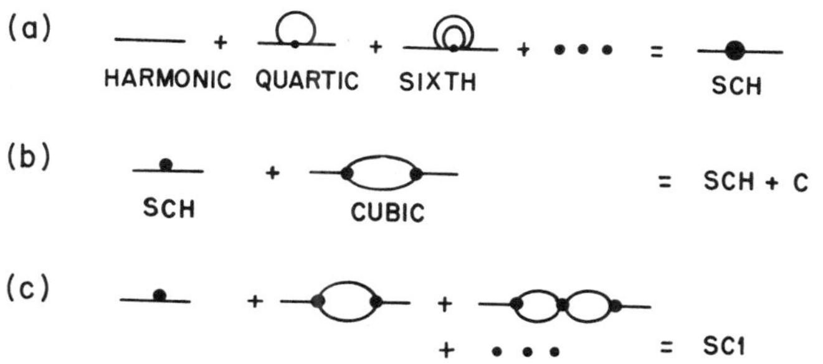

FIG. 3. (a) Sum of even anharmonic terms that leads to the SCH approximation. (b) The SCH approximation with the cubic anharmonic term added as a perturbation (SCH + C). (c) The sum of "bubble" terms that lead to the SC1 approximation.

where $F_1(\mathbf{Q}, \mathbf{q}\lambda) = e^{-W}[\mathbf{Q} \cdot \mathbf{e}^\lambda(\mathbf{q})](\hbar/2M\omega\lambda(\mathbf{q}))^{1/2}$ is the structure factor, and

$$A(\mathbf{q}\lambda, \omega) = \frac{8\omega_\lambda^2(\mathbf{q})\Gamma(\mathbf{q}\lambda, \omega)}{[-\omega^2 + \omega_\lambda^2(\mathbf{q}) + 2\omega_\lambda(\mathbf{q})\Delta(\mathbf{q}\lambda, \omega)]^2 + [2\omega_\lambda(\mathbf{q})\Gamma(\mathbf{q}\lambda, \omega)]^2} \tag{13.23}$$

is the one-phonon response function. Here Δ and Γ are the frequency shift and inverse lifetime ($\tau = \Gamma^{-1}$) of the phonon due to the cubic anharmonic term. The reader is referred to earlier reviews[47-50, 66, 67] for a full discussion. Briefly, Δ due to the cubic term is negative and reduces the phonon frequencies. The half-width Γ of the one-phonon group in Eq. (13.23) arises predominantly (for $T \ll \Theta_{DW}$) from the spontaneous decay of the phonon into two other phonons. The decay process does not, therefore, require the existence of thermal phonons and remains large even at $T = 0$ K. The SCP theory with the cubic anharmonic term added is often denoted the SCH + C approximation and is depicted diagrammatically in Fig. 3(b).

Goldman et al.[72] have added the additional anharmonic terms that they denote, SC1 approximation [see Fig. 3(c)]. The motivation is that, with these "bubble" terms added, the lowest-order SC1 free energy and SC1 phonon frequencies are consistent at $\mathbf{q} \to \mathbf{0}$. With the bubble terms included, the $S_1(\mathbf{Q}, \omega)$ is still given by Eqs. (13.22) and (13.23) but the phonon frequencies and Γ are increased somewhat.[73]

13.2.2.3. Short-Range Correlations. When the atoms have large-amplitude vibrations in wells having hard-core "boundaries" (Fig. 2), it is important to describe the hard core explicitly. Phrased differently, a Gaussian vibrational distribution, as displayed in Eq. (13.20), does not represent the real vibrational distribution accurately enough. To correct this, Nosanow[44] and Koehler[74] introduced the ground-state wave function

$$\psi = \left| \prod_{i \neq j} f(r_{ij}) \phi_G \right\rangle, \tag{13.24}$$

where ϕ_G is a Gaussian and $f(r) = \exp(-Kv(r)/4\varepsilon)$ is a "short-range correlation" function that vanishes when two atoms approach closely at small r. Here $v(r)$ is the interatomic potential and K is a variational parameter. The $f(r)$ is depicted in Fig. 4. Using Eq. (13.24), the usual SCH theory can be recovered with $v(r)$ in Eq. (13.19) replaced by a "softened" potential $\omega(r) = f^2(r)\{v(r) - \nabla^2[\log f(r)]\}$. In this approach Koehler and Werthamer[48, 50, 68, 69] introduced one-phonon excited states by operating on Eq. (13.24) with a one-phonon creation operator. The phonon frequencies were defined as the difference between the one-phonon excited state and the ground state.

A Brueckner T-matrix method may also be used.[46, 47, 49, 70, 75] In this case a "T-matrix" effective potential $T(r) = g(r)v(r)$ replaces $v(r)$ in Eq. (13.19), where $g(r)$ is obtained by solving a Bethe–Salpeter equation. Horner[47, 71] has

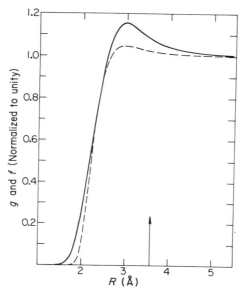

FIG. 4. The Nosanow–Jastrow function $f(r)$ (dashed curve), Eq. (13.24) and the short range function $g(r)$ (solid curve) obtained by solving the T matrix for bcc ^{3}He at $V = 21.46$ cm^3/mol.

refined these methods using the functional derivative technique. This leads to a more internally consistent theory, but one which is difficult to implement consistently in practice. All these methods are reviewed by Varma and Werthamer,[50] Koehler,[48] Horner,[47] and Glyde.[49]

It is fair to say that none of the treatments to date are entirely satisfactory. As we shall see, they do, however, provide a microscopic treatment of the dynamics, which agrees reasonably well with experiment. It is also not clear how to improve them in a simple and numerically tractable way. Certainly the microscopic theory of excitations in solid helium is substantially more complete and satisfactory than that for liquid ^{3}He or liquid ^{4}He.

13.2.3. Dynamic Form Factor

In crystals the $\rho(\mathbf{Q}) = \sum_l \exp[-i\mathbf{Q} \cdot (\mathbf{l} + \mathbf{u}_l)]$ in Eq. (13.1) is usually expanded in powers of $(\mathbf{Q} \cdot \mathbf{u})$, where $\mathbf{u}_l$ is the displacement of atom l from its lattice point $\mathbf{l}$. For an anharmonic crystal, this expansion is complicated.[76,77] It may be conveniently summarized as[54,76]

$$S(\mathbf{Q}, \omega) = S_0 + S_1 + S_{\text{INT}} + S_M. \qquad (13.25)$$

Here $S_0(\mathbf{Q})$ is the elastic scattering and $S_1(\mathbf{Q}, \omega)$, proportional to Q^2, represents the scattering that excites a single phonon only and is given by Eq. (13.22). Due to anharmonic coupling of excitations, the parts of $S(\mathbf{Q}, \omega)$

corresponding to single and multiphonon scattering (S_M) become coupled and interfere. The $S_{INT}(\mathbf{Q}, \omega)$, proportional to Q^3 in lowest order, represents the interference terms between single and multiphonon excitations. The S_M, proportional to Q^4 and higher order terms, contains the scattering that excites two or more phonons and does not contain a single phonon (as an intermediate state) at any point in the scattering. In a harmonic crystal $S_{INT} = 0$.

Ambegaokar, Conway, and Baym[76] (ACB) conveniently combine S_1 and S_{INT} as

$$S_P = S_1 + S_{INT}. \tag{13.26}$$

The S_P then contains all the terms that involve a single phonon in the scattering, either initially excited by the neutron or as an intermediate state in the scattering process. S_{INT} in Eq. (13.25) contributes intensity both within the sharp "one-phonon" peak (in the phonon group) and away from the peak in the "multiphonon" region. Similarly, $S_1(\mathbf{Q}, \omega)$ has a high-frequency tail reaching up to high ω in the multiphonon region. Since only the total $S(\mathbf{Q}, \omega)$ is observed, it is not possible to unambiguously separate S_1 from S_{INT} (and in some cases from S_2). Thus the multiphonon region, in addition to the one-phonon peak, contains very interesting physics when $\langle (\mathbf{Q} \cdot \mathbf{u})^2 \rangle \sim 1$.

Sum rules and the Debye–Waller factor may be used to determine when higher terms above S_1 are important in Eq. (13.25). $S(\mathbf{Q}, \omega)$ satisfies the f-sum rule (13.15). Both S_1 and S_P satisfy the ACB sum rule,[76]

$$\int d\omega\, \omega S_j(\mathbf{Q}, \omega) = \omega_R e^{-2W}[\hat{\mathbf{Q}} \cdot \mathbf{e}_1]^2, \qquad j = 1, P. \tag{13.27}$$

Here $\hat{\mathbf{Q}} = \mathbf{Q}/|Q|$ can be chosen parallel to the polarization vector $\mathbf{e}_1$ so that $[\hat{\mathbf{Q}} \cdot \mathbf{e}_1]^2 = 1$. Thus, Eqs. (13.27) and (13.26) show that $S_{INT}(\mathbf{Q}, \omega)$, the interference terms, do not contribute to the ACB (or the f-sum) rule. Their frequency dependence is such that

$$\int d\omega\, \omega S_{INT}(\mathbf{Q}, \omega) = 0. \tag{13.28}$$

As noted, the leading contribution to S_{INT} is proportional to Q^3. Wong and Gould[78] have identified a term in $S(Q, \omega)$ for liquid ^{4}He proportional to Q^3. Woo[79] has pointed out that there is no contribution to the f-sum rule from terms in $S(Q, \omega)$ proportional to Q^3 in liquid ^{4}He. Equation (13.28) suggests that Wong and Gould[78] and Woo[79] can all be correct. That is, if the Q^3 term in liquid ^{4}He proposed by Wong and Gould is like an interference term in the solid, it could contribute substantially to $S(\mathbf{Q}, \omega)$ but also satisfy Eq. (13.28) as shown by Woo.

Secondly, the ACB sum rule for S_1 or S_P contains a Debye–Waller factor,

$e^{-Q^2\langle u^2\rangle/3}$, compared with the f-sum rule. Thus in bcc ^{4}He, where $\langle u^2\rangle^{1/2} \simeq$ 1 Å^{-1}, $S_p/S \lesssim e^{-1}$ at $Q \gtrsim 1.5$ Å^{-1}. Multiphonon scattering will therefore dominate at $Q \gtrsim 1.5$ Å. In bcc ^{3}He, Glyde and Hernadi[55] find the phonon picture is still useful out to $Q \gtrsim 3$ Å^{-1} but 20–30 terms of $S(\mathbf{Q}, \omega)$ are needed in Eq. (13.25) to fulfill the f-sum rule. At $Q \gtrsim 3$ Å^{-1}, $S(\mathbf{Q}, \omega)$ in bcc ^{4}He is characteristic[36] of a gas of weakly interacting atoms as we discuss in Section 13.2.4.3.

13.2.4. Phonons and Single-Particle Excitations

13.2.4.1. *Phonon Energies.* Figure 5 shows the phonon dispersion curves observed by Minkiewicz *et al.*[36] in bcc ^{4}He at $V = 21$ cm^3/mol. Also shown are calculated curves by Glyde and Goldman[73] using the SC1 approximation to the SCP theory and a T-matrix treatment of SRC's. Considering that the SCP theory is a microscopic theory using only the interatomic potential, crystal volume and structure as input, the agreement with experiment is quite good. The agreement shown in Fig. 5 is typical of that obtained so far in low-density solid helium using a T-matrix method,[73, 80] the Nosanow–Jastrow treatment[68, 69] of short range correlations, or Horner's method.[71] The CBF frequencies defined by Koehler and Werthamer[68, 69] within the SCH + C approximation lie somewhat higher than the calculated

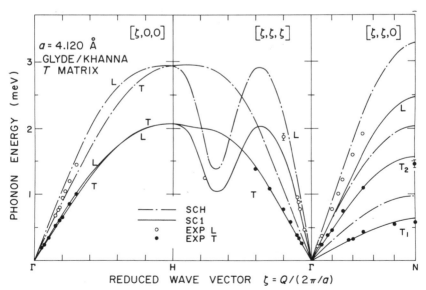

FIG. 5. Phonon frequencies observed in bcc ^{4}He at $V = 21$ cm^3/mol[36] and calculated in the SC1 approximation.[73]

SC1 values shown in Fig. 5. Although the "bubble" terms contained in the SC1 theory [see Fig. 3(c)] beyond the SCH + C approximation increase the phonon frequencies somewhat, the different treatments of short-range correlations lead to generally larger variations in the $\omega_{q\lambda}$, which tend to dominate this effect. It is interesting that different treatment of SRC's produce less variation in the higher order SC1 frequencies than in the lowest-order SCH frequencies.[73] This suggests that different methods of treating short-range correlations might give the same results if each is taken to high enough order in anharmonic theory.

The maximum phonon energy (~2-3 meV) in bcc ^{4}He is approximately twice the "maxon" energy in liquid ^{4}He at SVP. This reflects chiefly the steep volume dependence of the excitation energies as ^{4}He is compressed from the liquid ($V = 27.37$ cm^3/mol at 1 K and SVP) to the bcc phase.

The phonon frequencies in fcc ^{4}He at $V = 11.7$ cm^3/mol observed by Stassis et al.[38] are shown in Fig. 6. The calculations by Horner (see Ref. 38) clearly agree well. The phonon energies in fcc ^{4}He at $V = 11.7$ cm^3/mol are ~5 times higher than those in bcc ^{4}He (1 THz = 4.135 meV), again reflecting the steep volume dependence of phonon energies. The fcc ^{4}He is not nearly so anharmonic as the low-volume bcc ^{4}He. Indeed the SC1 phonon frequencies calculated by Goldman et al.[72] without SRC's (i.e., using only Gaussian vibrational distributions) also agree quite well with experiment, but not so well as those of Horner (see Ref. 38).

In fcc ^{4}He at $V = 9.03$ cm^3/mol, the agreement of theory with the observed frequencies of Eckert et al.[39] is even better. The SC1 phonon energies calculated by Goldman (see Ref. 39) lie generally somewhat above the observed values, while the SCH + C values of Collins and Glyde[56] lie generally somewhat below the observed values. In each case SRC's were not included and these do not appear to be required in ^{4}He compressed below

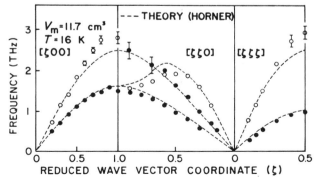

FIG. 6. Phonon frequencies in fcc ^{4}He as observed by Stassis et al.[38] and calculated by Horner (see Ref. 38). (From Stassis et al.[38])

10 cm^3/mol. The Lindemann ratio at 9.03 cm^3/mol, $\delta = 0.12$, is less than that ($\delta = 0.16$) in classical crystals[80–82] at their melting points.

13.2.4.2. *Lifetimes and Interference.* In Fig. 7(b), we show the phonon groups observed by Osgood et al.[36] in bcc ^{4}He at $\mathbf{Q} = (2\pi/a)(0.5, 0, 0)$ and $\mathbf{Q} = (2\pi/a)(1.5, 0, 0) \approx 2.3 \text{ Å}^{-1}$. These two Q's are identical points in reciprocal space separated by a reciprocal lattice vector $\tau = (2\pi/a)(1, 0, 0)$. Thus the two groups represent scattering from the same longitudinal phonon. If only the one-phonon part, $S_1(\mathbf{Q}, \omega)$, of $S(\mathbf{Q}, \omega)$ were being observed, the two groups should have the same shape (and approximately the same intensity). Figure 7(a) is an SCP calculation[55] of $S(\mathbf{Q}, \omega)$ for these two $\mathbf{Q}$'s in bcc ^{3}He at $V = 24$ cm^3/mol including $S_1 + S_{\text{INT}} + S_2$.

We note two points. First, the groups are broad, having a FWHM $(2\Gamma) \sim 1$–2 meV ≈ 10–20 K. These are typical phonon widths in solid helium. In fcc ^{4}He, although the phonon energies increase 5–6 fold, the widths remain similar to those shown in Fig. 7. Also, in fcc ^{4}He at 9.03 cm^3/mol the widths increase by only 10 to 15% between 22 and 38 K. The widths shown in Fig. 7 represent the $T = 0$ K values to within a few percent. In helium the Γ's are large and effectively independent of T because a single phonon can decay spontaneously into two other phonons via the cubic anharmonic coupling term.[48,49] The phonon group widths in bcc ^{3}He shown in Fig. 7 were calculated using

$$\Gamma(\mathbf{q}\lambda, \omega) = -\frac{\pi}{2} \sum_{1,2} |V(\mathbf{q}\lambda, 1, 2)|^2 \{(n_1 + n_2 + 1)\delta[\omega - (\omega_1 + \omega_2)]$$

$$+ (n_2 - n_1)[\delta(\omega + \omega_2 - \omega_1) - \delta(\omega - \omega_2 + \omega_1)]\}, \quad (13.29)$$

where $n_j = n_B(\omega_j)$ is the Bose function, the index j signifies $(q_j\lambda_j)$, and V is the cubic anharmonic coefficient. For $T \ll \Theta_{\text{DW}}$, this Γ is dominated by the temperature-independent term $\delta[\omega - (\omega_1 + \omega_2)]$, which represents the spontaneous decay of the phonon excited by the neutron, $(\mathbf{q}\lambda)$ into two other phonons 1 and 2.

The phonon half-widths Γ in solid helium are ~ 1000 times larger than those observed in liquid ^{4}He at low temperature ($T \approx 1.0$ K). The widths in liquid ^{4}He are also, in contrast, strongly temperature-dependent and increase from $\Gamma \approx 10^{-3}$ meV at 1.0 K to values comparable to those for the solid shown in Fig. 7 just below $T_\lambda = 2.17$ K. Since there is a single phonon dispersion curve only in liquid ^{4}He, the spontaneous decay of a phonon is kinematically excluded (except at $Q \lesssim 0.5$ Å^{-1}, where there is upward dispersion in the phonon energy).[83] Thus phonon decay in liquid ^{4}He is via a four-phonon process,[83] which requires the existence of a thermal phonon to scatter from the phonon excited by the neutron. The three-phonon

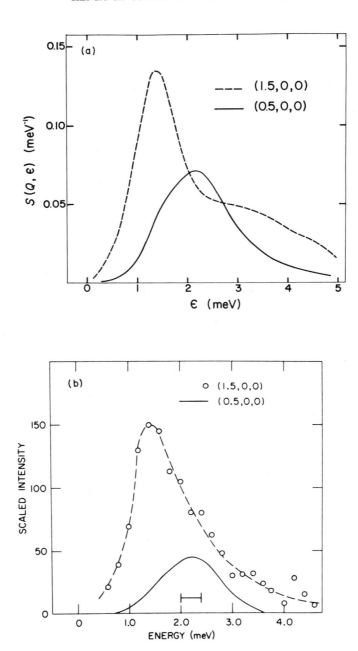

FIG. 7. Phonon groups (a) calculated in bcc ^{3}He[55] and (b) observed for bcc ^{4}He.[36]

spontaneous decay process is allowed in liquid ^{4}He only if the finite width of the phonons is taken into account.[84, 85] This does not, in any case, lead to a large Γ. Thus it is the spontaneous three-phonon decay process that leads to the large and nearly temperature-independent Γ's in solid helium.

Secondly, the scattered intensity at the two "equivalent" Q's in Fig. 7 clearly differs markedly. This tells us that, even at $Q \approx 2$ Å^{-1}, the higher order terms in addition to $S_1(\mathbf{Q}, \omega)$ contribute substantially to $S(\mathbf{Q}, \omega)$. Detailed calculations,[51, 54] such as those for bcc ^{3}He in Fig. 7, show that the interference term S_{INT} is responsible for the displacement of the peak positions shown in Fig. 7. The true single-phonon frequency probably lies at the midpoint between the peak positions shown in Fig. 7. The S_2 term leads to the "hump" at higher ω in bcc ^{3}He for $\mathbf{Q} = (2\pi/a)(1.5, 0, 0)$.

The interference terms contribute both within the sharp "one-phonon" part of the intensity and in the "multiphonon" region at higher ω.[54] For narrow phonon groups, the S_{INT} serves largely to raise or lower the intensity within the sharp "one-phonon" part, but cannot shift the peak position. Since $S_1(\mathbf{Q}, \omega)$ itself has high-frequency tails, which contribute in the "multiphonon" region, and S_{INT} contributes at all ω, it is generally impossible to isolate $S_1(\mathbf{Q}, \omega)$ from the total observed $S(\mathbf{Q}, \omega)$.

Since interference effects in solid helium have been discussed at length elsewhere,[25, 51, 54, 56] we do not elaborate on them here. We note only that there is equal evidence in liquid ^{4}He for something equivalent to one-multiphonon interference in $S(Q, \omega)$. For example, the observed[86] "one phonon" intensity, $Z(Q)$, in liquid ^{4}He shows variations with Q similar to those for the "one-phonon" intensity in solid ^{4}He cause by interference.[25, 54]

13.2.4.3. *Single-Particle Excitations.* At $Q \gtrsim 1.5$ Å^{-1}, Kitchens et al.[36] find that $S(\mathbf{Q}, \omega)$ ceases to show single-phonon structure in bcc ^{4}He. This means that the $S_M(\mathbf{Q}, \omega)$ terms in Eq. (13.25) are dominating the scattering. Indeed at $Q \approx 4.5$ Å^{-1}, Kitchens et al.[36] find that $S(\mathbf{Q}, \omega)$ does not display any collective character but rather is characteristic of scattering from weakly interactive atoms. At these Q values, the scattering is similar in the solid and the liquid. Essentially, the energy transferred by the neutron ($\approx \hbar\omega_R = \hbar^2 Q^2/2M \sim 6$ meV at $Q = 3$ Å^{-1}) exceeds the energy of the highest collective excitations (~ 2 meV) in bcc ^{4}He. The crystal can therefore no longer respond collectively at these higher energies. Rather, calculations by Horner[87] show that the peak position of $S(\mathbf{Q}, \omega)$ falls close to the free-atom recoil peak ω_R for $Q \gtrsim 3$ Å^{-1} in bcc ^{4}He. When high momentum and energy is transferred to an atom, its energy is too large for the surrounding atoms to respond collectively with the excited atom. The single excited atom has a kinetic energy that is larger than the potential energy due to its surrounding neighbors and therefore responds like an atom interacting only weakly with the remainder of the solid.

13.2.5. High Momentum Transfer

13.2.5.1. Impulse Approximation. Hohenberg and Platzman[88] originally proposed that scattering at large Q could be used to determine single-atom properties, particularly the momentum distribution $n(\mathbf{p})$. The basic idea is that at large Q, the momentum (impulse) transferred to a single atom from the neutron is much larger than the momentum transferred from interatomic interactions. At high enough Q the energy transferred to a single atom is greater than the interatomic potential energy and the atoms respond as if nearly free so that kinetic properties (e.g., $n(\mathbf{p})$, kinetic energy) are observed.

At what values of Q may we neglect the impulses imparted by interatomic forces? In solid helium, Monte Carlo[89,90] and phonon[91] calculations of the $\langle KE \rangle$ find that $\langle KE \rangle \approx \Theta_{DM}$ (i.e., $(\hbar^2/2M)k^2 \sim k_B\Theta_{DW}$). Since $\hbar^2/2M = 6.06$ K Å^2 and $\Phi_{DW} = 22.5$ K in bcc ^{4}He, this gives a root-mean-square (RMS) "wave vector" of atoms in the solid (due to interatomic interactions) of $k \sim 2$ Å^{-1}. Thus we might expect the "impulse approximation" (IA) to hold when Q is much greater than the k due to impulses from the surrounding atoms (i.e., when $Q \gtrsim 20$ Å^{-1}). Detailed calculations[92] find indeed that "final-state" interactions may be neglected at $Q \approx 30$ Å^{-1}, depending upon the accuracy required. For fcc ^{4}He where $\Theta_{DW} = 154$ K and $k \sim 5$ Å^{-1}, a $Q \gtrsim 50$ Å^{-1} will be required.[93] Essentially, the stronger the interatomic interactions, the higher the Q that is required for the IA to hold.[92,94]

The IA may be derived in two steps.[92,93] First, we approximate the coherent $S(\mathbf{Q}, \omega)$ in Eq. (13.1) by the incoherent $S_i(\mathbf{Q}, \omega)$. This ignores the terms in $\rho(\mathbf{Q}, t)$ and $\rho(-\mathbf{Q}, 0)$ of Eq. (13.1) that involve different atoms, referred to as the interatomic interference terms.[94] They are believed to be small when $Q \gg 2\pi/R \sim 2$ Å^{-1}. Since $S_i(Q) = 1$ at all Q, a signal that the interference terms are small is the point at which the observed $S(Q)$ approaches 1. This takes place at $Q \approx 8$ Å^{-1} in liquid ^{4}He. Although there has not been a conclusive determination of the Q at which $S(\mathbf{Q}, \omega)$ may be approximated by $S_i(\mathbf{Q}, \omega)$, a $Q \sim 10$ Å^{-1} is usually assumed.

Second, if we ignore the interactions of the single atom with the rest of the system (the final-state interactions) we obtain the IA from $S_i(\mathbf{Q}, \omega)$ as

$$S_{IA}(\mathbf{Q}, \omega) = \int d\mathbf{p}\, n(\mathbf{p})\delta\left(\omega - \frac{\hbar Q^2}{2M} - \frac{\mathbf{Q}\cdot\mathbf{p}}{M}\right). \quad (13.30)$$

The final-state interactions are completely negligible only in the limit $Q \to \infty$. Thus Eq. (13.30) is an approximation at finite Q.[92,95] Weinstein and Negele[96] and Reiter[97] have discussed counterexamples of systems having infinite forces (perfectly hard-core potentials or infinite impulses). In these special cases the final-state interactions are themselves infinite so the IA does not hold even at $Q \to \infty$. However, for realistic interactions the IA does hold at $Q \to \infty$. The S_{IA} may also be obtained from S_i in a short-time approximation,[88,92] hence the name impulse approximation.

There is currently much debate[98–103] on the values of Q at which $S(\mathbf{Q}, \omega)$ can be approximated by the IA result (13.30) for liquid ^{3}He and ^{4}He. This depends upon the accuracy needed. Where Eq. (13.30) is used to determine the condensate fraction n_0, high accuracy is needed. For example, since $n_0(T) \lesssim 13\%$, and n_0 within 10% accuracy is needed, the full $n(\mathbf{p})$ is needed to within $\sim 1\%$. Sears has recently discussed symmetrizing and scaling methods,[94] which can be used to reduce the contributions of final state interactions to $S(\mathbf{Q}, \omega)$. Detailed calculations[92] of $S_i(\mathbf{Q}, \omega)$ and $S_{IA}(\mathbf{Q}, \omega)$ in solid helium at low density (bcc and hcp) show that the IA holds well for $Q \gtrsim 30 \text{ Å}^{-1}$. If the symmetrizing procedures discussed by Sears are also employed, $n(\mathbf{p})$ can be determined[104] within 1% from $S_i(\mathbf{Q}, \omega)$ at $Q \gtrsim 30 \text{ Å}^{-1}$. At these high Q values, the response of liquid and solid helium should be similar. If only KE is needed, this may be obtained from the second moment of $S_i(\mathbf{Q}, \omega)$ as

$$\int_{-\infty}^{\infty} d\omega (\omega - \omega_R)^2 S_i(\mathbf{Q}, \omega) = \frac{4}{3\hbar} \omega_R \langle \text{KE} \rangle, \quad (13.31)$$

where $\omega_R = \hbar Q^2/2M$. In this case we need only assume $S(\mathbf{Q}, \omega) \approx S_i(\mathbf{Q}, \omega)$, which is almost certainly valid at $Q \gtrsim 10 \text{ Å}^{-1}$ in low-density helium.

13.2.5.2. Kinetic Energy in Solid Helium. Hilleke et al.[41] have recently explored hcp ^{4}He at $V = 18.20$ and $19.45 \text{ cm}^3/\text{mol}$ at $T = 1.70$ K using momentum transfers in the range $Q = 12.2$ to 22.5 Å^{-1}. The observed scattering intensity can be fitted by an isotropic Gaussian function within experimental error,

$$S(Q, \omega) = C \exp[-(\omega - \omega_R)^2/2\sigma^2]. \quad (13.32)$$

From Eq. (13.31) the $\langle \text{KE} \rangle$ can be obtained from the variance, $\sigma^2 = (4/3\hbar)\omega_R \langle \text{KE} \rangle$, of the Gaussian function in Eq. (13.32). In this way Hilleke et al.[41] obtained the first direct measurements of the atomic kinetic energies in solid ^{4}He: 34.3 ± 0.9 K at $18.20 \text{ cm}^3/\text{mol}$ and 31.1 ± 0.9 K at $19.45 \text{ cm}^3/\text{mol}$. Using the same method Sokol et al.[41] (see also Simmons and Sokol[42]) studied ^{4}He at $V = 21.7 \text{ cm}^3/\text{mol}$ in the liquid (4 and 1.80 K), bcc (1.70 K), and hcp (0.96 and 1.60 K) phases. Independent of the phase and temperature they originally deduced a $\langle \text{KE} \rangle = 19.5$ K, which has subsequently been revised[105] to 22 ± 3 K. These values are displayed in Fig. 8. To obtain Eq. (13.32) from Eq. (13.30), the $n(p)$ must also be Gaussian.

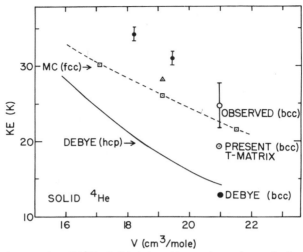

FIG. 8. Kinetic energy in solid ^{4}He. Solid circles with error bars, observed values in hcp ^{4}He;[41] open circle with error bar, bcc ^{4}He;[105] triangle, calculated Monte Carlo value in fcc ^{4}He;[90] squares with dashed line as guide to eye, Monte Carlo values in fcc ^{4}He;[89] solid curve, Debye model $\langle KE \rangle = (9/16)\,\theta_D$ in hcp ^{4}He; solid circle, Debye $\langle KE \rangle$ in bcc ^{4}He; open circle, anharmonic T-matrix $\langle KE \rangle$ in bcc ^{4}He.[91] (From Moleko and Glyde.[91])

Whitlock et al.[89] and Whitlock and Panoff[90] have made Monte Carlo (MC) evaluations of the ground state $\langle KE \rangle$ of fcc ^{4}He. As seen from Fig. 8 these MC $\langle KE \rangle$ values lie ~15% below the observed results in hcp ^{4}He at comparable density. At $V = 21.7$ cm^3/mol, Whitlock et al. calculated a $\langle KE \rangle$ of 23.4 and 22.4 K in the liquid and solid, respectively, which agrees well with the observed value[105] which is apparently independent of structure and phase and depends only on the volume. Also shown in Fig. 8 is a value calculated[91] for bcc ^{4}He using the SCP theory coupled with a T-matrix treatment of short-range correlations.

The $\langle KE \rangle$ provides a very direct and quantitative measure of the anharmonic nature of solid helium.[91] This may be seen as follows. First, a Debye model may be used to estimate the $\langle KE \rangle$ ($\langle KE \rangle = \tfrac{9}{16}\Theta_{DW}$). Using observed values of Θ_{DW}, we obtain kinetic energies that are approximately one half the observed values (see Fig. 8). Second, for a moderately anharmonic crystal, the vibrational distribution is approximately Gaussian. The observed Θ_{DW} can be used, via Eq. (13.17), to obtain an empirical value of the RMS amplitude $\langle u^2 \rangle$ for this distribution. The $\langle KE \rangle$ predicted using this Gaussian vibrational distribution is $\langle KE \rangle = 9\hbar^2/8M\langle u^2 \rangle = \Theta_{DW}/2$. This is also approximately one half the observed $\langle KE \rangle$. Thus a Gaussian vibrational distribution cannot be adjusted to fit both $\langle u^2 \rangle$ and the $\langle KE \rangle$. The SCH model (without short-range correlations) is an adjusted harmonic model. Thus the large ⟨KE⟩ demonstrates the presence of higher anharmonic contributions and short-range correlations beyond the SCH theory in solid helium and shows that the vibrational distribution cannot be Gaussian.

The large ⟨KE⟩ may be explained[91] in terms of anharmonic high-frequency tails in $S_1(\mathbf{Q}, \omega)$ and the one-phonon response function $A(\mathbf{q}\lambda, \omega)$ of Eq. (13.21). The ⟨KE⟩ may be expressed as

$$\langle \text{KE} \rangle = \frac{\hbar}{4N} \sum_{\mathbf{q}\lambda} \frac{1}{\omega_\lambda(\mathbf{q})} \int \frac{d\omega}{2\pi} \omega^2 A(\mathbf{q}\lambda, \omega). \quad (13.33)$$

Calculations[47-56] of $S_1(\mathbf{Q}, \omega)$ and $A(\mathbf{q}\lambda, \omega)$ do indeed have high-frequency tails that will contribute significantly[91] to the ⟨KE⟩. The ⟨KE⟩ calculation based on the T-matrix for bcc ^{4}He shown in Fig. 8 is obtained using Eq. (13.33). It agrees quite well with the MC and the observed value at $V = 21.7$ cm^3/mol while simpler models do not. Including further anharmonic terms could increase this ⟨KE⟩ if they further increased the magnitude of the high-frequency tails. Perhaps most importantly, the ⟨KE⟩ is a critical measure of the extreme "quantum" nature of solid helium. The anharmonic contributions from high-energy tails in $A(\mathbf{q}\lambda, \omega)$ (and therefore also in the momentum distribution) will also be included in MC calculations. There remains therefore an interesting discrepancy between theory and experiment at higher density.

Wong and Gould,[78] Kirkpatrick,[106] Family,[107] Talbot and Griffin,[108] Sears,[109] and others have shown that $S(\mathbf{Q}, \omega)$ in quantum liquids also has high-frequency tails. These tails are related to the hard core of the interatomic potential. For example, the ⟨KE⟩ in liquid ^{3}He (∼ 12-13 K) lies well above the Fermi energy (1.5 K). This demonstrates that $n(\mathbf{p})$ must have, as expected, a tail reaching beyond ε_F, which contributes significantly to the ⟨KE⟩. Thus, high-frequency tails in the dynamic response of helium due to the hard core of the interatomic potential seem common to all condensed phases.

13.2.6. Future Prospects

Benoit et al.[16] have recently observed a ($\frac{1}{2}$, 0, 0) Bragg peak in solid ^{3}He below the magnetic ordering temperature (∼ 1 mK). This peak is consistent with the expected uudd nuclear spin ordering. Further work to determine the magnetic structures of the other magnetic phases of solid ^{3}He would also be most interesting.

Solid ^{3}He and solid ^{3}He-^{4}He mixtures await further exploration by neutrons. With improvements in cryogenics and neutron facilities, the dynamics[55] and magnetic structures[110] of these solids may be revealed and the kinetic energies and atomic momentum distributions accurately determined. The advent of new spallation neutron sources with higher fluxes of epithermal neutrons should markedly advance the latter studies. These experiments will all be very difficult, but they will provide incisive new data to test theory along the lines discussed previously.

13.3. Liquid ^{4}He

13.3.1. Introduction

The study of excitations and interatomic correlations in liquid ^{4}He has a long and rich history. This is because at low temperature ^{4}He forms the prototype example in nature of a quantum liquid of interacting bosons. While the interaction between two ^{4}He atoms is simple and well known,[111] it is comparable in strength to the thermal and zero-point energies, making ^{4}He a complicated Bose liquid.

At $T_\lambda = 2.17$ K and SVP, ^{4}He undergoes a phase transition[19-21] from He(I) to He(II). In 1938 superfluidity was discovered in He(II) by Kapitza[112] and by Allen and Misener.[113] Shortly thereafter London[2] proposed that the λ transition in helium was the analog for a real liquid of the phenomenon of Bose-Einstein condensation in an ideal Bose gas. In a Bose gas of mass 4 particles at liquid ^{4}He density, a macroscopic fraction n_0 of atoms appear in the $p = 0$ state below a critical temperature $T_c = 3.09$ K and $n_0 \to 1$ as $T \to 0$ K. It can be readily shown that for $n_0 = 1$ the pair correlation function $g(r) = 1$ for all r. The He-He potential, however, has a steeply repulsive core,[111] which requires $g(r) = 0$ for $r < 2.5$ Å. Calculations for hard-sphere atoms[114] find $n_0 = 0.10$. In Section 13.3.3.4 we review neutron measurements that show that $n_0 \approx 0.13$ at $T \leq 1$ K, which agrees quite well with the best values of n_0 from Monte Carlo calculation, 0.09[115] and 0.113.[89] The relation between n_0 and superfluidity is, however, not entirely clear.

Landau[4] proposed that, since ^{4}He atoms are strongly interacting, only collective excitations involving many atoms can exist. These collective excitations are the "elementary excitations" of the fluid. He rejected the idea that single-particle excitations at low momentum were possible. By analogy with phonons, Landau proposed that the collective excitations would follow a sound-mode energy spectrum at low q,

$$\varepsilon(q) = cq, \tag{13.34}$$

where q is the excitation wave vector and, for a liquid, is interchangeable with Q. In addition he proposed a second, higher energy branch of the form $\varepsilon = \Delta + q^2/2\mu$. These excitations he called "rotons." To describe thermodynamic properties correctly, Landau later[5] moved these excitations from the origin to a finite momentum q_0,

$$\varepsilon(q) = \Delta + (q - q_0)^2/2\mu. \tag{13.35}$$

He then pictured the relations (13.34) and (13.35) as parts of a single continuous dispersion relation and noted:[5] "With such a spectrum it is of course impossible to speak strictly of rotons and phonons as of qualitatively different types of elementary excitations." From the experimental data then available, Landau[5] inferred the values $\Delta = 9.6$ K, $q_0 = 1.95$ Å^{-1}, and $\mu = 0.77 M$, where M is the ^{4}He atomic mass.

The phonon–roton dispersion relation from a recent compilation and analysis of experimental results by Donnelly et al.[116] is shown in Fig. 9 along with the form proposed by Landau[5] as an insert. A fit[116] to the spectrum (solid curve) yields $\Delta = 8.62$ K, $q_0 = 1.93$ Å^{-1}, and $\mu = 0.15 M$. From Figs. 10 and 11 we see that, in addition to the sharp "elementary excitation" peak, there is substantial structure in $S(Q, \omega)$ at higher ω. The now classic study by Cowley and Woods[86] showed that, for $Q \geq 0.7$ Å^{-1}, more than one half of the f-sum rule, Eq. (13.15), was taken up by this "multiphonon" scattering at high ω. We discuss the relative weight of the sharp component and its possible relation to the superfluid density, ρ_s, in some detail in Sections 13.3.2.1 and 13.3.3.3.

Bogoliubov[117] analyzed a gas of weakly interacting bosons at $T = 0$ K from first principles. With weak interactions n_0 remains large. In this limit he showed that the gas supported only collective excitations having an energy spectrum

$$\varepsilon(q) = [nv(q)q^2/M + q^4/4M^2]^{1/2}, \tag{13.36}$$

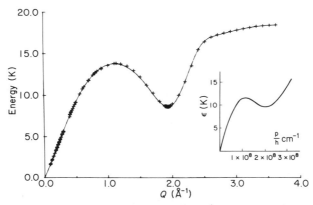

FIG. 9. The one-phonon dispersion relation for liquid ^{4}He at SVP and $T \leq 1.2$ K from a recent compilation of experimental results (crosses) by Donnelly et al.[116] The solid curve represents a cubic spline fit to these results. The inset shows the dispersion relation proposed by Landau[5] in 1947.

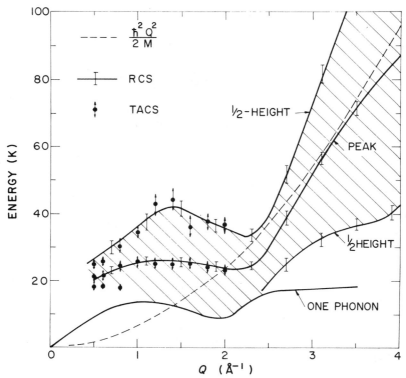

FIG. 10. The dispersion relation for the sharp one-phonon excitations in liquid ^{4}He at 1.1 K and SVP and, at higher energies, the upper and lower half-heights and the mean energy of the broad multiphonon peak in $S(Q, \omega)$. The dashed curve shows the dispersion relation for free ^{4}He atoms. (From Cowley and Woods.[86])

where $v(q)$ is the Fourier transform of the interatomic potential and $n = N/V$. At $q \to 0$, Eq. (13.36) has the form of Eq. (13.34) with $c = [nv(0)/M]^{1/2}$ and at high q Eq. (13.36) goes over into a free-particle spectrum, $\varepsilon = q^2/2M$. This confirmed Landau's[5] prediction at low q, for weakly interacting bosons at least, and at high momentum in Fig. 10 we see that the observed scattering is indeed centered near the free-particle energy. With appropriate choices[118] of $v(q)$ (e.g., $v(q) \approx \sin q/q$) the Bogoliubov spectrum qualitatively reproduces Landau's phonon–roton form. A central issue we explore in Sections 13.3.2.1 and 13.3.3.3 is whether the sharp "elementary excitation" component of $S(Q, \omega)$ depends on the existence of a condensate in ^{4}He. It does in Bogoliubov's weak interaction limit, which requires n_0 to be large. Landau[4] argued that the excitations would be "sharp" because, with dispersion of the shape shown in Fig. 9, an excitation cannot decay spontaneously into two

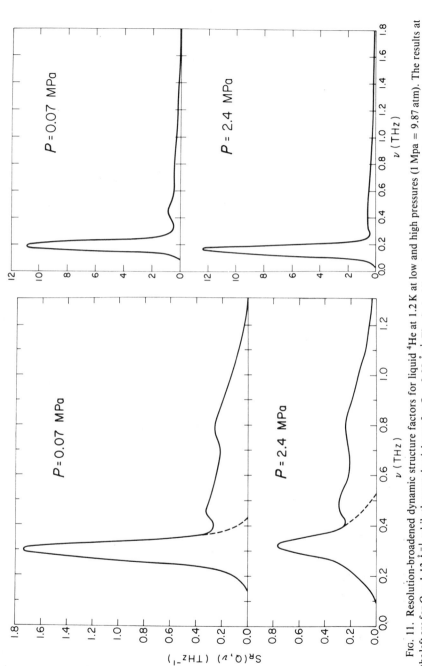

FIG. 11. Resolution-broadened dynamic structure factors for liquid ^{4}He at 1.2 K at low and high pressures (1 Mpa = 9.87 atm). The results at the left are for $Q = 1.13$ Å^{-1}, while those at the right are for $Q = 2.05$ Å^{-1}. The dashed lines show a plausible division of $S_R(Q, \nu)$ into "one-phonon" and "multiphonon" components. (From Svensson et al.[187])

other excitations (except at the upper end[119] and at low q, where there is upward dispersion—see Section 13.3.3.2). Phrased differently, are the condensate and sharp collective-excitation spectrum independent properties of a cold Bose liquid, or is the existence of a condensate responsible for the sharp excitations?

Feynman[120] derived microscopically the first expression for the phonon–roton energy spectrum appropriate to liquid ^{4}He. He argued that the excited state of the liquid containing a single excitation should be of the form $\Psi = \sum_l f(r_l)\phi$. Here ϕ is the ground-state wave function and $f(r_l)$ is a single-particle wave function for each atom involved in the excitation. He derived a variational equation for Ψ. This yielded a minimum energy eigenvalue of

$$\varepsilon(q) = \frac{q^2}{2M} S^{-1}(q) \tag{13.37}$$

for a function of the form $f(r) = e^{i\mathbf{q}\cdot\mathbf{r}}$. Since the density operator $\rho(r) = \sum_i \delta(r - r_i)$ has Fourier transform

$$\rho(\mathbf{q}) = \int d\mathbf{q}\, e^{-i\mathbf{q}\cdot\mathbf{r}} \rho(\mathbf{r}) = \sum_i e^{-i\mathbf{q}\cdot\mathbf{r}_i},$$

the excited state is

$$\Psi = \sum_i e^{i\mathbf{q}\cdot\mathbf{r}_i}\phi = \rho^+(q)\phi. \tag{13.38}$$

That is, the minimum energy excitation of wave vector q is a density fluctuation, having Fourier component $\rho(q)$, in agreement with the predictions of Landau. Feynman argued that, since bosons are indistinguishable, any new state that can be reached by a permutation or rotation of particles without a change in density must be identical to ϕ. Thus all excited states involve density changes. In an analogy with phonons, Feynman showed for $T = 0$ K, $\lim_{q \to 0} S(q) = q/2Mc$ so that as $q \to 0$, Eq. (13.37) assumes the phonon form of Eq. (13.34). Since $S(q)$ has a maximum at $q \approx 2$ Å^{-1} and $S(q) \to 1$ as $q \to \infty$, $\varepsilon(q)$ has a minimum in the "roton" region and $\varepsilon(q) \to q^2/2M$ at $q \to \infty$.

Feynman's variational phonon–roton energy lies substantially above the observed $\varepsilon(q)$ curve. Feynman and Cohen[121] proposed a more flexible variational wave function that simulated the interaction between phonons. This yielded a lower $\varepsilon(q)$ in good agreement with experiment at SVP. However, Padmore and Chester[122] find that the Feynman–Cohen spectrum does not describe the pressure dependence of the roton minimum at all well. Jackson and Feenberg[60,123] and others[61,124,125] have included the effects of phonon interactions more fully to obtain improved phonon energies and phonon lifetimes. The basic technique is the method of correlated basis functions[61] in which improved phonon wave functions are sought and matrix elements are evaluated to obtain excitation energies and lifetimes. This work provides, for example, a very precise form for $\varepsilon(q)$ at low q in terms of a power series in q. The progress up to 1975 is thoroughly reviewed by Woo[61] and Campbell.[126] Below we denote $\varepsilon(q)$ by $\omega(Q)$.

To describe all the observed scattering intensity, Miller et al.[127] proposed the form

$$S_S(Q, \omega) = Z(Q, \omega) + S_M(Q, \omega). \quad (13.39)$$

Here $Z(Q, \omega)$ describes the sharp, apparently single-excitation component usually written as $Z(Q, \omega) = Z(Q) \delta(\omega - \omega(Q))$, and S_M describes the remaining "multiphonon" component (called S_{II} by Miller et al.[127]). Equation (13.39) is intended to describe scattering from superfluid He(II) at $T \to 0$ K, as indicated by the subscript S in $S_S(Q, \omega)$. In general, both Z and S_M depend upon temperature. Cowley and Woods[86] have analyzed the observed scattering intensity in detail using Eq. (13.39) and identified the separate contributions of Z and S_M to $S(Q)$ and the f-sum rule, Eq. (13.15), as a function of Q. Detailed expansions of $Z(Q)$ and $S_M(Q, \omega)$ in powers of Q at low Q are reviewed by Woo.[61] While the form of Eq. (13.39) does not allow for possible interference between the single- and multiphonon scattering[25] (which can contribute to both Z and S_M), it provides a convenient framework for discussing $S(Q, \omega)$.

13.3.2. Theory

13.3.2.1. $S(Q, \omega)$ and the Condensate. In this section we review recent developments in the theory of excitations in liquid ^{4}He. The work prior to 1972 has been thoroughly reviewed by Woods and Cowley.[17] We also focus on those aspects since 1972 that relate to recent experiments. Ruvalds[128] and Zawadowski[129] have quite recently reviewed scattering involving two phonons or rotons (and bound states thereof), which contributes to $S(Q, \omega)$ at higher ω above the main peak (see Fig. 10). We therefore address this interesting component less extensively. Woo[61] has exhaustively reviewed the calculations of the phonon–roton $\omega(Q)$ using correlated basis function (CBF) methods. This includes an extensive discussion of the interactions between excitations. The CBF method may be regarded as an extension of the Bijl[130] and Feynman[120,131] approach, which focuses on determining the wave function of the elementary excited states. Interactions between these excited states are then calculated to determine corrections to the excitation energies and lifetimes of the excitations. We thus devote less attention to the evaluation of the precise form of $\omega(Q)$. Chester[132] has presented an excellent critical review of our understanding of liquid ^{4}He, which, in particular, places recent work in its historical perspective. Finally, Glyde[25] has emphasized the similarity between $S(Q, \omega)$ in liquid and solid helium, both calculated and observed, so that we devote little space to this here.

Current theories of $S(Q, \omega)$ may crudely be divided into two categories: (1) those that are microscopically based and (2) those that are more phenomenological and focus on the properties of the excitations given their existence. The first category is an extension of the pioneering work by Bogoliubov,[117] Hugenholtz and Pines,[133] Gavoret and Nozières,[134] Hohenberg and Martin,[135] Huang and Klein,[136] Ma and Woo,[137] Ma et al.,[138] Kondor and Szépfalusy,[139] Gould and Wong,[140] and Cheung and Griffin.[141] This work was reviewed by Woods and Cowley.[17] The approach uses Green function methods, relates the density–density response observed in neutron scattering to the single-particle excitations and includes an explicit account of the condensate. It is often denoted as the dielectric formulation of $S(Q, \omega)$. Since we wish to explore if (and how) $S(Q, \omega)$ depends on the superfluid density as suggested by Woods and Svensson,[142] we survey the results of this theory. The second category includes the calculation of phonon and roton energies, interactions and lifetimes by Bedell et al.[143, 144] Here the phonon–roton spectra were calculated using a semi-phenomenological polarization potential introduced intially by Aldrich et al.[145–147]. The properties of phonons and rotons are then investigated providing many very useful results. Much of this work has been reviewed by Pines.[148] Recent accurate calculations of $S(Q)$ and $g(r)$ by MC and CBF methods are discussed in Section 13.3.3.5.

The microscopic developments since 1972 have been made chiefly by Griffin and Cheung,[149] Wong and Gould,[78, 150] Szépfalusy and Kondor,[151] Wong,[152] and most recently Talbot and Griffin,[153, 154] Griffin,[155, 156] and Kirkpatrick.[106] We attempt to sketch the results, often without specific reference to the origin. The starting point is the time-ordered dynamic susceptibility (see Section 13.1.2.4), which may be separated into a "singular" and a "regular" part[134, 149, 151]

$$\chi(Q, \omega) = \chi_S(Q, \omega) + \chi_R(Q, \omega)$$
$$= \Lambda_\alpha(Q, \omega) G_{\alpha\beta}(Q, \omega) \Lambda_\beta(Q, \omega) + \chi_R(Q, \omega). \quad (13.40)$$

Here

$$\Lambda_\alpha(Q, \omega) = n_0^{1/2}[1 + \bar{P}_\alpha(Q, \omega)], \quad (13.41)$$

where n_0 is the condensate fraction and $G_{\alpha\beta}(Q, \omega)$ is the four-component, single-particle Green function.[17] The regular part is simply that part that does not involve $n_0 G_{\alpha\beta}$. This separation may be obtained as now explained: the product $\rho(Q)\rho(-Q)$ (see Eq. 13.2) contains four single-particle operators a_k.

If two of the a_k or $a_k^\dagger$ operate on the condensate state ($k = 0$), we obtain a factor n_0 (i.e., $a_0|\phi\rangle = n_0^{1/2}|\phi\rangle$). The remaining two operators in the product lead to a single-particle Green function (e.g., $G_{11}(Q, t) = -i\langle Ta_Q(t)a_Q^\dagger(0)\rangle$). The first role of the condensate is therefore to introduce in χ terms containing $n_0 G_{\alpha\beta}$. Terms of $\rho(Q, t)\rho(-Q, 0)$ containing one a_0 are proportional to $n_0^{1/2} G_{\alpha\beta}\bar{P}_\beta$ and "mix" single-particle and density–density excitations. Finally χ_R contains the terms in which all a_k's have $k \neq 0$ and should be similar above and below T_λ. Terms containing three or four a_0's require $Q = 0$ and do not contribute to scattering.

The separation in Eq. (13.40) is most suggestive. The $G_{\alpha\beta}$ is the single-particle Green function involving atoms outside the condensate having $k \neq 0$. The singular part χ_S is proportional to $G_{\alpha\beta}$. If $G_{\alpha\beta}$ has sharp singular peaks characteristic of "elementary excitations" these will be observed in $S(Q, \omega)$, provided n_0 is finite. In the limit of weak interactions considered by Bogoliubov where $n_0 \approx 1$, $\chi(Q, \omega)$ and $G_{\alpha\beta}(Q, \omega)$ become identical. When n_0 is small [$0 < n_0 < 0.13$ in He(II)] the general spectral structure of $\chi(Q, \omega)$ and $G_{\alpha\beta}(Q, \omega)$ could differ away from the "poles" of $G_{\alpha\beta}(Q, \omega)$ due to $\bar{P}_\alpha(Q, \omega)$, which couples the $G_{\alpha\beta}$ to pairs and higher numbers of excitations.[149] However, we expect the sharp peaks of G, identified as "quasiparticle" energies, to be observed in $S(Q, \omega)$, as indeed is observed in liquid ^{4}He (Fig. 11).

Thus, below T_λ where n_0 is finite, the poles of $S(Q, \omega)$ and of $G_{\alpha\beta}$ coincide. It is the single-particle G and its poles (if sharp) that are related to thermodynamic properties. From Eq. (13.40) and because of n_0, we are then able to use the excitation energies observed by neutron scattering to calculate thermodynamic properties. As we approach T_λ from below, the $G_{\alpha\beta}$ become thermally broadened. Under these circumstances $\bar{P}(Q, \omega)$ and $\chi_R(Q, \omega)$ might shift the peak position of $S(Q, \omega)$ away from that of $G_{\alpha\beta}(Q, \omega)$. The peaks of $S(Q, \omega)$ and $G_{\alpha\beta}(Q, \omega)$ might then differ and using the observed peaks to calculate thermodynamic properties could lead to disagreement with observed thermodynamic values. However, experience in the solid[25, 157] suggests that these interference and multiphonon terms, while changing the intensity substantially, shift the peak positions little. Above T_λ, where $n_0 = 0$ and $\Lambda = 0$, the first term of Eq. (13.40) vanishes and G cannot be observed in $S(Q, \omega)$. Above T_λ, there is no connection between $S(Q, \omega)$ and $G_{\alpha\beta}$, and we cannot use the peak positions of $S(Q, \omega)$ as excitation energies to evaluate thermodynamic properties.

The χ_R in Eq. (13.40) is often expressed in terms of an irreducible part $\tilde{\chi}$ (which contains no single interaction lines) in the form[78, 149–151] $\chi_R = \tilde{\chi}_R/[1 - v(Q)\tilde{\chi}_R] \equiv \tilde{\chi}_R/\varepsilon_R$. This equation may be taken as the definition of $\tilde{\chi}_R(Q, \omega)$ in terms of $v(Q)$ and $\chi_R(Q, \omega)$ and of the dielectric function ε_R. The G and Λ may also be expressed in terms of irreducible parts (e.g.,

$G = N/D$). Above T_λ, $\chi = \tilde{\chi}_R/\varepsilon_R$ and the density response has a pole at energy $\omega_1(Q)$, where

$$\varepsilon_R(q, \omega) = 0, \qquad T > T_\lambda. \tag{13.42}$$

The $\omega_1(Q)$ is the zero-sound mode (ZSM) energy above T_λ. Also, above T_λ the $G_{\alpha\beta}$ has a pole, denoted by $\omega_2(Q)$, where $D(Q, \omega_2) = 0$. As noted, for $T > T_\lambda$ there is no direct relation between χ and $G_{\alpha\beta}$ and we do not expect to observe $\omega_2(Q)$ in $S(Q, \omega)$.

For $T < T_\lambda$, Griffin and Cheung,[149] Szépfalusy and Kondor,[151] and Wong and Gould[78,150] show that χ and $G_{\alpha\beta}$ share the same common denominator

$$C = \bar{D}\varepsilon_R - \bar{\Lambda}v(Q)\bar{\Lambda}\bar{N}. \tag{13.43}$$

Here, C contains a "coupling" term $\bar{\Lambda}v(Q)\bar{\Lambda}\bar{N}$, which makes the denominator of χ and G the same and $\bar{\Lambda}^2$ is proportional to n_0. This is a more precise way of saying that χ and G have the same "sharp" peaks for $T < T_\lambda$. They[78,149-151] show that C has two zeros, one at $\bar{\omega}_1(q)$, which is close to ω_1, and the second at $\bar{\omega}_2(Q)$, which is close to ω_2:

$$C(Q, \bar{\omega}_1) = 0; \quad C(Q, \bar{\omega}_2) = 0; \quad T < T_\lambda. \tag{13.44}$$

The $\bar{\omega}_1$ and $\bar{\omega}_2$ are displaced slightly from ω_1 and ω_2 due to the small coupling term proportional to n_0. The relative strength or residue of these poles is not determined.

The dielectric formulation suggests[108] that the ZSM observed in $S(Q, \omega)$ is the ω_1 mode above T_λ and $\bar{\omega}_1$ below T_λ. The spectral distribution of intensity in $S(Q, \omega)$ will probably also differ significantly above and below T_λ. In this formulation it is the presence of a finite n_0 that makes the peaks of G and $S(Q, \omega)$ coincide below T_λ. We expect G to have sharp structure. In this sense we may say that the sharp peak of $S(Q, \omega)$ observed below T_λ is due to the presence of a macroscopic condensate. Certainly, we expect $S(Q, \omega)$ to differ above and below T_λ as indicated by the results of Woods and Svensson[142] (see Section 13.3.3.3). The data indicate that there is only one sharp component in $S(Q, \omega)$ below T_λ (see Fig. 11). Thus if $S(Q, \omega)$ really contains two peaks (at $\bar{\omega}_1$ and $\bar{\omega}_2$), as suggested by the dielectric theory, the weight of the $\bar{\omega}_2$ mode must be very small. As noted in Section 13.2.5.2, we expect $S(Q, \omega)$ to have high-frequency tails, due to the hard core of the interatomic potential, which are independent of statistics and n_0.

13.3.2.2. Polarization Potential Theory. Pines, Aldrich, Pethick, Bedell, and collaborators have developed[143-148,158,159] a polarization potential (PP) theory of liquid ^{4}He, ^{3}He, and ^{3}He-^{4}He mixtures. The theory takes the same basic form in all cases and attempts a unified treatment of liquid ^{4}He, ^{3}He, and mixtures. The PP theory has evolved and developed in time, beginning with a description of excitations and $S(Q, \omega)$ in ^{4}He by Aldrich and Pines[145,146] and in ^{3}He by Aldrich et al.[147] Bedell and Pines[158] extended the theory to describe scattering amplitudes and transport coefficients in liquid ^{3}He. More recently the PP theory has been applied to ^{3}He–^{4}He mixtures by Hsu et al.[159]

The theory contains three or four basic ideas and concepts. First, the dynamic susceptibility is expressed in the general random phase approximation (RPA) form $\chi = \chi_{sc}/(1 - V(q,\omega)\chi_{sc})$ discussed in Section 13.4.2.3. The χ_{sc} represents the screened response of a corresponding noninteracting fluid and is selected to be

$$\chi_{sc}(q,\omega) = \alpha_q \chi_{0*}(q,\omega) + (1 - \alpha_q)\chi_q^M. \quad (13.45)$$

Here χ_{0*} is the single-particle response having effective mass m_q^*, while χ_q^M represents multiparticle excitations. (We set $Q = q$ and $M = m$ to follow the notation of the PP theory.) Second, the interactions in ^{3}He and ^{4}He are assumed to be essentially the same and largely independent of statistics. Third, the RPA χ will have singularities and sharp peaks. These represent the collective excitations and resonant "modes" in the fluid, and the positions of the peaks as a function of q determine the excitation spectrum. Outside the peak region the weight of χ represents multiparticle hole or multiphonon excitations.

In ^{4}He, the PP interaction is written as

$$V(q,\omega) = f_q^s + f_q^v\left(\frac{\omega}{q}\right)^2. \quad (13.46)$$

Here f_q^s represents the interaction via density excitations and f_q^v that via current excitations. The f_q^s is the Fourier transform of the potential

$$f^s(r) = \begin{cases} a[1 - (r/r_c)^8], & r < r_c, \\ b[(r_c/r)^{12} - (r_c/r)^6], & r_c < r < r_t, \\ -[a_8/r^8 + a_6/r^6], & r_t < r. \end{cases} \quad (13.47)$$

The $f^s(r)$ is essentially a Lennard–Jones potential with a softened core and depends upon the parameters a, b, r_c, r_t. The radius of the core is set at $r_c = 2.68$ Å for all pressures in ^{4}He. The height of the core (a) is determined so that the $\chi(0,0)$ in Eq. (13.11b) satisfies the compressibility sum rule $[-\chi(0,0) = (nmc_1^2)^{-1}]$. The b is determined so that $f^s(r)$ is continuous at r_t. The f_q^v is expressed in terms of a q-dependent effective mass,

$$m_q^* = m + nf_q^v, \quad (13.48)$$

where $n = N/V$. At $q = 0$, f_0^v is determined so that m_3^* equals the observed thermodynamic ^{3}He mass, $m_3^* = m_3^0 + nf_0^3 = 3.1 m_3^0$. The q dependence of

f_q^v is determined so the resulting $S(Q, \omega)$ peak fits the observed phonon–roton curve. The α_q is determined by fits to experiment. The PP theory provides an excellent description of $S(Q, \omega)$ in liquid ^{4}He, particularly the "one-phonon" or "one-quasiparticle" part as will be discussed.

Bedell et al.[143, 144] present a comprehensive description of roton states, roton–roton interactions, and two-roton bound states, Their second paper,[143] particularly, reviews experiment and theory on rotons in some detail. As in the Landau–Khalatnikov (L-K) theory,[160] they treat rotons as quasi-particles (i.e., the sharp excitation in $S(Q, \omega)$ for $T < T_\lambda$ is equally described as a phonon or a quasiparticle). The interaction between two rotons (two q–p's) is then described in terms of a pseudopotential $f(r)$ of the polarization potential form, Eq. (13.47). Since the strength of this interaction has been shown[161] to be too strong to be described within the Born approximation (as used by L-K), they evaluate the interaction using the Bethe-Salpeter equation. Among many properties, they calculate the roton energy $\Delta(T)$ and inverse lifetime τ^{-1} due to this four phonon (q-p-q-p) scattering process. As in the L-K theory, the temperature dependence of $\Delta(T)$ and τ^{-1} comes primarily from $N_r(T)$, the number of thermal rotons available to take part in the scattering. For parabolic dispersion of the p-r curve near the roton minimum, $N_r(T) \approx \sqrt{T} \exp(-\Delta(T)/T)$.

In 1985, Manousakis et al.[162] have evaluated the condensate fraction n_0 and the momentum distribution $n(p)$ at $T = 0$ K. They used a variational wave function including pair and triplet correlations coupled with hypernetted chain summation techniques. They find $n_0 = 9.5\%$ at SVP in agreement with Green function Monte Carlo (GFMC)[90] results. The n_0 decreases with increasing pressure, though less rapidly than the GFMC values. Manousakis and Pandharipande[163] (MP) also evaluate the temperature dependence of $n(\mathbf{p})$ using CBF methods including pair, triplet, and momentum-dependent correlations in their basis functions. They particularly explore the low-momentum components of $n(\mathbf{p})$. Most recently, MP[164, 165] have calculated $S(Q, \omega)$ directly using CBF methods. Among other things, they present microscopic calculations of the multiphonon component of $S(Q, \omega)$ in liquid ^{4}He.

13.3.3. Experimental Results

13.3.3.1. Background. Liquid ^{4}He has been studied far more extensively by neutron scattering than any other material. Since the first measurements[3] in 1951, results have been reported from most of the neutron scattering centers of the world, virtually every type of neutron spectrometer has been employed in these studies, and the neutron energies have spanned the range from ultra-cold neutrons of 10^{-7} eV[166] to neutrons of nearly

20 eV.[167] In spite of this long history, exciting discoveries are still being made, and a surprising amount has been learned in the last decade primarily as a result of extensive and detailed studies of the temperature dependence of $S(Q)$ and $S(Q, \omega)$. We will concentrate largely on the results of these recent studies. For more extensive coverage of the earlier studies, the reader is referred to earlier articles,[17,18,25,168,169] particularly the reviews by Woods and Cowley[17] and Price.[18]

The transmission measurements of 1951 were soon followed by diffraction measurements such as those of Henshaw and Hurst[170] in 1953 and 1955. Then, stimulated by the proposal of Cohen and Feynman[171] that neutron scattering could be used to determine the dispersion relation for the elementary excitations, came the landmark inelastic scattering measurements of the late 1950s by Palevsky et al.,[172] Yarnell et al.,[173] and Henshaw,[174] which verified that superfluid ^{4}He did indeed have a dispersion relation of the "phonon–roton" form envisaged by Landau[5] (Fig. 9).

Following this early work, the inelastic scattering by liquid ^{4}He was studied in great detail throughout the 1960s, especially at Chalk River. The situation at the beginning of the 1970s is partially summarized by Fig. 10. The branch of sharp phonon–roton excitations was by then so well known that it is simply shown as a solid curve labeled "one phonon" (see also Fig. 9). Above the one-phonon branch, we see that there is a broad band of multiphonon scattering, which is centered at an energy somewhat higher than twice the energy Δ of the roton minimum for Q values up to about 2.3 Å^{-1}, and which then rises rapidly and becomes centered rather close to the dispersion relation for free ^{4}He atoms shown by the dashed curve. Although a great wealth of information already existed by this time, many important questions remained unanswered. For example: (1) Did the dispersion relation at low Q initially curve upward (anomalous dispersion) or downward (normal dispersion) from the velocity-of-sound line, Eq. (13.34)? (2) Why did there appear to be no dramatic change in $S(Q, \omega)$ on passing through the superfluid transition temperature T_λ? (3) Why did the width and the center position of $S(Q, \omega)$ at large Q oscillate with Q (see Figs. 21 and 22 of Ref. 86) rather than behave as expected from the impulse approximation, and further, why did $S(Q, \omega)$ at large Q show no evidence of a sharp central component arising from scattering by the zero-momentum atoms in the Bose condensate, as had been proposed by Hohenberg and Platzman?[88] Subsequent neutron scattering studies have provided answers for these as well as other important questions, as we shall now discuss. However, as we shall also see, they have left us with a different and even larger set of puzzles to be addressed by future studies.

13.3.3.2. Anomalous Dispersion. A knowledge of the detailed shape of the one-phonon dispersion relation at low Q is of crucial importance for understanding the interactions between excitations in liquid ^{4}He. If there is anomalous dispersion (i.e., if the phase velocity ω/Q exceeds the sound velocity for Q less than some value Q_c) then three-phonon decay processes can occur; otherwise, only much less probable four-phonon processes are allowed. In 1970, Maris and Massey[175] first proposed that there was anomalous dispersion in liquid ^{4}He. Since then, this topic has received a great deal of attention and numerous functional forms for the dispersion at low Q have been proposed; see, e.g., the articles by Maris,[83] Woo,[61] Aldrich et al.,[176] Svensson et al.,[177] Stirling et al.,[178] Donnelly et al.,[116] Stirling,[179] and Kirkpatrick.[180]

Indirect information about the form of the dispersion at low Q can be obtained from several types of measurements, but only an accurate determination of the dispersion relation by neutron inelastic scattering can give a direct answer. Although the measurements of Cowley and Woods[86] gave a hint of anomalous dispersion, the first reasonably convincing direct evidence came from the study of Svensson et al.[177] This study, which covered pressures up to 24 atm, indicated that the region of anomalous dispersion extended to $Q_c = 0.52$ Å^{-1} at SVP and that Q_c decreased with increasing pressure. A later study by Stirling et al.[178] gave results for pressures up to 20 bars and particularly good results for SVP. Stirling[179] has recently carried out even more accurate measurements at SVP and obtained the results shown in Fig. 12. These results show very clearly that there is anomalous dispersion for $Q \lesssim Q_c = 0.55$ Å^{-1}, substantiating the earlier value, $Q_c = 0.52$ Å^{-1}, of Svensson et al.[177] The inset shows that the maximum deviation from the sound velocity is just under 4%.

Neutron scattering measurements of the quality shown in Fig. 12 covering the complete range of pressure would be most valuable. The polarization potential calculations of Aldrich et al.,[176] which give a good description of the existing neutron results, as well as other types of measurements such as the ballistic-phonon propagation measurements of Dynes and Narayanamurti,[181] indicate that anomalous dispersion disappears entirely at pressures above about 20 bars. Kirkpatrick[180] has suggested that the anomalous dispersion in liquid ^{4}He at SVP should be substantially larger at 2 K than at 1.2 K. It would be very interesting to check this prediction experimentally, but it will undoubtedly be very difficult to determine precisely the dispersion at 2 K because of the substantial intrinsic widths of the phonon modes at this temperature (see Section 13.3.3.3).

13.3.3.3. $S(Q, \omega)$ in the Phonon–Roton Region and its Variation with Pressure and Temperature. Although the most detailed measurements of $S(Q, \omega)$ have been carried out at low temperatures ($T \approx 1.3$ K or lower) and SVP, there have also been numerous studies for other temperatures and pressures. The first study at higher P was carried out by Henshaw and

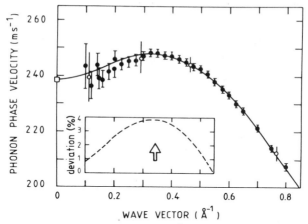

FIG. 12. Phonon phase velocities ω/Q for liquid ^{4}He at 1.2 K and SVP. The open square shows the sound velocity and the insert the deviation from this value. (From Stirling.[179])

Woods[182] and subsequent results have been presented by Dietrich et al.,[183] Woods et al.,[184, 185] Graf et al.,[186] Svensson et al.,[177, 187–189] Smith et al.,[190] Stirling et al.,[178] and Stirling.[191] Most of these studies were for $T \lesssim 1.3$ K, but some[185, 189] were for selected higher temperatures, and the study of excitations in the roton region by Dietrich et al.[183] covered wide ranges of both temperature and pressure.

As the pressure is raised at low temperature, the excitation energies increase for Q values up to nearly the roton minimum, while they decrease in the region of the roton minimum and above. Between SVP and the solidification pressure, the roton energy Δ and effective mass μ decrease by about 20% while the roton wave vector q_0 increases by about 6%. Depending on the Q value, the intensities of the sharp one-phonon peaks can either change greatly with pressure or hardly at all, as illustrated in Fig. 11. This figure shows the resolution broadened $S(Q, \omega)$ for the maxon (left) and roton (right) positions at low and high pressures obtained by Svensson et al.[187] While the roton peak increases slightly in intensity between low and high pressure (largely explained by the lowering of the roton energy), the maxon peak decreases in intensity by more than a factor of 2 and appears to broaden significantly as well. This dramatic intensity change has been interpreted[187] as indicating strong "anharmonic" effects in liquid ^{4}He, in particular interference between one-phonon and multiphonon processes. One would expect such effects to be strongest at the maxon position, where the one-phonon and multiphonon (generally dominated by two-roton contributions) energies are closest together (see Fig. 10), and to increase with increasing pressure since, as we have noted above, the maxon energy increases with pressure while the roton energy decreases. Detailed calculations[54] for bcc ^{4}He show that interference, due to the anharmonicity, between one-phonon and multiphonon scattering processes seriously distorts the observed one-phonon contributions to $S(Q)$ and the f-sum rule. The similarity of the observed one-phonon contributions for liquid ^{4}He and bcc ^{4}He (see Fig. 10 of Ref. 25) suggests that such interference effects are indeed important for liquid ^{4}He. As yet, however, there has been no precise formulation of anharmonic interference for a liquid. Since this intriguing aspect of liquid ^{4}He has recently been reviewed[25] by one of us, we will not discuss it further here.

The temperature variation of $S(Q, \omega)$ for liquid ^{4}He was investigated in some of the very earliest studies (Larsson and Otnes,[192] Yarnell et al.,[173] and Henshaw and Woods [182]). Subsequent work, either for a range of temperatures or at selected higher temperature values, has been presented by Cowley and Woods,[86] Dietrich et al.,[183] Woods et al.,[185, 193–196] Svensson et al.,[187, 189, 197] Dasannacharya et al.,[198] Woods and Svensson,[142] Sears et al.,[199] Tarvin and Passell,[200] Blagoveshchenskii et al.,[201, 202] Mezei,[203] Mezei and Stirling,[204] Pedersen and Carneiro,[205] Golub et al.,[166] and Stirling.[191]

Although the very early calculations by Bendt et al.[206] of thermodynamic properties using the excitation energies obtained by Yarnell et al.[173] (for $T \leq 1.8$ K) gave good agreement with the known experimental values, the results of neutron measurements closer to T_λ appeared to be inconsistent with the results of thermodynamic measurements. In particular, the intrinsic widths (inverse lifetimes) of the rotons appeared to be considerably larger than those calculated using the theory of Landau and Khalatnikov[160] with the parameters determined from viscosity measurements, and the roton energies Δ were considerably lower (typically 25% near T_λ) than the values inferred from thermodynamic measurements (see, e.g., the extensive compilation of results by Brooks and Donnelly[207]). Further, all of the studies prior to 1978 seemed to indicate that nothing dramatic happened to $S(Q, \omega)$ on passing through T_λ; rotons as well as other one-phonon excitations appeared to persist to temperatures well above T_λ. This was most puzzling since the "one-phonon" excitations had originally been thought to be associated with the superfluid phase. The dynamics of liquid ^{4}He appeared to be much more complex than had been anticipated and researchers began to question whether or not the presumed (but not at that time experimentally confirmed) existence of a Bose condensate below T_λ, which was generally believed to be the origin of superfluidity, was responsible for the unique, essentially δ-function excitations observed in superfluid ^{4}He at low temperatures. Perhaps these excitations were simply characteristic of a liquid at low temperatures? Subsequent measurements and analysis[142, 197] which we will now discuss, have, however, suggests that the "one-phonon" excitations may be unique to the superfluid phase of liquid ^{4}He.

The measurements of Svensson et al.[197] and Woods and Svensson[142] were the first to give evidence for a qualitative change in $S(Q, \omega)$ on passing through T_λ. In particular, they suggested that the "one-phonon" excitations disappeared at T_λ and hence really were a signature of the superfluid phase. Their results for the maxon wave vector ($Q = 1.13$ Å^{-1}) are shown in Fig. 13. The one-phonon peak, which dominates the scattering at low temperature (note the logarithmic scale), broadens and decreases in intensity as the temperature is raised but remains a clearly identifiable feature right up to 2.15 K, just 0.02 K below T_λ. At 2.27 K there is, however, no sign of this component and one is left with only the much broader peak characteristic of nonsuperfluid ^{4}He. Except at its leading (low-frequency) edge, where detailed balance effects are important, this broad peak then changes very

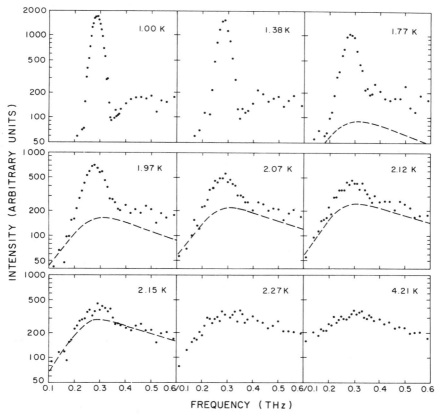

FIG. 13. Resolution-broadened dynamic structure factors for liquid ^{4}He for $Q = 1.13$ Å^{-1} and SVP. The dashed curves show the normal fluid components, i.e., the last term in Eq. (13.49). (From Woods and Svensson.[142])

little on further increasing the temperature. The well-defined peak in the scattering, which corresponds to one-phonon excitations, is thus observed only in the superfluid phase.

Woods and Svensson[142] also found that, for all $T < T_\lambda$, $S(Q, \omega)$ could be very well described as the sum of a "superfluid" component with a weight $n_s = \rho_s/\rho$ (ρ_s being the macroscopic superfluid density) and a "normal-fluid" component with a weight $n_n = 1 - n_s$, namely,

$$S(Q, \omega) = n_s S_s(Q, \omega) + n_n S_n(Q, \omega). \tag{13.49}$$

Here, as in Eq. (13.39), $S_s(Q, \omega)$ consists of a one-phonon peak (with a temperature-dependent width) plus a broad multiphonon component at higher ω, while $S_n(Q, \omega)$ consists of only the single broad peak characteristic of nonsuperfluid ^{4}He. In actual practice, $S_n(Q, \omega)$ is taken[142] to be the distribution observed at a temperature T^* just above T_λ (e.g., the 2.27 K distribution shown in Fig. 13) with its leading edge adjusted[142] slightly for each temperature so as to satisfy detailed balance. The dashed curves in Fig. 13 show the normal-fluid components, $n_n S_n(Q, \omega)$, obtained[142] via this procedure. While the one-phonon peak in $S_s(Q, \omega)$ can be described by the Miller et al.[127] form $Z(Q, \omega) = Z(Q) \delta[\omega - \omega(Q)]$ at very low temperatures ($T \leq 1$ K), at higher temperatures this peak exhibits substantial broadening. Woods and Svensson[142] found that the Lorentzian line shape

$$Z(Q, \omega) = \frac{Z(Q)}{\pi} \frac{\Gamma(Q, T)}{\Gamma^2(Q, T) + [\omega - \omega(Q)]^2} \tag{13.50}$$

gave a very good description of their results with the half-width $\Gamma(Q, T)$ taken to be the value for rotons given by the theory of Landau and Khalatnikov.[160] The strength $Z(Q, T)$ was determined as the integrated intensity of the one-phonon peak.

The integrated one-phonon intensities [Fig. 14(a)] obtained[142] by application of Eq. (13.49) are seen to be in excellent agreement with the quantity $n_s = \rho_s/\rho$ for all five values of Q that were studied. These values span the range from below the maxon to the roton minimum (see Figs. 9 and 10). From Fig. 14(b) we also see that the intrinsic widths of the one-phonon peaks (obtained by deconvoluting the Gaussian instrumental resolution assuming that the intrinsic line shapes were Lorentzians) fall, within experimental uncertainty, on a universal curve and agree very well with the width for rotons (dashed curve) calculated from the theory of Landau and Khalatnikov.[160] There is of course no reason to expect that Γ should be independent of Q and, in fact, in high-resolution measurements[204] at lower Q values, it is found to depend rather strongly on Q (see Fig. 17). There is also no reason to expect the Landau–Khalatnikov theory to be strictly correct at temperatures near T_λ.

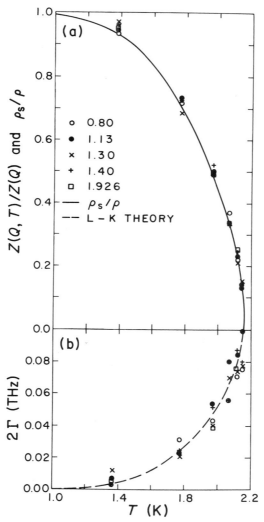

FIG. 14. (a) Normalized intensities of the one-phonon peaks in $S(Q, \omega)$ for liquid ^{4}He at SVP for the five Q values indicated and $n_s = \rho_s/\rho$ (solid curve). (b) The corresponding intrinsic widths of the one-phonon peaks and the width (dashed curve) for rotons calculated from the theory of Landau and Khalatnikov.[160] Results are based on an analysis in terms of Eq. (13.49). (From Woods and Svensson.[142])

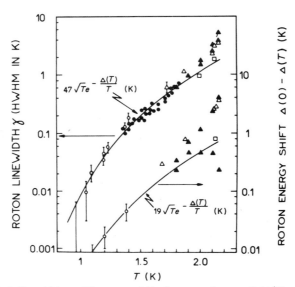

FIG. 15. Intrinsic linewidths and frequency shifts for rotons in superfluid ^{4}He at SVP. (From Mezei.[203])

Using the spin-echo technique, Mezei[203] has obtained very accurate roton linewidths and energy shifts at low temperatures ($T \leq 1.4$ K), where one would expect the Landau–Khalatnikov theory[160] to give the correct temperature variation. Figure 15, taken from Mezei's paper, shows that his results (open circles) indeed confirm the variation (solid curves) expected on the basis of the Landau–Khalatnikov theory. [Results very similar to the solid curves in Fig. 15 have been obtained by Bedell et al.,[143, 144] who have also presented calculations of Γ and $\Delta_0 - \Delta(T)$ for $P = 24.26$ bars and compared them with the experimental results of Dietrich et al.[183]] As in the Landau–Khalatnikov theory, Bedell et al. evaluate Δ and Γ assuming a four-phonon decay process, and the temperature variation of Δ and Γ arises predominantly from the changing population of thermal rotons needed for this decay process to occur. Other calculations of the temperature variation of Δ or Γ have been carried out by Titulaer and Deutch[208] and Roberts and Donnelly.[209] At higher temperatures, the Landau–Khalatnikov curves in Fig. 15 are in excellent agreement with the Raman scattering results of Greytak and Yan[210] (solid circles) and the results of Woods and Svensson[142] (open squares) obtained via Eq. (13.49), but not with the results of Dietrich et al.[183] (open triangles) or Tarvin and Passell (closed triangles).[200] These latter authors[183, 200] used the whole distributions to determine their roton parameters and hence, if the Woods–Svensson interpretation[142] is correct, their values will be seriously distorted by the presence of the broader normal-fluid component, which, at the roton position, is centered at a lower energy than the roton peak in $S_s(Q, \omega)$. Using the whole distribution thus gives roton widths and energy shifts that may be too large, as seen in Fig. 15. The energy shifts inferred by Dietrich et al.[183] and Tarvin and Passell[200] also violate a bound given by the theory of Bedell et al.[144] (see Figs. 2 and 4 of Ref. 144). Tarvin and Passell[200] have shown that the widths and especially the energies obtained by fitting to the whole distribution depend strongly on the functional form assumed for the line shape. [The two solid triangles for each temperature in Fig. 15 show their results for a Lorentzian line shape (smaller widths and larger shifts) and the harmonic oscillator line shape suggested by Halley and Hastings.[211]] This is exactly what one would expect from Eq. (13.49) since one is then fitting a single-peak line shape to a distribution that is the sum of two components characterized by different energies and linewidths. The one-phonon peak in the superfluid component, obtained by first subtracting $n_n S_n(Q, \omega)$, is, however, found to be a well-defined symmetric peak (see Fig. 2 of Ref. 142), which is very well described by the Lorentzian form, Eq. (13.50).

The interpretation of $S(Q, \omega)$ in terms of Eq. (13.49) has provided a solution to the puzzling problems mentioned earlier. Most importantly, it suggests that the excitations both above and below T_λ can differ. It has also enabled us to extract from the experimental distributions one-phonon energies and linewidths that, right up to T_λ, are in excellent agreement with the values inferred from thermodynamic measurements and calculated using the theory of Landau and Khalatnikov, even though this theory would not be expected to be strictly valid at such high temperatures. Further support for Eq. (13.49) comes from the fact that the analogous relationships

$$S(Q) = n_s S_s(Q) + n_n S_n(Q) \tag{13.51}$$

and

$$g(r) = n_s g_s(r) + n_n g_n(r) \tag{13.52}$$

that are implied by Eq. (13.49) are found[212, 213] to be very well satisfied by the best experimental results[213, 214] for $S(Q)$ and $g(r)$. As an example of the detailed description of $S(Q, \omega)$ provided by Eqs. (13.49) and (13.50), we show in Fig. 16 a comparison[215] of the observed and calculated values for $Q = 0.8 \text{ Å}^{-1}$ at $T = 2.12$ K. The agreement is obviously excellent. The description is not entirely adequate in all cases, however, because no provision has been made for allowing $\omega(Q)$ to vary with temperature. This shortcoming is especially serious for rotons, which exhibit a substantial variation of $\omega(Q)$ with temperature (see Fig. 15). Note, however, that when one follows the procedure[142] whereby $n_n S_n(Q, \omega)$ is first subtracted from the

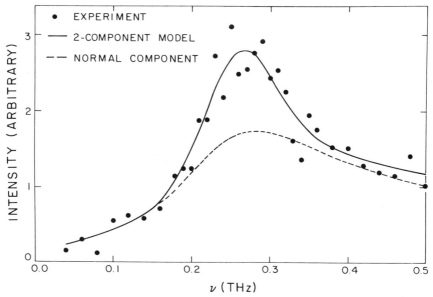

FIG. 16. $S(Q, \omega)$ for liquid ^{4}He for $Q = 0.8$ Å^{-1} at 2.12 K and SVP as observed in neutron scattering measurements (solid circles) and as calculated (solid curve) on the basis of Eqs. (13.49) and (13.50). The dashed curve shows the second term in Eq. (13.49). (From Woods and Svensson.[215])

observed $S(Q, \omega)$ and the remainder is then used to determine $\Gamma(Q, T)$ and $\omega(Q, T)$, this shortcoming plays no role. The main source of uncertainty then is in estimating $S_n(Q, \omega)$ for the temperature in question. It would be very useful to have additional theoretical work as well as detailed measurements of $S(Q, \omega)$ over a range of temperatures above, and especially just above, T_λ to help guide one in estimating $S_n(Q, \omega)$ at temperatures below T_λ, where it cannot be separately observed. Note also that $S_n(Q, \omega)$ must be determined from measurements just above T_λ since the central Rayleigh component, which appears in $S(Q, \omega)$ above T_λ, has been ignored. This component becomes increasingly important as $T \to 4.2$ K where, at low-Q values, $S(Q, \omega)$ is observed[195, 198] to have a distinct Rayleigh peak between the two Brillouin peaks.

Equation (13.49) was originally proposed by Woods and Svensson[142] simply as an empirical relationship. Early theoretical studies by Griffin[155, 216] and Griffin and Talbot[154] appeared to give direct support for Eq. (13.49), but in more recent studies[108, 153] these authors find that the poles in the first and second terms of Eq. (13.40) are strongly coupled and do not simply separate into parts proportionate to ρ_S and ρ_N as in Eq. (13.49). However, the region of strict validity of the theoretical calculations[108,153] is restricted to Q values considerably lower than those of the experiments so there may be no serious disagreement. Detailed studies at lower Q values as well as further theoretical calculations are required. Even after the shortcomings mentioned are corrected, Eq. (13.49) will probably turn out to be too simple to describe $S(Q, \omega)$ in complete detail for all Q and T. This relationship and the experimental results[142,197] on which it is based have, however, shown rather convincingly that well-defined one-phonon excitations are unique to He(II), and hence somehow related to the existence of a finite Bose condensate. This realization, coupled with recent theory,[108,153] need to be recognized to understand the dynamics of superfluid ^{4}He.

Mezei and Stirling[204] have studied $S(Q, \omega)$ for $0.3 \leq Q \leq 0.7$ Å^{-1} and $T \leq 1.7$ K under conditions of very high experimental resolution and obtained the intrinsic linewidths shown in Fig. 17. In contrast to the results at higher Q (Fig. 14), we see a marked Q dependence in the widths at low Q. The calculations of Talbot and Griffin[108] also indicate a marked dependence of the width on Q at low Q as do the early measurements of Cowley and Woods[86] (see Fig. 11 of Ref. 108). Note the drop in width between 0.5 and 0.6 Å^{-1} at 0.95 and 1.2 K in Fig. 17, and possibly also at 1.4 and 1.5 K. This is almost certainly a reflection of the fact that three-phonon decay processes, which are dominant in the region of anomalous dispersion at least at low temperature (for details see Maris[83]), are no longer allowed once one passes into the region of normal dispersion at $Q = 0.55$ Å^{-1} (see Fig. 12). Golub et al.[166] have determined the loss rate of ultra-cold neutrons due to interactions with superfluid ^{4}He over the temperature range 0.5 to 1.15 K, and found their results to be reasonably well described by calculations that

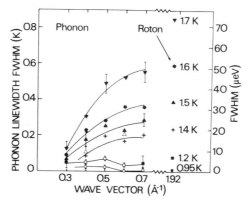

FIG. 17. Temperature dependence of intrinsic widths of rotons and low-Q phonons in liquid ^{4}He at SVP. (From Mezei and Stirling.[204])

include the effects of phonon absorption and phonon and roton scattering. If the energy spectrum of the neutrons upscattered by the liquid ^{4}He can also be determined, this should give much more detailed information about the processes involved in the interaction between the ultra-cold neutrons and the helium.

Before leaving this section, we briefly mention a few other studies of $S(Q, \omega)$ in the photon–roton region that have important implications for our understanding of the excitations and their interactions. One topic of long-standing interest is that of roton–roton interactions and, in particular, two-roton bound states. (For a discussion see Bedell et al.[143]) A direct comparison of the two-roton Raman scattering results of Murray et al.[217] with the very precise neutron scattering results of Woods et al.[218] for the roton region at 0.75 K and SVP shows that the interactions between two rotons with net $Q = 0$ is attractive, leading to a bound state with a binding energy[218] of 0.27 ± 0.04 K. On the other hand, the neutron measurements of Woods et al.[219] for $Q = 0.3$ Å^{-1} (the lowest Q for which the multiphonon scattering has yet been determined) gave a multiphonon peak that was well-defined, essentially symmetric, nicely separated from the one-phonon peak, and centered at an energy significantly above 2Δ (see also Ref. 187). From an analysis of their one-phonon results for Q values near the end point of the phonon–roton curve ($2.7 \leq Q \leq 3.3$ Å^{-1}), and also of the results of Graf et al.[186] and Svensson et al.[187] for the maxon position, Smith et al.[190] have concluded that, at least for $Q > 1$ Å^{-1}, roton–roton interactions are repulsive. Blagoveshchenskii et al.[202, 220] have, however, interpreted their neutron scattering results as indicating two-roton bound states for $Q \leq 1.65$ Å^{-1} with a maximum binding energy of 0.44 ± 0.07 K, even larger than the $Q = 0$ value given above. The question is clearly not yet resolved.

Detailed studies of the multiphonon component, $S_M(Q, \omega)$, of $S(Q, \omega)$ should in principle be able to give a wealth of information about the interactions between the one-phonon interactions. The measurements of Svensson et al.[187] (see Fig. 11) and Graf et al.[186] showed that there was definite structure in $S_M(Q, \omega)$. From theoretical studies such as those of Jackson,[221] Iwamato et al.,[222] Götze and Lücke,[223] and Manousakis and Pandharipande,[164, 165] one certainly expects that there should be such structure and that, in particular, one should be able to identify various features as arising from roton–roton, maxon–maxon, and maxon–roton processes. Stirling[191] has recently presented neutron scattering results for $S_M(Q, \omega)$ for $T = 1.27$ K and $Q = 1.13$ and 1.5 Å^{-1} at SVP, 15 and 25 bars, which show the clearest structure yet obtained. The structure varies substantially with Q as expected and is generally most marked at SVP. Most of the structure for $Q = 1.13$ Å^{-1} appears to arise from maxon–roton interactions, while roton–roton interactions appear to be more important for

$Q = 1.5 \text{ Å}^{-1}$. Further detailed studies of $S_M(Q, \omega)$ would be very valuable. Stirling[191] has also given new results for the Landau roton parameters, Δ, q_0, and μ, for $T = 0.05$ K at SVP and for $T = 0.9$ K at SVP, 15 and 24 bars.

$S_M(Q, \omega)$ also exhibits long tails extending to high frequencies as first emphasized by Woods et al.[219] who showed that even at $Q = 0.8 \text{ Å}^{-1}$ the tail extended to ≈ 70 K indicating that, although two-phonon processes appeared to be dominant at low Q, higher order processes were also important (two-phonon processes cut off below 40 K;—see Fig. 10). The existence of these weak high-frequency tails makes it very difficult to determine experimentally the complete $S(Q, \omega)$ so as to satisfy the necessary sum rules (see Ref. 187 for further discussion). Any error in determining the tails can have a very serious effect for studies where the higher frequency moments of $S(Q, \omega)$ are required. For example, to determine the effective pair potential in liquid ^{4}He by direct inversion of neutron scattering results, as attempted by Sears et al.,[199] the third moment $\int \omega^3 S(Q, \omega) \, d\omega$ must be known accurately. Theoretical studies by Wong and Gould,[78,150] Bartley and Wong,[157] Family,[107] Wong,[152] Talbot and Griffin,[108] and Kirkpatrick[106] have shown that the high-frequency tails of $S(Q, \omega)$ should exhibit an $\omega^{-7/2}$ dependence, and Wong[152] has demonstrated that this is indeed the case for the results of Woods et al.[219] for $Q = 0.8 \text{ Å}^{-1}$. Further detailed experimental testing of this prediction is extremely important since, if proven to be correct, it would be extremely helpful in establishing the tails of the distributions, something that has this far proven to be very difficult but is essential for many studies.

The whole area of determining and understanding $S(Q, \omega, P, T)$ for Q values in the phonon–roton region ($Q < 3.6 \text{ Å}^{-1}$) is clearly still a very fertile one for the fruitful interplay of theory and experiment. In spite of nearly 30 years of inelastic scattering measurements, there are many obvious ways in which neutrons can still make major contributions to our understanding of the one-phonon excitations and their interactions. Additional measurements of the type and quality of the ones discussed in this section are needed over the full range of pressures for the full range of superfluid temperatures, as well as at several temperatures above but close to T_λ. In this section we have deliberately emphasized the Woods–Svensson decomposition, Eq. (13.49), which has given us a greatly simplified and intuitively very appealing picture of $S(Q, \omega)$, resolved the puzzles from the earlier studies, and eliminated the long-standing disagreements between neutron and thermodynamic results. It has made us aware that the existence of well-defined one-phonon excitations is somehow related to Bose condensation and this, ultimately, may prove to be more important to a fundamental understanding of superfluid ^{4}He than the actual determination of the condensate fraction discussed in Section 13.3.3.4. We now have a reliable procedure for determining, from the experimental $S(Q, \omega)$, the energies Δ and intrinsic widths Γ of the one-phonon excitations right up to T_λ, where the excitations disappear. It will be very interesting to see how well Eq. (13.39) applies at higher pressures, and measurements of the type of Woods and Svensson[142] (Figs. 13, 14, and 16), Mezei[203] (Fig. 15), and Mezei and Stirling[204] (Fig. 17) at higher pressure should be a very high priority. While, as we have mentioned earlier, studies[186, 187] at higher pressures have given evidence for strong anharmonic effects in superfluid ^{4}He, Eq. (13.39), which works very well at SVP, makes no explicit allowance for anharmonic effects. It is obvious that we are going to need new theoretical studies in this area as well.

13.3.3.4. *S(Q, ω) at High Q and the Condensate.* At sufficiently high Q for the impulse approximation (see Section 13.2.5.1) to be essentially valid, $S(Q, \omega)$ will directly reflect the momentum distribution $n(\mathbf{p})$ of the atoms as indicated by Eq. (13.30). If a finite fraction, n_0, of the atoms in the superfluid phase [He(II)] are condensed into the zero-momentum state as originally proposed by London[2] (see Sections 13.1.1 and 13.3.1), then $n(\mathbf{p})$ can be written[100] as

$$n(\mathbf{p}) = n_0 \delta(\mathbf{p}) + (1 - n_0)n^*(\mathbf{p}), \qquad (13.53)$$

where $n^*(\mathbf{p})$ is the momentum distribution for the noncondensate atoms and the normalization is such that $\int n(\mathbf{p})\, d\mathbf{p} = \int n^*(\mathbf{p})\, d\mathbf{p} = 1$. Substituting Eq. (13.53) into Eq. (13.30) then gives an $S_{IA}(\mathbf{Q}, \omega)$, which consists of a δ-function condensate component at the recoil frequency, $\omega_R = \hbar Q^2/2M$, sitting on top of a broad noncondensate peak, which is symmetric about ω_R and whose width increases linearly with Q. The relative weight of the δ-function component will then give directly the condensate fraction n_0, while $n(\mathbf{p})$ for $p \neq 0$ can be determined from the noncondensate component via the relationship

$$pn(\mathbf{p}) = -\frac{1}{2\pi} \left\{\frac{\hbar Q}{M}\right\}^2 \frac{\partial}{\partial \omega} S_{IA}(\mathbf{Q}, \omega). \qquad (13.54)$$

[This expression applies only for measurements carried out at constant Q; for other types of scans, a much more complicated expression must be used (see, e.g., Ref. 98).]

Neutron inelastic scattering measurements of $S(Q, \omega)$ at large Q thus give the possibility of a direct experimental demonstration that a finite fraction of the atoms in superfluid ^{4}He are condensed into the zero-momentum state as required to confirm London's[2] intriguing and controversial proposal of 1938. This aim has been the main impetus for numerous studies of $S(Q, \omega)$ at high Q, but, in practice, it has turned out to be a very difficult task. Convincing evidence for a finite condensate fraction was only finally obtained[100] in 1982, 44 years after London's proposal and 16 years after

Hohenberg and Platzman[88] suggested that this method be used to determine n_0. (For recent reviews of the "condensate saga" see Refs. 224 and 225.)

Hohenberg and Platzman's[88] explicit proposal stimulated several experimental studies,[86, 98, 226–229] which, however, gave conflicting results with the inferred values of $n_0(T \approx 1.2$ K) ranging from 0.02 to 0.17. Much of this work was reviewed by Jackson[230] who concluded that the data were also consistent with a vanishingly small n_0. The work of Martel et al.[99] made it clear that essentially all of the previously inferred values were unreliable because of inadequacies in analysis, in particular the failure to appreciate the importance of final-state interactions and interference effects which have the consequence that the IA is not valid in the Q range of the measurements. It had of course been recognized earlier that the IA was not strictly correct; the results of Cowley and Woods[86, 226] for $4 \leq Q \leq 9$ Å^{-1} had shown that the peaks in $S(Q, \omega)$ were neither symmetric nor centered on ω_R and that their widths oscillated with Q rather than increasing linearly as expected from Eq. (13.30).

To take approximate account of the final-state interactions and interference effects, Martel et al.[99] replaced the δ function in Eq. (13.30) by a function $R(\mathbf{Q}, \omega)$, which was assumed to be a Lorentzian (i.e., describing a simple lifetime effect) with a full width at half-maximum (FWHM) given by

$$2\Gamma(\mathbf{Q}) = \rho(\hbar Q/M)\sigma(\mathbf{Q}), \qquad (13.55)$$

where ρ is the helium number density and $\sigma(\mathbf{Q})$ is the total ^{4}He–^{4}He atomic scattering cross section. This model gave a good quantitative description of the observed widths as illustrated in Fig. 18. Curve A shows the $2\Gamma(\mathbf{Q})$ values obtained from Eq. (13.55) with $\sigma(\mathbf{Q})$ taken from the atomic-beam measurements of Feltgen et al.[231] Note that it exhibits oscillations with the same periodicity and phase as those shown by the experimental widths. Curves B and B' are obtained by combining a Gaussian $n(\mathbf{p})$ with, respectively, a Lorentzian and a Gaussian $R(\mathbf{Q}, \omega)$ having a width given by curve A. Both give a good description of the observed widths leaving little doubt that the width oscillations are attributable to the final-state interactions and interference effects embodied in $\sigma(\mathbf{Q})$. Note that, for the Q range shown in Fig. 18, the condensate peak will, even under conditions of extremely high experimental resolution, have a width (curve A) approximately half that of the full $S(Q, \omega)$, and hence will not be directly observable as a separate sharper component in $S(Q, \omega)$.

In addition to clarifying why, and to what extent, the IA was not valid in the Q range of the measurements, Martel et al.[99] proposed a procedure for obtaining much more reliable estimates of $n(\mathbf{p})$ from the experimental distributions. Basically, this consisted of decomposing the observed $S(Q, \omega)$ into components that were symmetric and antisymmetric about ω_R and then

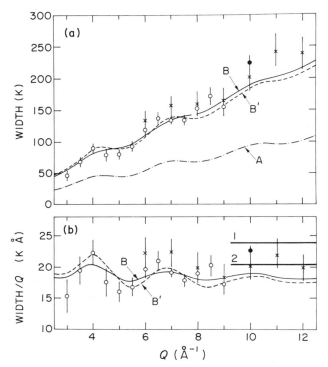

FIG. 18. (a) Intrinsic full widths at half-maximum (FWHM) of $S(Q, \omega)$ from the neutron measurements of Cowley and Woods[86] (open circles, $T = 1.1$ K) and Martel et al.[99] (solid circle and crosses, $T = 1.2$ K). Curve A is the quantity $2\Gamma(\mathbf{Q})$ calculated from Eq. (13.55) using the $\sigma(\mathbf{Q})$ results of Feltgen et al.[231] Curve B is the corresponding FWHM of $S(Q, \omega)$ calculated using a Gaussian $n(\mathbf{p})$ and a Lorentzian $R(\mathbf{Q}, \omega)$ (see text and Martel et al.[99]). Curve B' is the result of a similar calculation using a Gaussian $R(\mathbf{Q}, \omega)$. (b) The FWHMs shown in (a) divided by Q. The horizontal lines 1 and 2 show the asymptotic ($Q \to \infty$) results expected on the basis of the IA.[99] (Figure taken from Martel et al.[99])

using the symmetric part to obtain $n(\mathbf{p})$ via Eq. (13.54) (for details see Refs. 94 and 99). Tanatar et al.[104] have recently shown that this symmetrization procedure reduces the errors in $n(\mathbf{p})$ by a factor of ten for $Q = 20$ Å^{-1} in solid helium. This procedure was subsequently applied by Woods and Sears[232] to the earlier results of Cowley and Woods,[86] and by Sears et al.[100] to the results of new high-resolution measurements of $S(Q, \omega)$ for $4.0 \leq Q \leq 7.0$ Å^{-1} undertaken specifically to obtain information on the condensate. To further reduce the distortions caused by interference effects, these authors also averaged the $n(\mathbf{p})$ obtained from a range of Q values. The $n(\mathbf{p})$ obtained by Sears et al.[100] are shown in Fig. 19. Note that there is essentially no change in $n(\mathbf{p})$ between 4.27 and 2.27 K, but that on further cooling to below T_λ

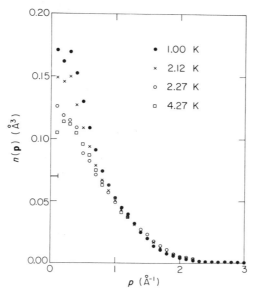

FIG. 19. The momentum distribution $n(\mathbf{p})$ for liquid ^{4}He at four temperatures at SVP. The horizontal bar shows the resolution HWHM. (From Sears et al.[100])

there is a marked increase in $n(\mathbf{p})$ at low p, which is a direct consequence of the appearance of a finite condensate fraction. The large increase at low p at 2.12 K (just 0.05 K below T_λ), approximately half the total increase observed at 1.00 K, is at first sight rather surprising but can be readily explained and is in fact conclusive evidence for a substantial condensate fraction. Its origin lies[100, 224, 225] in the singular behavior of $n^*(\mathbf{p})$ as $p \to 0$ described by

$$(1 - n_0)n^*(\mathbf{p}) = n_0(ap^{-2} + D + \cdots). \qquad (13.56)$$

Here $a = Mk_B T/8\pi^3\hbar^2\rho n_s$ and, as $T \to 0$, $D \to Mc/16\pi^3\hbar\rho\rho$, where c is the sound velocity (see Refs. 156 and 233 for additional details and references to earlier work). Note that the singular behavior of $n^*(\mathbf{p})$ only occurs if n_0 is finite. Because of the broadening by final-state interactions and instrumental resolution, the singular behavior of $n^*(\mathbf{p})$ gives an enhancement of $n(\mathbf{p})$ at low p for $T < T_\lambda$, which is indistinguishable from that given by the broadened condensate peak. The additional enhancement arising from the singular behavior of $n^*(\mathbf{p})$ is particularly large near T_λ where $n_s \to 0$, and this is the explanation of the large increase in $n(\mathbf{p})$ at small p for 2.12 K (Fig. 19). The observation[100] of this feature was thus conclusive evidence for the existence of a substantial condensate fraction—the long-sought convincing experimental demonstration that there really is a Bose condensate in superfluid ^{4}He. It was clearly of crucial importance to carry out measurements just above and especially just below T_λ.

To extract values of n_0 from the experimental $n(\mathbf{p})$, Sears et al.[100] used the relationship

$$n_0 = \varepsilon/(1 - \beta + \gamma), \qquad (13.57)$$

where ε and β are obtained by integrating, respectively, $n(\mathbf{p}) - n^*(\mathbf{p})$ and $n^*(\mathbf{p})$ from 0 to p_c, the cutoff momentum of the broadened condensate peak; $n^*(\mathbf{p})$ is taken to be the $n(\mathbf{p})$ observed at 2.27 K, and γ is the correction for the enhancement caused by the singular behavior of $n^*(\mathbf{p})$ as $p \to 0$. The problem of estimating γ, especially for temperatures near T_λ, is probably the main source of uncertainty in the procedure as has recently been emphasized by Griffin.[233] The values of n_0 for 1.00 and 2.12 K obtained by Sears et al. from the $n(\mathbf{p})$ of Fig. 19 are shown as solid circles in Fig. 20. The solid

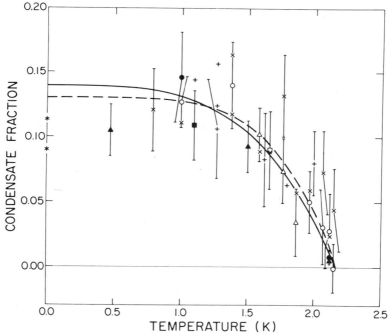

FIG. 20. Experimentally based values for the condensate fraction, $n_0(T)$, in superfluid ^{4}He at SVP from Sears et al.[100] (solid circles and square, and solid curve), Sears and Svensson[234] (open circles), Sears et al.[236] (open triangles), Robkoff et al.[237] (open square), Sears[240] (plusses and crosses), Mook[101] (solid triangles) and Campbell et al.[243] (dashed curve). The asterisks are from theoretical calculations for $T = 0$: upper,[89,241] lower.[115] (Figure from Svensson.[225])

square shows the value for 1.1 K obtained[100] by applying the same analysis procedure to the earlier $n(\mathbf{p})$ of Woods and Sears.[232] The solid curve shows the results of a fit[100] to these three values as well as the values shown by open symbols that were obtained earlier [234–237] from pair correlation functions via the method of Hyland et al.,[238] which will be discussed further in Section 13.3.3.5. There have been a great many theoretical estimates of $n_0(0)$ dating back to the work of Penrose and Onsager[239] in 1956, most values falling in the range 0.08 to 0.13. For a recent compendium see Sears.[240] Two of the better estimates are shown by asterisks in Fig. 20: see Refs. 89 and 241 for the upper and Ref. 115 for the lower. More recent calculations by Whitlock and Panoff[90] and Manousakis et al.[162] have given $n_0(0) = 0.092$, in excellent agreement with the lower asterisk in Fig. 20, while Puoskari and Kallio[242] have obtained $n_0(0) = 0.14$ in excellent agreement with the fit to the experimental values (solid curve). The difference between the two asterisks in Fig. 20 is typical of the difference that can result from small changes in the interatomic potential (see also Ref. 90). The difference between the theoretical values of $n_0(0)$ and the extrapolation from the experiments (solid curve) is well within the combined uncertainties. This was where we were[100] in the "condensate saga" in 1982, a rather pleasing position considering the previous uncertainties and disagreements.

Much more has been added to the saga since 1982. Campbell[243] has proposed a method for determining n_0 from the surface tension and obtained the results indicated by the dashed curve in Fig. 20. (Recently, Iino et al.[244] have analyzed new and highly accurate surface tension results using Campbell's method and obtained values for the condensate fraction in excellent agreement with those shown in Fig. 20.) Sears[240] has pointed out that n_0 can also be determined from the temperature variation of $\langle \mathrm{KE} \rangle$ via the relationship

$$n_0(T) = 1 - \langle \mathrm{KE}(T) \rangle / \langle \mathrm{KE}^*(T) \rangle, \qquad (13.58)$$

where KE* refers to the noncondensate atoms. From the $\langle \mathrm{KE}(T) \rangle$ values available from neutron inelastic scattering and neutron and x-ray diffraction measurements, he obtained the values of n_0 indicated by +'s and ×'s in Fig. 20. Since it was assumed that $\langle \mathrm{KE}^*(T) \rangle = \langle \mathrm{KE}(T_\lambda) \rangle$, these values are really upper limits on n_0, and they do in fact appear to be systematically higher than the other values especially near T_λ. The recent work of Ceperley and Pollock[245] also suggests that using Eq. (13.58) with $\langle \mathrm{KE}^*(T) \rangle = \langle \mathrm{KE}(T_\lambda) \rangle$ is only approximately correct. Mook[101] has also carried out a study very similar to that of Sears et al.[100] and obtained the results indicated by solid triangles in Fig. 20. From their very recent Monte Carlo discretized path-integral computations for liquid ^{4}He at finite temperatures, an extremely important advance on the theoretical front, Ceperley and Pollock[245] have obtained values of $n_0(T)$ for $T \geq 1$ K, which, except near T_λ where the results are not reliable because of the small size of their system (64 atoms), are also consistent with, though on average somewhat lower than, the experimental values in Fig. 20. It seems almost inconceivable that the good agreement between the different sets of experimentally based values (from neutron, x-ray, and surface tension measurements via five different methods of analysis), as well as the agreement with the theoretical values, can be fortuitous. Hence, in addition to being certain (from the $n(\mathbf{p})$ results of Fig. 19, as discussed) that a substantial condensate fraction exists in superfluid ^{4}He, we can be reasonably confident (from the results shown in Fig. 20) about its magnitude and temperature dependence at SVP.

Figure 21 shows a comparison of $n(\mathbf{p}, 1.00\text{ K})$ (solid circles) and the quantity $[1 - n_0(1.00\text{ K})]n(\mathbf{p}, 2.27\text{ K})$ (open circles). Aside from the distortions at small p, which are relatively small at 1.00 K and have been ignored in constructing Fig. 21, the latter quantity represents the noncondensate part of $n(\mathbf{p}, 1.00\text{ K})$, i.e., the last term in Eq. (13.53). The difference between the open and closed circles is thus a close approximation to the condensate part of $n(\mathbf{p})$, broadened, of course, by final-state interactions and the relatively small experimental resolution (FWHM shown by the horizontal bar on the figure). [The condensate component looks very large in Fig. 21 where $n(\mathbf{p})$ is plotted but, to obtain a numerical value, one must multiply by $4\pi p^2$ and integrate.] By deconvoluting the experimental resolution, one obtains an intrinsic momentum width, $\Delta p = \rho\sigma$ for the condensate peak, which gives a value for the effective cross section σ in very good agreement with the average value of $\sigma(\mathbf{Q})$ for the Q range (4–7 Å^{-1}) of the neutron measurements

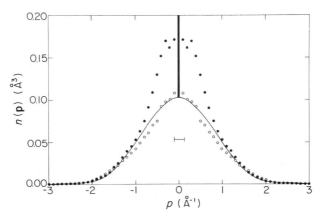

FIG. 21. A comparison of $n(\mathbf{p})$ (solid circles) and the quantity $(1 - n_0)n^*(\mathbf{p})$ (open circles) for 1.00 K. The horizontal bar shows the resolution FWHM. The solid curve is the calculated $n(\mathbf{p})$ of Whitlock et al.[89] for $T = 0$. (From Sears et al.[100])

known[231] from atomic-beam measurements. This agreement is further support for the procedure used by Sears et al.[100] to extract values of n_0 from the $n(\mathbf{p})$. (For additional details see Ref. 100.)

The solid curve in Fig. 21 (with the ideal δ-function condensate peak sitting on top of it) is the $n(\mathbf{p})$ for $T = 0$ obtained by Whitlock et al.[89] from a Monte Carlo calculation. It is clearly in quite reasonable agreement with the experimental distribution (open circles), through somewhat broader. (The difference is actually greater than it appears since the experimental distribution comes from measurements at 2.27 K and has not been corrected for the broadening by final-state interactions and experimental resolution.) For other recent comparisons of experimental and theoretical momentum distributions see Refs. 90, 115, 162, and 242. A very important feature of $n(\mathbf{p})$ for superfluid ^{4}He, which is not obvious from Figs. 19 and 21, is that it has an exponential tail at large $p(>1.5\ \text{Å}^{-1})$. This was first made clear by the work of Woods and Sears[232] (see Fig. 1 of Ref. 232) and also subsequently obtained by Whitlock and Panoff[90] from Monte Carlo calculations (see Fig. 5 of Ref. 90). In contrast, $n(\mathbf{p})$ for nonsuperfluid ^{4}He, as for all other materials, is Gaussian to a good approximation (see the recent review by Sears[246]). Griffin[233] has pointed out that a detailed study of the temperature dependence of the high-p tail of $n(\mathbf{p})$ for liquid ^{4}He should yield additional information about the condensate, but this has not yet been attempted.

Although the condensate component is readily identified in Fig. 21, it certainly does not stand out as a separate sharper component in $n(\mathbf{p})$ for 1.00 K or in the $S(Q, \omega)$ from which this $n(\mathbf{p})$ is obtained. This is because of the large broadening by final-state interactions and the small condensate fraction. Since $\sigma(\mathbf{Q})$, the quantity which governs this broadening, only decreases very slowly[231, 247] with increasing Q, one must go to very much larger Q to gain a marked advantage in the still unsuccessful quest to observe a separate sharper condensate peak directly in $S(Q, \omega)$—probably[225] to $Q \approx 100\ \text{Å}^{-1}$ and perhaps even substantially higher (see Ref. 225 for additional details). Such measurements can only be carried out using the high fluxes of epithermal neutrons from intense spallation sources.

Several measurements on liquid ^{4}He using spallation sources have already been reported. At Argonne National Laboratory (ANL), measurements on liquid as well as on polycrystalline solid (bcc and hcp) ^{4}He have been carried out [41, 42, 248, 249] over a wide range of densities for Q values up to $\approx 23\ \text{Å}^{-1}$ (see also Section 13.2.5.2). In most cases the distributions were observed to be Gaussian with very little dependence on temperature or phase for a given density. The $\langle KE \rangle$ values inferred from the measurements are, however, generally not in good agreement with the values obtained from Monte Carlo calculations.[89, 90] Simmons and Sokol[42] have recently pointed out that improvements in data analysis to take account of source asymmetries are required. One would hope that these improvements will largely eliminate the disagreements between theory and experiment. Preliminary values of n_0 inferred from the measurements at ANL have been included in a recent compilation by Manousakis et al. (see Fig. 9 and Ref. 2 in Ref. 162). Brugger et al.[103] have carried out measurements at Los Alamos National Laboratory at $Q \approx 83$ Å^{-1} at SVP. They obtained $\langle KE \rangle \approx 13$ K in good agreement with accepted values,[240] but their resolution was not high enough to detect any significant change between 4.2 and 1.2 K. At the KENS spallation facility in Japan, measurements[167,250] have been carried out at $Q \approx 150$ Å^{-1}, the largest value reported to date. The results show a change in line shape between 2.5 and 1.2 K that probably is attributable to the appearance of a Bose condensate below T_λ, but a preliminary analysis gives a value of n_0 about three times larger than the values in Fig. 20 (see Fig. 3 in Ref. 167). We will undoubtedly see numerous additional studies on liquid ^{4}He using spallation sources over the next few years.

Although we have, over the last decade, learned a large amount about the scattering at high Q in general, and about the condensate in particular, much work, both theoretical and experimental, remains to be done. We obviously need higher accuracy and this entails both better experiments and better analysis procedures. Higher accuracy would, for example, give us a better picture of how $n_0 \to 0$ at T_λ, and this is of high interest since $\sqrt{n_0}$ is the order parameter for superfluid ^{4}He. The value of 2β, the critical exponent[251] for n_0, can be inferred from the values of other critical exponents to be ≈ 0.70, but a direct determination is highly desirable. The large uncertainties of the present results for n_0 near T_λ do not permit an accurate determination of 2β, but it is interesting to note that the dashed curve in Fig. 20 corresponds to $2\beta = 0.71$. We also want to know n_0 at high densities, where it will be smaller because of the further depletion of the condensate by stronger interatomic interactions, and also whether it falls to zero continuously or discontinuously at the superfluid–solid interface. The x-ray measurements of Wirth et al.[252] suggest very little dependence of n_0 on density, but Mook[101] has interpreted his results as indicating a rather strong dependence on density. A summary of the results of several theoretical calculations of the density dependence of n_0, as well as preliminary results from the neutron measurements at Argonne National Laboratory mentioned previously, has recently been presented by Manousakis et al. (see Fig. 9 of Ref. 162). The best indications at present are that n_0 decreases only rather slowly with increasing density. There is an obvious need for additional and more accurate measurements.

The analysis procedures, many of which are based on simple physical models and sum-rule arguments, and even on intuition, are a major source of the present uncertainties. For example, Griffin[233] has recently discussed the problem of correcting for the singular behavior of $n^*(\mathbf{p})$ in the condensate studies and has reanalyzed the results of Sears et al.[100] and Mook[101] using a different procedure that gives considerably smaller values of n_0 than those shown by solid symbols in Fig. 20. Although, as he points out, his values must be regarded as lower limits on n_0, this work certainly indicates the need for an improved analysis procedure. We also need a better theory of final-state interactions. The assumption of Eq. (13.55) that the broadening by these interactions is describable as a simple lifetime effect, as originally proposed by Hohenberg and Platzman,[88] and used by Martel et al.[99] to successfully describe the previously puzzling oscillations in the width of $S(Q, \omega)$ (see Fig. 18), is only a rough first approximation. Kirkpatrick[106] has recently shown that the form used by Martel et al. is correct near the free-particle energy, $\hbar\omega_R$, but not well away from it. It is in fact well known (see, e.g., Sears[94]) that the broadening function $R(\mathbf{Q}, \omega)$ cannot be a simple positive function like a Gaussian or a Lorentzian, but must have oscillatory tails. New theoretical work on the detailed form of $R(\mathbf{Q}, \omega)$ would be most welcome. The scaling methods proposed by Sears[94] have not yet been applied to measurements for liquid ^{4}He, but it undoubtedly would be valuable to do so. There is at present high interest[92–94, 96, 97, 104, 106, 253–255] in the impulse approximation and its applicability to neutron scattering studies, and one would hope that this work will result in a much better understanding of final-state interactions leading to improved analysis procedures for the determination of $n(\mathbf{p})$, $\langle KE \rangle$, n_0, etc., from neutron scattering measurements at readily accessible Q values. The new work of Ceperley and Pollock[245] for finite temperature also gives the promise of the direct numerical calculation of various "problem" features—e.g., the singular behavior of $n^*(\mathbf{p})$ at low p provided that a sufficiently large system of particles can eventually be used in the computations.

Returning to experiments, it would now appear that, by using neutrons from intense spallation sources, there will be no real problem in reaching sufficiently high Q to have the possibility of observing a sharp central condensate component directly in $S(Q, \omega)$, but there is still some doubt about achieving the necessary high experimental resolution and counting statistics. On balance the prospects look rather good. While we await the necessary source intensity for these "ultimate" studies, it would be extremely valuable to carry out studies such as those of Sears et al.[100] and Mook[101] at sufficiently higher Q values (say 15–30 Å^{-1}) that final-state interactions and interference effects will be substantially less important. Such measurements, at the necessary high resolution, are certainly feasible with existing reactor and spallation sources.

13.3.3.5. Static Structure Factor and Pair Correlations. In comparison with the large number of x-ray diffraction studies (for references see Refs. 214, 237, 252, 256, and 257) and the vast number of neutron inelastic scattering studies, there have been relatively few neutron diffraction studies of liquid ^{4}He. The very early diffraction measurements of Henshaw and Hurst[170] were followed by Henshaw's studies[258, 259] of the effects of pressure and temperature on the atomic distribution. There was then a long gap until the measurements of Mozer et al.[260] for two temperatures at each of three densities, and a further gap until the highly accurate and very extensive measurements of Svensson and co-workers.[213, 214, 236, 261]

The values of $S(Q)$ obtained by Svensson et al.[214] for 11 temperatures at SVP are shown as solid circles and solid curves in Fig. 22. Open circles indicate the x-ray results of Gordon et al.[262] and triangles the compressibility

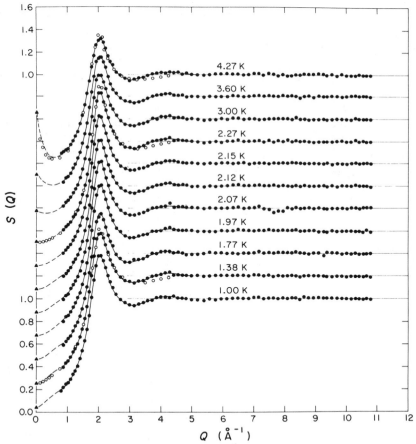

FIG. 22. The static structure factor $S(Q)$ of liquid ^{4}He for 11 temperatures at SVP. (From Svensson et al.[214])

limits. The dashed curves at low Q were obtained from the very accurate small-angle, x-ray results of Hallock.[263] In general, x rays have a distinct advantage at small Q because of the much narrower beams employed (typically 1 mm wide versus 5 cm for neutrons), while neutrons have a marked advantage at larger Q where the x-ray intensity falls off precipitously because of the atomic form factor. The x-ray results of Robkoff and Hallock[256] for seven temperatures at SVP are in rather good agreement with the neutron results of Fig. 22, except in the region of the principal peak where they are higher by $\approx 6\%$. This discrepancy has recently been discovered[264] to be, at least in major part, attributable to a systematic error in the x-ray measurements.

Note that, as a result of the low density and poorly defined short-range order arising from the large zero-point motion, the oscillations in $S(Q)$ for liquid ^{4}He have small amplitudes that decrease rapidly with increasing Q. There is no discernible structure beyond about $6\,\text{Å}^{-1}$ so the infinite-Q limit has effectively been reached in the measurements of Fig. 22. This allows a direct and very accurate normalization of the results and also eliminates the truncation problems that usually complicate the Fourier analysis required to determine $g(r)$ from $S(Q)$ via the relationship

$$g(r) = 1 + (2\pi^2 \rho r)^{-1} \int_0^\infty Q[S(Q) - 1] \sin Qr \, dQ, \qquad (13.59)$$

where ρ is the number density.

The temperature dependence of various features of the pair correlation functions obtained[214] from the results of Fig. 22 are shown in Fig. 23. Note that the amplitudes of the oscillations about the line $g(r) = 1$, which are a direct measure of the correlations between the atoms, exhibit the normal increase with decreasing temperature between 4.27 K and T_λ, but then there is a dramatic reversal and the amplitudes decrease continuously with further cooling in the superfluid phase, exhibiting a typical order–parameter type of behavior. This anomalous loss of spatial order on cooling, which is unique to superfluid ^{4}He, is also clearly seen in the recent results[212–214, 237, 252, 256, 257] for the temperature dependence of $S(Q)$ and, as pointed out by Henshaw,[258] was even evident in some of the very early neutron and x-ray results.[170, 258, 262] According to Cummings and co-workers,[238, 265, 266] this loss of spatial order is a direct consequence of the appearance of the zero-momentum Bose condensate below T_λ. To the extent that they are "in the condensate," atoms are completely delocalized because of the Heisenberg uncertainty principle, and hence might indeed be expected to no longer contribute to the spatial correlations that give rise to the oscillations in $g(r)$ and $S(Q)$. Cummings *et al.*[238] further proposed that values of the condensate fraction could be

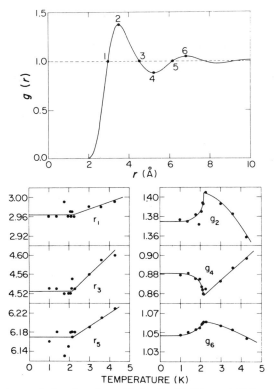

FIG. 23. The pair correlation function $g(r)$ for liquid ^{4}He at 1.00 K and SVP and the temperature variation of the heights of the first three extrema of $g(r)$ and the positions of the first three nodes of $[g(r) - 1]$. (From Sears and Svensson.[235])

determined from the temperature variation of $g(r)$ via the relationship

$$g(r) - 1 = (1 - n_0)^2[g_n(r) - 1], \qquad (13.60)$$

where $g_n(r)$ is the pair correlation function for the "noncondensate atoms." (In actual practice, $g_n(r)$ is usually taken[234, 235, 237, 252] to be $g(r)$ for a temperature T^* just above T_λ.) The tantalizing possibility of obtaining n_0 via Eq. (13.60) is one of the major reasons for the current high interest in $S(Q)$ and $g(r)$ for liquid ^{4}He.

Early attempts[238, 267] to obtain values of n_0 via Eq. (13.60) led to inconclusive results primarily because of the large uncertainties in the values of $g(r)$, but also partly because the reference temperatures T^* at which the $g_n(r)$ were determined were much too far above T_λ. The first diffraction results sufficiently accurate and complete to provide a critical test and application of Eq. (13.60) were those of Svensson et al.[214] This equation is only applicable at sufficiently large r ($\gtrsim 6$ Å, see Ref. 234) that the one-particle density matrix has effectively reached its asymptotic value, n_0, and, to be physically meaningful, it must give values of n_0 which are independent of r in this region. This requirement imposes stringent limitations on the behavior of $g(r)$ that can be tested experimentally. For example, the positions of the nodes of $g(r) - 1$ must be independent of temperature for $T < T_\lambda$. From Fig. 23 (see also Table I in Ref. 234) we can see that this is indeed the case. There is a substantial variation above T_λ, attributable to the density change of $\approx 17\%$ in this region, but below T_λ, where the density is constant to within $\approx 0.7\%$, the variations are small and apparently random, and easily attributable to experimental uncertainty. The constraints on the behavior of $g(r)$ imposed by Eq. (13.60) in fact appear to be entirely consistent with the experimental results (see Refs. 212, 234, and 235 for additional details.)

Open symbols in Fig. 20 show values of n_0 for SVP obtained[234-237] by application of Eq. (13.60). They are obviously in excellent agreement with the values given by the other methods discussed in Section 13.3.3.4. This agreement is independent evidence in support of Eq. (13.60), at least as an empirical relationship. Although Eq. (13.60) clearly has a lot of support from experiment, and seems to follow naturally from simple intuitive arguments,[212,235] considerable controversy[268] has surrounded the work[238,265,266] on which it is based. Cummings et al.[268] have effectively rebutted most of the criticisms leaving, it would appear, the main source of uncertainty to be in the estimation of $g_n(r)$ for $T < T_\lambda$, where it cannot be separately determined by experiments. As mentioned above, one usually assumes that $g_n(r, T) = g(r, T^*)$ where T^* is a temperature just above T_λ. It would be very valuable to have detailed theoretical investigations aimed at determining how $g_n(r)$ varies with temperature below T_λ. More systematic experimental studies over a range of temperatures above T_λ are also needed to provide guidance. The method of Cummings et al. for the determination of n_0 is most appealing because diffraction measurements are much less time consuming than the inelastic scattering measurements of $S(Q, \omega)$ at large Q discussed in Section 13.3.3.4, and the data analysis[214] needed to determine $g(r)$ is straightforward and reliable. It is only the final step of determining n_0 from the $g(r)$ that is, at present, somewhat uncertain. Additional theoretical attempts to justify or improve this method, as well as additional experimental tests, would thus be extremely valuable since, should the method prove to be reliable, it would allow us to obtain much more detailed information on n_0 and its variation with temperature and pressure than is likely to ever be obtained from inelastic scattering measurements. If, as seems highly probable, the results of Fig. 20 are correct, then, with hindsight, we have to conclude that Eq. (13.60), in spite of its possible shortcomings, gave us the first reliable values[234-237] of $n_0(T)$.

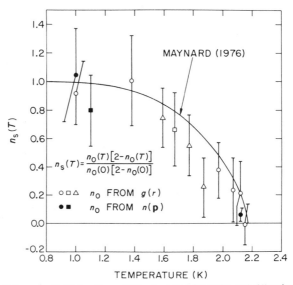

FIG. 24. Values of the superfluid fraction $n_s(T)$ calculated from Eq. (13.61) using the values of $n_0(T)$ indicated by the same symbols in Fig. 20. The solid curve shows experimental values of $n_s(T)$ from Maynard[270] for $T \geq 1.2$ K and from Brooks and Donnelly[207] for lower T. (From Svensson and Sears.[269])

As pointed out by Svensson et al.,[212] Eq. (13.60) can be combined with Eq. (13.52) to obtain an explicit relationship between $n_0(T)$ and the macroscopic superfluid density, $n_s(T)$, for all $T \leq T_\lambda$, namely,

$$n_s(T) = \frac{n_0(T)[2 - n_0(T)]}{n_0(0)[2 - n_0(0)]}. \tag{13.61}$$

Values of $n_s(T)$ obtained[269] via Eq. (13.61) from a subset of the $n_0(T)$ values in Fig. 20 are compared in Fig. 24 with the experimental $n_s(T)$ values of Maynard[270] supplemented at temperatures below 1.2 K by the values of Brooks and Donnelly.[207] Although Eq. (13.61) is not strictly correct in either of the limits $T \to 0$ and $T \to T_\lambda$ (see Ref. 212 for details), the overall good agreement in Fig. 24 indicates that it is sufficiently close to the mark to merit attention by theorists. In view of the enormous difficulty in obtaining accurate values of n_0 over a wide range of temperatures, and especially for temperatures near T_λ where n_0 is small, it would be a great advantage to have a reliable explicit relationship between n_0 and n_s. One could then combine values of n_s, which can be determined very precisely[270,271] over the entire temperature range, with accurate values of n_0 at low temperature, where it is largest and can be most reliably determined, to obtain, via Eq. (13.61) or an improved relationship, accurate values of n_0 for all temperatures.

An alternative explanation of the anomalous change in spatial order with temperature in superfluid ^{4}He has been proposed by Reatto and co-workers.[272] (See also the more recent papers by Puoskari and co-workers.[273]) These authors attribute the increase in height and sharpening of the principal peak of $S(Q)$ with increasing temperature simply to the increasing thermal population of rotons, the connection between rotons and the peak of $S(Q)$ being the obvious one of the Feynman dispersion relation, Eq. (13.37). They take no explicit account of the existence of the condensate. The results of their calculations of $\Delta S(Q, T) = S(Q, T) - S(Q, T_0)$, which use as input the experimental results for $S(Q)$ at $T_0 = 1.00$ K[214] and for the temperature dependence of the energies and intrinsic widths of rotons,[183, 200] are in approximate agreement[213, 214, 252, 272, 273] with experiment. The overall shape and amplitude of $\Delta S(Q, T)$ are reasonably well described, but the calculations give less sharpening of the main peak than observed and also indicate that the position Q_m of the peak changes considerably with T, in disagreement with experiment. Battaini and Reatto[272] have shown that better agreement on Q_m can be obtained if Feynman–Cohen rather than Feynman wave functions are used, i.e., if one takes account of backflow. Reatto and co-workers[272] suggest that their calculations are meaningful to temperatures somewhat above T_λ but, if the Woods–Svensson interpretation of $S(Q, \omega)$ for liquid ^{4}He discussed in Section 13.3.3.3, which indicates that rotons and other one-phonon excitations do not exist above T_λ, is valid, this suggestion is not correct. Their results for temperatures close to T_λ should also be treated with caution since, as discussed in Section 13.3.3.3, the values[183, 200] of the roton parameters used in their calculations may be seriously in error above ≈ 1.8 K because they were obtained from the entire observed distributions rather than just the superfluid component [see Eq. (13.49)].

The two competing explanations of the anomalous change in spatial order of superfluid ^{4}He lead one to expect different behavior as the density is increased. The smaller values of n_0 at higher density resulting from the presumed greater depletion of the condensate by stronger interactions would lead one to expect relatively smaller changes in spatial order if the explanation of Cummings and co-workers[238] is correct. In contrast, Reatto and co-workers[272] predict enhanced changes in spatial order at higher density since the roton gap Δ decreases with increasing density and hence the thermal population of rotons increases. Specifically, they find that $\Delta S_{\max}$, the maximum value of $\Delta S(Q, T)$, increases by about 60% for a 20% increase in density over that at SVP.

Wirth et al.[252] have presented extensive results from x-ray measurements for three different densities covering approximately the same range as the calculations of Reatto and co-workers.[272] The experimental values for $\Delta S_{\max}$ do not exhibit the substantial and systematic increase with density expected from the calculations. However, the experimental results for the lowest density are difficult to believe, since they do not follow the expected trend exhibited by the results for the two higher densities; relative to the latter results and the results of Svensson and co-workers[213,214] for SVP, they appear to be systematically high by ≈ 0.03-0.04. When they apply Eq. (13.60) to their results, Wirth et al. obtain values of $n_0(T)$, which, when plotted as a function of T/T_λ for each density, fall essentially on a universal curve contrary to expectations. This behavior has been interpreted[101,252] as evidence against the applicability of Eq. (13.60), at least at higher densities. We feel that this conclusion is premature. There appears to be a problem with at least some of the experimental results and, at any rate, the values of $n_0(0)$ obtained by Wirth et al.[252] are not seriously inconsistent with the values from numerous theoretical calculations and from neutron scattering measurements compiled by Manousakis et al. (see Fig. 9 of Ref. 162). In fact, the values of Wirth et al. for the two lower densities are in very good agreement with the preliminary results from the neutron measurements at ANL (see Ref. 2 of Ref. 162) over the same density region and give a nominal slope of $n_0(0)$ versus density consistent with that given by most of the theories. The ANL neutron results, while indicating a nominal dependence on density, are also, to well within the estimated errors, consistent with no variation of $n_0(0)$ with density. More, and especially better, measurements are needed before we can rule out or accept either the approach of Cummings and co-workers or that of Reatto and co-workers.

In addition to the interest from the point of view of the condensate, highly accurate experimental values of $S(Q)$ and $g(r)$ are often needed for the normalization of the results of inelastic scattering measurements[187,199] and are extremely important for testing the results of theoretical calculations. References to the theoretical work prior to 1980 as well as comparisons between their neutron diffraction results for $S(Q)$ and $g(r)$ and a selection of the best theoretical results existing at the time, both from calculations[274] based on variational wave functions of the Jastrow type and calculations[89] based on Green-function Monte Carlo (GFMC) methods, were given by Svensson et al.[214] The results from the GFMC calculations were in much better agreement with experiment, but they still underestimated the heights (by 3-4%) and overestimated the widths of the principal peaks of $S(Q)$ and $g(r)$. These discrepancies were believed to be largely attributable to the inadequacy of the Lennard-Jones pair potential used in the GFCM calculations, with some indication that many-body effects might also be important. The fact that the calculations were for $T = 0$ while the measurements were for $T = 1.00$ K was (and still is) believed to be of negligible importance.

In subsequent GFMC calculations, Kalos et al.[115] have tested several pair potentials and found that the HFDHE2 potential of Aziz et al.[111] gives by far the best agreement with experiment. The pair correlation function obtained by Kalos et al. using the HFDHE2 potential is compared in Fig. 25 with that from the neutron measurements of Svensson et al.[214] The agreement is essentially perfect. The agreement for $S(Q)$ (see Fig. 5 of Ref. 115) is equally good except right at the top of the principal peak, where the calculated values are low by about 3%. This is still within the combined uncertainties. As mentioned in Section 13.3.3.4, Ceperley and Pollock,[245] in an extremely important new advance, have carried out Monte Carlo discretized path-integral computations for liquid ^{4}He at finite temperatures ($\approx$1–4 K) at SVP, again using the HFDHE2 potential. The agreement (see Fig. 2 of Ref. 245) between their results for $g(r)$ at 2 K and the experimental results of Svensson et al.[214] is as perfect as the agreement shown in Fig. 25. Incidentally, Ceperley and Pollock note that the behavior of $g(r, T)$ expected on the basis of Eq. (13.60) is consistent with the results of their computations.

The quality of both the theoretical calculations and the experimental results has improved dramatically over the last decade. The level of agreement, at least at SVP, is now so good (Fig. 25) that at first sight there would appear to be little incentive for improvement. This superb agreement may,

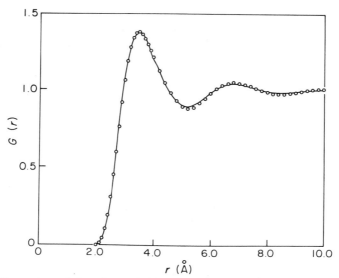

Fig. 25. Comparison of the pair correlation function for liquid ^{4}He at $T = 0$ (solid curve) from Green function Monte Carlo calculations[115] with that at $T = 1.0$ K (open circles) from neutron diffraction measurements.[214] (From Kalos et al.[115])

however, be somewhat fortuitous since the HFDHE2 potential does not agree[111] with all the two-body experimental data and, further, one would expect many-body forces to be of some importance in the liquid. The attempt of Sears et al.[199] to determine the effective pair potential directly from inelastic scattering results, while not giving a unique result, certainly indicated that many-body forces were important. When Kalos et al.[115] added three-body terms to their calculations they found, however, that the agreement with experiment worsened. The reason remains a puzzle, but it is clear that further investigation of many-body interactons is required. There is likely still some room for significant improvement in the calculations.

The situation at higher densities is not nearly as satisfactory. Robkoff and Hallock[275] have compared their x-ray results for higher densities with the calculations of Chang and Campbell[274] and Whitlock et al.[89] In general the agreement is not very good, but the same systematic error,[264] which caused the results of Robkoff and Hallock[256] for SVP to disagree with the neutron results,[214] as mentioned earlier in this section, probably affected their measurements at higher density as well, so this poor agreement should probably not be taken seriously at present.

New experimental results for $S(Q)$ and $g(r)$, of the quality and completeness of those of Svensson et al.[214] shown in Figs. 22 and 23, for several higher densities would be most valuable for testing the results of theoretical calculations of $S(Q)$ and $g(r)$, and to help us decide between the two competing explanations[238,272] of the anomalous change in spatial order with temperature in superfluid ^{4}He. For the latter purpose, even higher statistical precision would be desirable since the changes of $S(Q)$ and $g(r)$ with temperature are very small (typically[213,214,261] 2–5% at the position of the principal maximum).

13.3.4. Future Prospects

Because of the length of Section 13.3, a reflection of the fact that the quantity of experimental work on liquid ^{4}He far exceeds the total for solid helium, liquid ^{3}He, and liquid ^{3}He-^{4}He mixtures, much of the material that might have been included here has deliberately been placed in the individual subsections since we felt it would benefit the reader to have the detailed future needs appear in conjunction with the relevant background material. Here we will largely restrict ourselves to more general remarks, although a few specific items merit repetition.

We have learned a great deal about liquid ^{4}He over the last decade, primarily as a result of the detailed studies of the temperature dependences of $S(Q, \omega)$ and $S(Q)$ at SVP. The next logical step is to repeat many of these studies at several higher densities, preferably at constant density. Among the things we want to know are the detailed density dependence of the anomalous dispersion, the condensate fraction, $S(Q)$, $g(r)$, and $S(Q, \omega)$ in the phonon–roton region, and to what extent Eqs. (13.49) and (13.60) remain valid at higher density.

Although a major milestone in the condensate saga has been reached, as evidenced by the results in Fig. 20, we have still not achieved the original aim of observing a separate sharp condensate component directly in $S(Q, \omega)$. It now appears very hopeful that this aim will be achieved in some of the studies that are sure to be carried out at the new spallation sources over the next few years.

New theoretical work is as important as new experimental work. Several of the equations in Section 13.3.3, for example Eqs. (13.49), (13.51), (13.52), (13.60), and (13.61), are essentially empirical relationships, having strong support from experiment but no firm foundation in theory. Theoretical studies to justify or improve these relationships are urgently needed. In the application of these relationships and others such as Eqs. (13.53) and (13.58), it is necessary to make assumptions about how quantities like $S_n(Q, \omega)$, $S_n(Q)$, $g_n(r)$, $n^*(p)$, and $\langle KE^* \rangle$, which are characteristic of the normal fluid or the noncondensate atoms, vary below T_λ, where they cannot be separately observed in experiments. These assumptions are often a major source of uncertainty in the analysis, and we are in urgent need of theoretical guidance. In turn, additional measurements to determine the systematic variations of these quantities above, and especially just above, T_λ are needed to provide guidance for the theories.

Although we now have an explicit relationship, Eq. (13.61), between n_0 and n_s, the precise connection between Bose condensation and superfluidity is still not clear. We also now have strong experimental evidence that one-phonon excitations are unique to the superfluid phase, and hence somehow related to the existence of the condensate. From the point of view of the ultimate development of an all-encompassing theory of liquid ^{4}He, this is undoubtedly the most important experimental result of the last decade. The recent resurgence of interest in liquid ^{4}He on the part of theorists, and especially the new work[153-157] on microscopic theories for which the existence of the condensate is of crucial importance, will, we hope, soon lead us to a much better fundamental understanding. The extension to higher densities and larger systems of particles of the extremely important recent computations of Ceperley and Pollock[245] for finite temperatures is also sure to give exciting new results.

Even after 35 years of intensive study, there are a great many ways in which neutron scattering can, and undoubtedly will, make further major contributions to our understanding of liquid ^{4}He.

13.4. Liquid ^{3}He

13.4.1. Experimental Results

Liquid ^{3}He was first studied by inelastic neutron scattering in 1974 in the pioneering experiments of Scherm et al.[12] These experiments,[12,28,169,276,277] made using the high-flux reactor at the Institute Laue-Langevin (ILL) and time-of-flight methods, provided accurate values of the scattered intensity for $Q \gtrsim 1.5 \text{ Å}^{-1}$, which were compared with the predictions of several models of liquid ^{3}He. Sköld, Pelizzari, and collaborators[30,278–280] soon followed in 1976 with accurate measurements at low and intermediate Q at ANL, using the CP5 reactor and statistical chopper methods. Sköld et al.[30,278–280] were the first to show that collective excitations exist in ^{3}He at finite Q values $(0.4 \leq Q \leq 1.5 \text{ Å}^{-1})$. More recently Sokol et al.[281] have reported measurements at high Q using the spallation neutron source IPNS at ANL to investigate the atomic kinetic energy and momentum distribution. Mook[282] has recently reported scattering studies in the range, $4 \leq Q \leq 7 \text{ Å}^{-1}$. As we shall see, these difficult but rewarding experiments have stimulated a great amount of interest in the nature of excitations and their interactions in liquid ^{3}He.

All experiments observe the total dynamic form factor, Eq. (13.6), the sum of coherent scattering (S_c) from the density excitations and of spin-dependent scattering (S_I) from the spin-density excitations. The ratio (σ_i/σ_c) has the value[280] 0.20 ± 0.05. However, for $0.4 \leq Q \leq 1.0 \text{ Å}^{-1}$, $S_I(Q)$ appears to be 2–3 times greater than $S_c(Q)$ so that in this Q range the intensities in the two terms of Eq. (13.6) are comparable. At $Q > 2.5 \text{ Å}^{-1}$, $S_c(Q) \sim S_I(Q) \sim 1$ and Eq. (13.6) is dominated by $S_c(Q, \omega)$.

In Fig. 26 we show $S(Q, \omega)$ observed by Sköld and Pelizzari at $T = 40$ mK and $T = 1.2$ K. At $T = 40$ mK and $Q \leq 1.2 \text{ Å}^{-1}$, $S(Q, \omega)$ shows two peaks. That at lower energy is interpreted as the "paramagnon" or spin fluctuation resonance occurring in the spin-dependent S_I. This low-energy resonance has been proposed[23,283] as the explanation of many other properties of liquid ^{3}He and it is gratifying to observe it directly. The peak at higher energy ($E = 0.58$ to 1.05 in Fig. 26) is interpreted as the ZSM in the density excitations (S_c). For $Q \gtrsim 1.3 \text{ Å}^{-1}$ the paramagnon resonance and ZSM disappear and the scattering in both S_c and S_I is characteristic of that from weakly interacting single particle-hole (p-h) excitations or multiples of p-h excitations. In this picture, the incoming neutron excites a single atom from a state within the Fermi sea (leaving a hole) to a state above the Fermi sea (the particle excitation). The particle–hole pairs interact weakly and through this interaction multiple pairs can be excited.

At $T = 1.2$ K we see the scattered intensity differs little from that at 40 mK. This is most interesting since we expect Landau's[7,8] celebrated Fermi liquid theory to apply at $T \lesssim 40$ mK but not at 1.2 K. For $0.8 \leq Q \leq 1.2 \text{ Å}^{-1}$,

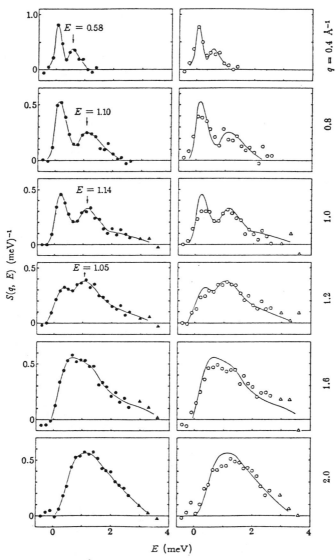

FIG. 26. $S(Q, E)$ for liquid ^{3}He observed by Sköld and Pelizzari[280] at $T = 40$ mK (left) and $T = 1.2$ K (right) ($q = Q$). The solid lines are fitted by eye to the $T = 40$ mK data. (Sköld and Pelizzari.[280])

we see that the "paramagnon" resonance broadens somewhat at 1.2 K. This is believed to be due to the thermal broadening of the single p–h excitation band. At $Q = 0.4$ Å^{-1}, the observed width of the paramagnon peak is limited by the instrument resolution and there may be thermal broadening between 40 mK and 1.2 K as yet undetected. There is also evidence in Fig. 26 of thermal broadening of the p–h excitations for $Q \geq 1.6$ Å^{-1}. The ZSM, however, does not appear to be thermally broadened. In the quantum limit, the absorption coefficient γ, for absorption of zero sound, which propagates when $\hbar\omega \gg 2\pi k_B T$, is[7,8] $\gamma = \gamma_{cl}[1 + (\hbar\omega/2\pi k_B T)^2]$, where γ_{cl} is the classical value. Assuming this applies at finite Q, and given a ZSM energy $\hbar\omega = 10$ K and $\gamma_{cl} \sim T^2$, we expect γ (and therefore the width of the ZSM) to be independent of temperature up to $T \approx \hbar\omega/2\pi k_B \approx 1.5$ K. This also suggests that the apparent effective mass m^* of the p–h excitations (into which the ZSM can decay) is similar at 40 mK and 1.2 K. Specific heat measurements[284] suggest that, at $Q = 0$, m^* drops rapidly from $m^* \approx 3m_0$ at $T \sim 10$ mK to $m^* \approx m_0$ at $T \approx 0.2$ K. If the p–h excitations responded with effective mass $m^* \approx m_0$ at 1.2 K, the ZSM would have disappeared at $Q \geq 0.5$ Å^{-1} due to decay to p–h excitations.

The energy of the ZSM is displayed in Fig. 27. We note that the observed energy extracted from $S(Q, \omega)$ depends somewhat upon how $S_I(Q, \omega)$ is interpreted.[280] The dashed line shows the extension of the zero-sound velocity. The shaded area shows the band of Q and ω in which single p–h excitations can be excited, assuming a Q-independent effective mass of $m^* = 3m_0$.

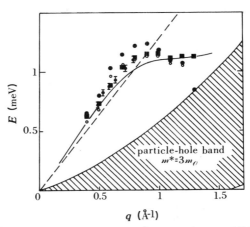

FIG. 27. Energy of the zero-sound mode in liquid ^{3}He. Experiments of Sköld and Pelizzari:[280] open circles, estimated peak positions; solid squares, obtained by fitting S_I to a Fermi gas model; solid circles without error bars, obtained by fitting S_I to a paramagnon model. Experiments of Hilton et al.:[276] solid circles with error bars. The solid curve is from a calculation by Aldrich and Pines,[147] and the dashed line indicates the zero-sound velocity.

At very low Q, ($Q \leq 0.01$ Å^{-1}), the density excitations should be well described by the sound phenomena predicted by Landau. Since neutrons are a high-frequency ($\omega \sim 10^{12}$ s^{-1}) probe, the excitations induced by neutrons should be collisionless, zero sound in which the atomic collision terms can be ignored. For $Q \lesssim 0.1$ Å^{-1} the average interaction between particles is strong relative to the energy transfer $\hbar\omega$. Thus the system responds entirely collectively and there is no weight of the response function in the noninteracting single p-h band. Pines[285] predicted that a response of this nature should continue up to $Q \sim k_F$, where k_F is the Fermi wave vector ($k_F = 0.78$ Å^{-1} at SVP). The data and subsequent calculations support this. The intensity in $S_c(Q, \omega)$ appears[147] predominantly in the ZSM, and excitation of independent p-h's is negligible (or small) up to $Q \sim 1$ Å^{-1}. If there were damping to single p-h excitations only, $S_c(Q, \omega)$ would be a delta function, $\delta(\omega - \omega_Q)$, at the ZSM energy ω_Q up to $Q \sim 1$ Å^{-1}. At low Q, multi-p-h excitation intensity grows[286] as Q^4 in $S_c(Q, \omega)$. The solid line in Fig. 27 is the ZSM energy calculated by Aldrich and Pines.[147] This shows upward dispersion of the ZSM (above the extension of $\omega_Q = c_0 Q$, where $c_0 = 192$ m/s is the zero-sound velocity), in agreement with experiment.

Figure 28 shows the FWHM of the ZSM. At $0.5 \leq Q \leq 1.3$ Å^{-1}, a typical width is ~ 0.5 meV, giving a mode lifetime $\tau \approx 4$ meV$^{-1} \approx 16 \times 10^{-12}$ s. At $Q \gtrsim 1.3$ Å^{-1} the ZSM disappears, presumably due to the rapid decay to single p-h excitation (see Fig. 27). If m^* were $m^* = m_0$, the single p-h band would reach above the dashed line shown in Fig. 27 at $Q \gtrsim 0.5$ Å^{-1}. The ZSM would then disappear by decay to p-h excitations at this Q value. The existence of the ZSM up to ~ 1.3 Å^{-1} is strong evidence that the effective mass is $m^* \approx 3m_0$ for $Q \lesssim 1.3$ Å^{-1}, although some distribution of m^* values may be possible in a more sophisticated model.

Outside the single p-h band, the ZSM can decay to multi-p-h excitations or to pairs or higher numbers of collective modes. The triangles in Fig. 28 show the width of the ZSM due to decay to pairs of p-h excitations calculated by Glyde and Khanna.[287] This width is generally below but consistent with the observed widths. Since the ZSM has upward dispersion for

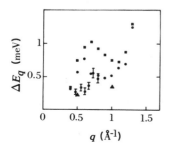

FIG. 28. Width ΔE of the zero-sound mode in liquid ^{3}He. Experiments of Sköld and Pelizzari,[280] symbols as Fig. 27, and of Hilton et al.,[276] open circles with error bars. The solid triangles show values calculated by Glyde and Khanna[287] due to decay to pairs of p-h excitations.

$0.3 \lesssim Q \lesssim 0.6$ Å^{-1}, decay to pairs of modes is possible. There may possibly be evidence of an increased width in this Q range in Fig. 28.

The spin-dependent $S_I(Q, \omega)$ may be understood roughly as follows. Measurements[23,283] of the static susceptibility $\chi_I(Q)$ show that the nuclear spins have a tendency toward ferromagnetic ordering. This means the spin-antisymmetric interaction entering $\chi_I(Q)$ and $S_I(Q, \omega)$ is negative. The negative interaction pushes the response within the p-h band to lower energies, turning it into a narrow resonance. Since the resonance lies in the p-h band, it is not a mode of the spin system. As Q increases, the spin-antisymmetric interaction becomes less important relative to the energies of the p-h excitations themselves and the peak broadens. From Fig. 26 we see that the peak position is quite constant but moves to slightly higher energies as Q increases beyond $Q \sim 1$ Å^{-1}. This suggests that the interaction is not greatly Q-dependent for $Q \lesssim 1$ Å^{-1}.

As we shall see, $S_I(Q, \omega)$ may be quite well described[287] within Fermi liquid theory with $m^* = 3m_0$ and a spin-antisymmetric interaction $F_0^a = -0.67$ at low Q. It may also be well described[288] within the paramagnon model[289] with a p-h mass $m^* = m_0$ and a larger interaction ($\bar{I} \sim 0.9$) (in our sign convention $\bar{I}$ is negative). However, we would expect the p-h excitations entering S_c and S_I to have the same m^* and the existence of a ZSM in S_c for $Q \lesssim 1.3$ Å^{-1} strongly suggests $m^* \sim 3m_0$.

Hilton et al.[277] have measured the inelastic scattering intensity at pressures of 10 and 20 bars as well as at SVP. The observed intensity shows no sharp features characteristic of collective excitations. As the pressure increases the scattering intensity moves to lower energy and decreases in magnitude. Since the thermodynamic ($Q = 0$) m^* increases with pressure, we would expect the p-h band to move to lower energy as the pressure increases. Thus S_I and the p-h component of S_c should move to lower energy. $S_c(Q)$ is also expected to decrease with pressure so that the total intensity would also decrease. If these were the only components observed, then we would indeed expect the intensity to decrease and move to lower energy. The ZSM energy, however, is predicted[287] to increase with pressure. The width of the ZSM is also predicted [287] to increase. Unless the ZSM disappears entirely, these predictions are not consistent with the observed[277] pressure dependence.

We now interpret these experiments in terms of models of $S(Q, \omega)$.

13.4.2. Interpretation

13.4.2.1. The Dynamic Susceptibility. Most numerical calculations of $S(Q, \omega)$ in liquid ^{3}He evaluate the dynamic susceptibility, Eq. (13.11a), in the form

$$\chi_{c,I}(Q, t) = -(i/V)\langle T[\rho_\uparrow(Q, t) \pm \rho_\downarrow(Q, t)][\rho_\uparrow^+(Q, 0) \pm \rho_\downarrow^+(Q, 0)]\rangle, \quad (13.62)$$

where T is the time-ordering operator and the plus (minus) sign is associated with χ_c (χ_I). This is done within a generalized RPA to χ,

$$\chi_{c,I}(Q, \omega) = \frac{\chi_0(Q, \omega)}{1 - V_{s,a}(Q, \omega)\chi_0(Q, \omega)}. \tag{13.63}$$

Here $\chi_0(Q, \omega)$ is the dynamic susceptibility for a suitable Fermi gas. The χ_0 may describe free fermions, fermions having an effective mass (χ_{0*}) or fermions screened by an average interaction (χ_{sc}). The single p–h part of χ_0 has the form of the Lindhard function[290]

$$\chi_0(Q, \omega) = \frac{\hbar}{V} \sum_k \frac{n(\mathbf{k}) - n(\mathbf{k} + \mathbf{Q})}{\hbar\omega - \varepsilon(\mathbf{k} + \mathbf{Q}) + \varepsilon(\mathbf{k}) + i\eta}, \tag{13.64}$$

where the $\varepsilon(k)$ may be $\hbar k^2/2m_0$, $\hbar k^2/2m^*$ or fully "dressed" single particle energies; χ_0 is the same for χ_c and χ_I since the cross term $\rho_\uparrow \rho_\downarrow$ in (13.62) vanishes for noninteracting fermions. The V_s and V_a are general Q- and ω-dependent spin-symmetric and spin-antisymmetric interactions, respectively [$V_{s,a} = \frac{1}{2}(V_{\uparrow\uparrow} \pm V_{\uparrow\downarrow})$]. Calculations that go beyond Eq. (13.63) are noted below.

Since Eq. (13.63) is often used, we sketch its derivation within the general theory of interacting Fermi liquids [see, e.g., Nozières[9] (Chapter 6) and Abrikosov et al.[10] (Sections 10, 16, 18, and 19)]. We note in particular the approximations needed to reach Eq. (13.63) and the meaning of $V(Q, \omega)$. The full $\chi(Q, \omega)$ may be written as a sum of individual particle-hole excitations

$$\chi(Q, \omega) = \frac{1}{V^2} \sum_{k_1 k_2} \int \frac{d\omega_1}{2\pi} \int \frac{d\omega_2}{2\pi} \chi(12, Q\omega). \tag{13.65}$$

In $\chi(12, Q\omega)$, index 1 denotes the particle and 2 denotes the hole. The general equation describing the propagation of each pair is

$$\chi(12, Q\omega) = \chi_0(1, Q\omega)(2\pi)^4 \Delta_{k_2, k_1-Q} \delta_{\omega_2, \omega_1-\omega} + \chi_0(1, -Q\omega) \int \frac{d\omega_3}{(2\pi)} \frac{d^3k_3}{(2\pi)^3}$$
$$\times V(k_1\omega_1, k_3\omega_3; k_1 - Q\omega_1, -\omega, k_3 + Q, \omega_3 + \omega)$$
$$\times \chi(32, Q\omega). \tag{13.66}$$

The first term, $\chi_0(1, Q\omega)$, represents free propagation of the pair without interaction ($2 = 1 - Q$). In the second term the pair interact and scatter into another pair. The V represents all possible pair scattering processes except those that have a single p–h pair as an intermediate state. That is, the complete interaction Γ is related to V by $\Gamma = V + V\chi_0\Gamma$, where $V\chi_0\Gamma$ is the process having a p–h (χ_0) intermediate state. This is the Bethe–Salpeter

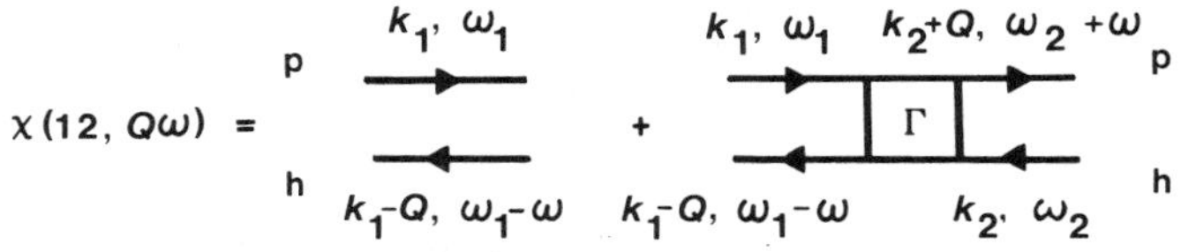

$$\chi(12, Q\omega) = \begin{array}{c} k_1, \omega_1 \\ \text{p} \longrightarrow \\ \text{h} \longleftarrow \\ k_1{-}Q, \omega_1{-}\omega \end{array} + \begin{array}{c} k_1, \omega_1 \quad k_2{+}Q, \omega_2{+}\omega \\ \longrightarrow \boxed{\Gamma} \longrightarrow \text{p} \\ \longleftarrow \quad \longleftarrow \text{h} \\ k_1{-}Q, \omega_1{-}\omega \quad k_2, \omega_2 \end{array}$$

FIG. 29. Diagrammatic representation of Eq. (13.66) for the propagation of a particle (p)-hole (h) excitation. The two parallel lines represent free propagation (χ_0) and Γ represents all possible interaction processes between the particle and hole.

equation in the p-h (χ_0) channel [Eq. (18.3) of Ref. 10 and Eq. (6.11) of Ref. 9]. If we take the long-wave limit ($Q \to 0$) of Eq. (13.66) we may show[9] that V reduces precisely to the interaction parameterized by Landau.

To obtain Eq. (13.63) we must make two approximations to the Fourier transform $\Gamma(r_1 t_1, r_2 t_2, r_3 t_3, r_4 t_4)$ of Γ, which will remove the dependence of V in (13.66) on $k_1 \omega_1$ and $k_2 \omega_2$. The first consists of setting $t_3 = t_1$ and $t_4 = t_2$. This makes the interaction local in time. With $t_1 = t_3$ the initial p-h pair in Fig. 29 disappears at the same instant, and with $t_2 = t_4$ the final p-h pair emerges at the same instant. This removes the dependence on ω_1 and ω_2. The second approximation consists of making the interaction local in space, i.e., we require $r_3 = r_1$ and $r_4 = r_2$ in Γ. This removes the dependence of Γ on k_1 and k_2. Then V reduces to $V(Q, \omega)$, and doing the integrals and sums as indicated in Eqs. (13.65) and (13.66), we find

$$\chi(Q, \omega) = \chi_0(Q, \omega) + \chi_0(Q, \omega)V(Q, \omega)\chi(Q, \omega) = \frac{\chi_0(Q, \omega)}{1 - V(Q, \omega)\chi_0(Q, \omega)},$$

which is the generalized RPA form. The impact of ignoring the dependence of V on $k_1 \omega_1$ and $k_2 \omega_2$ is difficult to assess. Calculations for the electron gas[291] suggest that fine structure in $S(Q, \omega)$ is ignored but the general overall shape is retained. Also, models of $V(Q, \omega)$ should represent all possible scattering processes except those that have a p-h as an intermediate state.

13.4.2.2. *Fermi Gas.* For a gas of noninteracting fermions[292] of mass m^*, χ reduces to Eq. (13.64) and $S_{0*}(Q, \omega) = -(n\pi)^{-1}[n(\omega) + 1](1 + \sigma_i/\sigma_c)\chi_{0*}''(Q, \omega)$. At $T = 0$ K and for $Q < 2k_F$,

$$\chi_{0*}''(Q, \omega) = -\left(\frac{dn}{d\varepsilon}\right)\left(\frac{\pi\hbar}{4x}\right) \times \begin{cases} y, & 0 < y < y_-, \\ 1 - (y - x^2)^2/4x^2, & y_- < y < y_+, \\ 0, & y_+ < y. \end{cases} \quad (13.67)$$

Here $x = Q/k_F$, $y = \hbar\omega/\varepsilon_F$, and $(dn/d\varepsilon) = 3n/2\varepsilon_F = m^*k_F/\pi^2\hbar^2$ is the density of states per unit volume at the Fermi surface. The $y_+ = x^2 + 2x$ and $y_- = |x^2 - 2x|$. For $Q > 2k_F$, χ_{0*}'' is the same as Eq. (13.67) except it is

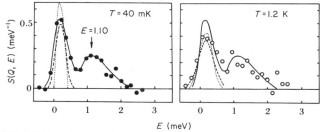

FIG. 30. $S_0^*(Q, \omega)$ (dotted curves) for $m^* = 3.1 m_0$ compared to the experimental results of Sköld and Pelizzari[280] (see Fig. 26) for $Q = 0.8$ Å^{-1}. The dashed curves show S_0^* folded with a resolution width of 0.25 meV.

zero for $y < y_-$. The y_+ is the upper edge of the p-h band and y_- the lower edge for $Q > 2k_F$ [i.e., $\omega_\pm = \omega_R(1 \pm 2k_F/Q)$]. At finite temperature, χ_0 is[293]

$$\chi_0^{\prime\prime}{}_*(Q, \omega) = -\left(\frac{dn}{d\varepsilon}\right)\left(\frac{\pi\hbar}{4x}\right) z \ln\left[\frac{1 + e^{\alpha_-}}{1 + e^{\alpha_+}}\right], \quad (13.68)$$

where $\alpha_\pm = [\mu/\varepsilon_F - (y \pm x^2)^2/4x^2]/z$ and $z = k_B T/\varepsilon_F$. Here α_- is the Stokes term and α_+ is the anti-Stokes term. For $T \to 0$ K, the chemical potential $\mu \to \varepsilon_F$ and Eq. (13.68) reduces to Eq. (13.67).

In Fig. 30 we compare $S_{0*}(Q, \omega)$, Eq. (13.68) calculated assuming an effective mass $m^* = 3.1\, m_0$ ($\varepsilon_F = p_F^2/2m^*$), with the observed $S(Q, \omega)$. The $S_{0*}(Q, \omega)$ is zero outside the p-h excitation band shown in Fig. 27. At low Q the observed paramagnon resonance lies within the p-h band but is peaked at low ω. The ZSM lies above the p-h band. This indicates that there is a strong negative interaction V_a in χ_I and a very strong positive interaction V_s in χ_c. At $Q \approx 2$ Å^{-1}, the apparent m^* in the observed intensity is reduced and, as we shall see, the interaction V_s in χ_c is negative at this Q value. S_{0*} (with $m^* \neq m_0$) violates the f-sum rule, giving

$$\int d\omega\, \omega S_{0*}(Q, \omega) = \frac{\hbar Q^2}{2m^*} = \frac{m_0}{m^*}\omega_R. \quad (13.69)$$

With $m_0/m^* < 1$, Eq. (13.69) tells us that the p-h band is pushed to lower energy. The larger m^*, the more the p-h band is pushed to lower ω.

13.4.2.3. *Interactions.* Akhiezer et al.[294] derived the general result [Eq. (13.6)] for interacting liquid ^{3}He. They examined low Q and assumed that $S_c(Q, \omega)$ was sharply peaked in a ZSM, $S_c(Q, \omega) = Z(Q)\delta(\omega - \omega_Q)$, as in ^{4}He. They approximated S_I by S_0.

Glyde and Khanna[287, 295] (GK) constructed an extremely simple model of $S(Q, \omega)$ at low Q ($Q \leq 1.5$ Å^{-1}) based on Landau theory.[7,8] As noted,

TABLE I. Landau Parameters for Liquid ^{3}He at SVP

Vol. (cm^3/mol)	F_0^s	F_1^s	F_0^a	F_1^a
36.83	10.07	6.04	−0.67	−0.67

Eq. (13.65) for the full $\chi(Q, \omega)$ reduces to the Landau transport equation for the response of the particle (and spin) density to an external field. In this limit V reduces to the Landau interaction[296] (in energy units)

$$\lim_{Q \to 0} V_s(Q, \omega) = \left(\frac{dn}{d\varepsilon}\right)^{-1} \left[F_0^s + \frac{F_1^s}{(1 + F_1^s/3)} \left(\frac{\omega}{Qv_F}\right)^2\right], \quad (13.70a)$$

$$\lim_{Q \to 0} V_a(Q, \omega) = \left(\frac{dn}{d\varepsilon}\right)^{-1} \left[F_0^a + \frac{F_1^a}{(1 + F_1^a/3)} \left(\frac{\omega}{Qv_F}\right)^2\right]. \quad (13.70b)$$

Here the F_L are the Lth angular momentum components of the interaction, and $L = 0$ and 1 only are retained. The F_L are listed in Table I for liquid ^{3}He at SVP. At $Q \to 0$, only excitations near the Fermi surface are possible so that in Eq. (13.66) $|k_1| = |k_2| = k_F$, $|\omega_1| = |\omega_2| = \varepsilon_F/\hbar$, and V depends only on Q and ω. The interaction of Eq. (13.70) is assumed to hold at finite Q and ω and extrapolation of $\chi_{c,I}$ to finite Q and ω is made using Eq. (13.63), with $\chi_{0*}(Q, \omega)$ having effective mass $m^* = m_0(1 + F_1^s/3) = 3.1\, m_0$.

The large positive value of F_0^s pushes $S_c(Q, \omega)$ to high frequency and into a delta-function peak at the ZSM. Since F_0^s is obtained by fitting to the observed first sound velocity $[-\chi_c(0, 0) = (dn/d\varepsilon)(1 + F_0^s)^{-1} = (nmc_1^2)^{-1}]$ this must predict the ZSM energy correctly at $Q \to 0$. The model provides no decay of the ZSM until it enters the p–h band. The calculated ZSM energy is shown in Fig. 31. At higher pressure the ZSM energy increases because F_0^s increases and the p–h band decreases in energy because m^* increases. The resulting $S_c(Q, \omega)$, which includes single p–h excitations only, satisfies the f-sum rule

$$\int d\omega\, \omega S_c(Q, \omega) = \frac{\hbar Q^2}{2m^*} \left(1 + \frac{F_1^s}{3}\right) = \omega_R. \quad (13.71)$$

To allow for multi-p–h excitation at higher Q without violating the f-sum rule, Glyde and Khanna (GK) modified the $V_s(Q, \omega)$ at higher Q.

Later GK[287] evaluated the decay of the ZSM to pairs of p–h excitations. This provides a decay mechanism for zero sound outside the single p–h band. From Fig. 28 we see that the calculated width ΔE of the ZSM due to this mechanism is comparable but somewhat less that the observed width. Also, since two p–h excitations are expected to be uniformly spread over Q and ω,

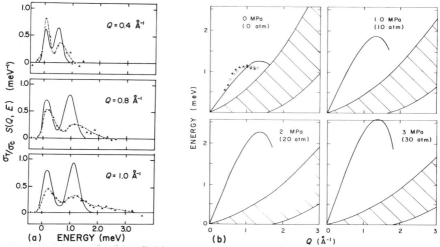

FIG. 31. (a) $S(Q, \omega)$ for liquid ^{3}He at SVP, and (b) the variation of the zero-sound-mode energy with pressure as calculated by Glyde and Khanna.[287]

we do not expect this mechanism to provide any strong Q dependence of ΔE. This mechanism predicts that the damping ΔE will increase with pressure.

Since F_0^a is negative, the $S_I(Q, \omega)$ is shifted to lower ω into a paramagnon peak as is observed. The first moment of S_I is $\hbar Q^2/2m^*(1 + F_1^a/3) \sim \omega_R/4$. Since spin current is not conserved, $S_I(Q, \omega)$ need not fulfill the f-sum rule. The resulting total $S(Q, \omega)$ is compared with observed values in Fig. 31.

As discussed in Section 13.3.2.2, Aldrich et al.[145,147] and Bedell and Pines[158,297,298] have developed a polarization potential (PP) theory for liquid ^{3}He using the basic RPA form, Eq. (13.63). In Eq. (13.63) the zero-order coherent χ is, as in ^{4}He ($q = Q$ to follow PP notation),

$$\chi_{sc}^c = \alpha_q \chi_{0*}(q, \omega) + (1 - \alpha_q)\chi_q^M, \tag{13.72}$$

where $\chi_{0*}(q, \omega)$ is the Lindhard function having a q-dependent effective mass m_q^*. The $\chi_{sc}^I = \chi_{0*}(q, \omega)$. The PP spin-symmetric and spin-antisymmetric interactions are

$$V_{s,a}(q, \omega) = f_q^{s,a} + g_q^{s,a}\left(\frac{\omega}{q}\right)^2. \tag{13.73}$$

In Eq. (13.73),

$$f_q^{s,a} = \tfrac{1}{2}[f_q^{\uparrow\uparrow} \pm f_q^{\uparrow\downarrow}], \tag{13.74}$$

where $f_q^{\uparrow\uparrow}$ and $f_q^{\uparrow\downarrow}$ are Fourier transforms of spin-dependent interactions, $f^{\uparrow\uparrow}(r)$ and $f^{\uparrow\downarrow}(r)$, both having the form of Eq. (13.47). The $f^{\uparrow\uparrow}(r)$ and $f^{\uparrow\downarrow}(r)$

differ only in the choice of the core radius r_c in Eq. (13.47). Aldrich and Pines[147] find the ZSM is best fitted if $r_c^{\uparrow\downarrow}$ is ≈ 3 Å at SVP, slightly larger than in ^{4}He. At $p \sim 21$ atm, $r_c^{\uparrow\downarrow} \approx 2.68 - 2.80$ Å. Bedell and Pines,[158,297] from fits to transport coefficients, determine a small difference

$$\delta = r_c^{\uparrow\uparrow} - r_c^{\uparrow\downarrow} = \begin{cases} 0.03 \text{ Å} & \text{at SVP} \\ 0.007 \text{ Å} & \text{at 21 atm,} \end{cases}$$

which reflects the effects of statistics. With $r_c^{\uparrow\uparrow}$ and $r_c^{\uparrow\downarrow}$ given, the parameter a in (13.47) is determined by requiring $(dn/d\varepsilon)f_{q=0}^{s,a} = F_{L=0}^{s,a}$. These relations are not quite obvious since f_0^s is the $q = 0$ limit of $f_q^s = \int d\mathbf{r}\, e^{i\mathbf{q}\cdot\mathbf{r}} f^s(r)$ and F_0^s is the $L = 0$ angular momentum component of F^s. The b and r_t in Eq. (13.47) are determined by continuity requirements as in ^{4}He.

The effective mass is $m_q^* = m_0 + n g_q^s$ with

$$g_q^s = g_0^s \left[1 - \theta(x - x_0)\left(\frac{x - x_c}{x_0 - x_c}\right)^2\right], \tag{13.75}$$

where $x = q/k_F$, $x_0 = 2.3$, and $x_c = 1.2$ at SVP and $x_c = 0.21$ at 21 atm. The g_0^s is determined to reproduce the observed m^* at $q = 0$ (i.e., $n g_0^s / m^0 = F_1^s/3$). The g_q^a appears to be set to the appropriate Landau parameter.

The PP theory reproduces Landau theory at $q = 0$ ($\alpha_0 = 1$). The theory then allows an extension to higher q and ω both via Eq. (13.63) and the interaction of Eq. (13.73). The softened core of $f(r)$ simulates the effects of pair collisions as in a Brueckner theory and induced interactions are embodied in the renormalized parameters, such as f_q, g_q, m_q^*. The theory has some especially attractive features. It predicts upward or anomalous dispersion in the ZSM in agreement with experiment (see Fig. 27), since f_q increases with q from $q = 0$ to $q \approx 1$ Å^{-1}. Also, since $x_0 = 2.3$ in Eq. (13.75), $m_q^* = m^*$ up to $q = 2.3 k_F \approx 1.7$ Å^{-1}. This is needed to keep the p–h band down at lower energy so that the ZSM remains above the p–h band. It enters the p–h band at $q \sim 1.3$ Å^{-1} as is observed (see Fig. 27). Outside the p–h band the ZSM is not damped in the PP theory and $V(Q, \omega)$ is entirely real. The PP theory provides an excellent description of transport properties and their pressure dependence. The attractiveness of the theory is somewhat blunted by the large number of parameters. The determination of these parameters is also discussed by Béal-Monod and Valls.[299]

Béal-Monod[288] has evaluated $S_I(Q, \omega)$ using the paramagnon model of spin fluctuations. This uses the form of Eq. (13.63) for χ_I with χ_0 given by Eq. (13.64) with $m^* = m_0$. The spin-antisymmetric interaction is reduced to a constant $V_a(Q, \omega) = (dn/d\varepsilon)^{-1}\bar{I}$. Although it has microscopic origins, the paramagnon model differs from Fermi liquid theory only in the choice of parameters ($m^* = m_0$, $V_a = \bar{I}$). The resulting $S_I(Q, \omega)$ fits that part of the observed intensity attributed to spin fluctuations very well. This seems due chiefly to using $m^* = m_0$, which spreads $S_I(Q, \omega)$ (the p-h band) to higher ω. Also the value $\bar{I} = 0.89$, obtained from fitting neutron data, agrees with the static ($\omega = 0$) value $\bar{I} = 0.91$. This suggests that V_a is not greatly Q-dependent up to $Q \sim 1.5$ Å^{-1}. One would expect the m^* in χ_0 for χ_I and χ_c to be the same.

Glyde and Hernadi[300] have calculated $S(Q, \omega)$ in the range $2 \lesssim Q \lesssim 5$ Å^{-1} using a microscopic model and the relation (13.63). The properties of ^{3}He are calculated within the Galitskii-Feynman-Hartree-Fock (GFHF) approximation beginning with the bare He-He atom potential. In this model the p-h interaction $V(Q, \omega)$ is approximated by the G-F T matrix. This takes account of pair collisions via the pair potential accurately, particularly the effects of the steep repulsive core. However, it neglects the components of the pair interaction taking place via the collective excitations. Since the liquid does not support collective excitations at $Q \gtrsim 1.5$ Å^{-1}, the latter component should be small for $Q \gtrsim 2$ Å^{-1}. The single particle energies ε_k are calculated using the G-F T matrix. The resulting ε_k, used in Eq. (13.64) to evaluate $\chi_0(Q, \omega)$, are complex. The total $S(Q, \omega)$ has a peak position and general shape that agrees well with experiment. However, the peak height is too small and the calculated $S(Q, \omega)$ extends up to higher ω than the observed $S(Q, \omega)$. Both these differences would be corrected if the effective mass in the calculated $S(Q, \omega)$ was increased to $m^* \approx 2$. The GFHF theory predicts $m^* \approx 1$ and comparison with experiment suggests the apparent effective mass is closer to 2 at $Q \approx 2$ Å^{-1}. The magnitude and sign of the GF $V_s(Q, \omega)$ appears to be approximately correct.

13.4.2.4. The Mori Formalism. $S(Q, \omega)$ has been described using the Mori[301] formalism by Lovesey,[302] Takeno and Yoshida,[303] Valls et al.,[304] and by Pathak and Lücke.[305] Early work is reviewed by Lovesey and Copley.[306] Generally the Mori formalism provides a framework in which $\chi(Q, \omega)$ can be related to empirical quantities [e.g., $S(Q)$, static properties, moments of $S(Q, \omega)$, and a relaxation function]. A reasonable semi-empirical description of excitations and $S(Q, \omega)$ is possible. To get damping of modes, models for the relaxation function are needed. The method does not, however, lend itself to testing specific microscopic pictures of excitations in liquid ^{3}He.

In this approach $\chi(Q, \omega)$, or related functions, are generally expressed in the RPA form, Eq. (13.63). It may be rewritten as [305]

$$V_{s,a}(Q, \omega) = -[\chi_{c,I}^{-1}(Q, \omega) - \chi_0^{-1}(Q, \omega)] \qquad (13.76)$$

and simply viewed as a definition of $V(Q, \omega)$. In this sense, $\chi(Q, \omega)$ may always be "rigorously" expressed in the form of Eq. (13.63). Models and approximations are then made to Eq. (13.76). First, a static approximation

($\omega = 0$) to $V(Q, \omega)$ can be made $[\chi(Q, 0) = -\chi(Q)]$ giving

$$V_s(Q, 0) = \chi_c^{-1}(Q) - \chi_0^{-1}(Q). \qquad (13.77)$$

Takeno and Yoshida[303] (TY) used Eq. (13.77) in Eq. (13.63) with $\chi_c^{-1}(Q)$ determined by fits to the observed $S(Q)$ with $\chi_0(Q)$ and $\chi_0(Q, \omega)$ given by Eq. (13.64) for $m^* = m_0$. This gives a ZSM having an energy in good agreement with experiment (see Fig. 32). However the ZSM lies immediately above the $m^* = m_0$ p-h band, which does not agree with experiment.

Next a dispersion relation can be derived for each $\chi_\alpha(Q, \omega)$ ($\alpha = c, I$) in Eq. (13.76) in terms of a relaxation function $M_\alpha(Q, \omega)$, which includes the possibility of decay or "relaxation" of collective excitations.[302-306] Takeno and Yoshida constructed a simple model of $M_\alpha(Q, \omega)$ to represent the decay of the ZSM to multi-particle-hole excitations. The resulting broad $S(Q, \omega)$ are shown in Fig. 32. These are typical of the $S(Q, \omega)$ obtained using model M_α, especially at high Q.[302, 306]

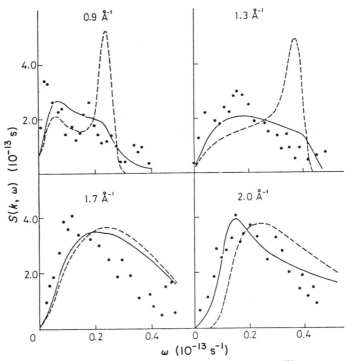

FIG. 32. $S(Q, \omega)$ for liquid ^{3}He calculated by Yoshida and Takeno:[303] dashed curve, static approximation, Eq. (13.77); solid curve, with inclusion of the decay of the zero-sound mode.

Using the Mori method, TY and Valls et al. show that the approximate forms of $V(Q, \omega)$ in Eqs. (13.70) and (13.73) correspond to a low-frequency limit of $V(Q, \omega)$. At higher ω, $V(Q, \omega)$ contains higher powers of ω beyond ω^2 and is in general complex. In the low-ω limit, TY derive the specific interactions of Eq. (13.70) by requiring that the $S_c(Q, \omega)$ and $S_I(Q, \omega)$ satisfy the compressibility ($\langle\omega^{-1}\rangle$) and f-sum rules at $Q \to 0$. Actually, S_I need not satisfy the f-sum rule and their V_a is therefore not quite correct. Valls et al.[304] also show, within the Mori method, what approximations are needed to arrive at the RPA, Eq. (13.63). Within the RPA, they find Q-dependent potential parameters very similar to those in the PP theory by fitting to the Landau parameters and the observed $S(Q)$.

13.4.3. High Momentum Transfer

Since the 1950s there has been enormous interest[60, 306–312] in evaluating the ground state energy E of liquid ^{3}He to test our understanding of interacting Fermi liquids. The most reliable results to date, using Green-function Monte Carlo (GFMC) methods,[308, 312] find $E = -2.3 \pm 0.1$ K, ~ 0.2 K above the observed[313] value $E = -2.47 \pm 0.01$ K. Since E is a sum of kinetic and potential parts, there is equal interest in having a separate experimental value of the $\langle KE \rangle$ to test theory.

Sokol et al.[281] have reported the first measurements of inelastic scattering intensity from liquid ^{3}He at high Q, $10 \leq Q \leq 16$ Å^{-1}. The observed intensity is fitted by a Gaussian of form Eq. (13.32) to within the experimental uncertainty. If the observed $S(Q, \omega)$ is well approximated by $S_i(Q, \omega)$, the variance σ^2 of the fitted Gaussian gives the second moment in Eq. (13.32). In this way Sokol et al. find $\langle KE \rangle = (3\hbar/4\omega_R)\sigma^2 = 8.1 \pm ^{1.7}_{1.3}$ K at SVP and $T = 0.2$ K. Since the spin-antisymmetric interaction is weak, $S_I \approx S_i$ at $Q \gtrsim 1.5$ Å^{-1}. The spin-symmetric interaction in ^{3}He and ^{4}He is similar so, as noted in Section 13.2.5.1, we expect $S_c \approx S_i$ at $Q \gtrsim 10$ Å^{-1}.

Lhuillier and Lévesque[310] (L–L) find a $\langle KE \rangle$ of 12–13.5 K using variational Monte Carlo methods. L–L included both pair and triplet correlations between atoms in their variational function. Manousakis et al.[309] find $\langle KE \rangle = 12.28$–$13.48$ K using combined variational and hypernetted chain summation methods. The lowest $\langle KE \rangle$ is obtained when pair, triplet, and momentum dependent correlations are included while the higher value is obtained for pair correlations only. The most recent GFMC value[312] is $\langle KE \rangle = 13$ K. The calculated $\langle KE \rangle$ appears to lie above the observed value. However, the quantity observed is the full width at half maximum W of the scattering intensity. The second moment σ^2 is obtained from W assuming $S(Q, \omega)$ is Gaussian ($\sigma^2 = W^2/8 \ln 2$). If $S(Q, \omega)$ is not simply a Gaussian but has high-frequency tails,[106–109] the σ^2 and the $\langle KE \rangle$ given by Eq. (13.31) could be significantly larger than that for a Gaussian.

The region of Q in which the IA, Eq. (13.30), holds well enough to determine $n(p)$ in liquid ^{3}He is as yet unknown and untested. For noninteracting fermions,

$$n(p) = \begin{cases} 1, & p < p_F, \\ 0, & p > p_F. \end{cases} \qquad (13.78)$$

The corresponding $S_0(Q, \omega)$ is then given by Eqs. (13.67) and (13.68), i.e., for $T = 0$ K and $Q > 2k_F$,

$$S_0(Q, \omega) = (1 + \sigma_i/\sigma_c)(3/4v_F Q)[1 - (\omega - \omega_R)^2/(v_F Q)^2]$$

for $\omega_- < \omega < \omega_+$, where $\omega_\pm = \omega_R(1 \pm 2k_F/Q)$, and zero elsewhere. This is also the IA result, Eq. (13.30), for $n(p)$ given by Eq. (13.78). $S_0(Q, \omega)$ is simply an "inverted parabola" centered at the recoil frequency ω_R.

For interacting atoms, $n(p) \sim 0.6$ for $p < p_F$ and $n(p)$ has a tail reaching up to high-momentum values. As noted in Section 13.2.5.2, this high-energy tail is a general characteristic of atoms interacting via a potential having a hard core and is common to solid and liquid helium. According to most calculations,[314] the step in $n(p)$ at p_F is reduced from 1 to ~ 0.5. The reduced step in $n(p)$ and the high-momentum tail broaden $S(Q, \omega)$ generally and especially in its "wings" where $S(Q, \omega)$ approaches zero [i.e., at $\omega \approx \omega_R(1 \pm 2k_F/Q)$]. If the IA, Eq. (13.30), holds precisely, the broadening is due solely to $n(p)$ and a measurement of $S(Q, \omega)$ can be used to determine $n(p)$. However, if Eq. (13.30) does not hold precisely, there can be broadening in $S(Q, \omega)$ due to interaction between the atoms via $V(Q, \omega)$ in Eq. (13.63) and the single-particle energies ε_k in Eq. (13.64), which will have an imaginary part (see Fig. 2 in Ref. 300). There can also be thermal broadening, but this can be made negligible if $k_B T < \varepsilon_F/10$.

A measurement of $n(p)$ is most interesting as a fundamental test of Fermi liquid concepts. However, it will require high precision and complete confidence in Eq. (13.30). This presents a challenge to experimentalists and theorists alike and much activity in this rewarding new area is expected.

13.4.4. Future Prospects

Measurement of $S(Q, \omega)$ in ^{3}He under pressure can provide incisive information. For example, calculations[287] of $S(Q, \omega)$ based on Fermi liquid theory (FLT) make clear predictions. Since m^* increases[22,315] with pressure, the p-h band is predicted[287] to move to lower energy. The $S_I(Q, \omega)$ should therefore move to lower energy and the paramagnon peak with it. The $V_a(Q = 0)$ is quite independent[22,315] of pressure. Thus a measurement could distinguish which effective m^* operates in $S_I(Q, \omega)$, the FLT m^* or the paramagnon theory $m^* = m_0$, which is independent of pressure. The ZSM energy is predicted to increase with pressure, as shown in Fig. 31. The width of the ZSM, due to decay to pairs of p–h excitations, is predicted[287] to increase with pressure. Measurements of the width as a function of pressure could distinguish between decay channels for the ZSM. The total scattering intensity is predicted to decrease with pressure, making the measurements more difficult.

Measurements of $S(Q, \omega)$ as a function of temperature would also be most interesting. For example, broadening of $S_I(Q, \omega)$ has been observed between 40 mK and 1.2 K. The temperature at which $S_I(Q, \omega)$ broadens depends upon the effective mass governing $S_I(Q, \omega)$—on the ratio $k_B T/\varepsilon_F$, where $\varepsilon_F = p_F^2/2m^*$. A measurement of the degree of broadening versus temperature could be used to infer $m^*(T)$. It would be most interesting to compare this apparent $m^*(T)$ with that observed in the specific heat.[284] Similarly up to 1.2 K, the damping of the ZSM does not appear to increase with temperature. It would be most interesting to determine at what temperature the ZSM begins to decay rapidly or disappear. This rapid decay would probably be due to decay to single p–h excitations, as the p–h band moves up in energy toward the ZSM. This could also be used to identify $m^*(T)$ and to distinguish between decay mechanisms for the ZSM. Current calculations[315–318] suggest that $m^*(\varepsilon)$ peaks quite sharply at the Fermi surface, $\varepsilon = \varepsilon_F$. Thus, as temperature increases and regions away from ε_F are sampled, the average effective mass should fall with temperature.

Measurements of $S(Q, \omega)$ at high Q hold promise of testing our microscopic ideas and models of ^{3}He. Precise values of $\langle KE \rangle$ can distinguish between ground-state theories of liquid ^{3}He and eventually those at finite temperature. Equally, measurements of $n(p)$ are most interesting, but, as in liquid ^{4}He, are likely to be hard won information. Direct comparison of observed and calculated values of $S(Q, \omega)$ may also be used to test microscopic models. These are models both of the single particle energies entering χ_0, Eq. (13.64), and of the interparticle interaction $V(Q, \omega)$. Comparison of calculated and observed $S(Q, \omega)$ directly at high Q, where interactions are relatively small compared to the scattering energies, may be the best region to begin to test microscopic calculations of excitations in liquid ^{3}He.

13.5. Liquid ^{3}He–^{4}He Mixtures

13.5.1. Introduction

Dilute solutions of ^{3}He atoms in liquid ^{4}He represent fascinating examples of interacting fermions and bosons. Such mixtures display a rich array of properties and constitute a field in themselves.[18, 20, 21, 60, 61, 128, 319–321] Here, we focus solely on excitations in mixtures with emphasis on excitations having wave vectors $Q \gtrsim 0.1$ Å^{-1}. There are just three reported studies of mixtures by inelastic neutron scattering [322-324] and several by light scattering.[325-327] We review these experiments and the theory that assists in the interpretation of neutron data.

The work on mixtures up to 1970 was reviewed by Ebner and Edwards.[319] This review still forms an excellent introduction to the subject. Baym and Pethick[321] have reviewed progress up to 1975 with emphasis on thermodynamic and transport properties. Woo[61] has reviewed microscopic theory up to 1975, while excitations, with emphasis on neutron and light scattering, have been reviewed by Price[18] and Ruvalds.[128]

13.5.2. Excitations and Neutrons

To introduce ^{3}He-^{4}He mixtures, we open with the simplistic picture of the mixture as two interpenetrating but independent fluids of ^{3}He and ^{4}He atoms. ^{3}He-^{4}He and ^{3}He-^{3}He interactions are ignored but ^{4}He-^{4}He interactions are fully retained. The concentration of ^{3}He atoms, $x = N_3/(N_3 + N_4)$, is small. In this picture, the ^{3}He atoms form a dilute gas of noninteracting Fermions with ^{4}He serving only to dilute the ^{3}He gas. In the limit $x \to 0$, the ^{3}He excitations are independent particle excitations of energy $\varepsilon = (\hbar k)^2/2m_3$. As far as ^{4}He is concerned, the ^{3}He serves only to dilute the pure ^{4}He. The ^{4}He excitations are therefore nearly pure ^{4}He phonon-rotons with a dispersion curve shifted somewhat, largely because of the decrease in density caused by the ^{3}He atoms.

The Fermi wave vector of the ^{3}He gas having concentration x is $k_F = (3\pi^2 nx)^{1/3} \approx x^{1/3} k_F(^3\text{He})$, where $k_F(^3\text{He}) \approx 0.8$ Å^{-1} is the Fermi wave vector of pure ^{3}He at SVP. Thus, for a 1% solution, $k_F \approx 0.17$ Å^{-1} and, for 6%, $k_F \approx 0.31$ Å^{-1}. For the concentrations, $x > 5$%, employed in the neutron studies,[322-324] k_F is thus not negligibly small. The ^{3}He excitations are therefore p-h excitations in which particles are excited out of the Fermi sea, leaving holes within it. These p-h excitations have energies centered at $\varepsilon = (\hbar k)^2/2m_3$ with an allowed band width of $2k_F$ parallel to the wave vector axis.

In the initial Landau-Pomeranchuk[328] (L-P) model for $x \to 0$, the ^{3}He atoms interact with the ^{4}He atoms but not with each other. This complicated ^{3}He-^{4}He interaction, which may be described in different ways and at different levels of sophistication,[61,321,328-335] modifies the ^{3}He atom energy both at rest and in motion. At low k, the ^{3}He energy is expanded in powers of k as[61,321] ($x \to 0$)

$$\varepsilon_3(k) = \mu_3 + \frac{(\hbar k)^2}{2m_3^*}. \qquad (13.79)$$

Fits to observed properties [61,132] give $\mu_3 = -2.8$ K and $m_3^* = 2.4 m_3$. With this effective mass, the Fermi energy is $\varepsilon_F \approx (3/2.4) x^{2/3} \varepsilon_F(^3\text{He})$, where $\varepsilon_F(^3\text{He}) \approx 1.5$ K is the pure ^{3}He value at SVP. Thus, at $x = 1\%$, $\varepsilon_F \approx 0.1$ K and, at $x = 6\%$, $\varepsilon_F \approx 0.3$ K. Low temperatures are therefore required to keep the Fermi gas degenerate ($k_B T \ll \varepsilon_F$). The lowest temperature studied[323] by neutrons is $T = 0.6$ K, so very substantial thermal broadening of the p-h excitation band in the observed $S(Q, \omega)$ is expected.

The dynamic form factor for ^{3}He-^{4}He mixtures is (see Section 13.1.2)

$$S(Q, \omega) = x(\sigma_{c3}/\sigma_{c4})[S_c(Q, \omega) + (\sigma_{i3}/\sigma_{c3}) S_I(Q, \omega)] + (1 - x) S_{44}(Q, \omega)$$
$$+ 2[x(1 - x)(\sigma_{c3}/\sigma_{c4})]^{1/2} S_{34}(Q, \omega). \tag{13.80}$$

The first term, which we call $S_{33}(Q, \omega)$, is proportional to x and represents the scattering from ^{3}He atoms alone. It is identical to that for pure ^{3}He, with now the ^{3}He atoms diluted by and interacting with ^{4}He as well as ^{3}He atoms. The factor $(\sigma_{c3}/\sigma_{c3}) = 4.34$ is the ratio of the ^{3}He and ^{4}He coherent scattering cross sections, and $(\sigma_{i3}/\sigma_{c3}) = 0.20$. The second term is as for pure ^{4}He and the cross term is

$$S_{34}(Q, \omega) \frac{1}{2\pi} \int_{-\infty}^{\infty} dt\, e^{i\omega t} \frac{1}{(N_3 N_4)^{1/2}} \langle \rho_3(Q, t) \rho_4(-Q, 0) \rangle. \tag{13.81}$$

Equation (13.80) is normalized so that $S(Q, \omega)$ reduces exactly to the ^{4}He $S(Q, \omega)$ for $x \to 0$.

Given $(\sigma_{c3}/\sigma_{c4}) = 4.34$ then at $x = 6\%$, the scattering intensity from ^{3}He is 25% of that from ^{4}He and the coefficient of S_{34} is 1. The intensity observed by Hilton et al.[323] at $Q \approx 1.5$ Å^{-1} is shown in Fig. 33. The broad peak at low energy corresponds to ^{3}He quasiparticle-hole (qp-h) excitations [chiefly $S_{33}(Q, \omega)$], while the sharp strong peak corresponds to the ^{4}He one-phonon resonance.

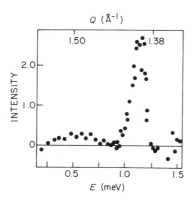

FIG. 33. Scattered intensity for a ^{3}He-^{4}He mixture containing 6% ^{3}He observed by Hilton et al.[323] at $T = 0.6$ K and a scattering angle of 87°.

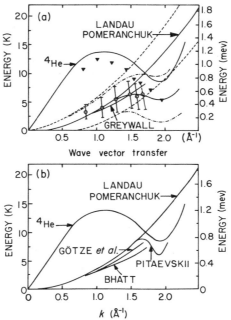

FIG. 34. Summary of results for ^{3}He qp-h excitation energies in dilute ^{3}He-^{4}He mixtures. (a) The centers (circles) of the ^{3}He peaks observed by Hilton et al.[323] at $T = 0.6$ K for $x = 6\%$. The error bars indicate the FWHMs of the peaks. The unlabeled solid curve and the dash–dot curves show, respectively, the center position and band edges calculated[323] using parameters given by Aldrich.[145] The curve labeled Greywall shows the result of a fit[345] to thermodynamic data. Also shown are the dispersion curves for pure ^{4}He (see Fig. 9) and the Landau-Pomeranchuk[328] result [Eq. (13.79)] for free quasiparticles having $m_3^* = 2.4 m_3$ with band edges (for $x = 0.06$) indicated by dashed curves. Triangles show the low-energy edge of the ^{4}He phonon–roton peaks observed by Hilton et al.[323] (b) The ^{4}He and Landau-Pomeranchuk curves as in the upper part plus the band-center positions given by the theories of Pitaevskii,[342] Bhatt,[348] and Götze et al.[349]

13.5.3. ^{3}He Excitations

In Fig. 34 we show the ^{3}He qp-h excitation spectrum observed by Hilton et al.[323] for $x = 6\%$. Circles indicate the center of the qp-h band. For $k \lesssim 1.2$ Å^{-1} the center of the qp-h band is well described by the L-P form, Eq. (13.79), for noninteracting ^{3}He qp's having $m_3^* = 2.4 m_3$. Since ^{3}He is lighter than ^{4}He, its kinetic energy will be larger, to lowest order[329] by (m_4/m_3). Due to its increased kinetic energy, ^{3}He occupies a larger volume in the fluid, by $(1 + \alpha)$ where[321] $\alpha = 0.28$. Thus a moving ^{3}He atom constitutes both a number-density and a mass-density excitation in the fluid. This density excitation creates (emits) density excitations (the ^{4}He phonon-rotons) and interacts with thermal phonons in the fluid.[61, 321, 328–335] Since there are few

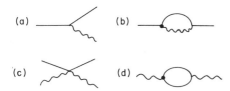

FIG. 35. Interaction processes; solid lines indicate ^{3}He quasi-particles, and wavy lines indicate ^{4}He phonons. (a) ^{3}He scattering leading to phonon emission. (b) Contribution of (a) to ^{3}He self-energy. (c) ^{3}He-phonon scattering. (d) Contribution of ^{3}He quasiparticle-hole creation process to phonon self-energy.

thermal phonon–rotons (p–r) at low temperature, the ^{3}He–phonon–roton emission process, depicted in Fig. 35(a), is expected to dominate.[61] The contribution to $\varepsilon_3(k)$ from p–r emission, depicted in Fig. 35(b), leads to the μ_3 and m_3^* in Eq. (13.79) at low k and to possible deviations from Eq. (13.79) at high k.

Thermodynamic properties[321] suggest that the ^{3}He–^{3}He interaction is weak. For example, the spin-symmetric Landau parameter, which enters S_c, is[321] $F_0^s \approx 0.20$. The spin-antisymmetric Landau parameter, which enters S_I, is [321] $F_0^a \approx 0.05$ to 0.10. Both S_c and S_I in Eq. (13.80) should then be well approximated by $S_0^*(Q, \omega)$, Eq. (13.67), for noninteracting fermions of mass $m_3^* = 2.4\, m_3$. Consistent with this picture, the neutron data show no evidence of a contribution from a ^{3}He–^{3}He interaction, but this may be masked by thermal broadening. Also, in this simple picture, S_0^* will violate the f-sum rule, Eq. (13.15), by a factor (m_3/m_3^*). This suggests that the ^{3}He-(p–r) interaction is energy dependent with a component $\propto \omega^2$. Tan and Woo[336] suggest that the ^{3}He-roton interaction is indeed energy dependent.

Around 1970, measurements of second[337] and fourth[338] sound, of ion mobilities,[339] and of the normal-fluid fraction[340] could not be explained using the L-P form for $\varepsilon_3(k)$ and a ^{4}He roton energy Δ unaffected by the presence of the ^{3}He. Brubaker et al.[337] proposed that $\varepsilon_3(k)$ dropped below the L-P form at high k (as if m_3^* increased with k). Initially, Esel'son et al.[338–340] proposed that Δ decreased with increasing x, but they later returned[341] to a departure of $\varepsilon_3(k)$ from the L-P form. The light scattering experiments of Woerner et al.[325] and of Surko and Slusher[329] showed no shift in Δ, even up to $x \approx 30\%$. This was confirmed by direct neutron scattering measurements of the full phonon–roton curve.[322,323] As a possible explanation, Pitaevskii[342] proposed that $\varepsilon_3(k)$ might have a small "roton-like" minimum at $k \approx 1.9\, \text{Å}^{-1}$ as indicated in the lower part of Fig. 34. This found some theoretical support.[343,344]

The neutron results[323] indicate that $\varepsilon_3(k)$ falls below the L-P value for $k \gtrsim 1.5\, \text{Å}^{-1}$ (Fig. 34) and that the qp-h band disappears as a well-defined excitation when it approaches the phonon–roton curve. The ^{3}He qp-h band would be expected to be repelled downward by the phonon–roton curve due to the interactions depicted in Figs. 35(a) and 35(c). The existing neutron data do not, however, establish whether or not this repulsion leads to a minimum in $\varepsilon_3(k)$. Greywall,[345] by fitting his high-precision measurements of the specific heat and of second sound, has obtained the $\varepsilon_3(k)$ marked Greywall in Fig. 34. This curve is consistent with the neutron results. Present thermodynamic data[345] therefore suggests a ^{3}He qp-h band, which lies below the L-P form for $k > 1$ Å^{-1}, but which has no minimum. Neutron data suggests the qp-h band becomes heavily broadened when it approaches the ^{4}He roton minimum. As yet, the distribution of intensity within the band is not well determined.

Bagchi and Ruvalds[346] and Ruvalds et al.[347] evaluated Fig. 35(c) in second-order perturbation theory with a vertex interaction deduced from the p-r curve in pure ^{4}He. They found a parabolic $\varepsilon_3(k)$ with $m_3^* \approx 2.2 m_3$. Bhatt[348] calculated the contribution to the ^{3}He self-energy, $\Sigma[k, \varepsilon_3(k)]$, due to Fig. 35(c) using a vertex fitted to reproduce the observed m_3^* at $k \to 0$ and reasonable assumptions about its momentum dependence. This yielded an $\varepsilon_3(k)$ that falls below the L-P curve as shown in Fig. 34. Bhatt obtained[348] a qp-h band that broadened and disappeared as a well-defined excitation for $k > 1.7$ Å^{-1}. The $\varepsilon_3(k)$ calculated by Götze et al.[349] is also shown in Fig. 34. More recent calculations by Lücke and Szprynger[350] find $\varepsilon_3(k) \propto k^2$ at low k but with deviations from quadratic dependence near the p-r curve. The calculations of Bhatt[348] and of Götze et al.[349] agree with experiment as well as can be expected, given the almost certain momentum and energy dependence[350] of the ^{3}He-(p-r) interactions.

13.5.4. ^{4}He Excitations

Rowe et al.[322] and Hilton et al.[323] have investigated $S(Q, \omega)$ for $0.8 \le Q \le 2.3$ Å^{-1} in the energy ($\hbar\omega$) range, where we expect the scattering to be predominantly from ^{4}He single phonons. In this Q and ω range, S_{44} and S_{34} of Eq. (13.80) contribute, probably roughly with equal intensity. At low x, S_{34} is expected to be dominated by scattering from ^{4}He, to contribute within the single p-r peak, and to distort the S_{44} peak shape.[31,323,350] However, since little was known about S_{34}, the observed intensity was interpreted as if from S_{44} alone and expressed as shifts (due to ^{3}He) of the p-r energy and lifetime from the values for pure ^{4}He.

The observed shift, $\delta\omega(Q)$, in p-r energy due to ^{3}He is small (see Fig. 36). At $Q \approx 1.9$ Å^{-1} $\delta\omega(Q)$ passes through zero. Below $Q \approx 1.9$ Å^{-1}, $\delta\omega(Q)$ is negative and above $Q \approx 1.9$ Å^{-1}, $\delta\omega(Q)$ is positive. This was explained by Hilton et al.[323] in terms of a shift in Q value due to the decreased density in the mixture, $Q_{\text{mix}} = Q_4 (n_4/n_{\text{mix}})^{1/3}$. At fixed Q, we are therefore sampling a different part of the p-r dispersion curve in the mixture than in pure ^{4}He. In an earlier experiment, Rowe et al.[322] found $\delta\omega(Q)$ small but positive everywhere.

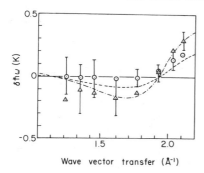

FIG. 36. Shifts in the ^{4}He phonon-roton energy $\delta\hbar\omega$ observed by Hilton et al.[323] at $T = 0.75$ K for ^{3}He-^{4}He mixtures containing 6% (dashed curve) and 12% (dot-dashed line) ^{3}He. The curves show the shifts predicted by shifting the wave vector (see text and Hilton et al.[323])

Beginning with pure ^{4}He, we expect deviations due to ^{3}He from three sources. First, the added ^{3}He dilutes the liquid and reduces its density. At lower density the interatomic forces are reduced. As the density is lowered in pure ^{4}He, for example, the maxon energy comes down and the roton energy increases while the roton wave vector decreases. At a wave vector $Q \approx 1.9$ Å^{-1} the single phonon energy is rather little changed[183] by changes in the density. By analogy, we expect $\omega(Q)$ to decrease for $Q \lesssim 1.9$ Å^{-1} as ^{3}He is added to pure ^{4}He, and to increase for $Q \gtrsim 1.9$ Å^{-1}. Lücke and Szprynger,[350] Hsu,[351] and Hsu et al.[159] have identified this as the largest factor affecting $\omega(Q)$, supporting the empirical relation proposed by Hilton et al.[323] Second, we expect that phonons can be annihilated by scattering from the ^{3}He, with the creation of ^{3}He particle-hole pairs. This is the same process that modifies the ^{3}He p-h energy band in Fig. 35(a). Its contribution to the p-r self-energy, shown in Fig. 35(d), is[159,351] ($q = Q$)

$$\Sigma(q, \omega_q) = nZ_q u_q^2 \chi(q, \omega_q). \tag{13.82}$$

Here u_q is the vertex describing the ^{3}He-(p-r) interaction, and $\chi(q, \omega_q)$ is the ^{3}He particle-hole susceptibility. The factor $Z_q u_q$ denotes the renormalized vertex, the large dot in Fig. 35(d). The real and imaginary parts of Eq. (13.82) are

$$\delta\omega_q = nZ_q u_q^2 \chi'(q, \omega_q) \tag{13.83a}$$

and

$$\Gamma_q = nZ_q u_q^2 \chi''(q, \omega_q), \tag{13.83b}$$

where, as in all treatments to date, u_q has been assumed to be real (which it almost certainly is not). Equation (13.83b) may also be interpreted as repulsion of the p-r curve due to hybridization with the ^{3}He qp-h excitation band. With a real u_q, Γ_q is zero everywhere except when q and ω_q lie in the ^{3}He p-h band. (The ^{3}He p-h band is defined by the region of q and ω space where $\chi'' \neq 0$.) The ω_q lie in the ^{3}He p-h band only near the roton minimum. Thus Ruvalds et al.,[347] who evaluated Eq. (13.83b), found a Γ_q that was sharply peaked near $q \approx q_R$ but zero elsewhere. Third, the phonons can simply scatter from the ^{3}He atoms. This process is depicted in Fig. 35(c). It contributes to damping at all q values and gives a shift $\delta\omega(q)$, which is expected to be comparable to that given by Eq. (13.83a).

In their original paper, Bartley et al.[352] evaluated Eq. (13.83a), i.e., the second contribution to $\delta\omega(q)$. They found $\delta\omega(q)$ positive for $q < q_c = 1.85$ Å^{-1} [$\omega(q)$ repelled upward above the ^{3}He p-h band] and $\delta\omega(q)$ negative for higher $q[\omega(q)$ repelled downward below the ^{3}He p-h band]. Here q_c is the value of q at which $\omega(q)$ crosses $\varepsilon_3(q)$. They did not evaluate the first contribution to $\delta\omega(q)$ and so their predictions disagreed with experiment in sign and magnitude. Bagchi and Ruvalds[346] evaluated Eq. (13.83a) and found an upper limit to its magnitude based on values of Γ_q obtained from Raman scattering. (They also evaluated the roton-^{3}He scattering contribution.) Ruvalds et al.[347] found Eq. (13.83a) had the same sign but a smaller magnitude than predicted by Bartley et al.[352]

More recently, Lücke and Szprynger[350] and Hsu[351] have also included the first contribution to the shift $\delta\omega_q$, i.e., that due to the density change caused by ^{3}He. For Eq. (13.83a) they find a sign in agreement with Bartley et al. but a much smaller magnitude. Since the first term dominates, the total $\delta\omega_q$ is negative for $q < 1.9$ Å^{-1} and positive for $q > 1.9$ Å^{-1}. The zero of $\delta\omega_q$ found by Hsu[351] at $q \approx 1.9$ Å^{-1} is a very satisfying feature of the polarization potential theory. This zero requires a zero in u_q. The u_q is an Aldrich-Pines pseudopotential[145] and the position of this zero is independent of reasonable variations of the parameters in the pseudopotential. Hsu[351] has added a very approximate evaluation of the third contribution to $\delta\omega_q$ bringing the final $\delta\omega_q$ closer to the observed values of Hilton et al.

The roton linewidth depends upon both x and T. This is expected since in the T and x ranges of the neutron studies the ratio $T/\varepsilon_F(x, T)$ is ≈ 1 and is changing rapidly with x and T. Thus, the shape of the ^{3}He qp-h band changes markedly between $0 \leq T \leq 1.5$ K and for $0 \leq x \leq 0.25$. Hsu has evaluated Eq. (13.83b) including its temperature dependence, but this gives generally too small an Γ. The damping of phonons and rotons due to scattering from ^{3}He atoms [Fig. 35(b)] is clearly important. Ruvalds has suggested that Γ_q due to Eq. (13.83b) will be zero unless $x > 2.5\%$ since, for $x < 2.5\%$, the phonon-roton curve does not intersect the ^{3}He qp-h band. A full calculation of Γ_q, including in particular the contribution from phonon-^{3}He scattering, remains an open problem.

Pedersen and Cowley[31] have calculated $S(Q, \omega)$ using the polarization potential theory and including the interference term $S_{34}(Q, \omega)$. They noted that the values of the PP interactions f_q^s and g_q^s in Eqs. (13.46) and (13.73) are approximately the same in pure ^{3}He and ^{4}He at the same density. They therefore assumed $f_q^{33} = f_q^{34} = f_q^{44}$ and $g_q^{33} = g_q^{34} = g_q^{44}$ for coherent scattering, with the interaction in the mixture depending only on the density. In this way they obtain a shift in the ^{4}He p-r energy with x that is small and in agreement with experiment. They find that the interference term $S_{34}(Q, \omega)$ is important in determining the p-r peak position and shape. They also predict substantial structure in the scattering intensity within the ^{3}He qp-h band.

Talbot and Griffin[353] have expressed the full $\chi(Q, \omega)$ in terms of its regular components and proper self-energies using the dielectric formalism discussed in Section 13.3.2. They prove in particular that, as conjectured initially by Franck[354] and now well known experimentally, the ^{3}He atoms do not contribute to superfluid flow in a mixture.

13.5.5. Future Prospects

Measurements of the ^{3}He qp-h excitations at very low temperature $[k_B T \ll \varepsilon_F(x)]$ to reduce thermal broadening would be most interesting. At low temperature the intensity within the band should be greater and the edges will be better defined. From the arguments given above, we do not expect the intensity within the band to be given by the free particle $S_{33}(Q, \omega)$. A measurement of this intensity will assist in determining the magnitude and frequency dependence of the ^{3}He-^{4}He and ^{3}He-^{3}He interactions. More precise measurements of the p-r line shapes as a function of x will assist in determining the form and magnitude of $S_{34}(Q, \omega)$. Precise values of the p-r line width away from the ^{3}He qp-h band would also provide important information on the decay processes beyond decay to single qp-h states. This is also a challenging area for improved calculations.

Finally, neutron scattering at high Q can determine the $\langle KE \rangle$ of the ^{3}He and ^{4}He atoms to test microscopic theories. Measurements of the $\langle KE \rangle$ in solid mixtures would be most interesting. Measurements of $n(p)$, while more difficult, will also be most rewarding in determining the interactions in ^{3}He-^{4}He mixtures. An additional challenge would be to determine the condensate fraction, n_0, for ^{3}He-^{4}He mixtures. As ^{3}He is added to liquid ^{4}He, lowering the density and hence decreasing the interatomic interactions that deplete the condensate, one would expect [355] n_0 to increase, at least initially. Exploring the pressure dependence of n_0 and the excitations would provide much new information on interactions.

References

1. F. London, "Superfluids," Vol. 2. Wiley, New York, 1954.
2. F. London, *Nature (London)* **141**, 643 (1938); *Phys. Rev.* **54**, 947 (1938).
3. L. Goldstein, H. S. Sommers, Jr., L. D. P. King, and C. J. Hoffman, *Proc. Int. Conf. Low Temp. Phys.* (R. Bowers, ed.), p. 88 (1951).

4. L. D. Landau, *J. Phys. USSR* **5**, 71 (1941).
5. L. D. Landau, *J. Phys. USSR* **11**, 91 (1947).
6. S G. Sydoriak, E. R. Grilly, and E. F. Hammel, *Phys. Rev.* **75**, 303 (1949).
7. L. D. Landau, *Zh. Eksp. Teor. Fiz.* **30**, 1058 (1956) [*Sov. Phys.—JETP (Engl. Transl.)* **3**, 920 (1957)].
8. L. D. Landau, *Zh. Eksp. Teor. Fiz.* **32**, 59 (1957) [*Sov. Phys.—JETP (Engl. Transl.)* **5**, 101 (1957)].
9. P. Nozières, "Theory of Interacting Fermi Systems." Benjamin, New York, 1964.
10. A. A. Abrikosov, L. P. Gorkov, and I. E. Dzyaloshinskii, "Quantum Field Theoretical Methods in Statistical Physics." Pergamon, Oxford, 1965.
11. D. D. Osheroff, R. C. Richardson, and D. M. Lee, *Phys. Rev. Lett.* **28**, 885 (1972); D. D. Osheroff, W. J. Gully, R. C. Richardson, and D. M. Lee, *Phys. Rev. Lett.* **29**, 920 (1972); A. J. Leggett, *Phys. Rev. Lett.* **29**, 1227 (1972).
12. R. Scherm, W. G. Stirling, A. D. B. Woods, R. A. Cowley, and G. J. Coombs, *J. Phys. C* **7**, L341 (1974).
13. D. G. Henshaw, *Phys. Rev.* **109**, 328 (1958).
14. M. Bitter, W. Gissler, and T. Springer, *Phys. Status Solidi* **23**, K155 (1967); F. P. Lipschultz, V. J. Minkiewicz, T. A. Kitchens, G. Shirane, and R. Nathans, *Phys. Rev. Lett.* **19**, 1307 (1967).
15. D. W. Osborne, B. M. Abraham, and B. Weinstock, *Phys. Rev.* **82**, 263 (1951).
16. A. Benoit, J. Bossy, J. Flouquet, and J. Schweizer, *J. Phys. Lett.* **46**, L923 (1985).
17. A. D. B. Woods and R. A. Cowley, *Rep. Prog. Phys.* **36**, 1135 (1973).
18. D. L. Price, in Ref. 19, Part II, p. 675.
19. "The Physics of Liquid and Solid Helium" (K. H. Bennemann and J. B. Ketterson, eds.), Part I. Wiley, New York, 1976; Part II, 1978.
20. J. Wilks, "The Properties of Liquid and Solid Helium." Oxford Univ. Press, London and New York, 1967.
21. W. E. Keller, "Helium-3 and Helium-4." Plenum, New York, 1969.
22. J. C. Wheatley, *Rev. Mod. Phys.* **47**, 415 (1975).
23. A. J. Leggett, *Rev. Mod. Phys.* **47**, 331 (1975).
24. D. Vollhardt, *Rev. Mod. Phys.* **56**, 99 (1984).
25. H. R. Glyde, *in* "Condensed Matter Research Using Neutrons" (S. W. Lovesey and R. Scherm, eds.), p. 95. Plenum, New York, 1984.
26. "75th Jubilee Conference on Helium-4" (J. G. M. Armitage, ed.). World Scientific, Singapore, 1983.
27. V. F. Sears, *J. Phys. C* **9**, 409 (1976).
28. W. G. Stirling, R. Scherm, P. A. Hilton, and R. A. Cowley, *J. Phys. C* **9**, 1643 (1976).
29. V. F. Turchin, "Slow Neutrons," p. 108. Isr. Program Sci. Transl., Jerusalem, 1965.
30. K. Sköld, C. A. Pelizzari, R. Kleb, and G. E. Ostrowski, *Phys. Rev. Lett.* **37**, 842 (1976).
31. K. S. Pedersen and R. A. Cowley, *J. Phys. C* **16**, 2671 (1983).
32. W. H. Keesom, "Helium," p. 180. Elsevier, Amsterdam, 1942.
33. C. Domb and J. S. Dugdale, *in* "Progress in Low Temperature Physics" (J. Gorter, ed.), Vol. 2, p. 338. North-Holland Publ., Amsterdam, 1957.
34. V. J. Minkiewicz, T. A. Kitchens, F. P. Lipschultz, R. Nathans, and G. Shirane, *Phys. Rev.* **174**, 267 (1968).
35. R. A. Reese, S. K. Sinha, T. O. Brun, and C. R. Tilford, *Phys. Rev. A* **3**, 1688 (1971); T. O. Brun, S. K. Sinha, C. A. Swenson, and C. R. Tilford, *in* "Neutron Inelastic Scattering," Vol. 1, p. 339. IAEA, Vienna, 1968.
36. E. B. Osgood, V. J. Minkiewicz, T. A. Kitchens, and G. Shirane, *Phys. Rev. A* **5**, 1537 (1972); T. A. Kitchens, G. Shirane, V. J. Minkiewicz, and E. B. Osgood, *Phys. Rev. Lett.*

29, 552 (1972); V. J. Minkiewicz, T. A. Kitchens, G. Shirane, and E. B. Osgood, *Phys. Rev. A* **8**, 1513 (1973).
37. J. G. Traylor, C. Stassis, R. A. Reese, and S. K. Sinha, *in* "Neutron Inelastic Scattering 1972," p. 129. IAEA, Vienna, 1972.
38. C. Stassis, G. Kline, W. Kamitakahara, and S. K. Sinha, *Phys. Rev. B* **17**, 1130 (1978).
39. J. Eckert, W. Thomlinson, and G. Shirane, *Phys. Rev. B* **16**, 1057 (1977).
40. W. Thomlinson, J. Eckert, and G. Shirane, *Phys. Rev. B* **18**, 1120 (1978).
41. R. O. Hilleke, P. Chaddah, R. O. Simmons, D. L. Price, and S. K. Sinha, *Phys. Rev. Lett.* **52**, 847 (1984); P. E. Sokol, R. O. Simmons, D. L. Price, and R. O. Hilleke, *Proc. Int. Conf. Low Temp. Phys., 17th* (U. Eckern, A. Schmid, W. Weber, and H. Wühl, eds.), Part II, p. 1213. North-Holland Publ., Amsterdam, 1984.
42. R. O. Simmons and P. E. Sokol, *in* "Neutron Scattering" (G. H. Lander and R. A. Robinson, eds.) p. 156. North-Holland Publ., Amsterdam, 1986. [Reprinted from *Physica B + C (Amsterdam)* **136B + C**, 156 (1986)].
43. L. H. Nosanow and G. L. Shaw, *Phys. Rev.* **128**, 546 (1962).
44. L. H. Nosanow, *Phys. Rev.* **146**, 120 (1966).
45. P. F. Choquard, "The Anharmonic Crystal." Benjamin, New York, 1967; T. R. Koehler, *Phys. Rev. Lett.* **17**, 89 (1966); H. Horner, *Z. Phys.* **205**, 72 (1967); N. Boccara and G. Sarma, *Physics (N.Y.)* **1**, 219 (1965).
46. R. A. Guyer, *Solid State Phys.* **23**, 413 (1969).
47. H. Horner, *in* "Dynamical Properties of Solids" (G. K. Horton and A. A. Maradudin, eds.), Vol. 1, p. 451. North-Holland Publ., Amsterdam, 1974.
48. T. R. Koehler, *in* "Dynamical Properties of Solids" (G. K. Horton and A. A. Maradudin, eds.), Vol. 2, p. 1. North-Holland Publ., Amsterdam, 1975.
49. H. R. Glyde, *in* "Rare Gas Solids" (M. L. Klein and J. A. Venables, eds.), Vol. 1, p. 382. Academic Press, New York, 1976.
50. C. M. Varma and N. R. Werthamer, in Ref. 19, Part I, p. 503.
51. H. Horner, *Phys. Rev. Lett.* **29**, 556 (1972).
52. N. R. Werthamer, *Phys. Rev. Lett.* **28**, 1102 (1972).
53. V. F. Sears and F. C. Khanna, *Phys. Rev. Lett.* **29**, 549 (1972).
54. H. R. Glyde, *Can. J. Phys.* **52**, 2281 (1974).
55. H. R. Glyde and S. I. Hernadi, *Phys. Rev. B* **25**, 4787 (1982).
56. W. M. Collins and H. R. Glyde, *Phys. Rev. B* **18**, 1132 (1978).
57. A. F. Andreev, *in* "Progress in Low Temperature Physics" (D. F. Brewer, ed.), Vol. 8, p. 67. North-Holland Publ., New York, 1982.
58. M. Roger, J. H. Hetherington, and J. M. Delrieu, *Rev. Mod. Phys.* **55**, 1 (1983).
59. F. A. Lindemann, *Phys. Z.* **11**, 609 (1911).
60. E. Feenberg, "Theory of Quantum Fluids." Academic Press, New York, 1969.
61. C.-W. Woo, in Ref. 19, Part I, p. 349.
62. K. A. Brueckner, *Phys. Rev.* **97**, 1353 (1955); **100**, 36 (1955); J. D. Jeukenne, A. Lejeune, and C. Mahaux, *Phys. Rep. C* **25**, 83 (1976).
63. E. Østgaard, *Phys. Rev.* **170**, 257 (1968); H. R. Glyde and S. I. Hernadi, *Phys. Rev. B* **28**, 141 (1983).
64. D. Lévesque, J.-J. Weis, and M. L. Klein, *Phys. Rev. Lett.* **15**, 670 (1983).
65. H. R. Glyde, *Can. J. Phys.* **49**, 761 (1971).
66. R. A. Cowley, *Rep. Prog. Phys.* **31**, 123 (1968); *Adv. Phys.* **12**, 421 (1963).
67. H. R. Glyde and M. L. Klein, *CRC Crit. Rev. Solid State Sci.* **2**, 181 (1971).
68. T. R. Koehler and N. R. Werthamer, *Phys. Rev. A* **3**, 2074 (1971).
69. T. R. Koehler and N. R. Werthamer, *Phys. Rev. A* **5**, 2230 (1972).
70. H. R. Glyde and F. C. Khanna, *Can. J. Phys.* **49**, 2997 (1971).

71. H. Horner, *J. Low Temp. Phys.* **8**, 511 (1972).
72. V. V. Goldman, G. K. Horton, and M. L. Klein, *Phys. Rev. Lett.* **24**, 1424 (1970).
73. H. R. Glyde and V. V. Goldman, *J. Low Temp. Phys.* **25**, 601 (1976).
74. T. R. Koehler, *Phys. Rev. Lett.* **18**, 654 (1967).
75. H. Horner, *Phys. Rev. A* **1**, 1713, 1722 (1970).
76. V. Ambegaokar, J. Conway, and G. Baym, in "Lattice Dynamics" (R. F. Wallis, ed.), p. 261. Pergamon, New York, 1965.
77. A. A. Maradudin and A. E. Fein, *Phys. Rev.* **128**, 2589 (1962).
78. V. K. Wong and H. Gould, *Ann. Phys. (N.Y.)* **83**, 252 (1974).
79. Ref. 61, p. 422.
80. V. V. Goldman, *J. Phys. Chem. Solids* **30**, 1019 (1969).
81. J. P. Hansen, *Phys. Rev.* **172**, 221 (1970).
82. L. K. Moleko and H. R. Glyde, *Phys. Rev. B* **30**, 4215 (1984).
83. H. J. Maris, *Rev. Mod. Phys.* **49**, 341 (1977).
84. A. J. Leggett and D. ter Haar, *Phys. Rev.* **139**, A779 (1965).
85. C. J. Pethick and D. ter Haar, *Physica (Amsterdam)* **32**, 1905 (1966).
86. R. A. Cowley and A. D. B. Woods, *Can. J. Phys.* **49**, 177 (1971).
87. H. Horner, *Proc. Int. Conf. Low Temp. Phys., 13th* (K. D. Timerhaus, W. J. O'Sullivan, and E. F. Hammel, eds.), Vol. II, p. 125. Plenum, New York, 1974.
88. P. C. Hohenberg and P. M. Platzman, *Phys. Rev.* **152**, 198 (1966).
89. P. A. Whitlock, D. M. Ceperley, G. V. Chester, and M. H. Kalos, *Phys. Rev. B* **19**, 5598 (1979).
90. P. A. Whitlock and R. M. Panoff, *Proc. 1984 Workshop High-Energy Excitations Condensed Matter* Los Alamos Nat. Lab., Publ. LA-10227-C, Vol. II, p. 430 (1984).
91. L. K. Moleko and H. R. Glyde, *Phys. Rev. Lett.* **54**, 901 (1985).
92. H. R. Glyde, *J. Low Temp. Phys.* **59**, 561 (1985).
93. B. Tanatar, G. C. Lefever, and H. R. Glyde, in Ref. 42, p. 187. [Reprinted from *Physica B+C (Amsterdam)* **136B+C**, 187 (1986).]
94. V. F. Sears, *Phys. Rev. B* **30**, 44 (1984).
95. V. F. Sears, *Phys. Rev.* **185**, 200 (1969).
96. J. J. Weinstein and J. W. Negele, *Phys. Rev. Lett.* **49**, 1016 (1982).
97. G. Reiter, in Ref. 90, Vol. II, p. 493.
98. H. A. Mook, *Phys. Rev. Lett.* **32**, 1167 (1974).
99. P. Martel, E. C. Svensson, A. D. B. Woods, V. F. Sears, and R. A. Cowley, *J. Low Temp. Phys.* **23**, 285 (1976).
100. V. F. Sears, E. C. Svensson, P. Martel, and A. D. B. Woods, *Phys. Rev. Lett.* **49**, 279 (1982).
101. H. A. Mook, *Phys. Rev. Lett.* **51**, 1454 (1983).
102. V. F. Sears, *Can. J. Phys.* **59**, 555 (1981).
103. R. M. Brugger, A. D. Taylor, C. E. Olsen, J. A. Goldstone, and A. K. Soper, *Nucl. Instrum. Methods* **221**, 393 (1984).
104. B. Tanatar, G. C. Lefever, and H. R. Glyde, *J. Low Temp. Phys.* **62**, 489 (1986).
105. P. E. Sokol, private communication.
106. T. R. Kirkpatrick, *Phys. Rev. B* **30**, 1266 (1984).
107. F. Family, *Phys. Rev. Lett.* **34**, 1374 (1975).
108. E. Talbot and A. Griffin, *Phys. Rev. B* **29**, 2531 (1984).
109. V. F. Sears, *Solid State Commun.* **11**, 1307 (1972).
110. M. Roger and H. R. Glyde, *Phys. Lett. A* **89A**, 252 (1982).
111. R. A. Aziz, V. P. S. Nain, J. S. Carley, W. L. Taylor, and G. T. McConville, *J. Chem. Phys.* **70**, 4330 (1979).

112. P. Kapitza, *Nature (London)* **141**, 74 (1938).
113. J. F. Allen and A. D. Misener, *Nature (London)* **141**, 75 (1938).
114. M. H. Kalos, D. Levesque, and L. Verlet, *Phys. Rev. A* **9**, 2178 (1974).
115. M. H. Kalos, M. A. Lee, P. A. Whitlock, and G. V. Chester, *Phys. Rev.* **B 24**, 115 (1981).
116. R. J. Donnelly, J. A. Donnelly, and R. N. Hills, *J. Low Temp. Phys.* **44**, 471 (1981).
117. N. N. Bogoliubov, *J. Phys. USSR* **11**, 23 (1947).
118. R. M. Brugger, *in* "Thermal Neutron Scattering" (P. A. Egelstaff, ed.), p. 54. Academic Press, London, 1965.
119. L. P. Pitaevskii, *Sov. Phys.—JETP (Engl. Transl.)* **9**, 830 (1959).
120. R. P. Feynman, *Phys. Rev.* **94**, 262 (1954).
121. R. P. Feynman and M. Cohen, *Phys. Rev.* **102**, 1189 (1956).
122. T. C. Padmore and G. V. Chester, *Phys. Rev. A* **9**, 1725 (1974).
123. H. W. Jackson and E. Feenberg, *Ann. Phys. (N.Y.)* **15**, 266 (1961); *Rev. Mod. Phys.* **34**, 686 (1962).
124. A. Bhattacharyya and C.-W. Woo, *Phys. Rev. Lett.* **28**, 1320 (1972).
125. A. Bhattacharyya and C.-W. Woo, *Phys. Rev. A* **7**, 204 (1973).
126. C. E. Campbell, *in* "Progress in Liquid Physics" (C. A. Croxton, ed.), p. 213. Wiley, New York, 1978.
127. A. Miller, D. Pines, and P. Nozières, *Phys. Rev.* **127**, 1452 (1962).
128. J. Ruvalds, *in* "Quantum Liquids" (J. Ruvalds and T. Regge, eds.), p. 263. North-Holland Publ., Amsterdam, 1978).
129. A. Zawadowski, in Ref. 128, p. 293.
130. A. Bijl, *Physica (Amsterdam)* **7**, 869 (1940).
131. R. P. Feynman, *in* "Progress in Low Temperature Physics" (C. J. Gorter, ed.), Vol. 1, P. 17. North-Holland Publ., Amsterdam, 1955.
132. G. V. Chester, *in* "The Helium Liquids" (J. G. M. Armitage and I. E. Farquhar, eds.), p. 1. Academic Press, London, 1975.
133. N. Hugenholtz and D. Pines, *Phys. Rev.* **116**, 489 (1959).
134. J. Gavoret and P. Nozières, *Ann. Phys. (N.Y.)* **28**, 349 (1964).
135. P. C. Hohenberg and P. C. Martin, *Ann. Phys. (N.Y.)* **34**, 291 (1965).
136. K. Huang and A. Klein, *Ann. Phys. (N.Y.)* **30**, 203 (1964).
137. S. K. Ma and C.-W. Woo, *Phys. Rev.* **159**, 165 (1967).
138. S. K. Ma, H. Gould, and V. K. Wong, *Phys. Rev. A* **3**, 1453 (1971).
139. I. Kondor and P. Szépfalusy, *Acta Phys. Hung.* **24**, 81 (1968).
140. H. Gould and V. K. Wong, *Phys. Rev. Lett.* **27**, 301 (1971).
141. T. H. Cheung and A. Griffin, *Phys. Rev. A* **4**, 237 (1971); A. Griffin, *Phys. Rev. A* **6**, 512 (1972).
142. A. D. B. Woods and E. C. Svensson, *Phys. Rev. Lett.* **41**, 974 (1978).
143. K. Bedell, D. Pines, and A. Zawadowski, *Phys. Rev. B* **29**, 102 (1984).
144. K. Bedell, D. Pines, and I. Fomin, *J. Low Temp. Phys.* **48**, 417 (1982).
145. C. H. Aldrich, III, Ph.D. Thesis, Univ. of Illinois at Urbana–Champaign, 1974.
146. C. H. Aldrich, III and D. Pines, *J. Low Temp. Phys.* **25**, 677 (1976).
147. C. H. Aldrich, III, C. J. Pethick, and D. Pines, *Phys. Rev. Lett.* **37**, 845 (1976); C. H. Aldrich, III and D. Pines, *J. Low Temp. Phys.* **32**, 689 (1978)
148. D. Pines, *in* "Highlights of Condensed Matter Theory," Course 87 of the Varenna Summer School, Soc. Italianna de Fisica, Bologna, Italy, p. 580 (1985).
149. A. Griffin and T. H. Cheung, *Phys. Rev. A* **7**, 2086 (1973).
150. V. K. Wong and H. Gould, *Phys. Rev. B* **14**, 3961 (1976); *Ann. Phys. (N.Y.)* **83**, 252 (1974).
151. P. Szépfalusy and I. Kondor, *Ann. Phys. (N.Y.)* **82**, 1 (1974).

152. V. K. Wong, *Phys. Lett. A* **61A**, 454 (1977).
153. E. Talbot and A. Griffin, *Ann. Phys. (N. Y.)* **151**, 71 (1983); *Phys. Rev. B* **29**, 3952 (1984).
154. A. Griffin and E. Talbot, *Phys. Rev. B* **24**, 5075 (1981).
155. A. Griffin, *Phys. Rev. B* **19**, 5946 (1979).
156. A. Griffin, *Phys. Rev. B* **30**, 5057 (1984).
157. D. L. Bartley and V. K. Wong, *Phys. Rev. B* **12**, 3775 (1975).
158. K. Bedell and D. Pines, *Phys. Rev. Lett.* **45**, 39 (1980).
159. W. Hsu, D. Pines, and C. H. Aldrich, III, *Phys. Rev. B* **32**, 7179 (1985).
160. L. D. Landau and I. M. Khalatnikov, *Zh. Eksp. Teor. Fiz.* **19**, 637 (1949).
161. See, e.g., J. Yau and M. J. Stephen, *Phys. Rev. Lett.* **27**, 482 (1971); I. Tüttö, *J. Low Temp. Phys.* **11**, 77 (1973).
162. E. Manousakis, V. R. Pandharipande, and Q. N. Usmani, *Phys. Rev. B* **31**, 7022 (1985).
163. E. Manousakis and V. R. Pandharipande, *Phys. Rev. B* **31**, 7029 (1985).
164. E. Manousakis and V. R. Pandharipande, *Phys. Rev. B* **33**, 150 (1986).
165. E. Manousakis, Ph.D. Thesis, Univ. of Illinois at Urbana-Champaign, 1985.
166. R. Golub, C. Jewell, P. Ageron, W. Mampe, B. Heckel, and I. Kilvington, *Z. Phys. B* **51**, 187 (1983).
167. N. Watanabe, in "Neutron Scattering in the 'Nineties," p. 279. IAEA, Vienna, 1985.
168. R. A. Cowley, *Proc. Conf. Neutron Scattering* (R. M. Moon, ed.), Vol. II, p. 935. CONF-760601-P2. Natl. Tech. Inf. Serv., Springfield, Virginia, 1976; also in Ref. 128, p. 27.
169. W. G. Stirling, *J. Phys., Colloq.* **39**(C6), 1334 (1978).
170. D. G. Henshaw and D. G. Hurst, *Phys. Rev.* **91**, 1222 (1953); D. G. Hurst and D. G. Henshaw, *Phys. Rev.* **100**, 994 (1955).
171. M. Cohen and R. P. Feynman, *Phys. Rev.* **107**, 13 (1957).
172. H. Palevsky, K. Otnes, K. E. Larsson, R. Pauli, and R. Stedman, *Phys. Rev.* **108**, 1346 (1957); H. Palevsky, K. Otnes, and K. E. Larsson, *Phys. Rev.* **112**, 11 (1958).
173. J. L. Yarnell, G. P. Arnold, P. J. Bendt, and E. C. Kerr, *Phys. Rev. Lett.* **1**, 9 (1958); *Phys. Rev.* **113**, 1379 (1959).
174. D. G. Henshaw, *Phys. Rev. Lett.* **1**, 127 (1958).
175. H. J. Maris and W. E. Massey, *Phys. Rev. Lett.* **25**, 220 (1970).
176. C. H. Aldrich, III, C. J. Pethick, and D. Pines, *J. Low Temp. Phys.* **25**, 691 (1976).
177. E. C. Svensson, P. Martel, and A. D. B. Woods, *Phys. Lett. A* **55A**, 151 (1975).
178. W. G. Stirling, J. R. D. Copley, and P. A. Hilton, in "Neutron Inelastic Scattering 1977," Vol. II, p. 45. IAEA, Vienna, 1978.
179. W. G. Stirling, in Ref. 26, p. 109.
180. T. R. Kirkpatrick, *Phys. Rev. B* **29**, 3966 (1984).
181. R. C. Dynes and V. Narayanamurti, *Phys. Rev. B* **12**, 1720 (1975).
182. D. G. Henshaw and A. D. B. Woods, *Proc. Int. Conf. Low Temp. Phys., 7th* (G. M. Graham and A. C. Hollis Hallett, eds.), p. 539. Univ. of Toronto Press, Toronto, 1961.
183. O. W. Dietrich, E. H. Graf, C. H. Huang, and L. Passell, *Phys. Rev. A* **5**, 1377 (1972).
184. A. D. B. Woods, E. C. Svensson, and P. Martel, *Phys. Lett. A* **43A**, 223 (1973).
185. A. D. B. Woods, P. Martel, and E. C. Svensson, in Ref. 178, p. 37.
186. E. H. Graf, V. J. Minkiewicz, H. Bjerrum Møller, and L. Passell, *Phys. Rev. A* **10**, 1748 (1974).
187. E C. Svensson, P. Martel, V. F. Sears, and A. D. B. Woods, *Can. J. Phys.* **54**, 2178 (1976).
188. E. C. Svensson, A. D. B. Woods, and P. Martel, *Phys. Rev. Lett.* **29**, 1148 (1972).
189. E. C. Svensson, W. G. Stirling, A. D. B. Woods, and P. Martel, *Proc. Conf. Neutron Scattering* (R. M. Moon, ed.), Vol. II, p. 1017. CONF-760601-P2. Natl. Tech. Inf. Serv., Springfield, Virginia, 1976.

190. A. J. Smith, R. A. Cowley, A. D. B. Woods, W. G. Stirling, and P. Martel, *J. Phys. C* **10**, 543 (1977).
191. W. G. Stirling, *Proc. Int. Conf. Phonon Phys., 2nd* (J. Kollár, N. Kroó, N. Menyhárd, and T. Siklós, eds.), p. 829. World Scientific, Singapore, 1985.
192. K.-E. Larsson and K. Otnes, *Ark. Fys.* **15**, 49 (1958).
193. A. D. B. Woods, E. C. Svensson, and P. Martel, *in* "Low Temperature Physics—LT14" (M. Krusius and M. Vuorio, eds.), Vol. 1, p. 187. North-Holland Publ., Amsterdam, 1975.
194. A. D. B. Woods, E. C. Svensson, and P. Martel, in Ref. 189, p. 1010.
195. A. D. B. Woods, E. C. Svensson, and P. Martel, *Can. J. Phys.* **56**, 302 (1978).
196. A. D. B. Woods, E. C. Svensson, and P. Martel, *Phys. Lett. A* **57A**, 439 (1976).
197. E. C. Svensson, R. Scherm, and A. D. B. Woods, *J. Phys. Colloq.* **39**(C6), 211 (1978).
198. B. A. Dasannacharya, A. Kollmar, and T. Springer, *Phys. Lett. A* **55A**, 337 (1976).
199. V. F. Sears, A. D. B. Woods, E. C. Svensson, and P. Martel, in Ref. 178, p. 23.
200. J. A. Tarvin and L. Passell, *Phys. Rev. B* **19**, 1458 (1979).
201. N. M. Blagoveshchenskii and E. B. Dokukin, *JETP Lett. (Engl. Transl.)* **28**, 363 (1978).
202. N. M. Blagoveshchenskii, E. B. Dokukin, Zh. A. Kozlov, and V. A. Parfenov, *JETP Lett. (Engl. Transl.)* **31**, 4 (1980).
203. F. Mezei, *Phys. Rev. Lett.* **44**, 1601 (1980).
204. F. Mezei and W. G. Stirling, in Ref. 26, p. 111.
205. K. S. Pedersen and K. Carneiro, *Phys. Rev. B* **22**, 191 (1980).
206. P. J. Bendt, R. D. Cowan, and J. L. Yarnell, *Phys. Rev.* **113**, 1386 (1959).
207. J. S. Brooks and R. J. Donnelly, *J. Phys. Chem. Ref. Data* **6**, 51 (1977).
208. U. M. Titulaer and J. M. Deutch, *Phys. Rev. A* **10**, 1345 (1974).
209. P. H. Roberts and R. J. Donnelly, *J. Low Temp. Phys.* **15**, 1 (1974); see also R. J. Donnelly and P. H. Roberts, *J. Low Temp. Phys.* **27**, 687 (1977).
210. T. J. Greytak and J. Yan, *Proc. Int. Conf. Low Temp. Phys., 12th, Kyoto, Jpn.* (E. Kanda, ed.), p. 89. Academic Publ., 1971.
211. J. W. Halley and R. Hastings, *Phys. Rev. B* **15**, 1404 (1977).
212. E. C. Svensson, V. F. Sears, and A. Griffin, *Phys. Rev. B* **23**, 4493 (1981).
213. E. C. Svensson and A. F. Murray, *Physica B + C (Amsterdam)* **108B+C**, 1317 (1981).
214. E. C. Svensson, V. F. Sears, A. D. B. Woods, and P. Martel, *Phys. Rev. B* **21**, 3638 (1980).
215. A. D. B. Woods and E. C. Svensson, unpublished work.
216. A. Griffin, *Phys. Lett. A* **71A**, 237 (1979).
217. C. A. Murray, R. L. Woerner, and T. J. Greytak, *J. Phys. C* **8**, L90 (1975).
218. A. D. B. Woods, P. A. Hilton, R. Scherm, and W. G. Stirling, *J. Phys. C* **10**, L45 (1977).
219. A. D. B. Woods, E. C. Svensson, and P. Martel, *in* "Neutron Inelastic Scattering 1972," p. 359. IAEA, Vienna, 1972.
220. N. M. Blagoveshchenskii, E. Dokukin, Zh. A. Kozlov, and V. A. Parfenov, *JETP Lett. (Engl. Transl.)* **30**, 12 (1979).
221. H. W. Jackson, *Phys. Rev. A* **4**, 2386 (1971); **8**, 1529 (1973).
222. F. Iwamato, K. Nagai, and K. Nojima, in Ref. 210, p. 189.
223. W. Götze and M. Lücke, *Phys. Rev. B* **13**, 3825 (1976).
224. E. C. Svensson, in Ref. 26, p. 10.
225. E. C. Svensson, in Ref. 90, Vol. II, p. 456.
226. R. A. Cowley and A. D. B. Woods, *Phys. Rev. Lett.* **21**, 787 (1968); A. D. B. Woods and R. A. Cowley, *in* "Neutron Inelastic Scattering," Vol. I, p. 609. IAEA, Vienna, 1968.
227. O. K. Harling, *Phys. Rev. Lett.* **24**, 1046 (1970); *Phys. Rev. A* **3**, 1073 (1971).
228. H. A. Mook, R. Scherm, and M. K. Wilkinson, *Phys. Rev. A* **6**, 2268 (1972).
229. L. Aleksandrov, V. A. Zagrebnov, Zh. A. Kozlov, V. A. Parfenov, and V. B. Priezzhev, *Zh. Eksp. Teor. Fiz.* **68**, 1825 (1975) [*Sov. Phys. JETP (Engl. Transl.)* **41**, 915 (1976)];

E. B. Dokukin, Zh. A. Kozlov, V. A. Parfenov, and A. V. Puchkov, *Pis'ma Zh. Eksp. Teor. Fiz.* **23**, 497 (1976) [*JETP Lett. (Engl. Transl.)* **23**, 453 (1976)]; *Zh. Eksp. Teor. Fiz.* **75**, 2273 (1978) [*Sov. Phys.—JETP (Engl. Transl.)* **48**, 1146 (1978)]; see also N. M. Blagoveshchenskii, I. V. Bogoyavlenskii, L. V. Karnatsevich, Zh. A. Kozlov, Yu. Ya. Milenko, V. A. Parfenov, and A. V. Puchkov, *Pis'ma Zh. Eksp. Teor. Fiz.* **37**, 152 (1983) [*JETP Lett. (Engl. Transl.)* **37**, 184 (1983)].
230. H. W. Jackson, *Phys. Rev. A* **10**, 278 (1974).
231. R. Feltgen, H. Pauly, F. Torello, and H. Vehmeyer, *Phys. Rev. Lett.* **30**, 820 (1973).
232. A. D. B. Woods and V. F. Sears, *Phys. Rev. Lett.* **39**, 415 (1977).
233. A. Griffin, *Phys. Rev. B* **32**, 3289 (1985); also in Ref. 42, p. 177. [Reprinted from *Physica B+C (Amsterdam)* **136B+C**, 177 (1986).]
234. V. F. Sears and E. C. Svensson, *Phys. Rev. Lett.* **43**, 2009 (1979).
235. V. F. Sears and E. C. Svensson, *Int. J. Quantum Chem., Quantum Chem. Symp.* **14**, 715 (1980).
236. V. F. Sears, E. C. Svensson, and A. F. Murray, unpublished work.
237. H. N. Robkoff, D. A. Ewen, and R. B. Hallock, *Phys. Rev. Lett.* **43**, 2006 (1979).
238. G. J. Hyland, G. Rowlands, and F. W. Cummings, *Phys. Lett. A* **31A**, 465 (1970); F. W. Cummings, G. J. Hyland, and G. Rowlands, *Phys. Kondens. Mater.* **12**, 90 (1970).
239. O. Penrose and L. Onsager, *Phys. Rev.* **104**, 576 (1956).
240. V. F. Sears, *Phys. Rev. B* **28**, 5109 (1983).
241. P. M. Lam and M. L. Ristig, *Phys. Rev. B* **20**, 1960 (1979).
242. M. Puoskari and A. Kallio, *Phys. Rev. B* **30**, 152 (1984).
243. L. J. Campbell, *Phys. Rev. B* **27**, 1913 (1983); *AIP Conf. Proc.* No. 103, p. 403 (1983).
244. M. Iino, M. Suzuki, and A. J. Ikushima, *J. Low Temp. Phys.* **61**, 155 (1985).
245. D. M. Ceperley and E. L. Pollock, *Phys. Rev. Lett.* **56**, 351 (1986).
246. V. F. Sears, *Can. J. Phys.* **63**, 68 (1985).
247. J. P. Toennies, *Faraday Discuss. Chem. Soc.* **55**, 129 (1973).
248. R. O. Hilleke, Ph.D. Thesis, Univ. of Illinois at Urbana–Champaign, 1983.
249. R. O. Simmons, in Ref. 90, Vol. II, p. 416.
250. N. Watanabe, in Ref. 90, Vol. I, p. 56.
251. B. D. Josephson, *Phys. Lett.* **21**, 608 (1966).
252. F. W. Wirth, D. A. Ewen, and R. B. Hallock, *Phys. Rev. B* **27**, 5530 (1983).
253. P. M. Platzman and N. Tzoar, *Phys. Rev. B* **30**, 6397 (1984).
254. G. Reiter and R. Silver, *Phys. Rev. Lett.* **54**, 1047 (1985).
255. G. Reiter and T. Becher, *Phys. Rev. B* **32**, 4492 (1985).
256. H. N. Robkoff and R. B. Hallock, *Phys. Rev B* **24**, 159 (1981).
257. R. B. Hallock, *Physica B+C (Amsterdam)* **109/110B+C**, 1557 (1982).
258. D. G. Henshaw, *Phys. Rev.* **119**, 9 (1960).
259. D. G. Henshaw, *Phys. Rev.* **119**, 14 (1960).
260. B. Mozer, L. A. De Graaf, and B. Le Neindre, *Phys. Rev. A* **9**, 448 (1974); see also R. D. Mountain and H. J. Raveché, *J. Res. Natl. Bur. Stand., Sect. A* **77A**, 725 (1973).
261. V. F. Sears, E. C. Svensson, A. D. B. Woods, and P. Martel, *At. Energy Can. Ltd.* [Rep.] **AECL-6779** (1979).
262. W. L. Gordon, C. H. Shaw, and J. G. Daunt, *J. Phys. Chem. Solids* **5**, 117 (1958).
263. R. B. Hallock, *Phys. Rev. A* **5**, 320 (1972).
264. R. B. Hallock, private communication.
265. F. W. Cummings, in "Statistical Mechanics" (S. A. Rice, K. F. Freed, and J. C. Light, eds.), p. 319. Univ. of Chicago Press, Chicago, Illinois, 1972.
266. F. W. Cummings, in "Cooperative Phenomena" (H. Haken and M. Wagner, eds.), p. 108. Springer-Verlag, Berlin and New York, 1973.

267. H. J. Raveché and R. D. Mountain, *Phys. Rev. A* **9**, 435 (1974).
268. A. Griffin, *Phys. Rev. B* **22**, 5193 (1980); G. V. Chester and L. Reatto, *Phys. Rev. B* **22**, 5199 (1980); A. L. Fetter, *Phys. Rev. B* **23**, 2425 (1981); F. W. Cummings, G. J. Hyland, and G. Rowlands, *Phys. Lett. A* **86A**, 370 (1981); H. B. Ghassib and R. Sridhar, *Phys. Lett. A* **100A**, 198 (1984).
269. E. C. Svensson and V. F. Sears, in "Frontiers of Neutron Scattering" (R. J. Birgeneau, D. E. Moncton, and A. Zeilinger, eds.), p. 126. North-Holland Publ., Amsterdam, 1986. [Reprinted from *Physica B + C (Amsterdam)* **137B + C**, 126 (1986).]
270. J. Maynard, *Phys. Rev. B* **14**, 3868 (1976).
271. D. S. Greywall and G. Ahlers, *Phys. Rev. A* **7**, 2145 (1973).
272. C. De Michelis, G. L. Masserini, and L. Reatto, *Phys. Lett. A* **66A**, 484 (1978); G. Gaglione, G. L. Masserini, and L. Reatto, *Phys. Rev. B* **23**, 1129 (1981); S. Battaini and L. Reatto, *Phys. Rev. B* **28**, 1263 (1983).
273. M. Puoskari, A. Kallio, and P. Pollari, *Phys. Scr.* **29**, 378 (1984); T. Chakraborty, A. Kallio, and M. Puoskari, *Phys. Rev. B* **33**, 635 (1986).
274. C. C. Chang and C. E. Campbell, *Phys. Rev. B* **15**, 4328 (1977).
275. H. N. Robkoff and R. B. Hallock, *Phys. Rev. B* **25**, 1572 (1982).
276. P. A. Hilton, R. A. Cowley, R. Scherm, and W. G. Stirling, *J. Phys. C* **13**, L295 (1980).
277. P. A. Hilton, R. A. Cowley, W. G. Stirling, and R. Scherm, *Z. Phys. B* **30**, 107 (1978).
278. K. Sköld and C. A. Pelizzari, in "Quantum Fluids and Solids" (S. B. Trickey, E. D. Adams, and J. W. Dufty, eds.), p. 195. Plenum, New York, 1977.
279. K. Sköld and C. A. Pelizzari, *J. Phys. C* **11**, L589 (1978).
280. K. Sköld and C. A. Pelizzari, *Philos. Trans. R. Soc. London, Ser. B* **290**, 605 (1980).
281. P. E. Sokol, K. Sköld, D. L. Price, and R. Kleb, *Phys. Rev. Lett.* **54**, 909 (1985).
282. H. A. Mook, *Phys. Rev. Lett.* **55**, 2452 (1985).
283. G. Baym and C. Pethick, in Ref. 19, Part II, p. 1.
284. D. S. Greywall, *Phys. Rev. B* **27**, 2747 (1983).
285. D. Pines, in "Quantum Fluids" (D. F. Brewster, ed.), p. 257. North-Holland Publ., Amsterdam, 1966.
286. D. Pines and P. Nozières, "The Theory of Quantum Liquids." Benjamin, New York, 1966.
287. H. R. Glyde and F. C. Khanna, *Can. J. Phys.* **58**, 343 (1980).
288. M. T. Béal-Monod, *J. Low Temp. Phys.* **37**, 123 (1979); *J. Magn. Mater.* **14**, 283 (1979).
289. S. Doniach and S. Engelsberg, *Phys. Rev. Lett.* **17**, 750 (1966); N. Berk and J. R. Schrieffer, *Phys. Rev. Lett.* **17**, 433 (1966); S. Doniach, *Proc. Phys. Soc., London* **91**, 86 (1967); W. F. Brinkman and S. Engelsberg, *Phys. Rev.* **169**, 417 (1968); see also Refs. 23 and 283.
290. J. Lindhard, *Mat.-Fys. Medd.—K. Dan. Vidensk. Selsk.* **28**(8), 1 (1954).
291. F. Green, D. Nielson, and J. Szymanski, *Phys. Rev. B* **31**, 2779, 2796 (1985).
292. L. van Hove, *Phys. Rev.* **95**, 249 (1954).
293. R. Jullien, M. T. Béal-Monod, and B. Coqblin, *Phys. Rev. B* **9**, 1441 (1977); F. C. Khanna and H. R. Glyde, *Can. J. Phys.* **54**, 648 (1976).
294. A. I. Akhiezer, I. A. Akhiezer, and I. Ya. Pomeranchuk, *Sov. Phys. JETP (Engl. Transl.)* **14**, 343 (1962).
295. H. R. Glyde and F. C. Khanna, *Phys. Rev. Lett.* **37**, 1692 (1976); *Can. J. Phys.* **55**, 1906 (1977).
296. A. Widom and J. L. Sigel, *Phys. Rev. Lett.* **24**, 1400 (1970); S. Babu and G. E. Brown, *Ann. Phys. (N.Y.)* **78**, 1 (1973).
297. K. Bedell and D. Pines, *Phys. Lett. A* **78A**, 281 (1980).
298. D. Pines, in "Recent Progress in Many-Body Theories" (J. G. Zabolitsky, M. de Llano, M. Fortes, and J. W. Clark, eds.), Lecture Notes in Physics, Vol. 142, p. 202. Springer-Verlag, Berlin and New York, 1981.

299. M. T. Béal-Monod and O. T. Valls, *J. Low Temp. Phys.* **56**, 585 (1984).
300. H. R. Glyde and S. I. Hernadi, *Phys. Rev. B* **29**, 4926 (1984).
301. H. Mori, *Prog. Theor. Phys.* **33**, 423 (1965); **34**, 399 (1965).
302. S. W. Lovesey, *J. Phys. C* **8**, 1649 (1975).
303. S. Takeno and F. Yoshida, *Prog. Theor. Phys.* **60**, 1585 (1978); F. Yoshida and S. Takeno, *Prog. Theor. Phys.* **62**, 37 (1979).
304. O. T. Valls, G. F. Mazenko, and H. Gould, *Phys. Rev. B* **18**, 263 (1978); O. T. Valls, H. Gould, and G. F. Mazenko, *Phys. Rev. B* **25**, 1663 (1982).
305. K. N. Pathak and M. Lücke, *Phys. Rev. B* **28**, 2468 (1983).
306. S. W. Lovesey and J. R. D. Copley, in Ref. 178, p. 3.
307. K. A. Brueckner and J. L. Gammel, *Phys. Rev.* **109**, 1040 (1958).
308. K. E. Schmidt, M. A. Lee, M. H. Kalos, and G. V. Chester, *Phys. Rev. Lett.* **47**, 807 (1981); M. A. Lee, K. E. Schmidt, M. H. Kalos, and G. V. Chester, *Phys. Rev. Lett.* **46**, 728 (1981).
309. E. Manousakis, S. Fantoni, V. R. Pandharipande, and Q. N. Usmani, *Phys. Rev. B* **28**, 3770 (1983).
310. C. Lhuillier and D. Lévesque, *Phys. Rev. B* **23**, 2203 (1981); D. Lévesque, *Phys. Rev. B* **21**, 5159 (1980).
311. E. Krotscheck, J. W. Clark, and A. D. Jackson, *Phys. Rev. B* **28**, 5088 (1983), and references therein.
312. R. M. Panoff, private communication.
313. T. R. Roberts, R. H. Sherman, and S. G. Sydoriak, *J. Res. Natl. Bur. Stand., Sect. A* **68A**, 567 (1964).
314. P. M. Lam, H. W. Jackson, M. L. Ristig, and J. W. Clark, *Phys. Lett. A* **58A**, 454 (1976); P. M. Lam, J. W. Clark, and M. L. Ristig, *Phys. Rev. B* **16**, 222 (1977).
315. B. L. Friman and E. Krotscheck, *Phys. Rev. Lett.* **49**, 1705 (1982).
316. E. Krotscheck and R. A. Smith, *Phys. Rev. B* **27**, 4222 (1983).
317. G. E. Brown, C. Pethick, and A. Zaringhalam, *J. Low Temp. Phys.* **48**, 349 (1982).
318. H. R. Glyde and S. I. Hernadi, *Phys. Rev. B* **28**, 141 (1983).
319. C. Ebner and D. O. Edwards, *Phys. Rep.* **2C**, 77 (1970).
320. I. M. Khalatnikov, "An introduction to the Theory of Superfluidity," Benjamin, New York, 1965; also in Ref. 19, Part I, p. 1.
321. G. Baym and C. Pethick, in Ref. 19, Part II, p. 123.
322. J. M. Rowe, D. L. Price, and G. E. Ostrowski, *Phys. Rev. Lett.* **31**, 510 (1973).
323. P. A. Hilton, R. Scherm, and W. G. Stirling, *J. Low Temp. Phys.* **27**, 851 (1977).
324. R. Scherm, W. G. Stirling, and P. A. Hilton, *J. Phys., Colloq.* **39**(C6), 198 (1978).
325. R. L. Woerner, D. A. Rockwell, and T. J. Greytak, *Phys. Rev. Lett.* **30**, 1114 (1973).
326. C. M. Surko and R. E. Slusher, *Phys. Rev. Lett.* **30**, 1111 (1973).
327. For review see M. J. Stephen, in Ref. 19, Part I, p. 307.
328. L. D. Landau and I. Ya. Pomeranchuk, *Dokl. Akad. Nauk. SSSR* **59**, 669 (1948); I. Ya. Pomeranchuk, *Zh. Eksp. Teor. Fiz.* **19**, 42 (1949).
329. G. Baym, *Phys. Rev. Lett.* **17**, 952 (1966).
330. J. Bardeen, G. Baym, and D. Pines, *Phys. Rev. Lett.* **17**, 372 (1966); *Phys. Rev.* **156**, 207 (1967).
331. W. E. Massey and C.-W. Woo, *Phys. Rev. Lett.* **19**, 301 (1967).
332. T. B. Davison and E. Feenberg, *Phys. Rev.* **178**, 306 (1969).
333. C.-W. Woo, H.-T. Tan, and W. E. Massey, *Phys. Rev. Lett.* **22**, 278 (1969); *Phys. Rev.* **185**, 287 (1969).
334. V. R. Pandharipande and N. Itoh, *Phys. Rev. A* **8**, 2564 (1973).
335. A. Fabrocini and A. Polls, *Phys. Rev. B* **30**, 1200 (1984).

336. H.-T. Tan and C.-W. Woo, *Phys. Rev. Lett.* **30**, (1973); see also Ref. 350.
337. N. R. Brubaker, D. O. Edwards, R. E. Sarwinski, P. Seligmann, and R. A. Sherlock, *Phys. Rev. Lett.* **25**, 715 (1970); *J. Low Temp. Phys.* **3**, 619 (1970).
338. N. E. Dyumin, B. N. Esel'son, E. Ya. Rudavskii, and I. A. Serbin, *Zh. Eksp. Teor. Fiz.* **56**, 747 (1969) [*Sov. Phys.—JETP (Engl. Transl.)* **29**, 46 (1969)].
339. B. N. Esel'son, Yu. Z. Kovdrya, and V. B. Shikin, *Zh. Eksp. Teor. Fiz.* **59**, 64 (1970) [*Sov. Phys.—JETP (Engl. Transl.)* **32**, 37 (1971)].
340. V. I. Sobolev and B. N. Esel'son, *Zh. Eksp. Teor. Fiz.* **60**, 240 (1971) [*Sov. Phys.—JETP (Engl. Transl.)* **33**, 132 (1971)].
341. B. N. Esel'son, V. A. Slyusarev, V. I. Sobolev, and M. A. Strzhemechnyi, *Pis'ma Zh. Eksp. Teor. Fiz.* **21**, 253 (1975) [*JETP Lett. (Engl. Transl.)* **21**, 115 (1975)]; V. I. Sobolev and B. N. Esel'son, *Pis'ma Zh. Eksp. Teor. Fiz.* **18**, 689 (1973) [*JETP Lett. (Engl. Transl.)* **18**, 403 (1973)].
342. L. P. Pitaevskii, *Proc. U.S.-Sov. Symp. Condensed Matter, Berkeley, Calif.*, 1973, unpublished.
343. C. M. Varma, *Phys. Lett. A* **45A**, 301 (1973).
344. M. J. Stephen and L. Mittag, *Phys. Rev. Lett.* **31**, 923 (1974).
345. D. S. Greywall, *Phys. Rev. Lett.* **41**, 177 (1978); *Phys. Rev. B* **20**, 2643 (1979).
346. A. Bagchi and J. Ruvalds, *Phys. Rev A* **8**, 1973 (1973).
347. J. Ruvalds, J. Slinkman, A. K. Rajagopal, and A. Bagchi, *Phys. Rev. B* **16**, 2047 (1977).
348. R. N. Bhatt, *Phys. Rev. B* **18**, 2108 (1978).
349. W. Götze, M. Lücke, and A Szprynger, *J. Phys., Colloq.* **39**(C6), 196 (1978); *Phys. Rev. B* **19**, 206 (1979).
350. M. Lücke and A. Szprynger, *Phys. Rev. B* **26**, 1374 (1982).
351. W.-C. Hsu, Ph.D. Thesis, Univ. of Illinois at Urbana-Champaign, 1984.
352. D. L. Bartley, J. E. Robinson, and V. K. Wong, *J. Low Temp. Phys.* **12**, 71 (1973); D. L. Bartley, V. K. Wong, and J. E. Robinson, *J. Low Temp. Phys.* **17**, 551 (1974).
353. E. Talbot and A. Griffin, *J. Low Temp. Phys.* **56**, 141 (1984).
354. J. Franck, *Phys. Rev.* **70**, 983 (1946).
355. W.-K. Lee and B. Goodman, *Phys. Rev. B* **24**, 2515 (1981).

14. CLASSICAL FLUIDS

Peter A. Egelstaff

Department of Physics
University of Guelph
Guelph, Ontario, Canada

14.1. Introduction to Atomic and Simple Molecular Fluids

14.1.1. Why Study Fluids with Neutrons?

The division of condensed matter into three states is readily understood if the interaction between two molecules is separated into short-ranged repulsive and long-ranged attractive parts and it is *assumed* that the sum of pair interactions is a good approximation to the total interaction energy in the dense matter. Van der Waals' famous equation enshrined these ideas, i.e.,

$$P = P_0 + a\rho^2, \tag{14.1}$$

where P and P_0 are, respectively, the pressures of the real fluid and a hypothetical fluid composed of molecules with repulsive interactions only, a is related to the area of the attractive part of the interaction, and ρ is the number density. In the years since van der Waals, the pair theory of fluids[1] has been developed in great detail, and the validity of the idea that the sum of pair interactions is a first approximation to the total was fully established. This idea may, of course, be based on the alternative approximation that the molecules (and atoms) are undeformable. Corrections to the latter approximation have been discussed for many years and theoretical results for the long-range corrections are widely used, but so far the short-range corrections are poorly understood. It is recognized, however, that the latter are more important than the former in many cases.

Two obstacles once stood in the way of progress in the understanding of fluids, but in recent years both have been removed. They were the need for an accurate knowledge of pair potentials for simple cases (expecially the noble gases) and for reliable methods for calculating the predictions of the pair theory of fluids. High-quality fitting procedures to fit the pair theory to *low-density* experimental data, also of high quality, removed the first obstacle for noble gas fluids, and modern computer simulation techniques removed the second. For other cases theoretical calculations, adjusted to fit bulk fluid data or solid-state data, have provided a first approximation for the pseudopotentials of liquid metals and for the simpler intermolecular potentials. Thus, today excellent first approximations to the microscopic properties of simple fluids (e.g., noble gases, liquid metals, and simple molecular fluids) may be obtained if the appropriate input is used and the correct theoretical procedures are followed.

For the first time, some of the serious limitations to the pair approximation for noble gases may be observed clearly; for example, in Fig. 1 the predicted and measured pressures of krypton as a function of density at room temperature[2] are plotted. For densities comparable to or larger than the critical density ($\sim 6 \times 10^{27}$ atoms/m^3 in this case) there are discrepancies of $\sim 30\%$. It may be shown that these are related to higher order modifications to the interactions at an interatomic distance comparable to the diameter of a krypton atom. The proper investigation of problems of this kind requires high-quality microscopic experimental data for many states, including both static and dynamic data. Comparison of such data with the predictions of

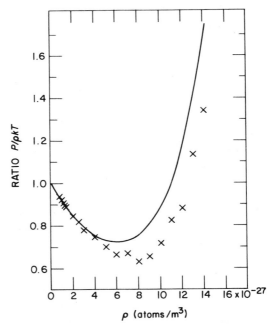

FIG. 1. Experimental pressures (in units of $\rho k_B T$) for krypton at room temperature as a function of density ρ (full line). The crosses are computer simulation calculations using the pair theory of fluids and an accurate pair potential. The differences are due to higher order interactions. (Redrawn from Ram and Egelstaff.[2])

modern computer simulations and modern theories[†] may provide a new physical insight into these problems and stimulate the building of new models or theories, which will account correctly and simultaneously for many discrepancies such as the one shown in Fig. 1. Since many-body potential effects of 10 to 20% are typical, data good to ~1% are desirable, which will test these models to about 10%.

The study of fluids covers both liquids and dense gases, which form one continuous state of matter since they are not separated by a phase boundary. By considering liquids and dense gases together, the discussion is spread over a wide range of state variables—pressure, volume, and temperature (P, V, T)—and extends over regions where the relative importance of the repulsive and attractive parts of the interaction varies considerably. At each P, V, T point a precise (~1%) determination of the static and dynamic pair correlation functions may be accomplished through radiation scattering experiments; the overall quality should be as good as or better than the pair theory predictions. It would be desirable to learn something of the triplet and higher correlation functions as well, but experimentally only integrals over these functions are available by measuring the thermodynamic derivatives of the pair functions. Consequently these data as a function of P, $V\,T$ should allow good derivatives to be determined at some points.

While these remarks apply to atomic fluids, much of their content applies also to simple molecular fluids (composed of, e.g., diatomic molecules). However, there are two complications. First, the intermolecular forces are both position and orientation dependent, so that measured properties such as $S(Q)$ are averaged over molecular orientations. Second, the pair interaction is less well known, and the low-density neutron scattering data are often employed to improve or test models of the pair interaction. The structure of the fluid at high density may differ from that at low density due to the influence of molecular shape on the packing of nearest neighbors. The study of molecular fluids is a complicated but rich field.

The choice of radiation for these programs is limited to neutrons or electromagnetic radiation. For static measurements (i.e., $d\sigma/d\Omega$), the theoretically attainable precision for nuclear correlation functions is about an order of magnitude better for neutrons than for x rays,[26] and so they are the obvious choice except for special cases. For dynamic measurements (i.e., $d^2\sigma/d\Omega\,d\omega$), a wide range of momentum transfer, with corresponding energy transfers, is attainable only with neutrons, since present techniques with electromagnetic radiation allow suitable energy transfers to be measured

---

[†] This objective should be contrasted with the objective of many recent theoretical papers, which is to test methods for pair theory calculations against the corresponding computer simulation.

only for long wavelengths and hence low momenta ($< 10^{-3}$ Å^{-1}). Moreover, electromagnetic radiation of all kinds is scattered by the electrons, whereas it is the center-of-mass (i.e., nuclear) correlations that are needed theoretically. Thus the experimentalist has almost no choice: in order to make the required measurements, neutron diffraction and inelastic scattering must be used in almost every situation.

Unfortunately, neutron fluxes are much lower, and less readily available, than fluxes of electromagnetic radiation. The consequent practical difficulties have prevented the inherent quality of neutron experiments from being exploited. Most of the published neutron experiments on fluids are of low quality relative to the ideal situation discussed earlier. Thus it is desirable to discuss the experimental and data reduction techniques for neutron measurements on fluids in some detail, and this is the purpose of the present chapter. If sufficient time, effort, and understanding are devoted to this field, the flux handicap can be reduced and the quality of the results improved.

While this chapter is concerned with simple fluids only, many of the arguments will apply also to more complicated systems, including solutions and alloys, as the need for higher precision is felt in the future. Ionic solutions are treated in detail in Chapter 15 by Enderby and Gullidge.

14.1.2. Preparations for S(Q) Work

In Fig. 2(b) two calculated structure factors $S(Q)$ for krypton at 297 K and $\rho = 6 \times 10^{27}$ atoms/m^3 are shown for the potentials drawn in Fig. 2(a). The region $0 < Q < Q_m$, where Q_m is the position of the principal peak of $S(Q)$, is sensitive to the attractive part of the potential and to the many-body interactions. The magnitude of the "forward" peak as $Q \to 0$ depends upon the compressibility and is therefore large in compressible gases but very small and difficult to measure properly for incompressible liquids. In contrast, the principal peak is much larger and easier to measure in liquids than gases. The region $Q > Q_m$ depends mainly upon the short-range (largely repulsive) part of the pair interaction, the amplitude of the oscillations depending on its steepness, and their period depending on the atomic diameter σ. A change in σ (when $r = \sigma$, $u(r) = 0$) will change the period of the oscillations in $S(Q)$. All the above points are well known and may be demonstrated by calculations as a function of density and of the shape of the pair potential, which will not be reviewed here. However, Fig. 2(b) illustrates the points that substantial changes in the potential may produce smaller relative changes in $S(Q)$.

In the case of molecular fluids similar remarks apply overall, but the fluid structure factor is modified by the single-molecule structure factor (i.e., the transform of the internal structure of the molecule). This has two effects: the intensity shows a general decrease from low Q to $Q \sim \pi/d$ (where d is

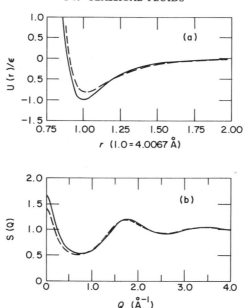

FIG. 2. (a) An accurate pair potential for krypton (full line) compared to the Lennard–Jones potential (dashed line). (b) Calculated structure factor $S(Q)$ for krypton at room temperature and $\rho = 6 \times 10^{27}$ atoms/m³ and the two pair potentials used for Fig. 2(a). (Redrawn from Ram and Egelstaff.[2])

an intramolecular distance), and superimposed on the oscillations of Fig. 2(b) are additional oscillations of wavelength $2\pi/d$. In cases where the internal structure of the molecule is of interest, the latter oscillations must be measured carefully, but when the liquid- or gas-like structure is required, the intramolecular structure factor is subtracted and the data reduced to the form shown in Fig. 2(b).

If an experimental test of the pair potential is required, it is necessary to study the low-density gas with the highest possible precision. But if the interpretation of the structure of a particular liquid is being investigated, several other approaches may be considered. For example, if a monatomic liquid or dense gas (e.g., krypton) is to be studied and the pair potential is well known, the task reduces to a study of the structural contribution of many-body forces. In this connection liquids and dense gases are more suitable samples than solids, as the distance between nuclei is a continuous variable (rather than confined to discrete lattice distances), and the density may be varied relatively easily. However, very precise measurements of $S(Q)$ are again required (i.e., ~1% or better). A different example is provided by liquid mixtures or compounds for which the structure is poorly understood, or perhaps for which almost no data exists. In such cases a crude pair correlation function, from which approximate coordination numbers and internuclear distances may be deduced, will be very useful. As the broader questions about this type of sample are answered, the structural questions become more sophisticated, so that the precision demanded of the experimentalist is increased. Thus at a subsequent stage in the investigation (particularly when isotopic substitution techniques are used), the precision needed will approach 1% again for meaningful comparisons with theoretical and other results.

For some of these experiments, the requirement is for 1% absolute precision, while for others it is sufficient to obtain a 1% measurement for one sample relative to another or to observe a difference between samples to about 1%. In all cases, however, it is necessary that the 1% figure apply to all values of Q measured. That is, the data must be correct to 1% (absolute or relative as required) as Q is varied from low to high values. If this requirement is met, a measurement relative to a suitable standard may be converted to an absolute result by an absolute normalization at one point only.

Any measurement to 1% precision is difficult, and in this case (see Fig. 2), the absolute value of a smoothly varying function over a wide range of the variable Q is required, so that the problems are especially difficult. These five observations are worth considering at the beginning of an experiment:

(1) Human mistakes are the biggest problem: forward planning and *design* can ensure that all features of a complicated apparatus work simultaneously and properly.

(2) Regular repetition of runs is necessary to check for flaws and to test for stable performance.

(3) Doubtful runs ruin accurate work, so they should be thrown away and these runs repeated.

(4) Standardization, background, calibration, and alignment measurements need to be interspersed by the instrument *user* between the data collection runs.

(5) Continuous thought during an experiment is necessary to minimize errors or to force them to cancel one another.

Although these matters are well understood by most experimentalists, the time required to implement them is rarely available. On central high-flux facilities the restrictions on instrument time for each user usually allows time for item (1) only, thus limiting the quality of the results. In contrast, the local lower flux facilities available to the same user may become more competitive if time is available to spend on items (2) to (5) as well as (1). It is clear that whatever the flux or the sophistication of the instrument, accurate and reliable measurements cannot be compressed into a very short time.

It is important to optimize a neutron diffractometer to give the highest counting rate consistent with the desired resolution (usually the broadest resolution function acceptable). The shape of $S(Q)$ suggests that the resolution ΔQ in Q may be wider at high Q than at low Q, and these quantities can be adjusted separately. The contribution to ΔQ from the spread in neutron wavelengths $\Delta\lambda$ in the incident beam is $\Delta Q = Q(\Delta\lambda/\lambda)$. For a conventional two-axis neutron diffractometer on a reactor (see Part A, Chapter 3), this is the main contribution to ΔQ at high Q. Thus if the subscript "H" denotes the high Q limit, then

$$(\Delta Q)_H \simeq Q_H \Delta\lambda/\lambda = 2 dQ_H \Delta\theta/\lambda, \qquad (14.2)$$

where the monochromator crystal spacing and Bragg angle are denoted by d and θ, respectively (and $\cos\theta \simeq 1$). If $\Delta\phi$ is the spread in scattering angle ϕ (from all sources including that shown in Fig. 12), it gives a contribution $(Q\cot\phi/2)\Delta\phi/2$ to ΔQ. Because of the behavior of $\cot\phi/2$, this term is important at low angles, and if $\Delta\phi \gg \phi d\Delta\theta$, the term proportional to $\Delta\lambda$ may be neglected, so that

$$(\Delta Q)_L \simeq Q_L \Delta\phi/\phi = 2\pi \Delta\phi/\lambda, \qquad (14.3)$$

where the subscript "L" denotes the low limit. Frequently Eqs. (14.2) and (14.3) may be adjusted independently and the ratio of these limits is

$$\frac{(\Delta Q)_H}{(\Delta Q)_L} = \frac{dQ_H}{\pi}\frac{\Delta\theta}{\Delta\phi}, \qquad (14.4)$$

which is usually ~ 3.

Once $\Delta\theta$ and $\Delta\phi$ are selected, the counting rate may be increased with the size of the instrument, since the rate is proportional to $A\Delta\alpha_i \Delta\theta \Delta\phi \Delta\alpha_f$ where A is the sample area and α_i and α_f are the ingoing and outgoing azimuthal angles. These quantities and the distances from source to sample to detector should be increased until the whole of a large source area is viewed and until the background from a very large area detector becomes significant, while keeping $\Delta\theta$ and $\Delta\phi$ fixed. To obtain this advantage, a large sample is required, and in many cases this is available. Similar optimization arguments may be applied to diffractometers that use other variables, but the basic conclusion—that for large sources, after fixing the resolution, the larger the instrument the higher the counting rate—is unchanged.

14.1.3. Preparations for $S(Q, \omega)$ Work

Many of the ideas discussed in the previous section apply to measurements of the inelastic scattering function $S(Q, \omega)$ as well. The resolution requirements are more complex, however, as resolution in Q and ω space must be considered, and simultaneous optimisation in Q and ω space is needed to maximize the counting rate. Since all structural effects are controlled by the pair distribution function $g(r)$ and hence the structure factor $S(Q)$, the Q-space resolution arguments are essentially the same as for $S(Q)$. Except for $Q < 0.1$ Å^{-1} for noble gas fluids and $Q < 1$ Å for liquid metals, the shape of $S(Q, \omega)$ is a single-peaked, bell-shaped curve when plotted as a function of ω at fixed Q. The full width in ω at half-maximum (FWHM) of $S(Q, \omega)$, will vary with Q in the way shown in Fig. 3 for a dense classical fluid. It is seen that FWHM for the total scattering function fluctuates above or about the FWHM for the self-scattering function. It is sufficient for many purposes, therefore, to discuss resolution requirements on the basis of the self-function. At high Q, the width of $S_s(Q, \omega)$ varies approximately as

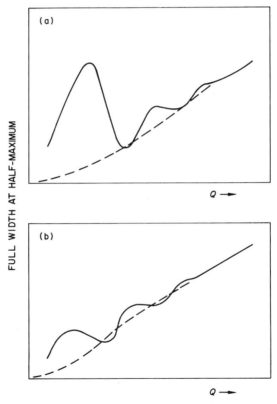

FIG. 3. Diagrams showing the (typical) variation of the full width at half-maximum (FWHM) of $S(Q, \omega)$ with ω. The full line corresponds to $S(Q, \omega)$ data, while the dashed line corresponds to $S_s(Q, \omega)$; (a) for an incompressible liquid such as lead and (b) for a compressible liquid such as argon.

$\Delta\omega \simeq 2Q\sqrt{kT/M}$ and becomes smaller at low Q, where $\Delta\omega \simeq 2DQ^2$ and where D is the diffusion coefficient. The change from one region to the other occurs in the neighborhood of the principal maximum of $S(Q)$ for dense liquids and at lower Q for lower densities.

It is difficult, although desirable, to design an instrument that has different energy resolutions at high and low Q. Usually this is done by a mechanical rearrangement such as changing Soller slits and monochromators on a crystal spectrometer or by changing flight paths and pulse and gate widths on a time-of-flight machine. The counting rate for a time-of-flight spectrometer with long flight paths may be maximized for a given resolution by reducing the pulse and gate widths to allow shorter flight paths and thus increase the detector solid angle to the limits set by the Q resolution. Similar optimization adjustments may sometimes be made for a three-axis crystal spectrometer. For example, if the incident and scattered wavelengths (λ_0 and λ_1, respectively) are approximately equal and $\phi \approx 90°$, then $(\Delta\omega)^2$ or $(\Delta Q)^2 \propto (\Delta\lambda_1)^2 + (\Delta\lambda_0)^2 \simeq$ constant, and the intensity $\sim (\Delta\lambda_0 \Delta\lambda_1)^2$. Thus the optimum intensity occurs for $\Delta\lambda_1 \approx \Delta\lambda_0$ in this case.

14.1.4. Comment on Theoretical Formulas

The theoretical formulas for an isotropic, homogeneous fluid in thermal equilibrium with its surroundings are[†]

$$\frac{d^2\sigma}{d\Omega\,d\omega} = b^2 \frac{k}{k_0} S(Q, \omega), \tag{14.5}$$

and

$$\frac{d\sigma}{d\Omega} = b^2 \int_{-\infty}^{E_0/\hbar} \frac{k}{k_0} S(Q_\omega, \omega)\,d\omega, \tag{14.6}$$

where

$$\frac{\hbar^2 Q_\omega^2}{2m} = 2E_0 - \hbar\omega - 2\cos\phi\sqrt{E_0(E_0 - \hbar\omega)}, \tag{14.7}$$

and

$$\frac{k}{k_0} = \left[1 - \frac{\hbar\omega}{E_0}\right]^{1/2}.$$

Equations (14.5)–(14.6) are sometimes referred to as formulae for the "no-scattering limit," because they are derived on the basis of single scattering

---

[†] In the present chapter **k** denotes a wave vector of the scattered neutron, denoted by $\mathbf{k}_1$ in Chapter 1, Part A.

without attenuation, that is, for a negligible sample. In addition the negligible sample is freely suspended in empty space, so that it evaporates. For a reasonable amount of sample in a container, several corrections need to be applied to the intensity in order to obtain the no-scattering limit, and these will be discussed in Sections 14.3 and 14.4. In addition, for fluids an accurate absolute scale is important when using Eqs. (14.5) and (14.6), and thus normalization techniques will be described also. Finally, these expressions assume detectors of uniform efficiency and very sharp resolution functions.

14.2. Data Collection

Several instruments have been described in Part A, Chapter 3, and here the use of diffractometers and inelastic scattering spectrometers for the study of fluids will be described. The discussion will be kept on a general basis and so be applicable to many different instruments. In order to complete a successful experiment yielding reliable absolute data, the user should measure the performance of the instrument at the outset.

14.2.1. Initial Measurements

The background counting rate of a detector arises from several origins. First, there is a background observed when the neutron source is switched off, due to radioactivity in the detector, noise in the electronic equipment, power fluctuations, etc. It should always be kept as a small component of the total background and should be checked each time the source is turned off: let this component be B_1. Second, there is "room background" from neutrons leaking through the shield around the source and neutrons escaping from a neighbor's apparatus. This background may be measured when the neutron source is on and the beam shutter for the instrument is closed. If this measurement is B_2, the room component is $B_2 - B_1$, and again this should be a small part of the total. Third, there is background from the neutron beam used by the instrument. Usually the beam striking the sample is defined by a cadmium diaphragm, and if this diaphragm is covered by 1-mm thick cadmium sheet, another background B_3 is measured. This background should be compared to two measurements: first, the background B_4 measured with the apparatus vacant (i.e., with no sample and no air in the beam), and second, the background B_5 measured with a cadmium sample (or with cadmium sheet wrapped around the real sample or with another suitable absorber). In a well-designed instrument $B_3 \simeq B_4 \simeq B_5$, that is, the beam background does not arise from neutrons scattered in the neighborhood of the sample. If there is a difference between these quantities, it is better to select B_5 as the background for use later, but, since it will be inconvenient to repeat this measurement on a frequent basis, the stability of the background should be determined by repeating measurements of B_3 or B_4 and perhaps of B_2. The difference $B_4 - B_5$ is due to neutrons passing through the sample and must be modified by attenuation factors described later.

All diffractometers record data as a function of a variable parameter, either angle-of-scatter or time-of-flight. In the former case, the angle ϕ_0 corresponding to the beam direction and the wavelength λ are treated as unknowns, while in the latter case the angle of scatter ϕ and a wavelength offset λ_0 are treated as unknowns. By making measurements of the position (ϕ_p or λ_p) of the Bragg peaks from a number of powdered samples with widely different d spacings (e.g., Cu, Pb, NaI, etc.), the two unknown parameters may be determined accurately. For angle scanning

$$\lambda = 2d\left[\sin\left(\frac{\phi_p}{2}\right)\cos\left(\frac{\phi_0}{2}\right) - \cos\left(\frac{\phi_p}{2}\right)\sin\left(\frac{\phi_0}{2}\right)\right], \quad (14.8)$$

and for time scanning

$$\lambda_p - \lambda_0 = 2d\sin(\phi/2). \quad (14.9)$$

These two equations for many d's are solved to find the two unknown quantities and their uncertainties. Repetition of this experiment at intervals provides one check of the stability of the instrument. For good measurements of this kind, ΔQ should be small, and this requirement may occasionally conflict with the larger ΔQ required by the optimization procedure of Section 14.1.2. In such cases ΔQ should be altered between the two measurements. Also, for pulsed sources λ_0 may depend on wavelength λ if the effective origin in the moderator varies with λ.

Additional calibrations are required for inelastic scattering spectrometers, which vary from instrument to instrument. For example, in the case of a multiangle, time-of-flight instrument, it is useful to measure the relative detector distances and geometries by using a small vanadium sample (e.g., a 5-mm diameter tube scattering 2% of the beam) placed at the sample center position. The flight time for elastically scattered neutrons to each detector and monitor is determined and compared. Then the same experiment is repeated with increasing diameter tubes up to the sample diameter and differences in these data used to determine geometrical factors influencing the time distribution, such as the different times-of-flight for scattering from different parts of the sample. In the case of a three-axis spectrometer, the resolution function of the analyzer is required and it may have an asymetrical shape (Part A, Chapter 3). The second-order contamination may be measured, and in the "constant Q" mode the spread in exact Q values may be added to the resolution function.

In many experiments on fluids a wide beam is used, and in order to determine precise cross sections, it is necessary to measure the beam profile and the position of the sample center relative to the profile. For most instruments this is accomplished readily by installing a tensioned nylon wire (1-mm diameter) at the sample position and traversing it across the beam. Counting rates on each detector (or detector position) are recorded as a function of position of the wire in the beam. Both vertical and horizontal traverses are made, and if necessary sample or diaphragm position adjustments are made. Because this method is quantitative, it provides data that may be used later in the correction programs (e.g., attenuation or multiple scattering programs). In this respect it may be contrasted to the simpler (nonquantitative) method of taking neutron photographs with a scintillator and a camera. This method is excellent for an initial setup, but is so nonlinear that good beam profiles cannot be obtained.

14.2.2. Vanadium Measurements

Spin-dependent nuclear scattering is the most convenient process to use for calibration purposes since, in the case of heavy target nuclei, the scattering is nearly elastic and isotropic. Of naturally occurring elements, vanadium is the most convenient, since it is almost monoisotopic (^{51}V), and the two scattering amplitudes for ^{51}V are of opposite sign and nearly equal amplitude. Thus the scattering by vanadium is almost all incoherent—and may be made completely so by a small, random alloying† addition. In special cases (e.g., gas samples) other calibration materials may be useful. For example, a low-pressure sample of ^{36}Ar (see Table I) may be employed if its state is well determined.

A vanadium metal foil of 0.25-mm thickness will scatter 1% of a neutron beam. A fairly compact sample (~3-mm thickness) covering part of the beam area only may be used if the beam profile is well known and the data correction programs include the effects of sample and counter geometry, beam profile, and beam divergence. Unfortunately, the most commonly used routines do not include these effects, and some experimentalists enhance the systematic errors by using a much thicker vanadium sample (up to 10-mm thickness or diameter, scattering about 33% of the beam) in order to increase the counting rate. To include geometric effects in the calibration, it is desirable to make a vanadium specimen of the same shape and scattering

---

†Some "random" alloys, e.g., Ti/Zr may be made of an incoherent composition and may be used as *subsidiary* standards, since the random areas of their mixtures can be determined only through comparison with vanadium.

Table I. Approximate Thickness to Scatter 10% of Thermal (25 meV) Neutron Beam in Transmission

Liquids and solids at room temperature	Thickness (mm)	Gases at 1 atm and room temperature	Thickness (m)
Light water	0.3	Argon 36	0.5
Heavy water	2.2	Nitrogen[a]	1.6
Lead	3.0	Krypton	5.2
Aluminium	12.0	Natural argon	58.0

[a] Because of the relatively large scattering from nitrogen, neutron beans and samples are often placed in either a vacuum or a natural argon atmosphere. For scattering purposes, 100-atm nitrogen can be held in a container of ~2 cm diameter.

power as the sample under study. For slab samples, several sheets of 0.25-mm vanadium foil may be stacked together with suitable spacing to give a specimen of the same dimensions and total scattering cross section as the sample. For matching cylindrical samples, a set of concentric vanadium tubes is best, or otherwise a length of foil may be wound into a spiral having the same diameter and scattering power as the sample. In the case of dense gases, these calibration specimens have substantial spaces between the layers of vanadium, while for liquids the sheets may be in contact with each other. The thickness of a vanadium specimen that scatters 10% of the beam is 2.5 mm, this is used for matching samples that have thicknesses of 2.5 mm or less. The magnitude of geometrical scattering effects (due to path length, angular definition, or beam profile effects) may be determined by comparison of the scattering patterns from vanadium specimens having the same transmission coefficient but different shapes and packing geometries, after making corrections with the standard correction routines. Discrepancies of the order of a few percent may occur.

Data from the vanadium specimen may be used to test the instrument's stability over the long term, by making runs of the same kind at regular intervals and comparing them. The total performance of the instrument may drift because of drifts in the high voltage, in amplifier components, in the line voltage, in detector cross talk, etc., or in mechanical changes such as twisting of girders as parts move, slight eccentricity changes, ambient temperature changes, slight relative movement of monitor and detector, etc. These drifts, which may show up in the long-term, can contribute a standard deviation (σ_2) to the final data. It can be observed by comparing the standard deviation (σ_3) calculated from the differences between a number of vanadium runs with the standard deviation (σ_1) due to the counting statistics of these runs. The deviation $\sigma_2 = \sqrt{\sigma_3^2 - \sigma_1^2}$ should be much less that σ_1 for an ideal instrument.

The size of σ_1 should be related to the final precision required (say σ_4) and, if the standard deviation from counting statistics on the sample is σ_5, it is usual to make $\sigma_5 \simeq 2\sigma_1$, so that it is the largest contribution to σ_4. Because most instruments scan through a range of angles or times, these standard deviations refer to data taken at each point along the scan. However, some experimentalists draw a smooth curve through vanadium data taken with $\sigma_1 > \sigma_5$ and use the smaller deviation $\sigma_1/\sqrt{n}$ (where n is the number of scanned points) to compare with σ_5. This procedure is deceptive since the smooth curve itself can be drawn with undulations of magnitude σ_1, and thus although the data is smooth the errors in $S(Q)$ are not reduced. The latter point is particularly important when position-sensitive detectors are used.

14.2.3. Data Collection of $S(Q)$ and $S(Q, \omega)$

Samples of the fluid being studied are held in a container with windows larger than the beam area and the state conditions (P and T) properly controlled. Window materials depend upon the range of P and T to be covered and possible corrosion by the fluid. If none of these parameters are important, the usual window materials are an aluminium alloy of about 0.5-mm thickness or a vanadium foil of about 0.1-mm thickness. Both entry and exit windows taken together will scatter ~1% of the beam, depending on the effect of the alloying components. In other cases (e.g., high pressures), the window material will be thicker and other alloys used so that the scattering may be much larger, but the amount scattered should lie within the range of validity of the correction programs (Section 14.3). Otherwise the beam area should be reduced, the sample partitioned, or analogous changes made.

Neutrons scattered by the sample and its container should be absorbed in cadmium, boron, gadolinium oxide, or similar absorbing materials except where they are traveling in the direction of the detector. For example, this may mean installing absorbers inside the container and conducting the ingoing and outgoing beams through absorbing apertures, tunnels, etc., in the immediate vicinity of the sample. Care should be taken to see that these do not scatter neutrons (e.g., through structural parts or other components of a mixture), since they will be assumed to be perfect absorbers in the data analysis. For multidetector assemblies, similar remarks apply to structures inside the detector shield, especially those designed to separate individual detectors.

Most samples yield approximately the same counting rate per unit area, because their thickness is adjusted to scatter the same fraction of the beam. Table I lists the thicknesses for various materials in order to scatter about 10% of the beam, and it will be seen that thicknesses vary widely. Either a cylindrical or a slab shape is employed, and a transmission geometry is used frequently, except for strongly absorbing samples. The sample is installed in one of the instruments described in Chapter 3 (Part A) and, since the sample environmental containment apparatus may be complex, adequate provision for it is usually made as well. This can turn the sample location into a small laboratory, and it needs to operate automatically for several days in the same manner as the instrument itself.

Counting times for a single run vary from a few hours to several weeks, depending on the incident flux and the value of σ_4 (see Section 14.2.2) required. Runs are repeated, used for the tests described previously, and then added together. Separate measurements are made for each state condition and for an evacuated container. Sometimes additional runs are made with the container filled with an absorber having the same neutron transmission coefficient as the sample. This avoids the attenuation coefficient correction to the container in complicated cases, such as large containers where Bragg reflections from the two sides come at slightly different apparent angles but overlap one another.

It will be seen from the discussion in this section that many runs are made in order to determine a single $S(Q)$ or $S(Q, \omega)$. The procedure adopted is geared to the precision (σ_4) required and to the data reduction routines that the user intends to employ. An economy of effort is possible when the same container and the same instrumental setup are used for a series of samples, for example, at different (PVT) states. In many experiments the total time available is limited, and therefore it is worthwhile optimizing the distribution of time between the different runs that contribute to the final $S(Q)$ or $S(Q, \omega)$. A simple calculation will show how to minimize the final statistical error, keeping the total time constant and varying the distribution between runs for any experimental situation [e.g., see Eq. (14.46)].

14.3. Data Analysis

14.3.1. Reduction of Observed Quantities to Differential Cross Sections

The data accumulated in the way described must be transformed to the "no-scattering limit" of Eqs. (14.5) and (14.6). Sears[3] treats this problem through neutron transport theory. This treatment may be applied to any sample if the meaning of $S(Q, \omega)$ is suitably generalized. He expresses his result by replacing $S(Q, \omega)$ in Eq. (14.5) with a new function $s(\mathbf{k}_0, \mathbf{k})$ where

$$s(\mathbf{k}_0, \mathbf{k}) = S(Q, \omega)H_1(k_0, k) + \sum_{j=1}^{\infty} s_j(\mathbf{k}_0, \mathbf{k}) \qquad (14.10)$$

and

$$s_j(\mathbf{k}_0, \mathbf{k}) = \left(\frac{\rho\sigma_s}{4\pi}\right)^{j-1} \int \cdots \int d\Omega_1\, dE_1 \cdots d\Omega_{j-1}\, dE_{j-1}$$
$$\times S(Q_1, \omega_1) \cdots S(\mathbf{Q}_j, \omega_j) H_j(\mathbf{k}_0, \mathbf{k}_1 \cdots \mathbf{k}_{j-1}, \mathbf{k}),$$

where $\mathbf{k}_0$ and $\mathbf{k}$ are the initial and final wave vectors, and j is an index denoting the number of times the neutron is scattered. In these expressions $\mathbf{k}$ is the outgoing wave vector irrespective of the number of times the neutron is scattered and σ_s is the total scattering cross section. The H_j are attenuation factors and involve a $(j + 2)$-dimensional integral in $\mathbf{r}$ space. If it is assumed that the ratio of j to $j - 1$ scattering depends only on $\mathbf{k}_{j-1}$, then

$$H_j(\mathbf{k}_0, \mathbf{k}_1 \cdots \mathbf{k}_{j-1}, \mathbf{k}) = (2/j) H_{j-1}(\mathbf{k}_0, \mathbf{k}_1 \cdots \mathbf{k}_{j-2}, \mathbf{k}) B(\mathbf{k}_{j-1})$$
$$= \frac{2^{j-1}}{j!} H_1(\mathbf{k}_0, \mathbf{k}) B(\mathbf{k}_1) \cdots B(\mathbf{k}_{j-1}), \quad (14.11)$$

which defines $B(\mathbf{k}_{j-1})$, and the factor $2/j$ is chosen to give the correct expression in the small sample (infinite) plane slab limit. Sears[3] shows that the factors for other shapes do not differ much from those for the infinite plane slab but that for thick or large samples there may be larger errors. With the further approximation

$$R_j(\mathbf{k}_0, \mathbf{k}) = \left(\frac{\rho\sigma_s}{4\pi\delta}\right)^{j-1} \int \cdots \int d\Omega_1\, dE_1 \cdots d\Omega_{j-1}\, dE_{j-1}$$
$$\times S(Q_1 \omega_1) \cdots S(\mathbf{Q}_j, \omega_j) B(\mathbf{k}_1) \cdots (\mathbf{k}_{j-1})$$
$$\simeq R_2(\mathbf{k}_0, \mathbf{k}) \quad (14.12)$$

(i.e., that all orders of scattered spectra have the same shape), he obtains from Eq. (14.10) the approximate result.

$$s(\mathbf{k}_0, \mathbf{k}) = H_1(\mathbf{k}_0, \mathbf{k})\{S(Q, \omega) + \Delta R_2(\mathbf{k}_0, \mathbf{k})\} \quad (14.13)$$

and

$$\left(\frac{d^2\sigma}{d\Omega\, d\omega}\right)_{\text{obs}} = \left(\frac{\sigma_s}{4\pi}\right) \frac{k}{k_0} s(\mathbf{k}_0, \mathbf{k}). \quad (14.14)$$

These equations are correct if

$$\Delta = \frac{\exp(2\delta) - 1}{2\delta} - 1, \quad (14.15)$$

where δ is the ratio of double to single scattering in the approximation that $S(Q, \omega) = \delta(\omega)$, i.e., elastic isotropic scattering. It is usual to drop this restriction on the definition of δ and to define it as the ratio of double intensity for any $S(Q, \omega)$ to single intensity in the isotropic approximation and to define Δ as the ratio of total multiple to single scattered intensity for this case. Then $\int R_2 \, dE$ is normalized to unity and Eq. (14.13) is used for all cases. The assumptions behind Eqs. (14.11) and (14.12) are not satisfactory in general and therefore Eq. (14.13) depends on the assumption that orders higher than the second are relatively unimportant, that is, $\delta \ll 1$.

Equation (14.13) agrees with earlier equations based on intuitive expectations and is correct in the no-scattering limit. However, it has not been tested fully against computer simulations for thicker samples: its range of validity is not known although from its derivation it cannot be valid for very thick samples. For this reason the term $\Delta R_2(\mathbf{k}_0, \mathbf{k})$ in Eq. (14.13) is kept small relative to $S(Q, \omega)$ by choosing appropriate dimensions for the sample. The fraction of the incident beam scattered by the sample and its container is between 10 and 20% for most experiments, and then Δ is of the same order. In addition to the possible errors in the makeup of Eq. (14.13), there are a number of approximations made in calculating H_1, Δ, and R_2. For this reason, too, the sample thickness should be kept small and the experimental setup adjusted to correspond roughly to what is being calculated. Finally, it should be pointed out that Eq. (14.13) is used for the interpretation of all kinds of radiation scattering experiments, and therefore programs developed for x-ray scattering (e.g., to calculate H_1) may be used for neutron scattering and vice versa.

14.3.2. Attenuation Corrections

The quantity $H_1(\mathbf{k}_0, \mathbf{k})$ is called an attenuation factor and is defined as

$$H_1(\mathbf{k}_0, \mathbf{k}) = \frac{1}{V} \int_{\text{Vol}} d\mathbf{r} \exp\{-\Sigma(\mathbf{k}_0)L(\mathbf{r}, \hat{\mathbf{k}}_0) - \Sigma(\mathbf{k})L(\mathbf{r}, -\hat{\mathbf{k}})\}, \quad (14.16)$$

where $\Sigma(\mathbf{k})$ is the total collision cross section per unit volume, i.e., a macroscopic quantity, and $L(\mathbf{r}, \mathbf{k})$ is the distance from the point $\mathbf{r}$ (in the sample) to a point on the surface in the direction of the unit vector $-\hat{\mathbf{k}}$. To be included in these definitions are averaging over the beam profile, the divergence of the incident and scattered beams in the definition of $\mathbf{k}_0$ and $\mathbf{k}$, are the container, and any other materials in the beam in the definition of $\Sigma(\mathbf{k}_1)$, $\Sigma(\mathbf{k})$ and the L's, and for the integral over the volume including only that part of the sample that can be seen by the detector. Not all of these points are covered in some of the published tables of these factors. For example, Paalman and Pings[4a] consider a cylindrical sample completely immersed in a uniform parallel beam while Kendig and Pings[4b] consider a beam less than the diameter of the cylinder and passing symmetrically through its center. In both cases there is no allowance for beam divergence before or after scattering. If the same approximation is used for plane slabs, the attenuation factor is a simple exponential function. Poncet[5a] considers a sample in a cylindrical container while Soper and Egelstaff[6] include a sample composed of several concentric cylinders, an arbitrary beam profile with arbitrary offsets with respect to a diameter, and an arbitrary detector profile. The possibility that the sample height is greater than the beam height and a general definition of H_1 for double scattering are included too. However all the above cases[4, 5a, 6] assume $|\mathbf{k}_0| = |\mathbf{k}|$ as for elastic scattering. In many laboratories these programs have been modified to allow for inelastic scattering, $|\mathbf{k}_0| \neq |\mathbf{k}|$, as was done by Poncet[5b] or Groome et al.,[7] for example, which requires calculating $\Sigma(k)$ separately for k and k_0. This quantity is

$$\Sigma(k) = 2\pi\rho \int_{-1}^{+1} \frac{d\sigma}{d\Omega}(k)\, d(\cos\phi) \qquad (14.16a)$$

and is a function of (E, ρ, T). It differs from the bound or free atom cross sections listed in tables but may be read from graphs of total cross section versus k. In general, experimental results or a theoretical model for $d\sigma/d\Omega$ are used in Eq. (14.16a). The integration over the sample in Eq. (14.16) introduces an angular dependence that depends on the shape and absorption in the sample. This may be pronounced if the sample is strongly absorbing.

Unfortunately, some laboratories employ attenuation factor programs that include only some of the features listed above; as yet programs involving all these features are not universally available. Moreover, insufficient work has been done on the precision of the numerical integration routines, so that for both these reasons the precise errors on attenuation factors are not well known. Even when they are calculated with high precision there are errors due to imprecise knowledge of $\Sigma(\mathbf{k})$ and $L(\mathbf{r}, \mathbf{k})$, etc., or due to cross section errors, composition or density errors, and dimensional errors. Since $H_1(\mathbf{k}_0, \mathbf{k})$ is a linear factor, a corresponding error is introduced into the final result for $S(Q, \omega)$.

In many cases the attenuation corrections are denoted, following Paalman and Pings,[4a] by the letter "A" with subscript "S" for sample or "C" for container. To derive $S(Q)$ they are obtained from the routines described with the assumption $|\mathbf{k}_0| = |\mathbf{k}|$ and depend on the angle ϕ although this is not indicated explicitly in many papers.

14.3.3. Multiple Scattering Corrections in Inelastic Scattering

The product [see Eq. (14.13)]

$$H_1(\mathbf{k}_0, \mathbf{k}) \cdot \Delta \cdot R_2(\mathbf{k}_0, \mathbf{k}) \qquad (14.17)$$

14. CLASSICAL FLUIDS

is known as the multiple scattering correction, and attempts to evaluate it have been made for many years. Because the beam area is large, many samples are thin (in the sense of Table I) in the direction of the beam only and are relatively thick in other directions, for example, at a large angle to the beam. In such cases the multiple scattering can be large and dominated by first scattering events in directions at right angles to the beam (or at the angle of inclination of a slab sample). To minimize this effect, the sample may be divided by absorbing spacers, or absorbing materials may be introduced into the sample, to obtain similar scattering properties in all directions. However, these procedures reduce the multiple scattering at the expense of reduced intensity and so are not employed frequently. Some samples (e.g., xenon) are strong absorbers naturally, and in these cases large area samples may be used without large multiple scattering. However, the absorption factor and its angular dependence are then more important than for other samples.

Semi-empirical methods of calculation were used by early experimentalists studying $S(Q, \omega)$. For example, Egelstaff and London[8] used a simple multigroup numerical approach, dividing the angular range into three groups and using a single energy transfer group. Brockhouse[9] used an empirical approach, by assuming that after two scattering events the energy spectrum was a Maxwellian at the sample temperature and hence independent of ϕ. These two general methods—numerical and empirical—have been continued to the present time. Later numerical methods assumed a theoretical form for $S(Q, \omega)$ and used numerical integration[10] or Monte Carlo simulation techniques[11] to obtain the correction. Brockhouse's[9] procedure was modified by Hawkins and Egelstaff,[12] who fitted a Maxwellian with an arbitrary temperature to the measured spectrum at low angles. They assumed that the spectrum of multiply scattered neutrons was independent of angle and hence given by the observed low-angle spectrum. This spectrum is dominated by the product in Eq. (14.17) except for $\omega \sim 0$. The variation of intensity with angle is taken from the $(d\Sigma/d\Omega)_{MS}$ calculation [Eq. (14.20)]. The best empirical method seems to be a generalization of this approach, in which some rough estimates are made of the range of low angles over which the product in Eq. (14.17) is dominant, and the observed spectrum is fitted to an arbitrary smooth function. This function is continued through the $\omega \simeq 0$ region by an ad hoc extension based on an estimate of the multiple elastic scattering. Such a method will work only if the multiple scattering is small and good low-angle data are available.

Sears[3] calculated $R_2(\mathbf{k}_0, \mathbf{k})$ by numerical integration using, in the case of the sample, a theoretical model for $S(Q, \omega)$. But this method does not take into account the container scattering, and to include this approximately, the sample and container may be "homogenized." That is, the total cross section

Σ_T and scattering cross section $\Sigma_S(Q, \omega)$ (given by the sum of the nuclear cross sections for all the atoms in the system) are divided by the total volume to give average cross sections per unit volume. These are used in a calculation for a hypothetical homogeneous system having the same external dimensions as the real system. Provided the multiple scattering is not large this method is acceptable.

The Bischoff et al.[11] Monte Carlo method for liquids was extended by Copley[13] to cover container scattering and several alternative geometries. This program is now used in many laboratories and some modifications have been incorporated by various users. Copley[14] describes one modification to speed up the calculation of elastic scattering, for example, multiple Bragg scattering from the container. This process leads to an elastic peak (of varying amplitude) at each angle of scatter and is an important correction if the container is thick. Nevertheless, evaluation of multiple scattering by this method is time-consuming and expensive. Some improvement may be possible by exploiting the advantages of modern parallel processors. This method is the only universal method and must be used where the multiple scattering is large. Again theoretical models for $S(Q, \omega)$, for sample and container respectively, are employed and may be improved by iteration on extensive experimental data.[69]

For solid or liquid samples, the multiple scattering varies slowly with thickness because the geometry is changed as the thickness is altered, and the two effects partially compensate. But for gaseous samples, the pressure may be changed without altering the sample shape and so produce a linear effect in the multiple scattering. This can be done with a gas having a large scattering cross section (e.g., hydrogen) and hence relatively low pressures so that there is no significant change in the scattering function $S(Q, \omega)$. In Fig. 4(a) the calculated functions for $\phi = 0.5°$ are shown, and above $1°$ the difference cannot be seen on this scale. Because the only change is a scale change in the absorption factor and in Δ [Eq. (14.13)], such data are a good experimental test of the multiple scattering calculations and a rare example in which the multiple scattering can be determined experimentally.

Figure 4(b) and (c) shows two examples in which the multiple scattering is very different. Data[15] taken with a 7.5-cm diameter pressure cell (walls of 3-mm aluminium alloy) containing 10 atm or 20 atm hydrogen gas at room temperature are shown in Fig. 4(b). The multiple scattering in the 20-atm sample is approximately twice that in the 10-atm case, so that after taking out the absorption factor the correction in the 10-atm case is equal to the difference between the two sets of data (except for double Bragg scattering in the container). This is true at all angles. Calculations using the empirical method suggest that the multiple scattering should be very small and this is seen here. The double Bragg scattering is larger relatively in the 10-atm

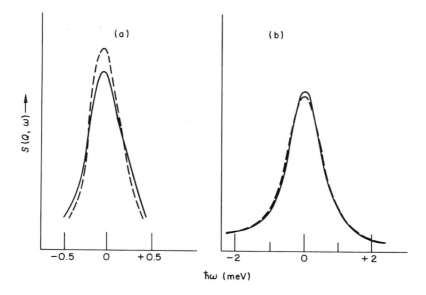
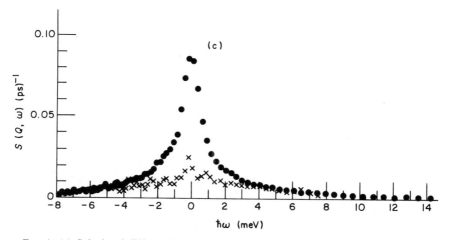

FIG. 4. (a) Calculated differential cross sections for hydrogen gas at 10 atm (full line) and 20 atm (dashed line) for $E_0 = 31$ meV and $\phi = 0.5°$. This difference cannot be seen on this scale for $\phi > 1°$. (b) Experimental distribution of scattered neutrons for hydrogen gas at room temperature and pressures of 10 atm (full line) and 20 atm (dashed line) and for $\phi = 9°$ and $E_0 = 31$ meV. Multiple scattering is given by the difference between these curves. (Redrawn from Egelstaff et al.[15]) (c) Differential cross sections for dense krypton gas at room temperature $E_0 = 12.5$ meV, and $\phi = 9°$. The solid circles are the experimental data and the crosses are Monte Carlo calculations of multiple scattering. (Redrawn from Egelstaff et al.[16])

sample and this can be seen, too. A different example is provided by experiments[16] on dense krypton gas using a thick-walled aluminum alloy pressure vessel (internal and external diameters of 25 and 43 mm, respectively). The data for a pressure of ~1000 atm ($\rho \simeq 14 \times 10^{27}$ atoms/m^3) at room temperature and an angle of 9° are shown in Fig. 4(c). Multiple scattering was calculated by the Monte Carlo method[13,14] and gas container double scattering was the most important effect. The calculations and observations at low angles agreed without any scale adjustments being required.

The instrumental resolution is not usually an important factor in multiple scattering work. If the resolution has been chosen to give a good measurement of the single scattering, then the broader multiple spectrum will not be greatly affected, and it is possible that the errors in its determination will be more significant than resolution corrections. Nevertheless, in the case of the Sears[3] method an additional numerical integration may be included readily, and Copley[13] has included the resolution in his Monte Carlo simulation program.

14.3.4. Multiple Scattering in Diffraction or Intensity Measurements

For diffraction experiments it is usual to evaluate a macroscopic cross section for multiple scattering incorporating all the components of the sample (e.g., sample, container, heater, etc.). For a simple sample, it is necessary to evaluate [see Eqs. (14.13) and (14.15)]

$$\left(\frac{d\Sigma}{d\Omega}\right)_{MS} = \frac{N\sigma_s}{4\pi} \int_0^\infty \frac{k}{k_0} H_1(\mathbf{k}_0, \mathbf{k}) \Delta R_2(\mathbf{k}_0, \mathbf{k}) f_0(E) \, dE, \qquad (14.18)$$

where $f_0(E)$ is the counter efficiency function normalized to unity for $|\mathbf{k}| = |\mathbf{k}_0|$. It is usual to assume that the $\mathbf{k}$ dependence of $H_1(\mathbf{k}_0, \mathbf{k})$ is weak so that

$$H_1(\mathbf{k}_0, \mathbf{k}) \simeq H_1(\mathbf{k}_0, \mathbf{k}_0) = A. \qquad (14.19)$$

The redefinition of A, following Eq. (14.15), and renormalization of R_2 leads to the result

$$\left(\frac{d\Sigma}{d\Omega}\right)_{MS} \simeq \frac{N\sigma_s}{4\pi} A\Delta \int_0^\infty \frac{k}{k_0} R_2(\mathbf{k}_0, \mathbf{k}) f_0(E) \, dE \simeq A\Delta\left(\frac{N\sigma_s}{4\pi}\right). \qquad (14.20)$$

In this equation the final step is based on the approximation $|\mathbf{k}_0| \simeq |\mathbf{k}|$, so that there is an additional term related to the inelasticity corrections (Section 14.4), which has been neglected. This term is a product of small quantities (i.e., of Δ and the inelasticity correction). However, if, for example, $\Delta \simeq 0.2$ and the inelasticity correction is ~0.2, the product is ~0.04 and a more refined treatment is needed to achieve accurate data. In such cases Eq. (14.18) must be evaluated correctly. For the papers cited below Eq. (14.20) was used in all cases. To generalize the result to a complex sample or geometry, the definitions of H_1 and R_2 are generalized.

In the case of slabs, multiple scattering results have been given by Vineyard,[17] Cocking and Heard,[18] Sears,[3] and Soper.[19] The latter offers the most flexibility and so covers a variety of experimental situations better than do the earlier papers. In the case of cylinders, results have been given by Blech and Averbach,[20] Sears,[3] and Soper and Egelstaff.[6] As described in Section 14.3.2 the latter paper covers the experimental situation with greater flexibility than do the former papers. The definition of Δ is broadened to include samples partially irradiated by a narrow beam. Also the numerical integration routines are improved.

Figure 5 compares results from Blech and Averbach,[20] Sears,[3] and Soper and Egelstaff,[6] and it can be seen that Blech and Averbach[20] and Soper and Egelstaff[6] disagree with Sears.[3] The magnitude of the change in δ due to changes in beam dimensions is shown also. All these results refer to isotropic scattering, and for small diameter cylinders or thin slabs δ is independent of ϕ. Figure 6 shows the angular dependence of multiple scattering for liquid lead as a slab[17] and for a hard sphere fluid in a cylindrical container.[6] The anisotropic scattering leads to a small variation of δ only, and since both bases are for dense liquids, this effect should be smaller for other fluids.

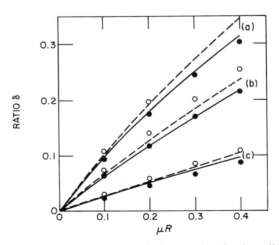

FIG. 5. Comparison of the ratios of double to single scattering (δ) using cylinders with ratios of radius to height of (a) 0.2, (b) 1.0, and (3) 5.0. The beam profile is uniform and the full line is for a fully immersed sample while the dashed line is for a beam width equal to half the sample diameter (from Soper and Egelstaff[6]). Solid circles are from Blech and Averbach,[20] while open circles are from Sears.[3]

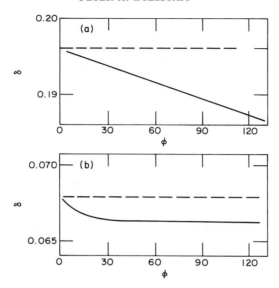

FIG. 6. (a) Multiple scattering ratio δ for a slab sample with μ times thickness = 0.132; isotropic scattering is shown by the dashed line and liquid lead near the triple point by the full line. (Data taken from Cocking and Heard.[18]) (b) Multiple scattering ratio δ for a cylindrical sample of a hard sphere fluid ($\rho\sigma^3 = 0.918$), the radius to height ratio is 0.1 and $\mu R = 0.063$, shown by the full line. For comparison the same sample for isotropic scattering is shown by the dashed line.

14.4. Inelasticity Corrections to Intensity Measurements

14.4.1. A General Formula for the Cross Section

The definition of $S(Q)$ is

$$\lim_{Q' \to Q} \int_{-\infty}^{+\infty} S(Q', \omega) \, d\omega = S(Q), \tag{14.21}$$

where the notation $Q' \to Q$ means that for the whole range of $-\infty < \omega < +\infty$, Q' must lie in the neighborhood of the given Q. Differential cross sections do not satisfy this condition because in this case Q' is given by Q_ω of Eq. (14.7). However, if the limit $\hbar\omega/E_0 \to 0$ is taken, Q_ω is seen to be a simple function of E_0 and ϕ, and the differential cross section is proportional to $S(Q)$. This limit is referred to as the static approximation. Thus an "inelastic correction" is needed, which is simply the difference between the quantity measured with $\hbar\omega/E_0 \neq 0$ and the ideal result for $\hbar\omega/E_0 \to 0$. Inelasticity corrections are straightforward in principle, although many papers label them complex and difficult. The principal reasons seem to be either that the wrong experimental conditions are used or that the wrong formula is used and that either or both errors are compensated by employing the wrong constants in order to get an approximate fit to the observations. This practice is still prevalent, even in papers where the data warrants a better treatment. An extensive literature exists, some of which is either misleading or incorrect, and therefore a lengthy treatment of the subject seems advisable. To begin this treatment, the cross-section formula for the scattered intensity will be discussed in a way that highlights its behavior. Detailed formulas will be given for atoms, and for molecules the reader will be referred to formulas in the literature.

In order to calibrate the instrument in absolute units, it is usual to deduce from vanadium data the experimental result for a perfectly elastic scattering substance so that the ratio of intensities of the sample to this ideal substance may be measured. The data for perfect elastic scattering will be denoted by the subscript "e" and the sample by subscript "s." For a time-of-flight experiment, the ratio of differential cross sections[21] (sample divided by elastic scatterer) corrected to the no-scattering limit is

$$\left[\frac{\sigma_e}{\sigma_s}\left(\frac{d\Omega}{d\sigma}\right)_e\right]\left(\frac{d\sigma}{d\Omega}\right) = \int_{-\infty}^{+\infty} N(k, k_e) F(Q_\omega, \omega) S(Q_\omega, \omega) \, d\omega \quad (14.22)$$

where

$$N(k, k_e) = \frac{\Phi(1/k_0) f(k) k}{\Phi(1/k_e) f(k_e) k_0}, \quad (14.23)$$

$\Phi(1/k) d(1/k)$ is the incident neutron spectrum, and $f(k)$ is the detector efficiency function. This quantity is unity, for example, if a slowing down dE/E spectrum and a $1/v$ detector are used. The factor k/k_0 comes from Eq. (14.5). In this type of experiment (see Part A, Chapter 3), the three variables k_0, k, and k_e are related by

$$(a/k_0) + (1/k) = (a + 1)/k_e, \quad (14.24)$$

where a is the ratio of incident to scattered flight paths. In this formula and those following, the continuous beam case may be obtained by letting $a \to \infty$, and therefore it will not be given separately. Because the experimental variable (e.g., $1/k$) is changed to energy transfer, the scattering function is multiplied[21] by a Jacobian $F(Q_\omega, \omega)$, which can be reduced to

$$F(Q_\omega, \omega) = \left.\frac{\partial(1/k_0)}{\partial(1/k_e)}\right|_\omega = \frac{a + 1}{a + (k_0/k)^3}, \quad (14.25)$$

where the derivative is calculated from Eq. (14.24) after eliminating k through the equation for energy conservation. Then k_0/k may be related to

Q and ω through the conditions for conservation of momentum and energy [Eq. (14.7)]. In terms of the natural nondimensional variable $p = 2m\omega/\hbar Q^2$,

$$\frac{k_0}{k} = \frac{1+p}{p\cos\phi \pm \sqrt{1-p^2\sin^2\phi}}, \qquad (14.26)$$

and this ratio is a double-valued function. However, because k_0/k is a positive real number, p is restricted by the conditions

$$\begin{aligned}-\cos\phi \le p \le +\cos\phi, &\qquad \phi < 90° \\ -1 \le p \le +1, &\qquad \phi > 90°.\end{aligned} \qquad (14.27)$$

(Note that as $k \to 0$, $p \to +1$, and as $k \to \infty$, $p \to -1$; the value of p may be estimated from $\hbar^2 Q^2/2m \simeq 2$ meV for $Q = 1$ Å^{-1}; and for constant Q the scattering function $S(Q, \omega)$ may be expressed as function of p with limits $\pm\infty$.) Thus $F(Q_\omega, \omega)$ has discontinuities and is zero outside the limits in Eq. (14.27). It acts as a sampling factor that determines how $S(Q, \omega)$ is sampled through the conditions in Eqs. (14.24) and (14.7). Also, due to the inclusion of this factor, the upper limit of $E_0/\hbar$ in Eq. (14.6) may be replaced by infinity in Eq. (14.22).

Thus the basic cross-section ratio of Eq. (14.22) has a simple structure: it is the product of the van Hove scattering function with a kinematic sampling factor and an experimental factor $N(k_0, k_e)$, and this product is integrated over all energy transfers. Moreover, the usual experiment with a two-axis crystal diffractometer (Part A, Chapter 3) is also included by this formula in the limit $a \to \infty$. In this steady-state source limit, $k_0 \to k_e$ from Eq. (14.24).

The behavior of k/k_0 as a function of p is shown in Fig. 7 for $\phi = 45°$ or $135°$ (this relation is determined by conservation laws only and so it applies to all experiments). To calculate an inelasticity correction equation, Eq. (14.26) can be expanded about $p = 0$ to give

$$k/k_0 = 1 - p(1 - \cos\phi) + (p^2/2)(1 - \cos\phi)^2 + \cdots, \qquad (14.28)$$

and this expansion is shown in the figure. It is a reasonable fit if $|p| \le 0.2$ at $135°$ or if $|p| \le 0.6$ at $45°$ and would be larger at smaller angles. In contrast, the range of k/k_0 fitted decreases with decreasing angle and would be $\sim 1 \pm \phi/4$ at low angles. To deduce $S(Q)$ from the integral in Eq. (14.22), it is desirable to design the experiment so that $k \approx k_0 \approx k_e$, that is, the k range is consistent with the expansion in Eq. (14.28). However, the significant range of p is determined by the properties of the sample, i.e., by the behavior of $S(Q, \omega)$. For example, in the case of simple diffusion, the range of p is proportional to the diffusion coefficient. If the experimentalist wishes to take good data integrated out to five half-widths in ω, then the range of p required is $\pm 10\, mD/\hbar$. For low angles this implies that $D < (0.3 \times 10^{-4})/\phi$

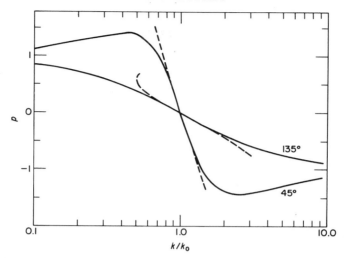

FIG. 7. The nondimensional ratio $p = 2m\omega/\hbar Q^2$ in a neutron scattering experiment as a function of k/k_0 for $\phi = 45°$ and $135°$ (full lines). An expansion for $k \simeq k_0$ [Eq. (14.28)] is shown by the dashed lines.

(cm^2 s^{-1}) and at $45°$ that $D < 3 \times 10^{-5}$ cm^2 s^{-1}, while at $135°$ it is necessary that $D < 10^{-5}$ cm^2 s^{-1}.

For these intensity measurements, the final wave number (k) is not known because only the wave number k_e is measured. Thus, an ad hoc experimental variable is adopted, i.e., the value of p for $Q = Q_e$, so that $p_e = \hbar\omega/2E_e(1 - \cos\phi)$. These quantities ($p$ and p_e) may be related by a somewhat involved equation after eliminating k and k_0 from Eqs. (14.7) and (14.24). Figure 8 shows two examples (for $\phi = 45°$ and $135°$) of this relationship. The odd behavior that occurs for p_e near the end of its positive range arises from the way the limit $\hbar\omega = E_0$ is approached in the experiment. This region should be avoided in order to obtain a meaningful integral (Eq. 14.22). To relate the integral to $S(Q)$ in a simple way, it is necessary that $p \approx p_e$ and that it may be expanded about $p = 0$ (i.e., $k = k_0$). In the case of $a \to \infty$, the ratio p_e/p becomes

$$\frac{p_e}{p} = 1 - \frac{\hbar\omega}{2E_e} + \left(\frac{\hbar\omega}{2E_e}\right)^2 \frac{\cos\theta}{2(1 - \cos\theta)} \left[1 + \frac{\hbar\omega}{2E_e} + \frac{5}{4}\frac{\hbar\omega}{2E_e} + \cdots\right]. \quad (14.29)$$

For the expansion used in Section 14.4.2, the first three terms are retained. The series in Eq. (14.29) truncated at this point is plotted in Fig. 8. It can be seen that a reasonable fit is obtained over a range similar to that fitted in Fig. 7. If a model for $S(Q, \omega)$ is available, the ranges of Q and ω over which $S(Q, \omega)$ has significant intensity may be estimated, and so the significant

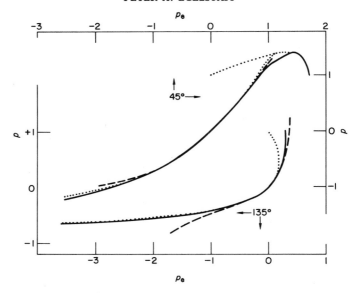

FIG. 8. The variation of $p_e = 2m\omega/\hbar Q_e^2$ with p (Fig. 7) for $\phi = 45°$ and $135°$ (full lines) and $a \to \infty$. An expansion for $p_e \simeq p$ is shown by the dashed lines [Eq. (14.29) truncated after three terms]. The dotted line shows p_e versus p for $a = 10$.

range of p may be estimated for the Q values of interest. From the condition that $p \simeq p_e$, the angles at which these Q's can be studied may then be found by comparison with Fig. 8, for example.

An examination of the series in Eq. (14.29) shows that two conditions must be satisfied in order that $p_e \simeq p$, namely,

$$\frac{|\hbar\omega|}{2E_e} \ll 1 \quad \text{and} \quad \left(\frac{\hbar\omega}{2E_0}\right)^2 \ll \frac{2(1-\cos\phi)}{|\cos\phi|}, \quad (14.30)$$

which are approximately the same as the conditions $(k/k_0)^2 \simeq 1$ and $p_e^2 \ll 2|\sec\phi|/(1-\cos\phi)$; better limits could be deduced by comparison with Fig. 7, for example. The first condition is related to the mean energy shift of $S(Q,\omega)$ and is an important constraint at high angles (say $\sim 90°$), while the second condition is related to the width of $S(Q,\omega)$ and is an important constraint at low angles. A rough guide may be obtained, however, by inserting the known frequency moments for $S(Q,\omega)$ for a classical monatomic fluid (atoms of mass M) into these conditions. From the moments

$$\bar{\omega} = \frac{\hbar Q^2}{2M} \quad \text{and} \quad \overline{\omega^2} = \frac{k_B T}{M} Q^2 + O\left(\frac{\hbar^2}{M^2}\right), \quad (14.31)$$

the conditions in Eq. (14.30) become

$$\frac{m}{M} \ll \frac{1}{1 - \cos \phi} \quad \text{and} \quad \left(\frac{m}{M}\right)^2 \ll \frac{2|\sec \phi|}{1 - \cos \phi}, \quad (14.31a)$$

and in addition,

$$\sqrt{\frac{k_B T}{E_e} \frac{m}{M}} \ll \frac{1}{\sqrt{1 - \cos \phi}} \quad \text{and} \quad \frac{k_B T}{E_e} \frac{m}{M} \ll 2|\sec \phi|. \quad (14.31b)$$

All four of these conditions must be satisfied for the expansion to be satisfactory. To see whether Eq. (14.31a) or Eq. (14.31b) is the more significant, these ratios must be compared:

$$(1 - \cos \phi)(m/M) \quad \text{and} \quad k_B T/E_e. \quad (14.31c)$$

It is clear that as $\phi \to 0$ then $k_B T/E_e \gg \phi^2 m/2M$, and so the second of the conditions in Eq. (14.31b) is always the most important, but for other angles various limits should be compared. In any case the conditions in Eq. (14.31) are approximations, and if a model $S(Q, \omega)$ is available, its predictions should be compared to the conditions in Eq. (14.30).

14.4.2. A Generalized Placzek Expansion

The Placzek[22] expansion consists of writing each of the factors of Eq. (14.22) as an expansion about $\hbar\omega = 0$. Since $(Q/Q_e)^2 = p_e/p$, the expansion for

$$\Delta Q^2 = Q^2 - Q_e^2 = [(p_e/p) - 1]Q_e^2 \quad (14.32)$$

has been given in Eq. (14.29). This may be used in a Taylor expansion of $S(Q, \omega)$ for $a \to \infty$. For other values of a, the expansion as far as the first three terms of Eq. (14.29) is

$$\frac{\Delta Q^2}{Q_e^2} = -\frac{\hbar\omega}{2E_e}\left(\frac{a-1}{a+1}\right) + \frac{1}{2}\left(\frac{\hbar\omega}{2E_e}\right)^2\left[\frac{\cos \phi}{1 - \cos \phi} + \frac{6a}{(a+1)^2}\right] + \cdots \quad (14.33)$$

This expression is then used in

$$S(Q_e^2 + \Delta Q^2; \omega) = S(Q_e^2; \omega) + \Delta Q^2 \frac{\partial S(Q^2, \omega)}{\partial Q^2}\bigg|_{Q = Q_e} + \cdots \quad (14.34)$$

If only the terms given in Eqs. (14.33) and (14.34) are employed, it may be shown[22] that only the two moments given in Eq. (14.31) will be required on taking the integral in Eq. (14.22). In this case the Placzek correction is $O(1/M)$ and is independent of the detailed dynamical behavior of the sample, i.e., it is valid for all materials and thermodynamic states. Moreover, the first-order quantum mechanical correction consists[22] of replacing $k_B T$ in Eq. (14.31) by $\frac{2}{3}\times$ (mean kinetic energy of the atoms). All other correction terms (including quantum mechanical ones) are $O(1/M^2)$ and in part depend[22] on the details of $S(Q, \omega)$, such as the mean square force. It is therefore desirable that these terms be kept small. In the case of atomic fluids the $O(1/M)$ term is a smooth function of ϕ, while part of the $O(1/M^2)$ depends on $g(r)$ and oscillates similarly to $S(Q)$. This is a further reason for keeping the $O(1/M^2)$ terms small. As a guide it may be assumed for coherent scattering that the ratio of the oscillatory part of $O(1/M^2)$ to the total $O(1/M^2)$ is roughly equal to $|S(Q) - 1|/S(Q)$ and that the ratio $O(1/M^2)/O(1/M)$ is roughly equal to $O(1/M)/S(Q)$. Then if the experimentalist would like the maximum part of $O(1/M^2) \simeq S(Q)/K$ and maximum oscillatory part of $O(1/M^2) \simeq |S(Q) - 1|/H$, where K and H are large numbers, the above conditions imply that $K \simeq H$ and that the maximum $O(1/M) \simeq S(Q)/\sqrt{K}$. This relates the maximum size of the Placzek correction to a precision quotient K. It also implies that the series is equal to the sum of a few terms, which is a condition for its convergence.[23] Thus for data good to 1% ($K \simeq 100$), the experiment should be designed so that the $O(1/M)$ Placzek correction is $\leq 10\%$.

The behavior of the sampling factor as a function of p is shown in Fig. 9 for angles of 90° and 30°. If $\phi \geq 90°$ it is single-valued, while for $\phi < 90°$ it is double-valued when $p > 1$. In addition the magnitude of p is restricted by the conditions in Eq. (14.27). If $S(Q, \omega)$ is such that part of it falls outside the observable p range, that part will not be seen experimentally. If in addition $S(Q, \omega)$ is reasonably smooth and roughly bell shaped, this would make good measurements impossible, because of both the unobservable regions and the large corrections needed when the functions p_e and F (Fig. 8 and 9) have unusual shapes. However, if $S(Q, \omega)$ divides (as a function of ω at constant Q) into two or more reasonably well-separated regions, it is possible to adjust the experimental conditions so that a region of $S(Q, \omega)$ near $\omega = 0$ falls on the central part of the curves in Fig. 9 (i.e., $F \simeq 1$, and p near zero), while other regions fall outside the observable p range. In this case the expansion in Eq. (14.33) may be used for the observable part of $S(Q, \omega)$, which will be denoted by $J(Q, \omega)$. Such cases occur frequently for molecular fluids that are at temperatures well below that required to excite vibrational levels (e.g., for nitrogen at room temperature the lowest vibrational state is at an energy of $11k_B T$). If, in addition, the neutron energy is not large enough to excite the lowest vibrational state, the initial and final states of the molecules are the ground state. Other transitions, which may be seen in the complete $S(Q, \omega)$, are excluded by the experimental conditions and so, in this example, $J(Q, \omega)$ is the scattering function for $0 \rightarrow 0$ transitions. Thus to use the Placzek expansion it is necessary first that $J(Q, \omega)$

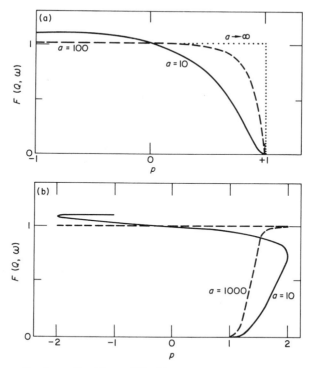

FIG. 9. The sampling factor for (a) $\phi = 90°$ with $a = 10$, 100, and ∞ and (b) for $\phi = 30°$ with $a = 10$ and 1000, as a function of the nondimensional variable $p = 2m\omega/\hbar Q^2$.

be confined to a region satisfying the conditions in Eq. (14.30) and second that the function $[S(Q, \omega) - J(Q, \omega)]$ either be zero (as for atomic fluids) or lie outside the observable range (as for some molecular fluids). In what follows only functions $J(Q, \omega)$ satisfying these conditions will be considered, and the notation $J(Q, \omega)$ will imply that these conditions are satisfied.

Thus an expansion for the central portion of the sampling factor is required and it is

$$F(Q_\omega, \omega) = 1 - \frac{3}{a+1}\left(\frac{\hbar\omega}{2E_e}\right) - \frac{3}{a+1}\left(\frac{\hbar\omega}{2E_e}\right)^2\left[1 + \frac{3}{2}\frac{a-1}{a+1}\right] + \cdots. \tag{14.35}$$

When this expansion is substituted into Eq. (14.22), the function $S(Q, \omega)$ must be replaced by $J(Q, \omega)$.

The function $N(k, k_e)$ may be expanded if $\Phi(1/k)$ is specified. It will be assumed for purposes of illustration that a pulsed source having a slowing down (dE/E) spectrum is used, so that $\Phi(1/k) \propto k$ [if a E^{-n} spectrum is used, additional terms proportional to $(1 - n)$ appear in the expansion below]. Putting $f(k) = 1 - \exp(-u/k)$, we get

$$N(k, k_e) = \frac{k(1 - \exp - u/k)}{k_e(1 - \exp - u/k_e)} \tag{14.36a}$$

$$= 1 - \frac{Aa}{a + 1}\left(\frac{\hbar\omega}{2E_e}\right) - \frac{1}{2}\left[\frac{a^2B}{(a + 1)^2} - \frac{3aA}{(a + 1)^2}\right]\left(\frac{\hbar\omega}{2E_e}\right)^2 + \cdots \tag{14.36b}$$

The expression in Eq. (14.36a) is the same as obtained from Eq. (14.23) when $k_0 = k_e$, and so covers the steady-state source as well if $a \to \infty$ in Eq. (14.36b). Here, A and B are detector constants related to the detector absorption coefficient u by

$$A = 1 - \left[\frac{1}{\varepsilon_e} - 1\right]\ln\left[\frac{1}{1 - \varepsilon_e}\right],$$

$$B = A + \left[\frac{1}{\varepsilon_e} - 1\right]\ln^2\left[\frac{1}{1 - \varepsilon_e}\right], \tag{14.37}$$

where

$$\varepsilon_e = f(k_e) = 1 - \exp(-u/k_e).$$

These constants are chosen so that the coefficients in Eq. (14.36b) reduce to A and $B/2$ when $a \to \infty$, and that for a $1/v$ detector $A = B = 0$ and for a "black" detector $A = B = 1$. The coefficients employed by other authors are related to A and B in the way shown in Table II.

At this stage the function $J(Q, \omega)$ may be Taylor expanded about Q^2 [Eq. (14.34)] and Eq. (14.33) substituted for ΔQ^2; it will be noted that $J(Q, \omega)$

TABLE II. Relationships between Different Detector Coefficients

Authors	Coefficients[a]	
Yarnell et al.[b]	$C_1 = \dfrac{1 + A}{2}$	$C_3 = \dfrac{1 + B - 2A}{4}$
Blum and Narten[c]	$k_0\varepsilon_1 = A - 1$	$k_0^2\varepsilon_2 = 2 - A - B$
Powles[d]	$\alpha_d = 2(1 + A)$	$\beta_d = -2(1 - 2A + B)$
Powles[d]	$F_1 = A - 1$	$F_2 = 2A - A^2 - B$
		$_2F = 2 - A - B$

[a] A and B are defined in Eq. (14.37).
[b] *Phys. Rev. A* **7**, 2130 (1973).
[c] *J. Chem. Phys.* **64**, 2804 (1976).
[d] *Mol. Phys.* **36**, 1161, 1181 (1978).

is even in Q and thus a function of Q^2. Finally, inserting this expansion and Eq. (14.35) and (14.36b) into Eq. (14.22) gives

$$S_{obs}(\phi, E_e) = \left[\frac{\sigma_e}{\sigma_s}\left(\frac{d\Omega}{d\sigma}\right)_e\right]\left(\frac{d\sigma}{d\Omega}\right)_s = \int_{-\infty}^{+\infty} J(Q_e, \omega)\,d\omega$$

$$-\frac{(Aa+3)MJ_1}{(a+1)2E_e} - \frac{2m(a-1)\dot{J}_1(1-\cos\phi)}{\hbar(a+1)} + \frac{m\cos\phi\dot{J}_2}{2E_e}$$

$$+\frac{6am\dot{J}_2(1-\cos\phi)}{2E_e(a+1)^2} + \frac{m(Aa+3)\dot{J}_2(1-\cos\phi)}{(a+1)E_e}$$

$$-\frac{\hbar^2 J_2(a^2B + 3(5a-1) - 9aA)}{8E_e^2(a+1)^2} + \frac{2m^2(1-\cos\varphi)^2\ddot{J}_2}{\hbar^2} + \cdots,$$

(14.38)

where dots denote $\partial/\partial Q^2$ and

$$J_n = \left[\int_{-\infty}^{+\infty} \omega^n J(Q, \omega)\,d\omega\right]_{Q=Q_e}. \quad (14.39)$$

The term in $\ddot{J}_2$ is $O(1/M^2)$ in the following regime (ii), but is $O(1/M)$ for rigid molecules in regime (i). The difference

$$P(\phi, E_e) = S_{obs}(\phi, E_e) - \int_{-\infty}^{+\infty} J(Q_e, \omega)\,d\omega \quad (14.40)$$

is known as the Placzek correction, and if subtracted from $S_{obs}(\phi, E_e)$ gives the zero moment J_0. To use Eq. (14.38) requires a calculation of the moments J_n for $n = 1$ and 2, and the monatomic case has been given in Eq. (14.31). Through the condition of detailed balance these moments may be shown to involve the same function. Another way to use Eq. (14.40) is to perform numerical integrations using a model $S(Q, \omega)$ having a number of correct moments.

Egelstaff and Soper[23] divide experiments into two regimes:

$$\begin{aligned}\text{Regime (i)} \quad & S(Q, \omega) - J(Q, \omega) \neq 0, \\ \text{Regime (ii)} \quad & S(Q, \omega) - J(Q, \omega) = 0.\end{aligned} \quad (14.41)$$

They give expressions for $P(\phi, E_e)$ and $a \to \infty$ for monatomic fluids in regime (ii) and for homonuclear and heteronuclear diatomic molecular fluids in both regimes (i) and (ii). It is clear that the moments calculated from Eqs. (14.39) and (14.41) will be different in the two regimes and the appropriate moments for each regime are given in their paper. For regime (i) they used the model of a molecular fluid with vibrational zero-point energy but inaccessible excited states. Placzek[21] gave expressions for $a \to \infty$ in regime (ii) only. He derived expressions for the moments S_n (i.e., S replaces J in Eq. 14.39), which are used in most later papers. He also worked out some higher order terms in the expansion and gave expressions for the total cross section. Blum and Narten[24] and Powles[25] made calculations for several types of molecules but were inconsistent in their use of the two regimes; that is, their moments are not consistent with either of the conditions in Eq. (14.41). Before using the formulas of Blum and Narten[24] and Powles,[25] it is necessary to go through the derivations and insert consistent expressions for the J_n.

In the case of a monatomic fluid, Eq. (14.40) becomes

$$P(\phi, E_e) = -\frac{2ym}{M}\left[\frac{aA + a + 2}{a + 1}\right] + \frac{m}{M}\frac{k_B T}{2E_0}\left\{1 - 2y + \frac{4y(Aa + 3)}{a + 1}\right.$$
$$\left. - \frac{2y[a^2 B - 6a(A + 1) - 3(5a - 1)]}{(a + 1)^2}\right\} + \cdots, \qquad (14.42)$$

where $y = (1 - \cos\phi)/2 = \sin^2\phi/2$. As an example[26] of the use of this expression Fig. 10 shows absolute and relative corrections for krypton at room temperature. Data for two wavelengths are shown and it can be seen that the curves are very similar showing the strong angular dependence of this effect. Moreover, the largest correction occurs for the shorter wavelength, and this is generally true. Because the shorter wavelength gives larger Q, $|S(Q) - 1|$ is smaller and so the relative correction rises sharply with decreasing wavelength. At very high Q, a correction equal in magnitude to $S(Q)/\sqrt{K}$ is much greater than $|S(Q) - 1|$.

14.4.3. Approximate Methods of Correction

The Placzek method is the only accurate general method and the formulas in Egelstaff and Soper[23] are easy to use, so that experiments should be set up to satisfy the conditions in Eq. (14.30). Nevertheless, the correction will never be large in well-considered experiments because its maximum size is $\sim S(Q)/\sqrt{K}$ (e.g., if $K \simeq 100$, the maximum correction is $\sim 10\%$). Thus approximate methods may work too although, because it is first necessary to check them against the general method, their usefulness is limited. If the only available experimental data violate Eq. (14.30), then approximate methods must be used.

Wick[27] has expanded $S(Q, \omega)$ about the recoil energy $(\hbar^2 Q^2/2M)$, and his method was used by Egelstaff and Soper[23] to obtain a general correction for either the self-part of $S(Q, \omega)$ or for incoherent scattering. [Note that the formula II-12 in Egelstaff and Soper[23] should be multiplied by the ratio of counter efficiencies for E_e and for the recoil energy, and formula II-12 minus 1 is denoted by $W(\phi, E_e)$]. To extend Wick's method to coherent scattering

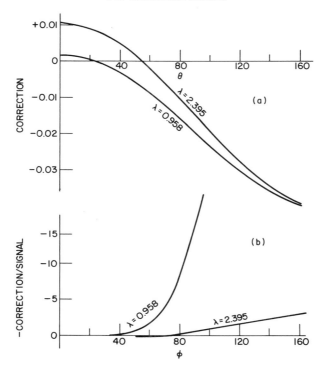

FIG. 10. (a) The inelasticity correction for krypton at 297 K for two wavelengths (2.395 and 0.958 Å), and detectors of 2.5-cm diameter ^{3}He at 10 atm. Note that the correction is nearly independent of λ for angles >40° and that at these angles it is slightly *larger* for the *shorter* wavelength.[26] (b) The ratio of the inelasticity correction in (a) to the envelope of $|S(Q) - 1|$—called the signal. These curves demonstrate the importance of restricting the angular range, especially for short wavelengths. (Redrawn from Egelstaff.[26])

requires the quantum mechanical coherent second moment, which depends on the structure of the fluid, among other difficulties. This formula is useful when the interference correction $O(\hbar/M^2)$ may be neglected, which is expected to be true if the difference between $P(\phi, E_e)$ and $W(\phi, E_e)$ is not large. In this case $W(\phi, E_e)$ will be a better correction for light elements, and the example of liquid lithium will be reviewed. Ruppersberg and Reiter[28] and Olbrich *et al.*[29] have made precise measurements on lithium, but their largest correction is ~20% and they employ $P(\phi, E_e)$ with $M/m \simeq 9.5$ rather than 7. It may be shown by expanding $W(\phi, E_0)$ that increasing the mass in $P(\phi, E_e)$ makes the two curves similar, although the additional mass varies with angle bringing the ratio M/m up by approximately 1.5 for large angles. Figure 11(a) compares $P(\phi, E_e)$ and $W(\phi, E_e)$ for lithium at 573 K using $\lambda = 0.695$ Å, and it can be seen that the difference grows with angle. In this case about half

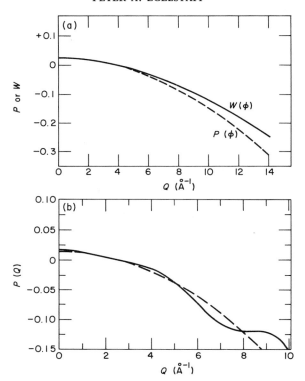

FIG. 11. Comparison of calculations of the inelasticity correction. (a) The inelasticity correction calculated for liquid lithium at 573 K and $\lambda = 0.695$ Å, according to the Placzek expansion [Eq. (14.42)], dashed line, and the Wick expansion [Egelstaff and Soper,[23] Eq. (II-12)], full line, denoted by $P(\phi)$ and $W(\phi)$, respectively. Note that these corrections are the same for $Q < 5$ Å^{-1} ($\phi < 32°$). (b) The Placzek correction for nitrogen at 297 K and $\lambda = 1.22$ Å with the de Gennes interference term (full line) and without this term (dashed line).

the scattering is incoherent and so the interference correction may be

$$\sim \frac{|S(Q) - 1|}{S(Q)} \frac{|W(\phi) - P(\phi)|}{2}.$$

Fortunately, this may be neglected over the range of Q shown. A larger error arises from the use of the empirical mass ratio 9.5 in $P(\phi)$ instead of using $W(\phi)$ with the lithium mass, and possibly the difference between these two results indicates underlying errors in the data ($\sim 4\%$).

For molecular fluids ad hoc formulas are used widely in the literature. For example, Deraman et al.[30] made careful measurements on nitrogen gas at room temperature and $\lambda = 1.22$ Å, where the initial and final states are the ground state, a good example of regime (i). However, they set to zero an interference term correction, originally pointed out by de Gennes.[31] Formulas for the de Gennes correction taken from Placzek[23] have been applied to this case, and the total correction with and without the de Gennes term is shown in Fig. 11(b). It can be seen that the interference correction may be neglected, except for the larger Q values (6–8 Å^{-1}). Over this range oscillations occur that are evident in the experimental data (Fig. 2 in Ref. 30) and that should be fitted to obtain the correct intramolecular term (which was not required in this work). The dashed line in Fig. 11(b) is the self-term correction, and in place of this Deraman *et al.*[30] used the Placzek correction for a rigid diatomic molecule, with an arbitrary effective mass determined by fitting to the data (they included a Q-dependent factor that was unimportant so that this mass was constant). The fitted value was 20% larger than the molecular mass and this difference is a possible indication of their underlying errors (~6%). In both of these examples (taken from the recent literature as examples of well-designed experiments), it was unnecessary to use an effective mass; the correct formulas with the proper target mass could have been used with equally satisfactory results. Both groups used relatively large scale factor adjustments (Section 14.5) and consequently the underlying errors (4% and 6%, respectively, at the high Q end of the data) may indicate that these scale factors are not constant (in Olbrich *et al.*[29] the data is bent to fit the renormalization conditions, see Section 14.5.2). For other examples in the literature, however (e.g., water), ad hoc adjustments have been made to parameters in the Placzek formula to secure a rough fit to observed data when the correction exceeds the safe limit [Eq. (14.30)]. For such poorly designed experiments, the results can be checked through numerical integration of Eq. (14.22) using only an accurate model of $S(Q, \omega)$ for the fluid being studied.

Inelasticity corrections are sometimes based on Eq. (14.22) and models for $S(Q, \omega)$ especially if the correction is large. However, the model must include the correct frequency moments to high order, if it is to be an improvement over the Placzek expansion. Therefore most of the models used to fit liquid or gaseous experimental $S(Q, \omega)$ data are unsatisfactory for this purpose. A model that has been used to test $P(\phi, E_e)$ calculations for molecular fluids is the free molecule. Exact formulas for a noninteracting rigid molecule may be written out and some modifications for zero point motion included.[23] But this model does not include the intermolecular terms and so is not valid for large corrections; if it is used for large corrections some approximation to the intermolecular terms must be used, for example, Vineyard's[32] approximation for $S_d(Q, \omega)$. Another model is that for the harmonic solid with a Bravais lattice. Exact formulas are available, and in this case the important high-energy transfers may be similar to those in the dense liquid. However, this model has not been exploited so far, perhaps because of fears that the interference terms may be overemphasized.

Finally, there is the possibility of using an experimentally determined $S(Q, \omega)$ in Eq. (14.22) or of finding $S(Q)$ by integrating an experimental $S(Q, \omega)$ if the inelasticity correction is very large. In either case the errors (see Section 14.5) on $S(Q, \omega)$ are such that the error on $S(Q)$ normally exceeds that from the Placzek or Wick corrections. Moreover, the data collection time is much longer for $S(Q, \omega)$ work. It is thus better to readjust the experimental conditions to keep the Placzek or Wick corrections within acceptable limits than to make inelastic scattering measurements covering $S(Q, \omega)$ in detail.

14.5. Evaluation of $S(Q)$ and $S(Q, \omega)$

14.5.1. Reduction of Vanadium Data

The measurements with a vanadium sample described in Section 14.2.2 for a diffraction experiment will be denoted by $I_V(\phi, k_e)$ and a similar measurement with a cadmium or another absorbing sample replacing the vanadium will be denoted by $I_{Cd}(\phi, k_e)$; in each case the data are normalized by the beam monitor readings. It will be assumed that these samples are not inside close fitting containers, and that a no-sample run has been taken that will be denoted by $I_0(\phi, k_e)$. At angles between 45° and 135° it is expected that $I_{Cd}(\phi, k_e) \simeq I_0(\phi, k_e)$, and that the difference between these two quantities may be neglected. However, the difference $I_0(\phi, k_e) - I_{Cd}(\phi, k_e)$ should be examined as a function of angle especially at low angles, to detect neutrons that have passed through the sample location and go on to generate a background by scattering from parts of the apparatus beyond the sample. Beam "scrapers" and baffles are installed in many instruments to eliminate such effects, but if they are not completely effective a difference is expected at low angles. This difference must be modified by an absorption factor before being subtracted from other readings, and in the following it will be assumed that this has been done. Often, the counting rate must be sacrificed to obtain completely clean conditions at low angles.

Then from Section 14.3 the measured intensity is related to the vanadium differential cross section $(d\sigma/d\Omega)_V$ by

$$I_V(\phi, k_e) - I_{Cd}(\phi, k_e) = C(\phi, k_e) A_V N_V \left[\left(\frac{d\sigma}{d\Omega}\right)_V + \frac{\Delta\sigma_s}{4\pi} \right], \quad (14.43)$$

where $C(\phi, k_e)$ is a calibration factor including the monitor efficiency and the detector geometry. N_V is the effective number of vanadium nuclei in the beam, given by[6]

$$N_j = \rho_j \int_{V_j} P(dV_j) D(dV_j) \, dV_j, \quad (14.44)$$

where ρ_j is the number density in the sample, $P(dV_j)$ and $D(dV_j)$ are the beam and detector profiles, respectively, and dV_j is a volume element for the jth component of the sample (in this case j is vanadium). From Section 14.4 the cross section of vanadium is

$$\left(\frac{d\sigma}{d\Omega}\right)_V = \frac{\sigma_s}{4\pi}[1 + P(\phi, k_e)], \tag{14.45}$$

where $P(\phi, k_e)$ is given by Eq. (14.42). The bound atom cross section $\sigma_s = 5.05 \pm 0.05$ b and the coherent cross section is 0.6% of σ_s and so is less than the error in the total cross section (in principle this error could be reduced in the future). Thus the absolute value of the calibration factor $C(\phi, k_e)$ may be determined to $\sim 1\%$ accuracy if the other quantities are well known. The uncertainties in N_V may be significant and thus $C(\phi, k_e)$ may differ from that for other samples unless the precautions described in Section 14.2.2 have been taken. For parallel-sided uniform slab samples the error in N_V should be less than for cylindrical samples. In some cases[24] the uncertainty can be reduced to $\sim 1\%$, but for many experiments the proper $C(\phi, k_e)$ is determined to about 5% only. This error is a systematic one and should generate the same systematic error in all $S(Q)$ measurements for which it is used. Therefore it may be reduced by any one of the other normalization procedures discussed in Section 14.5.2. The relative error in $C(\phi, k_e)$ as a function of ϕ or k_e can arise from errors in A_V, Δ, or $P(\phi, k_e)$, but these may be made small by good experimental design. Another relative error is the statistical error on each channel, namely,

$$I_V - I_C = \left(\frac{T_V}{M_V} - \frac{T_C}{M_C}\right) \pm \sqrt{\frac{T_V}{M_V^2} + \frac{T_C}{M_C^2}}, \tag{14.46}$$

where T_V and T_C are the total counts per channel recorded for vanadium and cadmium respectively and M_V and M_C are the corresponding flux monitor counts (which are assumed to be much larger). To make optimum use of the time available the cadmium sample need be run only until $(T_C/M_C^2) \ll (T_V/M_V^2)$.

If the instrument has been optimized in the way described in Section 14.1.2, the detectors and sample will be large, and an average angle ϕ should be calculated. As a guide Fig. 12 shows a median angle ϕ' given by

$$\cos \phi' = l \cos \phi \cdot (l^2 + d^2)^{-1/2} \tag{14.47}$$

where l is the sample to detector distance (between centers), and d is one quarter of the sum of detector and sample heights. For example, if $d = 0.1$ m and $l = 2$ m, then $\phi' = 4.15°$ when $\phi = 3°$. The angle ϕ' may be used in calculating Q with sufficient precision in most cases. Equation (14.47) is

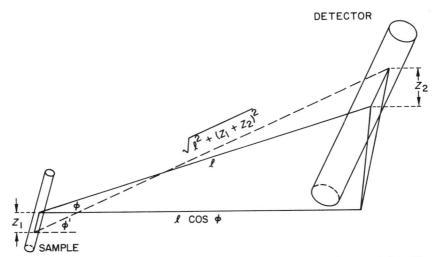

FIG. 12. Diagram to illustrate an approximation to the mean angle of scatter (ϕ') and its relationship to the angle (ϕ) used by most instruments.

useful for estimating limits of d for low-angle work, since the difference $2(\phi' - \phi)$ is approximately the spread in ϕ due to sample and counter height.

In the case of inelastic scattering measurements by the time-of-flight method, Eq. (14.43) is replaced [using Eq. (14.14)] by the following equation, after converting time-of-flight to ω. The intensity I is then the observed intensity times $\hbar t^3/t_0^2 E_0$.

$$I_V(\phi, k, k_0) - I_{Cd}(\phi, k, k_0) = N_V V(\phi, k, k_0) \frac{kf(k)}{k_0 f(k_0)} H_1(k_0, k)$$

$$\times \{S(Q_\omega, \omega) + \Delta R_2(k_0, k)\} \frac{\sigma_s}{4\pi}. \quad (14.48)$$

(This function should be folded with the resolution etc.—Part A, Chapter 3.) If the vanadium spectrum is integrated over k for fixed k_0 (which is measured in this experiment), the results reduce to Eq. (14.43) and $C(\phi, k_0)$ can be determined as before. This method will be satisfactory if $P(\phi, k_0)$ is small. Otherwise, or as an alternative, the elastic peak in the vanadium spectrum may be integrated and Eq. (14.48) integrated for this part of $S(Q, \omega)$ only [i.e., $e^{-2W}\delta(\omega)$]. Because the frequency distribution of a solid at low frequencies varies as the square of the frequency, the inelastic scattering under the elastic peak is a curve almost parallel to the ω axis and so may be subtracted by inspection in most cases. Here, $R_2(k_0, k)$ will consist of a broad part that may be subtracted in the same way and an elastic part that must be calculated by one of the procedures described in Section 14.3.3. Then this method will give $C(\phi, k_0)$ again, which may be compared to the previous result as a check on the approximations made. Both methods avoid resolution problems by the integration procedure. The function $C(\phi, k_0)$ is employed in the ideal equation

$$_0I_e(Q, \omega) = \left(\frac{N\sigma_s}{4\pi}\right)_V C(\phi, k_0)\, \delta(\omega) \tag{14.49}$$

to calibrate future results, where $_0I_e$ means the intensity for a perfect elastic scattering sample in the no-scattering limit. That is, $C(\phi, k_0)$ is the area of the spectrum seen by the detector if $S(Q) = 1$, the detector efficiency is constant, and the no-scattering limit is taken. Also, since $C(\phi, k_0)$ is an instrumental constant, its variation with time is a measure of instrumental stability.

In the case of experiments using a three-axis spectrometer, several modes of operation are possible, but the most convenient is to hold E constant and vary E_0, ϕ in order to keep Q constant (within limits set by the stepping intervals on E_0 and ϕ). For any energy transfer $\hbar\omega = E_0 - E$, the angle ϕ and the initial angle ϕ_i are related by

$$\cos \phi = \left(\cos \phi_i + \frac{\hbar\omega}{2E}\right)\left(1 + \frac{\hbar\omega}{E}\right)^{-1/2}. \tag{14.50}$$

This method has the advantage that the analyzer Bragg angle is not altered so the detector efficiency is fixed. However, small resolution changes occur (see Part A, Chapter 3) as the spectrum is measured and for this reason, and also to limit potential problems with orders, it is desirable to have $(\hbar\omega)_{max} \lesssim E/5$. The intensity is given by Eq. (14.48), except that $f(k)/f(k_0)$ for the detector is replaced by the same ratio for the beam monitor. If the monitor is a $1/v$ counter, this factor cancels the factor k/k_0. It is then possible to proceed with the two methods described and obtain a calibration factor $C(\phi, k)$ for this technique, which will be used in Eq. (14.49) in place of $C(\phi, k_0)$.

For some experiments it may be desirable to replace vanadium by another material for which $d\sigma/d\Omega$ or $d^2\sigma/d\Omega\, d\omega$ is well known. A particular case occurs for gases, since the geometry of the vessel may be included in the calibration by replacing the sample by another gas. For example, Powles et al.[33] in measuring $S(Q)$ for nitrogen calibrated their results against oxygen. Apart from its cost, Ar^{36} at a low pressure may be used too.

14.5.2. Static Structure Factor Data

Experimental data are taken for a sample inside a container and for the empty container, and background runs are taken and treated in the way described in Section 14.5.1. Then Eq. (14.43) is used for both sets of data, namely,

$$I'_g(\phi, k_e) = I_g(\phi, k_e) - I_{C_d}(\phi, k_e)$$
$$= C(\phi, k_e)\left[A_g N_g\left(\frac{d\sigma}{d\Omega}\right)_g + \left(\frac{d\Sigma}{d\Omega}\right)_{MS}\right], \quad (14.51)$$

where the subscript "g" means scattering target, and

$$\left(\frac{d\Sigma}{d\Omega}\right)_{MS} = \Delta_g \sum_j \left(\frac{\sigma_s}{4\pi}\right)_{gi} A_{gj:g} N_{gj} \quad (14.52)$$

is a macroscopic cross section for multiple scattering. This equation has been written as a sum over several components (sample + container + structures, etc.), which form the complete target, although usually $(d\sigma/d\Omega)_{MS}$ is computed directly and the sum need not be evaluated explicity. The factor $A_{gj:g}$ means scattering by the component T_j and attenuation by the whole target g; it depends on ϕ and E_0 as discussed in Section 14.3.2. If SC denotes a full container and C an empty one, the differential cross section for the sample is given by

$$N\left(\frac{d\sigma}{d\Omega}\right)_S = \frac{1}{A_{S:SC}}\left[\frac{I'(\phi, k_e)}{C(\phi, k_e)} - \left(\frac{d\Sigma}{d\Omega}\right)_{MS}\right]_{SC}$$
$$- \frac{1}{A_{C:C}}\left[\frac{I'(\phi, k_e)}{C(\phi, k_e)} - \left(\frac{d\Sigma}{d\Omega}\right)_{MS}\right]_C \frac{A_{C:SC}}{A_{S:SC}}. \quad (14.53)$$

Here, $(d\sigma/d\Omega)_S$ is the quantity obtained from the experiment, and it is converted to $S(Q)$ using the results of Section 14.4. So far all cross sections have been the sum of coherent and incoherent scattering, and as $S(Q)$ is proportional to the bound atom coherent cross section, these terms are separated at this point. Thus

$$S(Q) = \frac{4\pi}{\sigma_c}\left(\frac{d\sigma}{d\Omega}\right)_S - \left[\frac{\sigma_i}{\sigma_c} + \frac{\sigma_i + \sigma_c}{\sigma_c} P(\phi, k_e)\right]_S. \quad (14.54)$$

It will be seen from Eq. (14.53) and (14.54) that, if $(d\sigma/d\Omega)_{MS}$ is small, the errors on the number of atoms and on the cross sections of the sample material and the vanadium are directly related to errors appearing in $S(Q)$. These will amount to several percent in most cases, and great care is required if they are to be reduced to $\sim 1\%$. Therefore the structure factor obtained from Eq. (14.54) is renormalized using some other criterion. Because this criterion will be used for a single value of Q (or of r if transformed), it does not modify the angular dependence included in the vanadium data. Moreover, for any set of measurements covering successively several (PVT) states of a fluid, using the same container, only a single correction factor is needed. If, after this correction, the data as a function of Q, or as a function of (PVT) with simple dependence on N_S, do not satisfy the normalization rules, then other kinds of experimental error are important (e.g., scattering by impurities such as hydrogen). Some experimentalists try to reduce these errors as well by repeating the normalization procedure, but this is equivalent to altering the experimental observations and so the results should be reexamined in each case.

Several properties of $S(Q)$ may be used to establish a basis for renormalization, and the most commonly used ones are summarized here:

(1) $S(Q)$ is a smooth function of Q and is positive everywhere: this rule is used to reduce statistical fluctuations by suitably averaging points or to interpolate data over narrow regions where large corrections (e.g., due to the major Bragg reflections from a thick container) have spoiled the data.

(2) $$S(Q) \to \rho \kappa_B T \chi_T [(\sum b_j)^2 / \sum (b_j^2)] \text{ as } Q \to 0, \qquad (14.55)$$

where χ_T is the isothermal compressibility, b_j the bound atom coherent amplitude for the jth nuclear species, and the factor $(\sum b_j)/\sum b_j^2$ is used so that $S(Q) \to 1$ as $Q \to \infty$. Since $0 \le \chi_T \le \infty$, the $Q \to 0$ limit varies widely. If $0.2 < S(0) < 2$ and if good low Q data have been taken, it will be useful. The lower limit arises through the need to have a large enough quantity to measure, and the upper limit through the need to avoid a steeply varying function at low Q. Sometimes, this rule can be used to check σ_i in Eq. (14.54) or the constants in Eq. (14.53) (as, e.g., in Ref. 28).

(3) $S(Q)$ approaches a theoretical result for the largest range of Q measured (this may be either unity or a result depending on the short-range structure). This is the most widely used renormalization rule; in many cases it is the only rule needed and usually it is more reliable than either (2) or (4). In practice it is used to calculate a cross section and so normalize $d^2\sigma/d\Omega\, d\omega$ [rather than $S(Q)$] at large angles. Ideally when both (2) and (3) can be used, the same normalization should be obtained. Sometimes this is not so, and this indicates that the errors have accumulated to produce a nonconstant normalization factor.

(4) In r space, $g(r) \simeq 0$ for small distances which we denote by $r < r_c$ so that

$$-2\pi^2 \rho = \int_0^\infty Q^2 [S(Q) - 1]\, dQ \frac{\sin Qr}{Qr} dQ \quad \text{for} \quad r < r_c, \qquad (14.56)$$

where for atomic fluids $r_c \approx 2$ Å, while for molecular fluids r_c is less than or equal the minimum bond length. This rule has been used extensively in the x-ray field, mainly to improve the separation of the elastic ($k = k_0$) component from other components in the data and to overcome the inherent normalization problems with x rays. Since neither of these difficulties occurs with neutron scattering, it is not necessary to use this rule unless there is an unusual problem with the data. Moreover, because it requires a good knowledge of $S(Q)$ for all Q (especially at large Q), some assumption is needed to use it. In principle this rule is sensitive to the normalization factors, but this sensitivity makes it sensitive to errors as well. In this connection the importance of good data for the total cross section and σ_c and σ_i should be emphasized. Some experimentalists have used this rule to alter their observations to give $g(r)$ a good appearance at low r, but there seems to be little to recommend this practice. An alternative approach is to renormalize the data by rule (3), and to report deviations from rule (4) without making any further adjustments.

The latter approach was used by Teitsma and Egelstaff[34] for krypton gas at $\rho = 2.88 \times 10^{27}$ atoms/m^3 and $T = 297$ K, with the result shown in Fig. 13. Figure 13(a) shows the fit after the renormalization through rule (3), i.e., multiplying $S(Q)$ by a factor of 1.01, and Fig. 13(b) shows the resulting $g(r)$ with no adjustments to fit rule (4). These data are of good quality because the procedures described in the earlier sections of this chapter had been followed, although the 1% change was small fortuitously, because it is less than the total uncertainty in bound atom cross sections (2%). In Teitsma and Egelstaff,[34] the renormalization constants are reported explicitly, but in other papers[28-30, 33] covering accurate studies they are included in fictitious constants (e.g., for cross sections, masses, attentuation coefficients, etc.) used in place of real constants in order to satisfy rules (2) to (4). This practice makes it difficult for a reader to establish exactly what changes have been made to accomplish the renormalization. It seems, however, that in Ruppersberg and Reiter[28] and Olbrich et al.[29] on liquid lithium a correction of ~11% was used and in Powles et al.[33] on nitrogen gas a correction of ~7% was used. These corrections are larger than found for krypton mainly because the experimental and interpretive procedures described in earlier sections of this chapter were not followed in full detail, and in the case of lithium because of the large size of the incoherent scattering correction. Nevertheless, it is desirable to employ the best values of real constants and to renormalize by additive or multiplicative arbitrary constants that are then reported explicitly. The use of real constants and explicit reporting of renormalization changes is advised, because it enables different experimental procedures to be compared and so may lead to improvements in the future. This remark applies to both $S(Q)$ and $S(Q, \omega)$.

The structure factors for krypton, lithium, and nitrogen are not very

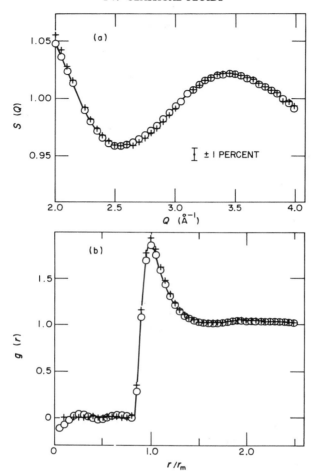

FIG. 13. (a) $S(Q)$ for krypton gas[34] at 297 K and $\rho = 2.88 \times 10^{27}$ atoms/m³ (circles and line) compared to a calculation (crosses) based on the potential (full line) of Fig. 2. The close agreement is indicated by the 1% marker.[34] (b) The pair correlation function for krypton gas at 297 K and $\rho = 2.88 \times 10^{27}$ atoms/m³ (about 100 atm) (circles and line from Teitsma and Egelstaff[34]), compared to a calculation (crosses) based on the potential (full line) of Fig. 2. (Redrawn from Teitsma and Egelstaff.[34])

sensitive to the Q-space resolution of the instruments employed. The reason is that, in optimizing the counting rate, the resolution has not been relaxed so far that Bragg reflections are broadened excessively. This is partly because Bragg reflections are useful in calibrating the wavelength, partly because they must be subtracted with the container scattering and partly to meet the needs of other users of the instruments. In such cases a resolution correction may be made through a Taylor expansion formula; i.e.

$$S(Q) = S_{obs}(Q) - \frac{1}{2} \frac{\partial^2 S_{obs}(Q)}{\partial Q^2} \frac{\int Q^2 R(Q)\, dQ}{\int R(Q)\, dQ} + \cdots, \qquad (14.57)$$

where $R(Q)$ is the Q-space resolution function (assumed to be symmetrical). This correction is greatest near the principal peak in $S(Q)$ because the derivative is largest there.

14.5.3. Dynamic Structure Factor Data

This discussion is similar to that given above except that the starting point is Eq. (14.48) rather than Eq. (14.43). Then the generalization of Eqs. (14.51)–(14.53) to the case of inelastic scattering is straightforward [where I' is the difference shown on the left-hand side of Eq. (14.48)]:

$$\frac{f(k_0)}{f(k)} I'_g(\phi, k, k_0) = C(\phi, k_z) \left[A_g N_g \left(\frac{d^2\sigma}{d\Omega\, d\omega} \right)_g + \left(\frac{d^2\Sigma}{d\Omega\, d\omega} \right)_{MS} \right] \qquad (14.58)$$

and

$$\left[N \left(\frac{d^2\sigma}{d\Omega\, d\omega} \right)_{obs} \right]_S = \frac{1}{A_{S:SC}} \left[\frac{I'(\phi, k, k_0) f(k_0)}{C(\phi, k_z) f(k)} - \left(\frac{d^2\Sigma}{d\Omega\, d\omega} \right)_{MS} \right]_{SC}$$
$$- \frac{1}{A_{C:C}} \left[\frac{I'(\phi, k, k_0) f(k_0)}{C(\phi, k_z) f(k)} - \left(\frac{d^2\Sigma}{d\Omega\, d\omega} \right)_{MS} \right]_C \frac{A_{C:SC}}{A_{S:SC}}, \qquad (14.59)$$

where the attenuation factors $A_{T_j:T}$ are now dependent on k (for time-of-flight data) or k_0 (for three-axis data) and k_z means k_0 (time-of-flight) or k (three-axis). In these equations the differential cross sections for multiple scattering are assumed to have been calculated without the counter efficiency included (i.e., for a black counter). However, if this efficiency is included in the calculation the factor $f(k_0)/f(k)$ should be moved outside the square brackets in Eq. (14.59). This factor refers to the detector for time-of-flight or to the monitor for three-axis experiments. In the case of time-of-flight measurements, a complete package of data reduction programs has been described by Verkerk and van Well.[35a]

The measured quantity is, approximately,

$$\left(\frac{d^2\sigma}{d\Omega\, d\omega} \right)_{obs} = \int_{-\infty}^{+\infty} R(\omega - \omega') \frac{k}{k_0} \{ \bar{b}^2 S_c(Q, \omega') + b_i^2 S_i(Q, \omega') \}\, d\omega', \qquad (14.60)$$

where $R(\omega)\, d\omega$ is the resolution function in ω space and has unit area. This is an approximation because the Q-space resolution is not shown (but see

Section 14.5.2), and because the measurement is not in ω space in most cases. The latter point is only significant when the resolution broadening is large.

A useful method of correcting a large quantity of data for the resolution function $R(\omega)$ has been described by Verkerk,[35b] based on the fact that the transform of Eq. (14.60) is a product. The observed function [i.e., Eq. (14.60)] is represented as the sum of a set of equispaced readings. A discrete Fourier transformation is made and the most significant set of Fourier components is chosen. The Fourier components of the observed function are divided by those of the resolution function and an inverse discrete transformation made. A limited number of these Fourier components are summed to give a good approximation to $S(Q, \omega)$. The important step is the choice of the significant components, and the way of doing this and estimating the errors on the deconvoluted data are described in Verkerk's paper.[35b] This method treats relatively large resolution corrections, but when the correction is small the simpler Taylor expansion method may be used successfully [i.e., Eq. (14.57) with ω replacing Q].

The sum rules

$$\int_{-\infty}^{+\infty} S_c(Q, \omega) \, d\omega = S(Q); \qquad \int_{-\infty}^{+\infty} S_i(Q, \omega) \, d\omega = 1, \qquad (14.61)$$

and for a classical fluid (either coherent or incoherent),

$$\int_{-\infty}^{+\infty} \omega S(Q, \omega) \, d\omega = \frac{\hbar Q^2}{2M}; \qquad \int_{-\infty}^{+\infty} \omega^2 S(Q, \omega) \, d\omega = \left(\frac{k_B T}{M}\right) Q^2 + 0\left(\frac{\hbar^2}{M^2}\right), \qquad (14.62)$$

may be used to check the normalization $S(Q, \omega)$ when the data has been listed at constant Q (see p. 453). If $S(Q)$ has been determined by the methods described above, Eq. (14.61) is the most useful for this purpose. In fact the whole of the renormalization discussion given for $S(Q)$ may be employed for $S(Q, \omega)$ through the use of Eq. (14.61). However, if the wings of $S(Q, \omega)$ have been determined accurately, Eq. (14.62) is a useful check as it is more sensitive to the wings that to the intense central portion of $S(Q, \omega)$. In their extensive study of coherent and incoherent inelastic scattering by liquid argon, Sköld et al.[36] used these tests. Their results are shown in Fig. 14 for the first three moments. Figures 14(a) and (d) show the ratios of experimental to theoretical values before multiple scattering and resolution corrections have been applied, while (b) and (e) show the data after these two corrections were applied. Because the experimental data are truncated before $\omega \to \infty$, an extrapolation to $\omega \to \infty$ was based on a theoretical model. Figures 14(c) and (f) show the ratios after this extrapolation, and the similarity to 14(b) and (e) shows that it is relatively unimportant. The standard deviation of

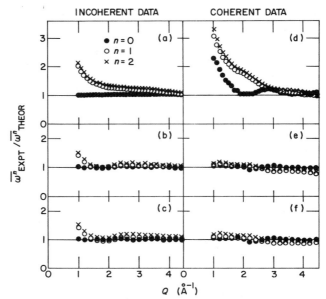

FIG. 14. A comparison of experimental and theoretical frequency moments [ω^n, Eqs. (14.70) and (14.71)] for liquid argon: (a) and (d) are data before multiple scattering and resolution corrections; (b) and (e) are after these corrections; and (c) and (f) are after extrapolation of the experimental data to $\omega \to \infty$. (Redrawn from Sköld et al.[36])

these moment ratios from unity was calculated from the 18 points shown in the figure and was 0.02, 0.12, and 0.19 for the incoherent zero, first, and second moments, respectively, and 0.05, 0.10, and 0.11 for the coherent data. The larger values for the first and second moments stem partly from the region $Q < 1.4 \text{ Å}^{-1}$, where the effect of the corrections is large.

It is sometimes useful to present $S(Q, \omega)$ as a family of curves at constant ω or constant Q. Provided the ω stepping is the same for each spectrum (as it is in time-of-flight data and may be chosen to be in three-axis data), the constant ω plot consists of one $+\omega$ and one $-\omega$ point for each spectrum and Q calculated from Eq. (14.7). The $\pm\omega$ points may be reduced from the symmetry of the scattering function:

$$e^{\hbar\omega/k_BT}S(Q, -\omega) = S(Q, +\omega). \qquad (14.63)$$

This rule, which is a useful test of the data, was deduced from the detailed balancing of scattering cross sections and was used in reactor physics calculations prior to the introduction of the van Hove function. However, for a classical fluid the $\pm\omega$ differences are negligible (i.e., $e^{\hbar\omega/k_BT} \simeq 1$). Constant Q plots may be obtained from the three-axis data taken in the way described in Section 14.5.1, and the $\pm\omega$ sides of the curves may be averaged using Eq. (14.63). In the case of time-of-flight data, an expansion of Eq. (14.7) for $\hbar\omega \ll E_0$ is useful [compare Eq. (14.29)]:

$$Q = Q_e\left[1 + \frac{\hbar\omega}{4E_0} + \cdots\right]. \tag{14.64}$$

For this case a Taylor expansion of $S(Q, \omega)$ shows that when $E_0 \gg k_B T/2AS(Q)$,

$$\hat{S}(Q_e, \omega) \simeq \frac{\hat{S}(Q, -\omega) + \hat{S}(Q, +\omega)}{2} + \cdots, \tag{14.64a}$$

where $\hat{S}$ is the symmetrized form of $S(Q, \omega)$, so that averaging $\pm\omega$ data for constant θ gives constant results. If the data involves larger ratios of $\hbar\omega/4E_0$, then $S(Q, \omega)$ is deduced from constant ω plots interpolated to the chosen value of Q. Frequently, the time-of-flight data are "binned," that is, suitably summed to give ω bins of constant width, in order to assemble convenient constant-ω plots. Both constant Q and constant ω plots are used in the literature for making comparisons with theoretical models or with computer simulations. It is, of course, equally satisfactory to compare with the measured cross sections, and it is sometimes argued that this is a more direct way to assess the agreement between experiment and theory.

Equation (14.59) shows that the observed function is strongly dependent on the counter efficiency function $f(k)$. In time-of-flight experiments, high-efficiency detectors are used and the efficiency as a function of k is not readily calculated with high precision. For this reason such detectors are calibrated against others that in turn are calibrated against low-efficiency detectors whose efficiency can be determined accurately (e.g., boron ion chambers, special types of scintillators, etc.). However, for the three-axis spectrometer used in the mode described earlier, only the monitor efficiency as a function of k_0 is required, and this is known for some monitors (i.e., those having well-defined absorption cross sections and good pulse height distributions). This is an advantage in some situations. If a large number of comparable $S(Q, \omega)$'s are required, e.g., when P, V, T, or concentration are varied, the three-axis spectrometer suffers from two disadvantages relative to the time-of-flight instruments. These are that the instrumental parameters are mechanically scanned point by point, rather than electronically scanned, and that the data collection time for a "complete" $S(Q, \omega)$ is longer for equal source strengths.[†] A "complete" $S(Q, \omega)$ does not cover all of the (Q, ω)

[†] This is the generally accepted view, but comparison of specific instruments may lead to opposite conclusions in some circumstances.

plane but a range wide enough to embrace the regions of interest only. When that region is narrow and a few states are to be covered, the advantages of point-by-point constant Q work and a good detector efficiency function may offset these disadvantages.

14.5.4. The Assessment of Quality in Experimental Data

The description of the experimental procedures and precautions has emphasized the need to look after each detail if good quality results ($\sim 1\%$ precision) are to be obtained. A substantial effort is required to do this; thus it is unwise to follow each step in detail without discovering whether it is necessary and so some method of assessing quality at each of several stages is desirable. Both $S(Q)$ and $S(Q, \omega)$ are smooth functions whose absolute magnitude is important, and the assessment of functions of this kind is always difficult. However, this task is simplified if such a stage-by-stage assessment is followed. To do this it is worth assessing the following six areas, to determine whether the experimental procedures and the data at various stages are consistent both internally and externally. It will be seen that a good assessment is difficult to do.

(1) The statistical precision required in the raw data corresponds to about 10^5 counts per point for $S(Q)$ or 10^6 counts per spectrum for $S(Q, \omega)$, and about 5×10^5 counts per angle for vanadium data if the final precision, after all checks and tests are complete, is to be $\sim 1\%$. Exceptional cases in which small differences are sought (e.g., between isotopes or states) may require larger totals. These totals should be broken into several smaller figures to check for stability, consistency, etc. In addition there should be some redundancy in the number of Q points or the number of spectra determined. These checks on stability, reproducibility, and consistency allow an assessment to be made of the instrument and the way it is being used.

(2) Each step in the elaborate procedures described in Sections 14.1 to 14.5 (or each assumption in the theoretical arguments) should be evaluated quantitatively in the circumstances of the proposed experiment: the error in altering or omitting a step may then be estimated. These steps do not carry equal weight for all experiments, so that some optimization of effort can be made as a result of this quantitative evaluation.

(3) The comparisons with normalization rules, moment theorems, etc., are made using real constants before any adjustments are made to the original data. They are examined in the light of the analysis of the experiment carried out under (2). It is important that the outcome of these comparisons is consistent with (2) for each test employed.

(4) An internal self-consistency test may be made between different sets of data taken as a function of P, V, and T changes, or for the same P, V,

and T but different containers, or for neutron wavelength changes, etc. These are relative comparisons in which the weighting of one of the factors considered under (2) can be altered.

(5) Intercomparison of different sets of high-quality results are very useful, but are possible only if each experimentalist publishes a clear and self-consistent account. For this reason most of these comparisons have been between data taken by the same group at different times. For this purpose long-term record keeping is needed, and then since many groups study the same material over many years (e.g., Li, N_2, Kr), this test can be useful.

(6) When comparisons are made with good quality theoretical predictions, the deviations should be reasonable on physical grounds. This applies to both the scattering functions and any difference functions that are available. In this test, as in others, subjective conclusions are unreliable and so some quantitative estimates are important.

14.6. Comparison of Experiment and Theory

14.6.1. Computer Simulations

In the hands of an experimentalist, the computer simulation of fluids is an important tool for planning experiments and for the interpretation of data. Except at low densities it is the only method available to calculate $S(Q)$ and $S(Q, \omega)$ for an arbitrary (P, V, T) state. With the advent of inexpensive computers capable of handling molecular dynamics and programs for the evaluation of molecular dynamics data, the coupling of neutron scattering and computer simulation techniques in this field is rapidly becoming an essential feature of experimental work. The method of molecular dynamics simulation has been described fully in the literature.[37,38] Various precautions must be taken to obtain reliable data, but these will not be reviewed here. The main limitations are the size of the system used in the simulation and the length of time for which it is followed. Many simulations fail to meet acceptable criteria for one or more of these parameters. For example, the simulation of $S(Q)$, either by calculating Fourier components or by calculating $g(r)$ and transforming, is done on a box of side L so that data for $Q < 4\pi/L$ cannot be determined (frequently this limit is 0.5 Å^{-1}). In the case of $S(Q, \omega)$, poor statistics at long times are usually a limitation, so that the computed $S(Q, \omega)$ spectra will be less precise and less extensive than the experimental ones. Also, with the high data rates from modern neutron instruments, the number of angles, energies, and states covered experimentally will be greater than the number of simulations available from present facilities in many laboratories. In spite of these limitations, the advantages of linking the two techniques on a routine basis are substantial.

The simulation is based on a pair potential that is an approximation to the interaction between an isolated pair of the atoms or molecules composing the real fluid. For noble gases excellent pair potentials have been determined by fitting models to low-density experimental data.[39] It is important to use these potentials or a good approximation to them. If a two-parameter potential is used for convenience, it is better to use the "effective atomic potential" (EAP) rather than the Lennard–Jones potential, since it is much closer to the known shape of the pair interaction. The EAP is

$$u(r) = \varepsilon \left[2\left(\frac{r_m}{r}\right)^{12} - 2\left(\frac{r_m}{r}\right)^{9} - \left(\frac{r_m}{r}\right)^{6} \right], \qquad (14.65)$$

where ε and r_m are the depth and position of the minimum, respectively (e.g., $201 k_B$ and 4.01 Å, respectively, for krypton). Since the r^{-12} form is steeper than the noble gas potential, the principal error occurs for $u(r) \to \infty$ at small r. On the scale of Fig. 2, the difference between Eq. (14.65) and the realistic potential could not be resolved, whereas the difference between Eq. (14.65) and the Lennard–Jones potential is quite clear. In the case of diatomic molecular fluids, a model potential is often used as the real potentials are not well known. Then comparisons with neutron data can be useful in improving the model to the extent that the model equations permit realistic adjustments to be made. For some molecular fluids *ab initio* potentials are available and are better than the models (e.g., HCl^{40}).

Computer simulation results are used in theoretical work to test liquid models, for example, integral equations for $S(Q)$, or generalized hydrodynamic, memory function or continued fraction models for $S(Q, \omega)$. Because the pair potential used in the simulation is known, some of the parameters in the model are known, and so approximations made for other parameters or in the structure of the model may be tested. For several models the parameters are related to the frequency moments of $S(Q, \omega)$; in pair theory the zero, second, and fourth moments may be calculated without difficulty. But if the model is compared to experimental data, not only is the potential not known exactly but the fourth and higher moments include terms depending on many-body forces. Moreover, the model is less reliable than the simulation that has been used to test it. Thus the direct use of models for comparison with experimental data is not satisfactory. However, in order to reduce the cost of simulation studies, it is possible to interpolate between different sets of computer data using such models with ad hoc parameters. Thus if the model parameters are fitted to the available computer data, the parameters may be interpolated and $S(Q, \omega)$ recalculated.

14. CLASSICAL FLUIDS

Because it is expensive to extend a simulation of $g(r)$ to large r or $I(Q, t)$ to large t, it is necessary to extrapolate the computer data before transforming. Some extrapolation procedures are based on models. The sensitivity of the resulting $S(Q)$ or $S(Q, \omega)$ to the model must then be assessed in each case.

The result of a simulation will be close to the experimental result and can be used in any estimates of experimental design parameters required beforehand, for example, in estimating resolution widths, ω and Q ranges to be studied, magnitude of changes with P, V, or T, importance of container scattering, etc. It can also be used in calculating the attenuation coefficients [Eq. (14.16a)] and as an input to the multiple scattering calculations (Section 14.3.3). If the inelasticity correction is large, it could be used also as a model as described in Section 14.4.3, although extensive data are required for this purpose and for the multiple scattering calculation.

Finally, computer simulation results are compared to the $S(Q)$ or $S(Q, \omega)$ data, and this chapter will be concluded with three examples of pair theory simulations compared to experimental data on atomic fluids. In the first, a liquid metal, the pair potential idea is imperfect as a metal is a two-fluid mixture of ions and electrons. However, at any state an approximate pair potential—the state-dependent pseudopotential—is used to represent the full many-body interaction and to calculate approximately the structure and dynamics at the atomic level. This is a good test of the pseudopotential picture. In the second example, the pair potential is well known (e.g., for noble gas atoms) and differences between theory and experiment are related to defects in the pair theory. The expansion of the potential energy as the sum of pair, three-body, four-body, etc., interactions is an approximation that has not been tested. The pair term is known to approximate experimental data to first order, but it is not known whether the series converges to the experimental result. It may be possible to examine this question experimentally. Alternatively, if the validity of the series is assumed, it may be possible to test models for the many-body forces (three-body or higher), the form of which are unknown for distances on the order of molecular or atomic dimensions. The major contribution to bulk properties by these forces arises from contributions over this range of distances.[41] This fact is ignored in most of the theoretical literature on bulk properties. Neutron scattering investigations of this kind may stimulate better theories of bulk properties.

The third example covers the attempts to identify the fundamental modes of motion in a fluid at arbitrary frequencies and wave numbers. There have been many contributions to this field from a number of opposing points of view, and it is still a controversial subject. Advances in neutron scattering experiments and in theoretical methods coupled with computer simulations are adding a further chapter to this topic.

14.6.2. Liquid Rubidium

An excellent series of experiments to determine $S(Q, \omega)$ for liquid rubidium at 315 K and 1 atm were carried out by Copley and Rowe.[42] A comparable molecular dynamics study was made by Rahman[43] using a volume-dependent pair potential proposed by Price et al.[44] This potential is consistent with the lattice dynamics of solid rubidium.

Copley and Rowe[42] used a sample container consisting of 32 slabs of liquid, each 2.54 cm wide and 0.22 cm high and 10.16 cm long, in an aluminum alloy container with 0.012 cm thick walls. Boron nitride dividers were used to separate the slabs of liquid and so reduce multiple scattering. This sample was used for time-of-flight experiments with an incident energy of 33 meV and angles $2.4 \leq \phi \leq 14.4°$. In another experiment they used a cylindrical aluminum alloy holder 1.68 cm (o.d.) and wall thickness 0.045 cm. The liquid was separated by boron nitride spacers (0.09 cm thickness) into several layers 1.68 cm high. Using a neutron energy of 4.94 meV, spectra were recorded by the time-of-flight method for $17.4 \leq \phi \leq 118.8°$. Then, using an energy of 33 meV, spectra were recorded for $26.4 \leq \phi \leq 108°$. In the former experiment $\Delta Q \approx 0.01$ Å^{-1}, and in the latter experiment $\Delta Q \approx 0.05$ Å^{-1} due to the spread in angles. The experimental procedures were similar to those described above.

For the molecular dynamics simulation, Rahman[43] used 500 particles in a box that gave a density of 1.502 g/cm^3 compared to 1.46 g/cm^3 in the experiment. The temperature was 319 K compared to 315 K, and these

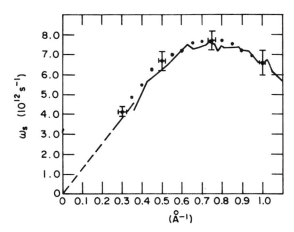

FIG. 15. The measured peak positions (ω_s) in $S(Q, \omega)$ for liquid rubidium at 320 K[42] compared to a molecular dynamics simulation[43] using a pseudo-pair-potential based on solid-state data.[44] The dashed line is an extrapolation of $\omega = cQ$. (Redrawn from Copley and Rowe[42] and Rahman.[43])

differences illustrate that there are problems in obtaining MD and experimental results at exactly the same (T, V) state (disregarding the pressure, which is different in this theory). The structure factors $S(Q)$ for the two cases agreed to within a factor 1.04 and the difference was attributed[42] to errors in the normalization. In addition the experimental peak height at $Q = 1.55$ Å^{-1} was several percent higher than the simulation after making this correction, but this is probably within experimental and computing errors.

For $Q < 1$ Å^{-1} the scattering function $S(Q, \omega)$ showed some structure, having a central peak ($\omega = 0$) and broad side peaks (at $\omega = \omega_s$). A similar shape was found in the simulation, and a plot of ω_s for $0.3 \leq Q \leq 1$ Å^{-1} is given in Fig. 15. Excellent agreement was found. In Fig. 16(a) the area of $I(Q, t)$ [i.e., $S(Q, 0)$] is shown, and in Fig. 16(b, c) several plots of $I(Q, t)$ are

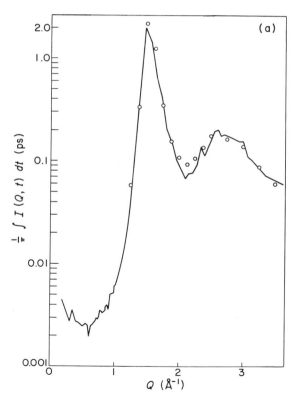

FIG. 16. The intermediate scattering functions for liquid rubidium for (a) $\omega = 0$ and (b) and (c) for Q's as indicated. The full lines are MD results and the circles are data from Copley and Rowe.[42] (Redrawn from Rahman.[43])

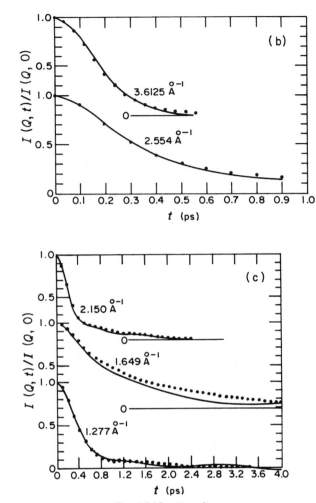

FIG. 16 (*Continued*)

given (circles are experimental results and the full line is the MD simulation). The overall agreement is again excellent for $1.25 \leq Q \leq 3.6 \text{ Å}^{-1}$, except at $Q_m = 1.65 \text{ Å}^{-1}$ where a better simulation is needed. Thus if a volume-dependent pseudopotential is deduced from solid-state data, the structure and dynamics [i.e., $S(Q)$ and $S(Q, \omega)$] of this liquid metal are described satisfactorily. It is expected that the volume dependence may be extended to some other states (Mountain[45]), and comparisons of simulations with data for $S(Q)$ along the coexistence curve have been made successfully.

14.6.3. Dense Krypton Gas

Krypton has been the subject of a series of papers[16,34,46,47] on the properties of the gas as a function of density at room temperature. The structure factor at low densities was measured[34] using a two-axis diffractometer and a neutron wavelength of 2.395 Å. Gas pressures corresponded to 10 to 200 atm. The experimental procedures were those described in Sections 14.1 to 14.5, and the numerous precautions and the detailed steps in the data analysis were followed. This led to accurate data for $0.3 \leq Q \leq 4.0$ Å^{-1}, and a good normalization (Fig. 13). The Monte Carlo method of computer simulation was used with a pair potential[48] deduced from bulk properties. Figure 17 shows $S(Q)$ for $Q < 2$ Å^{-1} and several densities: the circles and line are the experimental result and the crosses are the simulation. At the lowest densities the agreement between the two results is excellent, and as $|S(Q) - 1| < 0.025$ this is a good check on the procedures. At higher densities there are discrepancies for $Q < 1$ Å^{-1}, which increase as ρ^2. Teitsma and Egelstaff[34] relate the difference between the measured and simulated $S(Q)$ to the three-body potential term in its density expansion.

The dynamic structure factor was measured over a higher density range using the time-of-flight method.[16] Because a relatively thick-walled aluminum alloy pressure vessel was used (38 mm i.d. with 50 mm o.d. for 200 to 400 atm and 25 mm i.d. with 43 mm o.d. for 400 to 1000 atm), the scattering from the vessel was relatively more significant than in the other experiments discussed in this chapter. Attenuation corrections were calculated in the normal way, and the multiple scattering was calculated for the real geometry using Copley's Monte Carlo program.[13,14] Elastic multiple scattering is important in this case (see Fig. 4). A series of runs were taken covering 12 densities between 6.06 and 13.84×10^{27} atoms/m^3, and a number of tests were made to check the stability of the spectrometer during the course of the measurements. Molecular dynamics runs have been done for some densities.[47,49] At $\rho = 10.7 \times 10^{27}$ molecules/m^3 and $T = 297$ K, two simulations were done for 250 and 500 atoms, and were found to be in reasonable agreement. The averages of these results for several values of Q are shown in Fig. 18 (dashed line) compared to a smooth line through the experimental data. At the principal maximum in the structure factor $Q = 1.78$ Å^{-1}, there is excellent agreement without any further renormalization. At $Q = 0.8$ Å and in the neighborhood of 3 Å^{-1} there is less satisfactory agreement, and the experimental results are narrower than predicted. This suggests that the many-body potential terms act in such a way that the relevant atomic motions are slowed down.

14.6.4. The Normal Modes of a Fluid

There have been many ideas and experiments about the eigenmodes of a fluid, but it is still an incomplete and somewhat controversial subject. These

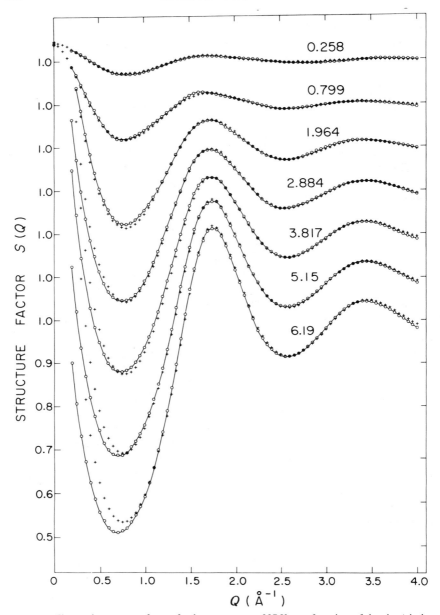

FIG. 17. The static structure factor for krypton gas at 297 K as a function of density (circles and line) compared to a Monte Carlo simulation (crosses) using a true pair potential. The number against each curve is the density in units of 10^{27} atoms/m^3. (Redrawn from Teitsma and Egelstaff.[34])

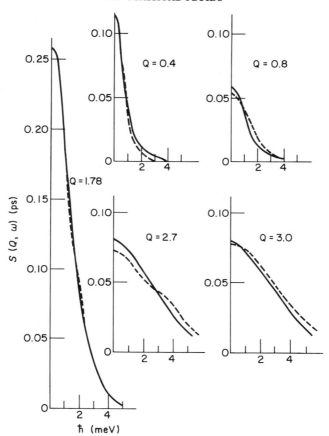

FIG. 18. The dynamic structure factor for krypton gas at 297 K and $\rho = 10.7 \times 10^{27}$ atoms/m^3 for several values of Q. The full lines are drawn through the data and the dashed line is a molecular dynamics simulation using a true pair potential. (Redrawn from Egelstaff et al.[47])

modes are well defined as $Q \to 0$ or $\to \infty$ (i.e., sound modes and free particle modes, respectively), but at intermediate Q's a suitable representation for them is not known. The reason is that they are not properly defined quantum states, but rather an amalgam of many highly excited states. It is likely that only an approximate definition will be possible eventually. Because the well-defined modes (at $Q \to 0$ or ∞) yield sharp peaks in $S_c(Q, \omega)$, it is argued sometimes that for true modes this should occur also at intermediate Q, in analogy with the observations on single crystals. However, this is too restrictive: that it is not a necessary condition may be seen from the observation that the well-defined modes of a polycrystalline sample do not always lead to peaks in $S_c(Q, \omega)$. Another possible argument is that a theoretical model in which the mode frequency is well defined may be fitted to $S_c(Q, \omega)$, thus finding the theoretical parameters for the mode. However, this is not direct evidence for its existence, and even a good determination of the parameters must be accompanied by an interpretive theory.

At long wavelengths in the hydrodynamic regime, the modes (longitudinal) that contribute to $S(Q, \omega)$ are sound wave modes centered at $\omega = \pm cQ$, where c is the velocity of sound, and a heat or entropy fluctuation mode centered at $\omega = 0$. Each mode contributes a Lorentzian in ω and, if $Q \to 0$, the function $S(Q, \omega)$ consists of three sharp peaks. As Q is increased these peaks broaden and eventually merge into one peak, centred at $\omega = 0$ if the recoil energy can be neglected. For noble gas fluids near the triple point they merge for $Q \simeq 0.1\ \text{Å}^{-1}$, while for some metals they merge for $Q \simeq 1\ \text{Å}^{-1}$. This region has been discussed through physical models, such as the viscoelastic theory; early work in this field has been discussed by Frenkel[50] and later work by Boon and Yip.[51] During the 1960s many models were published that attempted to describe $S(Q, \omega)$ in the range where the hydrodynamic modes overlap, and they were used to extend calculations to a few inverse angstroms. These included models based on solids,[52,53] on generalizations of hydrodynamics,[54-56] on collective variables,[57,58] and on continued fraction expansions,[59,60] among others. In each case $S(Q, \omega)$ has poles (denoted by $\pm \omega_s$), which are extensions of the normal hydrodynamic modes. Since for $Q \neq 0$ each pole is an amalgam of "yet to be defined" states covering a range of different frequencies, it is not likely that the width in $S(Q, \omega)$ reflects the lifetime of one mode.

The mean positions (ω_s) of these propagating modes correspond roughly with the maxima in $\omega S(Q, \omega)$ or $\omega^2 S(Q, \omega)$ because of their relationship to the imaginary part of $G(r, t)$ for a classical fluid. The latter function is related to the response of the fluid to an external potential and its Fourier transformation is

$$S_\text{I}(Q, \omega) = \{S(Q, \omega) - S(Q, -\omega)\}/2. \qquad (14.66)$$

Since this is an odd function of ω, the mean energy of the response spectrum will be given by

$$\bar{\omega}_\text{I} = \frac{\int_0^\infty \omega S_\text{I}(Q, \omega)\, d\omega}{\int_0^\infty S_\text{I}(Q, \omega)\, d\omega} \qquad (14.67\text{a})$$

$$\simeq \frac{\int_0^\infty \omega^2 \tilde{S}(Q, \omega)\, d\omega}{\int_0^\infty \omega \tilde{S}(Q, \omega)\, d\omega} \qquad (14.67\text{b})$$

for a classical fluid. [When $Q \to \infty$ or $T \to 0$, $\bar{\omega}_\text{I} \to \hbar Q^2/2MS(Q)$ and for $Q \to 0$, $\bar{\omega}_\text{I} \to cQ$]. The step between Eqs. (14.67a) and (14.67b) is made by the substitution $S(Q, \omega) = e^{\hbar\omega/2k_BT}\tilde{S}(Q, \omega)$ and $\sinh \hbar\omega/2k_BT \simeq \hbar\omega/2k_BT$ for a classical fluid. If $\omega^2\tilde{S}$ and $\omega\tilde{S}$ have single maxima (denoted by ω_p) at similar frequencies, then $\bar{\omega}_I \simeq \omega_p$. This is approximately the response frequency and consequently must be approximately the frequency (ω_s) of normal modes for that choice of Q. Because all the models satisfy the frequency moments of $S(Q, \omega)$, their propagating mode frequencies (ω_s) must also correspond roughly to ω_p. Clearly, these simple arguments would not be successful for ω_s near zero.

It is important to measure ω_s versus Q, and it was noted by the experimentalists using these models for trial calculations that ω_s corresponds approximately to the peak position in a cold neutron time-of-flight or wavelength spectrum. This differential cross section is proportional to

$$\frac{d^2\sigma}{d\Omega\, d\lambda} \alpha \left[1 - 2\frac{\hbar\omega}{E_0} + \left(\frac{\hbar\omega}{E_0}\right)^2 \right] S(Q, \omega), \quad (14.68)$$

and therefore for low E_0 and negative ω (neutron energy gain) it has a peak at $\omega_p \sim \omega_I \sim \omega_s$. Thus an easy way to determine (approximately) the poles of $S(Q, \omega)$ was to plot peak positions from cold neutron time-of-flight data.[61] Alternatively, the data could be fitted by adjusting the parameters in a model[62, 63] to find ω_s. In terms of the position (Q_m) of the principal maximum of $S(Q)$, these results for ω_s rose from $Q = 0$ to a maximum at $\sim Q_m/2$ and fell to zero at $\sim Q_m$, then rose again at higher Q, and the various methods of finding ω_s agreed reasonably well (near Q_m, $\omega \sim 0$ in Eq. (14.68) so that only the fitting method could be used). It was found that the whole spectrum could be fitted by these models significantly better than with simpler models [e.g., based on a Gaussian shape for $S(Q, \omega)$].

Since, within the framework of the model, $S(Q, \omega)$ had been separated into partial functions with peaks at $\pm\omega_s$, the observed peaks in the cross section [see Eq. (14.68)] and the deduced ω_s data were often referred to as "peaks in $S(Q, \omega)$."[64] However, in these experiments distinct peaks in $S(Q, \omega)$ (with a measureable width[62, 64]) were observed only for metals at the maximum frequencies near $Q_m/2$. Thus this observation implied[56] that peaks in $S(Q, \omega)$ itself would be seen for metals if $Q < Q_m/2$ and the right experiments were done (see Section 14.6.2), but that for other cases (noble gas fluids in the measureable Q range or metals where $Q > Q_m/2$), no distinct peaks would be observed if the full $S(Q, \omega)$ were measured. This was confirmed by excellent experiments on neon,[65] argon,[36] and rubidium[42] in which $S(Q, \omega)$ was measured properly. Thus the notation "peaks in $S(Q, \omega)$" for the whole ω-Q relationship was misleading and led to confusion. Moreover, the idea of giving a meaning to ω_s beyond the range at which distinct peaks were seen in $S(Q, \omega)$ was opposed by a number of authors who felt that unless a real peak was observed, ω_s was a theoretical parameter lacking physical meaning.

This alternative interpretation was expressed clearly in several papers[42, 65, 66] that rejected the conclusions of the earlier experiments. The significance of ω_s at high Q beyond the range at which peaks are seen in $S(Q, \omega)$ has remained an unresolved question.

Recently, de Schepper and Cohen[67] showed that the function $S(Q, \omega)$ derived from the generalized Enskog equation[51] of the kinetic theory of hard sphere fluids is given approximately by the sum of three Lorentzians whose coefficients depend on Q (only) for $Q\sigma < 12$. They proposed that the frequencies $\pm\omega_s$ had meaning for $Q\sigma < 12$, even when the Lorentzians overlapped to give a single peak at $\omega = 0$ for $S(Q, \omega)$. They further proposed that the same analysis should work for other fluids, and emphasized $\omega^2 S(Q, \omega)$ because the heat mode is suppressed in the current–current correlation function. This picture would fail if $S(Q, \omega)$ could not be fitted by three Lorentzians with ω-independent coefficients. While the earlier formulas for $S(Q, \omega)$[54–60] and the recent one[67] have essentially the same content, the latter feature is a test not proposed previously. However, several memory function models (see, e.g., Ailawadi et al.[58]) used a Q-dependent relaxation time as the only fitted constant and could be fitted to experimental data successfully.

De Schepper et al.[68] made computer simulations for the Lennard–Jones fluid with $\rho\sigma^3 = 0.692$ and $kT/\varepsilon = 0.97$, and obtained smooth single-peaked $S(Q, \omega)$ functions for $1 < Q\sigma < 15$. They showed that these data could be fitted with the sum of three Lorentzians over the whole range.

Van Well et al.[69] measured $S(Q, \omega)$ by the time-of-flight method for liquid argon at four different pressures between 20 and 400 bars along the 120 K isotherm employing the precautions described earlier and obtained excellent data. They found that $S(Q, \omega)$ was a smooth, single-peaked function of ω for $0.4 < Q < 4\,\text{Å}^{-1}$, and that it could be fitted accurately by the sum of three Lorentzians with ω-independent coefficients. Figure 19 shows ω_s as a function of Q derived[70] from the experimental and simulation data. The shape is similar to that seen in the older experiments but the precision and quality are much better. There is a good match between the Lennard–Jones fluid for this choice of parameters and liquid argon at this state. These authors deduce that propagating eigenmodes in a classical atomic fluid can be studied out to $4\,\text{Å}^{-1}$ even though distinct peaks in $S(Q, \omega)$ are not seen. A gap (i.e., where $\omega_s = 0$) occurs at $Q\sigma \sim 6$ ($Q \approx 2\,\text{Å}^{-1}$), which is clear in both sets of data, and here the modes are overdamped. Such an effect was observed in the earlier (but less extensive) experiments on metals and was interpreted as evidence for overdamped transverse modes,[71] thought to be the liquid analog of umklapp processes in solids, although this interpretation is controversial. Since it is expected that $\omega_s \simeq \omega_p$ except when $\omega_s \simeq 0$, it is interesting to compare the value of ω_p deduced from the maxima in

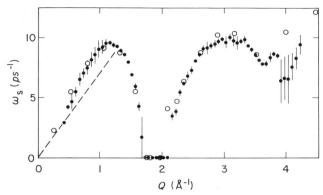

FIG. 19. The deduced peak positions (ω_s) in $S(Q, \omega)$ for liquid argon at 120 K (solid circles) compared to a molecular dynamics simulation (open circles) using a Lennard–Jones pair potential. (Courtesy of de Graaf et al., unpublished.)

$\omega^2 S(Q, \omega)$ with the data of the kind shown in Fig. 19. Sköld et al. measured $S(Q, \omega)$ for liquid argon at 85.2 K and 1 atm. and obtained the data for ω_p, while Rahman[72] made a molecular dynamics simulation for the Lennard–Jones fluid at 95 K and his results for ω_p are in reasonable agreement with the measured values. Van Well and de Graaf[73] evaluated ω_p from the data of de Schepper et al.[68] and van Well et al.,[69] and their results are shown in Fig. 20. There is good agreement between the measured and simulated values and also between the results for ω_s (Fig. 19) and ω_p (Fig. 20) when ω is different from zero. Thus both the model fluid and the real fluid exhibit similar behavior; the data may be analyzed simply, but the interpretation

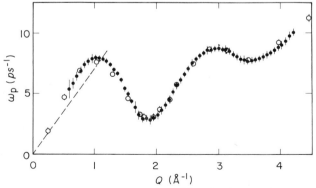

FIG. 20. The positions (ω_p) of the maxima in $\omega^2 S(Q, \omega)$ for liquid argon at 120 K (solid circles) compared to a computer simulation using a Lennard–Jones pair potential (open circles). (Courtesy of de Graaf et al., unpublished.)

of the constants in the formula as physically meaningful parameters is not accepted universally.

The proper description of normal modes of motion for a classical fluid is not yet clear or even whether such a description is possible if $Q \neq 0$, and so the interpretation of $S(Q, \omega)$ is not clear either and the subject is open to controversy. However, the interaction of theoretical results, computer simulations, and neutron scattering data may help to resolve these questions in the future.

Acknowledgment

I would like to thank Drs. John Copley, John Dore, Leo de Graaf, Henner Ruppersberg, Alan Soper, Peter Verkerk, and Ad van Well for a number of very helpful and constructive comments that were of assistance preparing this chapter.

References

1. Egelstaff, P. A. (1967). "An Introduction to the Liquid State." Academic Press, New York.
2. Ram, J., and Egelstaff, P. A. (1984). *J. Phys. Chem. Liq.* **14**, 29.
3. Sears, V. F. (1975). *Adv. Phys.* **24**, 1.
4a. Paalman, H. H., and Pings, C. J. (1962). *J. Appl. Phys.* **33**, 2635.
4b. Kendig, A. P., and Pings, C. J. (1965). *J. Appl. Phys.* **36**, 1692.
5a. Poncet, F. (1977). I.L.L. Rep. 77P015S and 139S.
5b. Poncet, F. (1978). I.L.L. Rep. 78P087S and 199S.
6. Soper, A. K., and Egelstaff, P. A. (1980). *Nucl. Instrum. Methods* **178**, 415.
7. Groome, L. J., Teitsma, A., Egelstaff, P. A., and Mook, H. A. (1980). *J. Chem. Phys.* **73**, 1393.
8. Egelstaff, P. A., and London, H. (1957). *Proc. R. Soc. London, Ser. A* **242**, 374.
9. Brockhouse, B. N. (1958). *Nuovo Cimento, Suppl.* **9**(1), 45.
10. Slaggie, E. L. (1967). *Nucl. Sci. Eng.* **30**, 199.
11. Bischoff, F. G., Yeater, M. L., and Moore, W. D. (1972). *Nucl. Sci. Eng.* **48**, 266.
12. Hawkins, R. K., and Egelstaff, P. A. (1975). *Mol. Phys.* **29**, 1639.
13. Copley, J. R. D. (1974). *Comput. Phys. Commun.* **7**, 289; **9**, 59, 64 (1975); **20**, 459 (1980).
14. Copley, J. R. D. (1981). *Comput. Phys. Commun.* **21**, 431.
15. Egelstaff, P. A., Gläser, W., Eder, O. J., Polo, J. A., Renker, B., and Soper, A. K., to be published.
16. Egelstaff, P. A., Gläser, W., Litchinsky, D., Schneider, E., and Suck, J. B. (1983). *Phys. Rev. A* **27**, 1106.
17. Vineyard, G. (1954). *Phys. Rev.* **96**, 93.
18. Cocking, S., and Heard, C. R. T. (1965). *U.K. At. Energy Res. Establ., Rep.* **AERE-R5016**.
19. Soper, A. K. (1983). *Nucl. Instrum. Methods* **212**, 337.
20. Blech, I. A., and Averbach, B. L. *Phys. Rev.* **137**, A1113.
21. Powles, J. G. (1973). *Mol. Phys.* **26**, 1325, eq. 4.13 with changes of notation.
22. Placzek, G. (1952). *Phys. Rev.* **86**, 377.
23. Egelstaff, P. A., and Soper, A. K. (1980). *Mol. Phys.* **40**, 553, 569.

24. Blum, L., and Narten, A. H. (1976). *J. Chem. Phys.* **64**, 2804.
25. Powles, J. G. (1978). *Mol. Phys.* **36**, 1161.
26. Egelstaff, P. A. (1983). *Adv. Chem. Phys.* **53**, 1.
27. Wick, G. C. (1954). *Phys. Rev.* **94**, 1228.
28. Ruppersberg, H., and Reiter, H. (1982). *J. Phys. F* **12**, 1311.
29. Olbrich, H., Ruppersberg, H., and Steeb, S. (1983). *Z. Naturforsch. A* **38A**, 1328.
30. Deraman, M., Dore, J. C., and Powles, J. G. (1984). *Mol. Phys.* **52**, 173.
31. de Gennes, P. G. (1959). *Physica (Amsterdam)* **25**, 825.
32. Vineyard, G. (1958). *Phys. Rev.* **110**, 999.
33. Powles, J. G., Dore, J. C., and Osae, E. K. (1980). *Mol. Phys.* **40**, 193.
34. Teitsma, A., and Egelstaff, P. A. (1980). *Phys. Rev. A* **21**, 367.
35a. Verkerk, P., and van Well, A. A. (1984). I.R.I. (Delft) Rep. 132-84-06.
35b. Verkerk, P. (1982). *Comput. Phys. Commun.* **25**, 325.
36. Sköld, K., Rowe, J. M., Ostrowski, G., and Randolph, P. D. (1972). *Phys. Rev. A* **6**, 1107.
37. Hansen, J. P., and McDonald, I. R. (1976). "Theory of Simple Liquids." Academic Press, New York.
38. Haile, J. M. (1980). "A Primer on the Computer Simulation of Atomic Fluids by Molecular Dynamics." Clemson Univ. Press, Clemson, South Carolina.
39. Aziz, R. (1983). *In* "Rare Gas Solids" (M. L. Klein and J. A. Venables, eds.), Vol. 3. Academic Press, New York.
40. Votava, C., and Alrichs, R. (1981). *Proc. Jerusalem Symp. Quantum Chem., Biochem., 14th*.
41. Meath, W., and Aziz, R. (1984). *Mol. Phys.* **52**, 225.
42. Copley, J. R. D., and Rowe, J. M. (1974). *Phys. Rev. Lett.* **32**, 49; *Phys. Rev. A* **9**, 1656 (1974).
43. Rahman, A. (1974). *Phys. Rev. Lett.* **32**, 52; *Phys. Rev. A* **9**, 1667 (1974).
44. Price, D. L. (1971). *Phys. Rev. A* **4**, 358 ; with parameters found in Price, D. L., Singwi, K. S., and Tosi, M. P., *Phys. Rev. B* **2**, 2983 (1970).
45. Mountain, R. D. (1978). *J. Phys. F* **8**, 1637.
46. Egelstaff, P. A., Teitsma, A., and Wang, S. S. (1980). *Phys. Rev. A* **22**, 1702.
47. Egelstaff, P. A., Salacuse, J., Schomers, W., and Ram, J. (1984). *Phys. Rev. A* **30**, 374.
48. Barker, J. A., Watts, R. O., Lee, J. K., Schafer, T. P., and Lee, Y. T. (1974). *J. Chem. Phys.* **61**, 3081.
49. Ullo, J. J., and Yip, S. (1984). *Phys. Rev. A* **29**, 2092.
50. Frenkel, J. (1946). "Kinetic Theory of Liquids." Oxford Univ. Press, London.
51. Boon, J. P., and Yip, S. (1982). "Molecular Hydrodynamics." Academic Press, New York.
52. Egelstaff, P. A. (1962). *U.K. Energy Res. Establ., Rep.* **AERE-4101**.
53. Singwi, K. (1964). *Phys. Rev.* **136**, A969.
54. Kadanoff, L. P., and Martin, P. C. (1963). *Ann. Phys. (N.Y.)* **24**, 419.
55. Ruijgrok, Th.W. (1963). *Physica (Amsterdam)* **29**, 617.
56. Egelstaff, P. A. (1966). *Rep. Prog. Phys.* **29**, 333.
57. Zwanzig, R. (1967). *Phys. Rev.* **156**, 190.
58. Ailawadi, N. K., Rahman, A., and Zwanzig, R. (1971). *Phys. Rev. A* **4**, 1616.
59. Sears, V. F. (1970). *Can. J. Phys.* **48**, 616.
60. Lovesey, S. W. (1973). *J. Phys. C* **6**, 1856.
61. Cocking, S. J., and Egelstaff, P. A. (1965). *Phys. Lett.* **16**, 130; Chen, S. H., Eder, O. J., Egelstaff, P. A., Haywood, B. C. G., and Webb, F. J. *Phys. Lett.* **19**, 269 (1965).
62. Cocking, S. J. (1967). *Adv. Phys.* **16**, 189.
63. Larsson, K. E. (1968). "Neutron Inelastic Scattering," Vol. 1, p. 397. IAEA, Vienna.
64. Egelstaff, P. A. (1967). *Adv. Phys.* **16**, 147.

65. Buyers, W. J. L., Sears, V. F., Lonngi, P. A., and Lonngi, D. A. (1975). *Phys. Rev. A* **11**, 697.
66. Copley, J. R. D., and Lovesey, S. W. (1975). *Rep. Prog. Phys.* **38**, 461.
67. de Schepper, I. M., and Cohen, E. D. (1982). *J. Statist. Phys.* **27**, 223; see also de Schepper, I. M., Cohen, E. D., and Zuilhof, M. J., *Phys. Lett. A* **101A**, 339; **103A**, 120 (1984).
68. de Schepper, I. M., van Rijs, J. C., van Well, A. A., Verkerk, P., de Graaf, L. A., and Bruin, C. (1984). *Phys. Rev. A* **A29**, 1602.
69. van Well, A. A., Verkerk, P., de Graaf, L. A., Suck, J.-B., and Copley, J. R. D. (1984). I.R.I. (Delft) Rep. 132-84-07; also to be published.
70. de Schepper, I. M., Verkerk, P., van Well, A. A., and de Graaf, L. A. (1984). *Phys. Rev. Lett.* **50**, 974.
71. Egelstaff, P. A. (1970). *In* "Current Problems in Neutron Scattering." p. 51. CNEN, Rome.
72. Rahman, A. (1967). *Phys. Rev. Lett.* **19**, 420.
73. van Well, A. A., and de Graaf, L. A. (1984). Private communication.

15. IONIC SOLUTIONS

John E. Enderby* and P. M. N. Gullidge

H. H. Wills Physics Laboratory
University of Bristol
Bristol BS8 1TL

15.1. Introduction

This is a review of how we learn about the behavior of ions in aqueous solution by applying the techniques of neutron scattering. We begin by focussing attention on a liquid that consists of a salt M_pX_q dissolved in heavy water, although the methods we describe apply equally well to nonaqueous solvents such as methanol, N,N' dimethyl formamide, etc. Here, M is to be regarded as the more electropositive element and will therefore form positively charged cations; similarly, X will exist in solution as negatively charged anions. The concentration scale used will be the molality, that is the number of moles of the anhydrous salt M_pX_q per kilogram of solvent.

The problems that such solutions present to scientists interested in obtaining a microscopic description of ionic behavior can be appreciated by reference to Fig. 1. It should be thought of as a "representative" snap shot, so that from a series of such pictures (taken, for example, at intervals of 10^{-13} s) one could deduce a typical or time-average spatial distribution of ions and water molecules. This distribution can be characterized in the example given by ten pair distribution functions denoted by $g_{\alpha\beta}(r)$. It is convenient to group the ten distribution functions in terms of the information they contain. Thus, in the example illustrated in Fig. 1, g_{OO}, g_{OD}, and g_{DD} describe the water-water coordination; g_{MM}, g_{XX}, and g_{MX} describe the ion-ion coordination; g_{MO} and g_{MD} describe the cation-water coordination; and g_{XO} and g_{XD} describes the anion-water coordination.

If neutrons are incident on such a liquid, the amplitude of the scattered waves is proportional to

$$\sum b_\alpha \sum_{i(\alpha)} \exp[-i\mathbf{Q}\cdot\mathbf{r}_i(\alpha)],$$

---

* Present address: Institut Laue Langevin, Grenoble, France, CEDEX 38042.

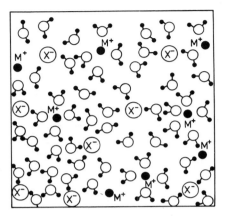

FIG. 1. A microscopic picture of an aqueous solution containing cations (M^+), anions (X^-), and water molecules.

where $\bar{b}_\alpha$ is the neutron coherent scattering length and $\mathbf{r}_i(\alpha)$ denotes the position of the ith nucleus of α type. The mean intensity

$$I(Q) = \sum_\alpha \sum_\beta \bar{b}_\alpha \bar{b}_\beta \left\langle \sum_{i(\alpha)} \sum_{j(\beta)} \exp i\mathbf{Q} \cdot [\mathbf{r}_j(\beta) - \mathbf{r}_i(\alpha)] \right\rangle \qquad (15.1)$$

can, in principle, be obtained from the experimental procedure described by Egelstaff in Chapter 14 of this volume.[1] In practice, as Egelstaff points out, a range of corrections has to be applied to the observed intensity $I_{obs}(Q)$ if $I(Q)$ is to be obtained reliably, not the least of which concerns the fact that the static approximation implied by Eq. (15.1) is a poor one for liquids that contain hydrogen or deuterium. The departure from the static approximation gives $I_{obs}(Q)$ a characteristic droop at high Q, the elimination of which involves the so-called Placzek corrections (cf. Chapter 14).

If for the moment we ignore all of these corrections, it is straightforward to demonstrate the connection between $I(Q)$ and $g_{\alpha\beta}(r)$. The partial structure factor $S_{\alpha\beta}(Q)$ is defined by

$$S_{\alpha\beta}(Q) = 1 + \frac{4\pi N}{QV} \int dr [g_{\alpha 3\beta}(r) - 1] r \sin Qr,$$

or

$$g_{\alpha\beta}(r) = 1 + \frac{V}{2\pi^2 Nr} \int dQ (S_{\alpha\beta}(Q) - 1) Q \sin Qr,$$

where $N = \sum_\alpha N_\alpha$ and N_α is the number of α-type species contained in a

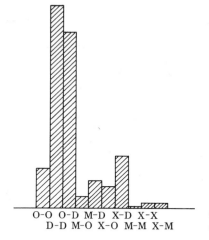

FIG. 2. The weighting of the various contributions to the total neutron pattern for a 4.35-m solution of $NiCl_2$ in D_2O.

sample with volume V. We also define the atomic fraction c_α of the α-species by $c_\alpha = n_\alpha/N$.

These definitions allow us to express $I(Q)$ as

$$I(Q) = N[c_\alpha(\bar{b}_\alpha)^2 + F(Q)],$$

where

$$F(Q) = \sum_\alpha \sum_\beta c_\alpha c_\beta \bar{b}_\alpha \bar{b}_\beta [S_{\alpha\beta}(Q) - 1].$$

Thus $F(Q)$, the quantity in principle accessible from a total scattering pattern, is a weighted average of ten individual structure factors. The difficulty in interpreting such a pattern of ion–water and ion–ion correlation function in aqueous solution can be appreciated by reference to the bar chart shown in Fig. 2. For practical purposes a total pattern can yield information only about $S_{DD}(Q)$ and $S_{OD}(Q)$ or their transforms $g_{DD}(r)$ and $g_{OD}(r)$, and that assumes that the multiple scattering and Placzek corrections can be handled satisfactorily. Moreover, as c_M is reduced, the vital information about ion–ion and ion–water correlations becomes increasingly buried in the total scattering pattern. We introduce for later use the quantity $F_0(Q)$, which is the total structure factor corrected for multiple scattering, self-absorption, etc., but not for the Placzek effect.

15.2. The Method of Differences

The neutron "first-order" difference method[2] allows one to gain direct information about the detailed arrangement of the water molecules around the ions in aqueous solutions. The quantity that is central to the method is the algebraic difference of $F_0(Q)$ from two samples that are identical in all respects except that the isotopic state of the cation M (or the anion X) has been changed; this quantity, denoted $\Delta_M(Q)$ [or $\Delta_X(Q)$], is the sum of four partial structure factors $S_{\alpha\beta}(Q)$ weighted in such a way that only those relating to ion-water correlations are significant. Explicity,

$$\Delta_M(Q) = A_M[S_{MO}(Q) - 1] + B_M[S_{MD}(Q) - 1] + C_M[S_{MX}(Q) - 1]$$
$$+ D_M[S_{MM}(Q) - 1],$$

$$\Delta_X(Q) = A_X[S_{XD}(Q) - 1] + B_X[S_{XD}(Q) - 1] + C_X[S_{XM}(Q) - 1]$$
$$+ D_X[S_{XX}(Q) - 1] \quad (15.2)$$

where

$$A_M = c_M c_O \bar{b}_O \Delta b_M; \quad B_M = c_M c_D \bar{b}_D \Delta \bar{b}_M,$$
$$C_M = c_M c_X \bar{b}_X \Delta b_M; \quad D_M = c_M^2[(\bar{b}_M)^2 - (\bar{b}_{M'})^2],$$

where $\Delta b_M = \bar{b}_M \cdots b_{M'}$, the difference in scattering length for the two isotopic compositions of the cation. The coefficients A_X–D_X are similarly defined.

Two properties of $\Delta(Q)$ are of particular importance in the present context. First, the Placzek distortions of $F_0(Q)$ are essentially eliminated so that a difference distribution function $\bar{G}(r)$ can be determined *directly* from

$$\bar{G}(r) = \frac{V}{2\pi^2 Nr} \int dQ\, \Delta(Q) Q \sin Qr. \quad (15.3)$$

In terms of the distribution functions $g_{\alpha\beta}(r)$, it follows at once that

$$\bar{G}_M(r) = A_M(g_{MO} - 1) + B_M(g_{MD} - 1) + C_M(g_{MX} - 1) + D_M(g_{MM} - 1) \quad (15.4)$$

and

$$\bar{G}_X(r) = A_X(g_{XO} - 1) + B_X(g_{XD} - 1) + C_X(g_{MX} - 1) + D_X(g_{XX} - 1). \quad (15.5)$$

Second, A and B are much greater than C and D so that the method in practice yields a high-resolution measurement of an appropriate combination of g_{MO} and g_{MD} or g_{XO} and g_{XD}.

The "second-order" difference method allows one to gain direct information about ion–ion correlations.[3] Let us again consider a salt $M_p X_q$ dissolved in D$_2$O and let $\Delta_{M_1}(Q)$ and $\Delta_{M_2}(Q)$ represent the two first-order differences for three solutions with M in the isotopic state M, 'M, and "M. Similarly, let $\Delta_{X_1}(Q)$ and $\Delta_{X_2}(Q)$ to be the corresponding quantities for isotopic substitutions of the anion X. It follows from Eqs. (15.1) and (15.2) that a linear combination of $\Delta_{M_1}(Q)$ and $\Delta_{M_2}(Q)$ will yield directly $S_{MM}(Q)$. In a similar way $S_{XX}(Q)$ can be derived from $\Delta_{X_1}(Q)$ and $\Delta_{X_2}(Q)$. The cross term $S_{MX}(Q)$ can be obtained as follows. Consider four samples whose isotopic state can be represented by

$$M_q X_p, \quad 'M_q X_p, \quad M_{q'} X_p, \quad \text{and} \quad 'M_{q'} X_p.$$

Let Δ_M^X represent the first-order difference for the first two samples and $\Delta_M^{'X}$ the difference for the third and fourth samples. Straightforward manipulation of Eqs. (15.1) and (15.2) yields

$$S_{MX}(Q) = \frac{\Delta_M^X - \Delta_M^{'X}}{2(p/q)c_M^2(\bar{b}_M - \bar{b}_{M'})(\bar{b}_X - \bar{b}_{X'})} + 1.$$

In real space, the three ion–ion distribution functions can now be obtained through

$$g_{\alpha\beta}(r) = \frac{V}{2\pi^2 Nr} \int dQ\, [S_{\alpha\beta}(Q) - 1]Q \sin Qr$$

by analogy with Eq. (15.3) and with α, β = M or X.

15.3. Experimental Aspects of the Method

In order to carry out the experiments, several conditions must be met. To begin with, a suitable pair of isotopes must be available, with scattering amplitudes that are known and are sufficiently different to give a statistically significant $\Delta S(Q)$. In Table I, a partial list of isotopes that have been or will be used in solution work is given. It is clear from Table I that representative experiments on a wide variety of aqueous solutions can be undertaken.

Each of the samples must be identical in all respects except for the isotopic state of either M, X, or both. This means that the molality of each solution must be equal to within $\pm 0.05\%$. Activity or electrical conductivity methods are usually able to measure the concentration to the required accuracy. Similarly, the ratio of deuterium to hydrogen (D/H) in the solution must be known to comparable accuracy. Infrared spectroscopy has been found to be the preferred method analysis at high D/H ratios—the normal situation for aqueous solution studies. The isotopic ratio for M or X must be determined

TABLE I. Ions and Possible Isotopes

Valence	Ion	Possible isotopes			Scattering lengths known
1	Li^+	^6Li	^7Li		Yes
1	K^+	^{39}K	^{41}K		Yes
1	$(ND_4)^+$	^{14}N	^{15}N		Yes
1	Ag^+	^{107}Ag	^{109}Ag		Yes
1, 2	Cu^+, Cu^{2+}	^{63}Cu	^{65}Cu		Yes
2	Mg^{2+}	^{24}Mg	^{25}Mg		Yes
2	Ca^{2+}	^{40}Ca	^{44}Ca		Yes
2	Ni^{2+}	^{58}Ni	^{60}Ni	^{62}Ni	Yes
2	Ba^{2+}	^{138}Ba	^{132}Ba		Yes
2	Sr^{2+}	^{86}Sr	^{88}Sr		Yes
2	Zn^{2+}	^{64}Zn	^{68}Zn		Yes
2	Sn^{2+}	^{102}Sr	^{122}Sn		Yes
2	Hg^{2+}	^{200}Hg	^{202}Hg		No
2, 3	Fe^{2+}, Fe^{3+}	^{56}Fe	^{57}Fe		Yes
2, 3	Cr^{2+}, Cr^{3+}	^{52}Cr	^{53}Cr		Yes
3	Nd^{3+}	^{142}Nd	^{144}Nd		Yes
3	Dy^{3+}	^{162}Dy	^{164}Dy		Yes
-1	Cl^-	^{35}Cl	^{37}Cl		Yes
-1	I^-	^{127}I	^{129}I [a]		No
-1	$(NO_2)^-, (NO_3)^-$	^{14}N	^{15}N		Yes
-1	$(CN)^-, (SCN)^-$	^{14}N	^{15}N		Yes
-1	$(ClO_3)^-, (ClO_4)^-$	^{35}Cl	^{37}Cl		Yes
-2	$(SO_3)^{2-}, (SO_4)^{2-}$	^{32}S	^{33}S		Yes
-2	$(CO_3)^{2-}$	^{12}C	^{13}C		Yes

[a] Slightly radioactive.

to at least $\pm 0.1\%$, so that the value of b can be established. This can be achieved by mass spectrometry, provided the instrument is suitable for thermal ionization of solid samples. As regards the b value for a particular nuclide, the most recent compilations by L. Koester (private communication) are especially valuable (see also Appendix to Part A of this volume).

Once the sample conditions have been met, the statistical precision required for the raw neutron data corresponds to approximately 10^6 counts per point over a Q range of 0.4–14.0 Å^{-1}. In practice there are very few reactor facilities where this range of Q and statistical accuracy can be achieved. The D4B instrument at the ILL Grenoble is well suited for these experiments. Other facilities are located at HFIR (Oak Ridge National Laboratory, Tennessee) and the ISIS (RAL, Chilton, Oxon, U.K.).

The data reduction techniques follow closely those described by Egelstaff.[1] Fortunately, the taking of differences, though making severe demands on the statistics, actually reduces the problems associated with systematic errors. We have already referred to the *Placzek corrections*, which continue to present serious problems for hydrogenous liquids and which essentially disappear in the first-order difference. The fact that the two samples are identical in size, shape, and density and are almost identical in terms of neutron cross sections means that multiple scattering corrections are also very small for the difference function. An empirical demonstration of the lack of sensitivity of the first-order difference to correction procedures, methods of sample preparation, and instrumentation is that $\Delta_{Cl}(Q)$ obtained by the ORNL group and the Bristol group are essentially the same.[4] On the other hand, $F(Q)$ obtained by different groups for the same liquid are frequently not consistent reflecting, as Egelstaff points out, the importance of the systematic errors.

15.4. Some Results

15.4.1. Introduction

This book is concerned with methodology rather than experimental results, but it is useful to illustrate what can be achieved by modern techniques. We shall therefore take a few examples for illustrative purposes only. For a more complete review, Enderby,[5] Copestake et al.,[6] Ichikawa et al.,[7] Neilson and Skipper,[8] and Enderby and Neilson[9] should be consulted.

15.4.2. Cation–Water Correlations

So far, the ions that have been studied by the method of differences are Li^+, K^+, Ag^+, ND_4^+, Cu^{2+}, Ca^{2+}, Nd^{3+}, and Dy^{3+}. In general, the results accord with computer simulation studies although there are detailed differences. For example, the concentration dependence found for the tilt angle θ for Ni^{2+} is, as yet, not properly explained (Table II). The form of $\Delta_{Ni}(Q)$

TABLE II. Ni^{2+}–Water Coordination

Ion and solute	Molality	Ion–oxygen distance (Å)	Ion–deuterium distance (Å)	θ (deg)[a]	Coordination number
Ni^{2+}					
$NiCl_2$	0.086	2.07 ± 0.03	2.80 ± 0.03	0 ± 20	6.8 ± 0.8
	0.46	2.10 ± 0.02	2.80 ± 0.02	17 ± 10	6.8 ± 0.8
	0.85	2.09 ± 0.02	2.76 ± 0.02	27 ± 10	6.6 ± 0.5
	1.46	2.07 ± 0.02	2.67 ± 0.02	42 ± 8	5.8 ± 0.3
	3.05	2.07 ± 0.02	2.67 ± 0.02	42 ± 8	5.8 ± 0.2
	4.41	2.07 ± 0.02	2.67 ± 0.02	42 ± 8	5.8 ± 0.2

[a] θ is the tilt angle.

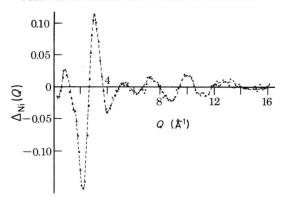

FIG. 3. $\Delta_{Ni}(Q)$ for a 4.35-m solution of NiCl$_2$ in D$_2$O.

and $G_{Ni}(r)$ shown in Figs. 3 and 4 is typical for a wide variety of cations, although K$^+$ and ND$_4^+$ do not show the characteristic double peak in $\bar{G}_M(r)$. This reflects the weakly hydrating character of these cations, the water molecules in the first shell being extremely labile.

15.4.3. Anion–Water Correlations

An extensive study of the chloride ion has been made by several authors,[4,5] and the new and surprising result is the stability of the Cl$^-$–D$_2$O coordination in respect of counter ion, ionic strength, and temperature (Fig. 5 and Table III). It is known from both NMR[10] and quasi-elastic neutron studies[11] that the Cl$^-$–D$_2$O lifetime is very short (ca. 5×10^{-12} s). Neilson and Enderby[3] have proposed a resolution of the paradox in terms of the specific nature of the Cl$^-$–D$_2$O interaction.

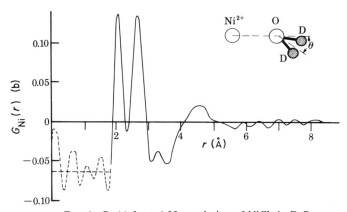

FIG. 4. $G_{Ni}(r)$ for a 4.35-m solution of NiCl$_2$ in D$_2$O.

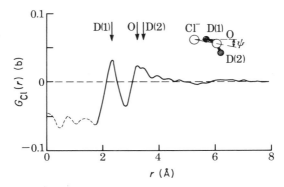

FIG. 5. $G_{Cl}(r)$ for a 4.35-m solution of NiCl$_2$ in D$_2$O.

TABLE III.   Cl$^-$-Water Coordination[a]

Solute	Molality	$r_{ClD(1)}$ (Å)	r_{ClO} (Å)	$r_{ClD(2)}$ (Range) (Å)	ψ[b] (Range) (deg)	Coordination number
LiCl	3.57	2.25 ± 0.02	3.34 ± 0.04	—	0	5.9 ± 0.2
LiCl	9.95	2.22 ± 0.02	3.29 ± 0.04	3.50 ± 3.68	0	5.3 ± 0.2
NaCl	5.32	2.26 ± 0.03	3.20 ± 0.05	—	0–10	5.5 ± 0.4
RbCl	4.36	2.26 ± 0.03	3.20 ± 0.05	—	0–10	5.8 ± 0.3
CaCl$_2$	4.49	2.25 ± 0.04	3.25 ± 0.04	3.55 ± 3.65	0–06	5.8 ± 0.2
NiCl$_2$	4.35	2.29 ± 0.02	3.20 ± 0.04	3.40 ± 3.50	6–11	5.5 ± 0.4
NiCl$_2$	3.00	2.23 ± 0.03	3.25 ± 0.05	—	0.6	5.5 ± 0.4
BaCl$_2$	1.10	2.24 ± 0.04	3.26 ± 0.05	—	0.6	6.2 ± 0.4

[a] Reference for Table 2, Enderby and Neilson.[9] Here, ψ is the tilt angle shown in Fig. 5.

15.4.4. Ion–Ion Correlation Functions

Two systems have been considered in detail: a 4.35 m solution of NiCl$_2$[3] and a 14.9 m solution of LiCl.[6] The r-space form of these data are shown in Figs. 6 and 7, and a summary of the ion–ion coordination parameters for NiCl$_2$ solution is given in Table IV. A full discussion of the real-space structure that can be deduced for solutions of NiCl$_2$ from these results has been given by Neilson and Enderby.[3]

The 14.9 m LiCl in water was chosen for investigation because it is intermediate between the pure molten salt and the NiCl$_2$ solution where there is sufficient water to satisfy the Ni^{2+} hydration. The pair correlation function $g_{ClCl}(r)$ shows two peaks (Fig. 7). The first at $r = 3.75 \pm 0.05$ Å is characterized by a coordination number of 2.3 ± 0.3 chloride ions surrounding a central chloride ion; the second peak at 6.38 ± 0.03 Å has a coordination

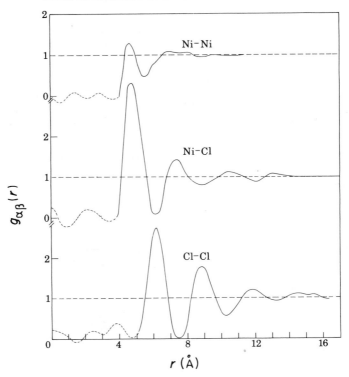

FIG. 6. The three ion–ion pair distribution functions for a 4.35-m solution of NiCl$_2$ in D$_2$O.

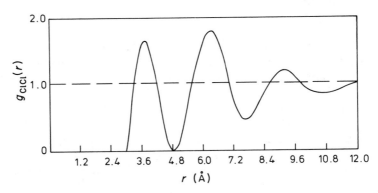

FIG. 7. The chloride–chloride pair distribution function for a 14.90-m solution of LiCl$_2$ in D$_2$O.

TABLE IV. Ion–Ion Distribution Functions in 4.35 m NiCl$_2$ in D$_2$O

Pair distribution function	Cut-off distance (r_0) (Å)	Position of first maximum (Å)	Position of first minimum (r_m) (Å)	Coordination number[a]
Ni–Ni	4.1 ± 0.1	4.7 ± 0.2	5.6 ± 0.2	
Ni–Cl	3.9 ± 0.1	4.6 ± 0.2	6.1 ± 0.2	5.8 ± 0.3 (Cl$^-$ around Ni^{2+})
Cl–Cl	5.0 ± 0.1	6.1 ± 0.2	7.5 ± 0.3	8.5 ± 0.3

[a] Obtained by integrating $4\pi(N/V)c_\alpha g(r)r^2$ between the limits $r_0 < r < r_m$.

number of 10. The origin of these features can be readily understood once it is realized that the lithium coordination consists of approximately three water molecules and three chloride ions.[6] The peak at 6.38 Å arises in the same way as the corresponding one in the NiCl$_2$ solution—through chloride ions binding to water molecules in the first hydration sphere. The inner peak arises from the direct interaction of chloride ions with Li$^+$ and is not observed in NiCl$_2$ solutions because of the complete hydration of Ni^{2+}. We expect this peak to be at a distance similar to the corresponding one in the molten salt, which is indeed the case[6] since $\bar{r}_{ClCl}$ is 3.86 Å.

This work has shown that in highly concentrated solutions structural features reflecting both the molten salt and the dilute solution coexist. It follows that changes in the ion–ion structure factor as the concentration of water is reduced will not be reproduced by models that disregard the molecular nature of water and treat it as a dielectric continuum.

15.5. The Kinetics of Water Exchange

15.5.1. Introduction

The exchange of water molecules between the hydration shell of cations or anions and "bulk" water is of considerable significance to solution chemists and can be regarded as the fundamental reaction for ions in solution. We have already seen in Section 15.4.4 how the stability of the cation–water conformation is closely linked with the formation of cation–anion complexes. Water exchange is also of significance in understanding the basic mechanism of redox reactions.

The lifetime of the complex τ_m can be defined as the mean time a given water molecule exists in the complex before changing places with a water molecule outside the complex. In the chemical literature τ_m is taken to refer to oxygen exchange rather than hydrogen exchange, which may be much faster. For example, τ_m for Al^{3+} is 10^5 s as determined by ^{17}O NMR; the

TABLE V. Mean Residence Time of Water Molecules in the First Hydration Shell

Ion	τ_m (s)^a	Coordination number	Ionic radius (Å)
Li$^+$	10^{-11}	6	0.68
Cu^{2+}	10^{-10}	4–6	0.72
Mg^{2+}	10^{-6}	6	0.66
Ni^{2+}	10^{-5}	6	0.69
Ga^{3+}	10^{-4}	6	0.62
Al^{3+}	10	6	0.51
Cr^{3+}	10^6	6	0.63

^{a} Rounded values.

rate constant for proton loss for Al(H$_2$O)$_6^{3+}$ is, however, 10^5 s^{-1}, which corresponds[12] to a lifetime of ~10^{-5} s.

The wide range of τ_m values found for aquoion has been commented on elsewhere, and currently accepted values, together with the coordination number, are displayed in Table V. These data, which are largely taken from the book by Burgess,[12] can be understood in a general way in terms of ionic size and charge. This is not the whole story, however, as a closer inspection of Table V reveals. For example, Cu^{2+} and Ni^{2+} have the same charge and essentially the same metal–oxygen distance whereas τ_m differs by three orders of magnitude.

The lifetimes shown in Table V are in general derived from NMR data, and for the more labile aquoions, the interpretation of the experimental data in terms of τ_m becomes increasingly sensitive to the input parameters of the models used. For $\tau_m < 10^{-7}$ s, the technique of high-resolution quasi-elastic neutron scattering offers new possibilities and gives new insights, as we shall now show.

15.5.2. Theory

The incoherent cross section of the hydrogen nucleus is so large that $S_i(Q, \omega)$ for aqueous solutions is dominated by the hydrogen contribution, other terms being sufficiently small that they may be neglected in the data analysis. Knowledge of $S_i(Q, \omega)$ thus leads directly to the proton dynamics, which, provided that observation times τ_{ob} are long compared with those characteristic of vibrational and rotational motions, may be identified with those of the water molecule itself. The usually accepted lower time limit for diffusive behavior is 10^{-11} s, which clearly satisfies this constraint.[11] In addition, the experimental conditions must obey

$$DQ^2\tau_c \ll 1,$$

where τ_c is the rotational correlation time for a water molecule and D is the translational diffusion coefficient. Typically in aqueous solutions $D \simeq 10^{-9}$ m^2 s^{-1} and $\tau_c \simeq 10^{-11}$ s so that Q must be $\ll 1$ Å, a severe constraint, which, however, can be satisfied by the technique of neutron back scattering.

For a proton obeying the diffusion equation, the correlation function $G_s(r, t)$ has a Gaussian form, and $S_i(Q, \omega)$ is a Lorentzian given by

$$S_i(Q, \omega) = \frac{(1/\hbar)DQ^2}{(DQ^2)^2 + \omega^2}.$$

This is the form to be expected when τ_m is short relative to τ_{ob}, so that during this observation time, any proton can sample the entire range of environments present in the solution. We refer to this situation as the "fast-exchange limit." When the opposite is the case (the "slow exchange limit"), and $\tau_m > \tau_{ob}$ of one aquoion (which in practice is always the cation) $S_i(Q, \omega)$ takes the form

$$S_i(Q, \omega) = (1/\hbar)\{x_1 D_1 Q^2/[(D_1 Q^2)^2 + \omega^2] + x_2 D_2 Q^2/[(D_2 Q^2) + \omega^2]\},$$

where x_i and D_i ($i = 1, 2$) are, respectively, the atomic fractions and diffusion coefficients of the two proton populations. The key aspects of the analysis introduced by Hewish et al.[11] are that one of the diffusion coefficients (D_1) is equal to that of the cation (D_{ion}), a quantity accessible from independent tracer measurements, and that the relative weighting of the two Lorentzians is known from the first-order difference method. Thus the fitting of a curve to the data involves only a single parameter D_2.

15.5.3. Experimental Procedure

Experiments have been carried out on the IN10 spectrometer at the ILL, an instrument with micron-electron-volt resolution on which measurements were taken in the range 0.1 Å $< Q < 0.3$ Å, at equal intervals of 0.04 Å. The incident wavelength was 6.27 Å, the monochromator was Si(111), and the experimental resolution function at each value of Q was obtained from the incoherent scattering from a sample of vanadium. To obtain the necessary energy resolution of about 1 μeV corresponding to $\tau_{ob} \simeq 10^{-9}$ s, the scattering angle for both the monochromator and the analyzer (also Si(111)) was fixed at 180° (i.e., in the back-scattering position). By this means the spread in wavelength resulting from the angular spread in the beam was minimized according to

$$|d\lambda/\lambda| = \cot \theta \, d\theta.$$

In order to scan the incident beam through a range of energies from +15 to −15μeV, the monochromator was mounted on a carriage that underwent oscillatory motion and use was made of the Doppler shift for this purpose. The principles of the back-scattering technique for high-resolution neutron spectroscopy have been fully described by Birr et al.[13]

The samples used were contained in a flat can, ~2 mm thick, with windows of 0.01 cm tantalum sheet. A version of a Monte Carlo simulation program due to Johnson was modified for the quasi-elastic region with micron-electron-volt resolution so that corrections for multiple scattering could be made. This program enables the ratio $J_1^*/J_{TOT} = R^*$ to be calculated, where J_{TOT} represents the total flux of scattered neutrons for some given (Q, ω) and J_1^* represents the flux of singly scattered neutrons that would be observed in the absence of absorption.

A theoretical curve $S(Q, \omega)$ was fitted to the experimental results by using a least-squares optimization algorithm, which compared the function

$$\int \frac{S(Q, \omega')R(Q, \omega - \omega')\,d\omega'}{R_1^*(Q, \omega)}$$

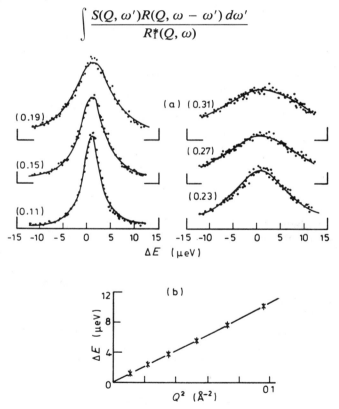

FIG. 8. (a) The neutron spectra for a 3-m solution of CsF in H_2O fitted by a single Lorentzian with Q values shown in brackets (in Å^{-1}). (b) The half-width of the Lorentzians as a function of Q^2.

with the observed spectrum $I(Q, \omega)$, where $R(Q, \omega)$ is the appropriate measured resolution function. The unknown diffusion coefficient can then be obtained by plotting the fitted Lorentzian width as a function of Q^2 and finding the slope.

15.5.4. A Test Experiment for the Fast-Exchange Limit

An example of the type of test experiments carried out by Hewish et al.[11] is afforded by the study of a 3-m solution of CsF in water. This is a complex liquid, but characterized by a proton population that is in the fast-exchange limit. It is to be expected, therefore, that

(i) $S_i(Q, \omega)$ at each value of Q will be a single Lorentzian,

(ii) the half-width of the Lorentzians should be a strictly proportional function of Q^2, and

(iii) the value of the *mean* diffusion coefficient $\bar{D}_{H_2O}$ predicted by these data should agree exactly with that derived from NMR spin-echo techniques, because both sets of data refer to the fast-exchange limit.

Inspection of Fig. 8 shows that (i) and (ii) are indeed satisfied. The slope of half-width versus Q^2 yields a value of $\bar{D}_{H_2O}$ of $1.69 \pm 0.15 \times 10^{-9}\,\text{m}^2\,\text{s}^{-1}$, which agrees, within experimental error, with the spin-echo value of $1.62 \pm 0.10 \times 10^{-9}\,\text{m}^2\,\text{s}^{-1}$.

15.5.5. Slow Exchange

Three cations, which are hydrated in the slow exchange limit, Mg^{2+}, Ni^{2+}, and Nd^{3+}, have been studied by the method. As an example we show in Fig. 9 the neutron spectra for a 3-m solution of $NiCl_2$. These spectra were corrected for multiple scattering in the same way as those for the test solutions discussed in Section 15.5.4. In this case, however, a single Lorentzian does not fit the data and this shows, as expected, that there are at least two relevant diffusion coefficients on the IN10 time scale. Similar conclusions were reached[14] for Mg^{2+} and Nd^{3+}.

The next approximation is to use a two-state model and equate D_2 with D_0, the diffusion coefficient of pure water. The two-state model assumes that the solvent molecules can be treated in terms of "free" water and "bound" water. On the IN10 time scale D_1 can be identified with D_{ion} and reliable values of this quantity as a function of concentration are available from the tracer measurements due to Stokes et al.[15] The results illustrated in Fig. 10 show that this model leads to gross disagreement between theory and experiment and demonstrates directly that the dynamical character of water molecules other than those in the first coordination shell are influenced by the presence of the cations. Hewish et al.[11] conclude that the two-state model,

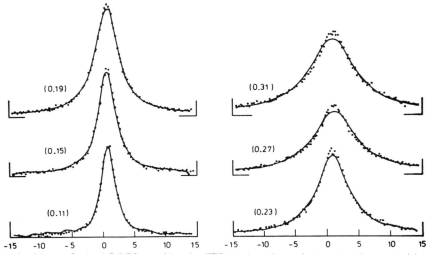

Fig. 9. An attempt to fit the data for a 3-m solution of NiCl$_2$ with a single Lorentzian.

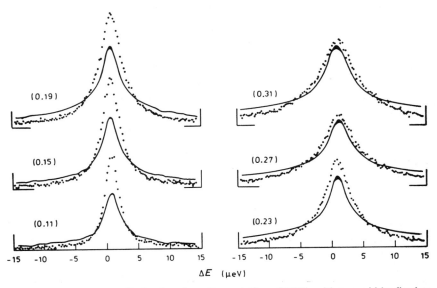

Fig. 10. An attempt to fit the data for a 3-m solution of NiCl$_2$ with two widths fixed to correspond to D_0 and D_{ion} and with six water molecules bound to Ni^{2+} and the rest considered free.

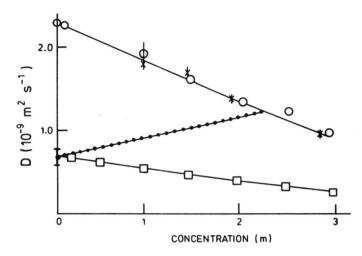

FIG. 11. Values for D_{ion} (□) as determined by Stokes et al.[15] and values for D_2 (∗) determined directly from the quasi-elastic neutron spectra compared with those derived (○) indirectly from the $\bar{D}_{H_2O}$ measurements of Stokes et al.,[15] on the assumption that $D_1 \equiv D_{ion}$ and that the coordination number is six. The effective second-shell diffusion constant obtained by Hewish et al.[11] from D_{ion} and D_2 (●).

frequently used in discussions of water dynamics in aqueous solution,[16] is unsatisfactory.

The results of further analysis in which D_2 was obtained explicitly from the experimental data are shown in Fig. 11. The concentration dependence of D_2 is of particular significance as Friedman[17] has emphasized. Detailed tests have shown that the weighting of the narrow Lorentzian characterized by D_2 must correspond to the primary hydration number of six; thus water molecules in the second and subsequent coordination spheres are in fast exchange with bulk water molecules, and although Hewish et al. proceeded to the next level of approximation and assumed a two-state model for second shell and bulk water molecules, more fundamental theories of D_2 and its concentration dependence are now being developed.[17] It will therefore be of considerable interest to measure D_2 for a range of aqueous solutions and, in particular, to determine both its magnitude and concentration dependence as a function of ionic charge and radius.

15.6. Conclusions

The method of isotopic substitution has enabled a very detailed picture to be obtained for the static structure of ionic solutions. It is difficult to see how any other method could be used to deduce the structural information now available to scientists interested in a fundamental description of these systems. Similarly the identification of water molecules whose dynamical properties are different from those associated with ionic hydration but do not correspond to those of free water has been made possible by high-resolution quasi-elastic neutron spectroscopy.

Acknowledgments

We wish to acknowledge the continued support of the Science and Engineering Research Council for ionic solution work and to thank the present and past members of the aqueous solutions group for helpful discussions and for performing many of the experiments referred to in this chapter. J. E. Enderby wishes to thank NATO for the award of a travel grant (RG/125/80), which allowed some of the ideas discussed in this chapter to be worked out in collaboration with Professor Harold Friedman.

References

1. P. A. Egelstaff, Chapter 14 in this volume.
2. A. K. Soper, G. W. Neilson, J. E. Enderby, and R. A. Howe, *J. Phys. C* **10**, 1793 (1977).
3. G. W. Neilson and J. E. Enderby, *Proc. R. Soc. London, Ser. A* **390**, 353 (1983).
4. S. Biggin, J. E. Enderby, R. L. Hahn, and A. H. Narten, *J. Phys. Chem.* **88**, 3634 (1984).
5. J. E. Enderby, *Annu. Rev. Phys. Chem.* **34**, 155 (1983).
6. A. P. Copestake, G. W. Neilson, and J. E. Enderby, *J Phys. C.* **18**, 4211 (1985).
7. K. Ichikawa, Y. Kameda, T. Matsumato, and M. Misawa, *J. Phys. C* **17**, L725 (1984).
8. G. W. Neilson and N. Skipper, *Chem. Phys. Lett.* **114**, 35 (1985).
9. J. E. Enderby and G. W. Neilson, *Rep. Prog. Phys.* **44**, 593 (1981).
10. H. G. Hertz, *in* "Water: A Comprehensive Treatment" (F. Franks, ed.), Vol. 3, p. 301. Plenum, New York, 1973.
11. N. A. Hewish, J. E. Enderby, and W. S. Howells, *J. Phys. C* **16**, 1777 (1983).
12. J. Burgess, "Metal Ions in Solutions." Horwood, Chichester, England, 1978.
13. M. Birr, A. Heidemann, and B. Arefeld, *Nucl. Instrum. Methods* **95**, 435 (1971).
14. P. Salmon, Ph.D. Thesis, Univ. of Bristol, 1986.
15. P. H. Stokes, S. Phang, and R. Mills, *J. Chem. Solids* **8**, 489 (1979).
16. T. Sakuma, S. Hoshing, and Y. Fujii, *J. Phys. Soc. Jpn.* **46**, 617 (1979).
17. H. L. Friedman, *Chem. Scr.* **25**, 42 (1985).

16. COLLOIDAL SOLUTIONS

Sow-Hsin Chen and Tsang-Lang Lin

Nuclear Engineering Department
Massachusetts Institute of Technology
Cambridge, Massachusetts 02139

16.1. Introduction

16.1.1. Various Colloidal Systems of Interest

The term colloid generally refers to a system of semimicroscopic particles or finely subdivided bulk matter that has a linear dimension between 10 and 10,000 Å dispersed in a continuous medium. The dispersion medium could be a gas, liquid, or solid. Fog, smoke, foam, milk, paints, etc., are examples of colloidal systems that are frequently encountered in daily life. Owing to their small size, the colloidal particles are characterized by a very large surface-to-volume ratio. This property is important for many chemical reactions to take place at the interface between the particles and the medium. Besides the size, the particle shape and the particle–particle interactions are also important properties of a colloidal system, since they determine the stability of the suspension.

Suspensions in liquid media are the most important type of colloidal systems. Based on the interactions between the colloidal particles and the solvent, one may classify the colloidal systems into three main categories: lyophobic (solvent hating), lyophilic (solvent loving), and association colloids, the latter of which consist of aggregates of amphiphilic molecules. In this chapter we are concerned exclusively with the latter two categories. Many macromolecules such as proteins and polymers are easily dissolved into solutions. In these cases, the colloidal particles are just the macromolecules. Since these macromolecules are soluble and form a true solution, they are lyophilic colloids. In biochemistry and molecular biology, one is interested in colloids in aqueous solutions since most of the reactions take place in an aqueous environment. Amphiphilic molecules, such as surfactants, have a water soluble polar head group at one end and a water insoluble aliphatic tail connected to the head group on the other end. At a sufficient concentration of the amphiphilic molecules in solution, aggregates called micelles can form.

As the concentration of the amphiphilic molecules is increased, the initially globular micelles may turn into cylindrical or disklike micelles.[1,2] Many colloidal systems are stabilized by an addition of amphiphilic molecules, which are then adsorbed onto the particle surfaces and form a protective layer against coagulation. The stability of some systems may depend on the electrostatic repulsion between the colloidal particles when the colloids carry charges on their surfaces. The formation of micelles and the capability of the micelles to solubilize materials that are otherwise insoluble in the solvent are the most prominent features of the amphiphilic molecules. For example, the detergency involves such processes as the forming of micelles to remove oil or dirt from solid surfaces. Other examples include the transport of lipid molecules in the bloodstream by serum lipoproteins, which are in the form of micelles; the absorption of corrosive products in motor oil by formation of microemulsions; the use of emulsions in the photographic processes; microencapsulation of drugs in microemulsions in the pharmaceutical industry; and applications of microemulsions in tertiary oil recovery in the petroleum industry. In general, we may define the association colloids as continuous dispersions in water or oil containing aggregates of amphiphilic molecules of one or more than one kind. Fig. 1 illustrates schematic models (conventional pictures) of the primary types of aggregates of the amphiphilic molecules. Determination of their size, shape, structure, and interactions will be our major concern in this chapter.

16.1.2. Thermodynamic Theory of the Formation and Growth of Association Colloids

Amphiphilic molecules are known to form aggregates, called micelles and vessicles, when they are dissolved in aqueous solutions in sufficiently large concentrations. As one increases the concentration of the amphiphilic molecules from zero, certain macroscopic properties, such as surface tension or osmotic compressibility of the solutions, are found to change abruptly over a narrow range of the concentration.[3] This suggests that there is a concentration above which some sorts of aggregates would form with a high probability and that these aggregates dramatically affect the transport and thermodynamic properties of the solution. This threshold concentration of the amphiphilic molecules is called the critical micellar concentration (CMC). In an aqueous environment there is a core in the aggregate that consists of a liquidlike arrangement of the hydrophobic part of the molecules (namely, the hydrocarbon tails) so that they are effectively shielded from the solvent when the molecules aggregate. The effective attractive tail–tail interaction, derived from the indirect effect due to the hydrophobic interaction, lowers the free energy of the system[4] (i.e., amphiphiles plus water), thus the

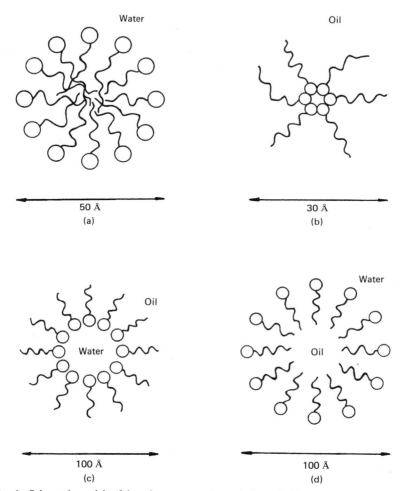

FIG. 1. Schematic models of the primary types of association colloids: (a) micelle, (b) inverted micelle, (c) water-in-oil microemulsion, and (d) oil-in-water microemulsion. Typical sizes are indicated.

hydrophobic interaction is in favor of formation of the aggregates. On the other hand, the electrostatic and the steric repulsive interactions between the polar head groups, which are confined to lie on the surface of the aggregates, will act to increase the free energy of the system. In addition to this, formation of ordered aggregates from otherwise dispersed disordered monomers in solution will cause the entropy of the system to decrease. Thus these two effects combined will not favor formation of the aggregates.

One may use a general thermodynamic argument in combination with some geometrical considerations to qualitatively explain the tendency of the self-organization of amphiphilic molecules in an aqueous environment without having to go into discussions of highly complicated intermolecular forces. This approach is called "the principle of mutually opposing forces" by Tanford.[4]

Chemical potentials of an isolated molecule or a monomer and a molecule in an aggregate of N molecules (a micelle) can be expressed in terms of the standard chemical potentials μ° as

$$\mu_1 = \mu_1^\circ + k_B T \ln X_1, \qquad (16.1)$$

$$\mu_N = \mu_N^\circ + (k_B T/N) \ln(X_N/N), \qquad N = 2, 3, 4, \ldots, \qquad (16.2)$$

where X_1 and X_N are the mole fractions or the concentrations of molecules in the monomer state and in an aggregate of N molecules (a N-mer), respectively. The standard chemical potentials μ_1° and μ_N° are the mean interaction free energies per molecule when the molecule is in the monomer state and in the N-mer state, respectively. The term $(k_B T/N) \ln/(X_N/N)$ is due to the entropic effect of the dispersion. These expressions apply only to dilute solutions where the interactions between monomers and aggregates can be neglected.

At thermodynamic equilibrium, the chemical potentials for each molecule, whether it is in the monomer state or in the N-mer state, are equal. We therefore equate Eqs. (16.1) and (16.2) to obtain

$$\frac{X_N}{N} = (X_1)^N \exp\left(-\frac{N(\mu_N^\circ - \mu_1^\circ)}{k_B T}\right). \qquad (16.3)$$

Equation (16.3) can be interpreted physically as (probability of formation of an N-mer) = (probability of having N monomers at the same location) $\times$ (a Boltzmann factor of having this N-mer formed from N monomers).[5-9]

16.1.2.1. Existence of CMC. Equation (16.3) allows us to understand mathematically the existence of the phenomena of CMC. Empirically, one knows that for many amphiphilic molecules of reasonable chain lengths (say from C_6 to C_{14}), aggregates tend to form with certain minimum sizes $N \cong M$. For example, for lithium dodecyl sulfate (LDS) with a chain length $l = 16.7$ Å (C_{12}), $M \cong 53$.[10] One can plausibly argue that μ_N° must decrease to a minimum value at $N = M = 53$. So that

$$\alpha_N = (\mu_1^\circ - \mu_N^\circ)/k_B T \qquad (16.4)$$

is positive and appreciably different from zero (say about 8) only when N is near M. Then with a constraint that the total molar fraction of the amphiphiles in solution must be X, we have

$$X_1 + X_M = X. \tag{16.5}$$

These two equations can be used in conjunction with Eq. (16.3) to write

$$X = X_1 + M(X_1 e^{\alpha_M})^M, \quad M \gg 1, \tag{16.6}$$

$$X_M = M(X_1)^M e^{M\alpha_M}. \tag{16.7}$$

Since M is large for small values of X, Eq. (16.6) shows that $X \cong X_1$ and the second term on the right-hand side of the equation is negligible. As one increases X, X_1 will also increase until at certain concentration called the CMC, at which X_1 reaches a threshold value $X_{1\mathrm{CMC}}$ the second term in Eq. (16.6) begins to dominate. Beyond this concentration, further increase in X can be accommodated by a very small increase of X_1 because the second term in Eq. (16.6) can take up the increase easily due to a large value of M. Equation (16.5) would then show that X_M will essentially increase linearly with X and the CMC can be conveniently defined to be an amphiphile concentration at which $X_M = X_1$. Substituting this equation into Eq. (16.6) we can show approximately that

$$\ln[X_{\mathrm{CMC}}/2] = -\alpha_M. \tag{16.8}$$

We therefore see that the value X_{CMC} depends only on having a maximum value of α_M at a certain value of M. This also indicates that the sharpness of the CMC phenomena depends on having a large M. This can be shown to be possible for amphiphilic molecules of longer chain lengths in the next section.

16.1.2.2. Existence of a Minimum Micellar Size and Aggregation Number. Because of the principle of mutually opposing forces, a micellar aggregate tends to forms a close-packed, semidisordered liquidlike core consisting of the hydrocarbon tails, each having a well-defined steric volume. The partially ionized head groups would then distribute themselves uniformly on the surface of the hydrocarbon core and try to maintain an optimum surface area per head group a_H.

According to the empirical rules of Tanford,[4] the steric volume of the hydrocarbon tail depends on the carbon number N_C as

$$v = 27.4 + 26.9 N_C. \tag{16.9}$$

From this equation $v = 350 \text{ Å}^3$ when $N_C = 12$. The fully stretched chain length can be estimated from

$$l_0 = 1.5 + 1.265 N_C, \qquad (16.10)$$

which gives $l_0 = 16.7$ Å for $N_C = 12$.

Denoting the minimum aggregation number by M, we have a volume conservation relation

$$Mv = \tfrac{4}{3}\pi R_0^3, \qquad (16.11)$$

and a surface area conservation relation

$$M a_H = 4\pi R_0^2, \qquad (16.12)$$

where R_0 is the radius of the hydrocarbon core. Equation (16.11) states that the volume of the hydrocarbon core is equal to the sum of the steric volumes of each hydrocarbon tail in the core. Equation (16.12) asserts that the surface area per head group is equal to the optimum value a_H.

For a minimum size aggregate to form, one requires that

$$R_0 \cong l_0. \qquad (16.13)$$

Then a_H can be calculated from Eqs. (16.11) and (16.12) as

$$a_H = \frac{3v}{R_0} = \frac{3v}{l_0} = \frac{3 \times 350 \text{ Å}^3}{16.7 \text{ Å}} = 62.9 \text{ Å}^2 \quad \text{for} \quad N_C = 12. \qquad (16.14)$$

From Eq. (16.12) M is then given by

$$M = \frac{4\pi R_0^2}{a_H} = \frac{4\pi l_0^3}{3v} = 56 \quad \text{for} \quad N_C = 12. \qquad (16.15)$$

Experimentally, it is observed that for the case of [LDS] = 1 gr/dl at 37°C in D_2O without salt added, $M = 53$,[10] which is consistent with Eq. (16.15) if $v = 364.3 \text{ Å}^3$, with an experimental error of 5%. This value of v is in agreement with 350 Å^3 calculated from the empirical formula in Eq. (16.9).

16.1.2.3. *Growth of the Micellar Aggregate.* As the surfactant concentration increases beyond CMC, it is observed that the mean aggregation number $\bar{n}$ gradually increases when no salt is added to the solution. However, with addition of salts the increase of $\bar{n}$ is often very rapid. It is often argued in the literature that the minimum spherical micelle with radius $R_0 = l_0$ tends to grow into cylindrical shape (rodlike structure) when salt is added. The reason for this is that an addition of salt (electrolyte) tends to screen out the Coulomb interaction between head groups on the surface of a micelle. Thus the optimum surface area per head group a_H decreases. A cylinder has less surface-to-volume ratio than a sphere. In fact it can easily be shown that the ratio ($v/l_0 a_H$) is equal to $\frac{1}{3}$ for a sphere and equal to $\frac{1}{2}$ for a cylinder.

For LDS micelles, we found that to a good approximation micelles grow from the minimum sphere into prolate ellipsoids.[10] In this case, due to the packing constraint, the semiminor axis of the ellipsoid must be equal to $b = l_0$. Thus we have a volume conservation relation

$$\bar{n}v = \tfrac{4}{3}\pi l_0^2 a, \tag{16.16}$$

which shows that the semimajor axis a is related uniquely to the mean aggregation number $\bar{n}$.

Missel et al. formulated a thermodynamic theory of the growth of micelles made from alkaline salts of alkyl sulfates along the line of sphere-to-rod transition.[11] They tested this theory with an extensive light scattering measurement of the hydrodynamic radius of those micelles.[11]

16.1.3. DLVO Theory of Interactions between Colloidal Particles.

Colloidal dispersions are in general not stable, which is due mainly to the van der Waals attractive forces between the particles. However, when they are charged, stability can be achieved by the Coulombic repulsive forces between the charged particles. Alternatively, one can adsorb surfactant or polymer molecules onto the particle surface, which provide physical barriers against coagulation.[12] Historically, Derjaguin, Landau, Verwey, and Overbeek (DLVO) were able to develop a theory to explain the stability of some colloidal systems in terms of the repulsive interaction potential between colloidal particles.[13] The interaction potential between colloidal particles considered in the DLVO theory consists of three parts:

(1) an effective hard-core repulsion with a diameter σ representing the excluded volume effect,
(2) a long-range van der Waals attraction, and
(3) an electric double-layer repulsion.

The van der Waals attractions arise from the dipolar interactions between individual molecules. The first type is the interaction between two permanent dipoles μ_1 and μ_2, and the statistically weighted potential is given by Keesom:[14]

$$V(r) = -\tfrac{2}{3}\mu_1^2 \mu_2^2 / k_B T r^6. \tag{16.17}$$

The second type, called the Debye force, is due to interaction of a permanent dipole and a dipole induced on the other particle by it. With a polarizability α and a dipole moment μ, the potential is given by[15]

$$V(r) = -2\alpha\mu^2/r^6. \qquad (16.18)$$

The third type is the interaction between two induced dipoles called *London dispersion forces*. The dipoles arise from fluctuations in the distributions of electrons. The potential is given by[16]

$$V(r) = -\tfrac{3}{4}\alpha^2 h\nu/r^6, \qquad (16.19)$$

where $h\nu$ is a characteristic amount of excitation energy.

All these forces have the same dependence on the spatial distance r and can be summed to give a total potential

$$V(r) = -\beta r^{-6}. \qquad (16.20)$$

Except for strong dipoles like H_2O, the attractive forces are dominated by the London forces, which are additive. By assuming additivity, the attractive potential between two identical spheres of diameter σ at the center-to-center distance r is given by

$$V(r) = -\frac{A}{12}\left[\frac{1}{x^2-1} + \frac{1}{x^2} + 2\ln\frac{x^2-1}{x^2}\right], \qquad (16.21)$$

where $x \equiv r/\sigma$ and A is the Hamaker constant[17] defined by $A \equiv \pi^2 n^2 \beta$, and n is the number of molecules per unit volume. It is worth noting that the Hamaker constant depends on the type and number density of a particular molecule but not on the size of the colloidal particle. Table I shows some typical values of the Hamaker constant A estimated from

$$A = \tfrac{4}{2}\pi\gamma' d, \qquad (16.22)$$

where γ' is the measured dispersion component of surface tension and d is the intermolecular spacing.[18] For particles dispersed in a liquid medium, the effective Hamaker constant is given by

$$A = (\sqrt{A_2} - \sqrt{A_1})^2, \qquad (16.23)$$

TABLE I. Hamaker Constants Calculated from Eqs. (1.3)–(1.6)

	A (erg)	d (Å)
Heptane	1.05×10^{-13}	2.2
Dodecane	9.49×10^{-14}	1.8
Eicosane	2.07×10^{-13}	2.6
Polystyrene	2.20×10^{-13}	2.3
Water	2.43×10^{-13}	3.3

where A_2 and A_1 are the Hamaker constants for the particles and the medium, respectively. For example, for Dodecane oil droplets in water, the Hamaker constant can be estimated from values listed in Table I to be

$$A = (\sqrt{9.49 \times 10^{-14} \text{ erg}} - \sqrt{2.43 \times 10^{13} \text{ erg}})^2$$
$$= 3.42 \times 10^{-14} \text{ erg},$$

which is of the order of $k_B T$ at room temperature.

Repulsive interactions between charged particles are commonly encounted in colloid science. Each charged particle dispersed in a dielectric medium is surrounded by a layer of the electrolyte ions of opposite sign. These charges combine with the surface charge layer on the macroion to form a double layer.

According to the DLVO theory, when the double layer is thin (i.e., when $k \equiv \kappa\sigma > 6$), the interaction potential of two spherical particles is given by

$$V_R = \frac{4Z^2e^2}{\varepsilon\sigma(2+k)^2} \ln(1 + e^{-k(x-1)}), \quad (16.24)$$

where

$$k = \left(\frac{8\pi N_A e^2 I}{10^3 \varepsilon k_B T}\right)^{1/2} \sigma = \text{Debye screening constant,} \quad (16.25)$$

$$I = \frac{1}{2} \sum Z_i^2 M_i = \text{ionic strength.} \quad (16.26)$$

M_i is the molar concentration of i-type ions in the medium, z_i the valency of the ions, ε the dielectric constant of the medium, N_A Avogadro's number, and Z is the number of charges on the particle surface.

This approximate expression was derived by summing over contributions from infinitesimal parallel rings in the two colloidal particles.

When the double layer is thick (i.e., $\kappa\sigma < 6$), the interaction potential can be found by solving the Poisson–Boltzmann equation in the linearized form using the Debye–Hükel approximation. The resultant potential between two macroions is given by

$$V_R = \frac{4Z^2e^2}{\varepsilon\sigma(2+k)^2} \frac{e^{-k(x-1)}}{x}. \quad (16.27)$$

The interaction potential can also be expressed in terms of the surface potential ψ_0 related to the surface charge Z by

$$\psi_0 = \frac{4Ze}{\varepsilon\sigma(2+k)}. \quad (16.28)$$

Recently, Medina-Noyola and McQuarrie[19] adopted a more systematic integral equation approach. Within the primitive model of electrolyte solutions, they solved the three-component Ornstein–Zernike equations in the mean spherical approximation for the case of a pair of charged large spherical particles in a dielectric medium containing two types of small ions (namely the counter ions and coions) of arbitrary size and concentration. They obtained the effective interaction potential between the large particles from the potential of mean force. This is justified because the concentration of large particles is assumed to be vanishingly low. When the size of the small ions is negligibly small compared to that of the large ions, their result agrees exactly with the expression given in Eq. (16.27). The potential in Eq. (16.27) is therefore more general than the original DLVO derivation may suggest.[13] In particular, the screened coulomb form of the potential is valid in a more general sense.

This approach can also be extended to include effects of the finite size of the ions. It was found that the repulsive potential is higher than that given by Eq. (16.27) at separations greater than one ion diameter but it is greatly reduced at separations less than one ion diameter when concentrations of the electrolytes are high (e.g., 1 M). This results in an effective negative (attractive) potential at short distance when the electrolyte concentration is high. It would be of great interest to test this point experimentally.

In general, to a first approximation, the repulsive potential is well

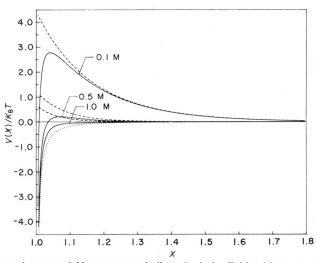

FIG. 2. The interaction potential between two similar spherical colloids with constant surface charge $Z = 20$; $A = 12 \times 10^{-14}$ erg. Dashed curves show the repulsive potential at electrolyte concentrations of $0.1 M$, $0.5 M$, and $1.0 M$. The dotted curve is the van der Waals attraction potential with the Hamaker constant set equal to $1 K_B T$, $T = 298.16$ K. Solid curves are the total potential. The diameter of the colloids is set equal to 50 Å.

represented by Eq. (16.27). Together with the van der Waals attractive forces, the total potential function may have very different features depending on the relative strength of attractive and repulsive forces and the range of these forces. Figure 2 shows a series of interaction potentials at constant surface charge but at different electrolyte concentrations. When the electrolyte concentration is increased, the repulsive forces are effectively screened out and the interaction is dominated by van der Waals attraction. We shall see this effect in the experimental section.

16.2. Small-Angle Neutron Scattering Technique

16.2.1. Complementarity of SANS, SAXS, and LS

In this chapter we shall often use abbreviations such as small-angle neutron scattering (SANS), small-angle x-ray scattering (SAXS), and light scattering (LS). Modern light scattering techniques include both static light scattering (SLS) and dynamic light scattering (DLS). In SLS one measures the intensity of light scattered by a system of particles at some range of scattering angles. DLS is often called quasi-elastic light scattering (QELS), where one measures spectral distribution of scattered light at various angles. The recent practice has been to use a time-domain technique called photon correlation spectroscopy (PCS), which directly measures the time correlations of the fluctuating intensity of light scattered from a system of colloidal particles undergoing Brownian motions.

The comprehensive reference of SAXS techniques is a recent monograph by Glatter and Kratky,[20] but an earlier pioneering work of Guinier and Fournat[21] is still very much in use. A large amount of literature exists on LS. Two monographs, one by Chu[22] and the other by Berne and Pecora[23] summarized the literature up to 1976. More recent progress has been reported in two summer school proceedings, one edited by Chen et al.[24] and the other by Degiorgio et al.[25] Chen et al.[24] covers some SANS work but a more complete review of SANS up to about 1975 can be found in an article by Jacrot.[26a] A very much more up-to-date review of SANS and its application to studies of association colloids has just been done by Chen.[26b]

As we mentioned in Section 16.1.1, sizes of aggregates in association colloids generally range from about 20 Å up to 10,000 Å. In order to get information with regard to size and size distribution, the wavelength of the radiation used in the scattering experiment should match the size range of interest. The basic geometrical consideration of the scattering of waves by a particle would show that the fundamental variable in a scattering experiment is the so-called Bragg wave vector defined by

$$\mathbf{Q} = \mathbf{k}_0 - \mathbf{k}_1, \qquad (16.29)$$

where $\mathbf{k}_0$ and $\mathbf{k}_1$ are, respectively, the incident and scattered wave vectors of the radiation. In a small-angle scattering experiment from an isotropic medium containing heavy particles, one measures predominantly the elastic and quasi-elastic scatterings for which $|\mathbf{k}_0| = |\mathbf{k}_1| = 2\pi/\lambda$, where λ is the wave length of the radiation in air.

Consequently the $\mathbf{Q}$ vector has a magnitude

$$Q = (4\pi/\lambda)n \sin(\theta/2), \qquad (16.30)$$

which we shall call the Bragg wave number. In Eq. (16.30) θ is the scattering angle and n the index of refraction of the medium. For light $n = 1.33$ in water, but for x rays and neutrons n is very close to unity.

A fundamental theorem in the theory of the scattering of waves by an extended object relates the real-space density distributions of the scattering object to the Q-space scattered intensity distribution in terms of a Fourier transform.[20] It follows from this theorem that the characteristic size in the real-space R is reciprocally related to the characteristic width of the intensity distribution in Q space. Therefore, for a typical aggregate size of R, one needs to do a scattering experiment that spans a Q range an order of magnitude on each side of the value $Q_0 = 2\pi/R$. Thus, at the lower end where $R \simeq 20$ Å, one should use x rays or neutrons where $\lambda = 1\text{-}20$ Å. At the upper end, where $R \simeq 10,000$ Å, one should use light with $\lambda = 4000\text{-}8000$ Å. It is also a consequence of this theorem that the maximum spatial resolution achievable in a scattering experiment is given in terms of the maximum wave number covered in the measurement, $Q_{\max}$, by

$$\Delta R = \pi/Q_{\max}. \qquad (16.31)$$

This means that in a typical SAXS and SANS measurement where $Q \leq 0.6$ Å^{-1}, the maximum spatial resolution achievable is no more than 5 Å; whereas in a typical light scattering experiment $Q \leq 0.003$ Å^{-1}, the resolution is no better than 1000 Å.

Another important consideration in a scattering experiment is the nature of the interaction between the radiation and the particles in the medium. Small-angle scattering is usually dominated by the coherent scattering of waves by the particles. The relevant quantity that comes into the description of the scattering cross section is the bound coherent scattering length, $\bar{b}_l$, of the lth nucleus constituting the particle in the case of SANS. In the case of SAXS, the corresponding quantity is the x-ray scattering amplitude in the forward direction, $f_l(0)$, of the lth atom. Numerical values for these constants are given in Table II for various elements usually encountered in association colloids. As one can see from the table, $\bar{b}_l$ is more or less the same order of magnitude for all nuclei except for the signs. On the other hand, $f_l(0)$ is always positive, and its magnitude increases linearly with the atomic

TABLE II. Scattering Properties of Some Elements[a]

Element	$\bar{b}^b$ [10^{-12} cm]	σ_i^c [10^{-24} cm^2]	$f_x(0)^d$ [10^{-12} cm]
Hydrogen	−0.3742	79.90	0.28
Deuterium	0.6671	2.04	0.28
Lithium	−0.190	0.91	0.84
Carbon	0.6648	0.001	1.69
Nitrogen	0.936	0.49	1.97
Oxygen	0.5805	0.000	2.25
Sodium	0.363	1.62	3.09
Magnesium	0.5375	0.077	3.38
Phosphorus	0.513	0.006	4.23
Sulphur	0.2847	0.007	4.5
Chlorine	0.9579	5.2	4.8
Calcium	0.49	0.03	5.6

[a] V. F. Sears, "Thermal Neutron Scattering Lengths and Cross Sections for Condensed Matter Research." Atomic Energy of Canada Ltd. AECL-8490, 1984.

[b] Coherent neutron scattering length.

[c] Incoherent neutron scattering refers to that part of the scattering of neutrons by individual nuclei in a sample that does not result in a constructive or destructive interference effect. Therefore, the incoherent scattering is essentially isotropic in all directions and can be regarded as a Q-independent background in the forward direction.

[d] X-ray scattering amplitude in the forward direction.

number of the element. From this table it can be concluded immediately that for x rays the scattering powers of hydrocarbons and of water are not significantly different. But for neutrons the scattering power of the hydrogen-containing molecules can be made to vary greatly by substitution of the protons by deuterons. For light scattering, the quantity equivalent to the $\bar{b}_l$ is $(2\pi/\lambda^2)n(\partial n/\partial \rho)_T$, where n is the index of refraction of the particles with respect to the medium and ρ the number density of the colloidal particles. The magnitude of this quantity is usually not large, since $(\partial n/\partial \rho)_T$ is rather small.

Although in many ways SANS is similar to the well-known SAXS technique, it has three unique features that distinguish it from SAXS and LS and can be used with great advantage in the study of association colloids:

(1) *The relative ease in measuring the absolute scattering cross section.* (See Section 16.2.2.) There are relatively simple procedures to calibrate the detector efficiency and to normalize the data to an absolute unit as we shall subsequently discuss. Owing to the short wavelength of thermal neutrons, the scattering intensity at small angles is not contaminated by scattering due to the sample container imperfections and to dust particles, which are always present in aqueous solutions. These problems are always the most difficult

to handle in small-angle LS. Furthermore, most samples of interest are sufficiently transparent to thermal neutrons when contained in a 1-mm (or up to several millimeters) quartz cell. One usually keeps the transmission to higher than 80% by varying the cell thickness and under such conditions multiple scattering is negligible.[27,28]

(2) *The possibility of varying the coherent scattering length density of the solvent continuously and by a large amount by simply changing the D_2O/H_2O ratio of the solvent.* In the case of association colloids, the isotope effect is usually not important[29] but there are exceptions.[30] This way one can utilize contrast matching between the solute and the solvent in an experiment.

(3) *The possibility of a selective deuteration of a particular functional group in an amphiphilic molecule to enhance the spatial resolution of a measurement.* We shall describe later an experiment of this kind for the study of the chain conformation of the hydrocarbon tail in a micellar core.

For SAXS the contrast variation technique is rather difficult to apply without changing the solvent chemistry drastically. Since aggregate formation results from a delicate balance between the hydrophobic interaction and the charge repulsion on the surface of a micelle, a small change in solvent condition can induce a large change in the state of an aggregate. In general, the SAXS technique is rather difficult to apply to the determination of the structure of association colloids because there is no contrast between the solvent and the hydrocarbon part of the aggregate. X rays are primarily scattered by the polar part of the amphiphilic molecules. It may become, however, a powerful tool for the study of the structure of the head group region and also the counter-ion distribution and hydration. This involves using the anomalous dispersion technique, making use of the sharp K-edge absorption of x rays by metal ions and should be possible when synchrotron x-ray sources become widely available. The technique is equivalent to the contrast variation technique in SANS and involves no change in the property of the solvent. In this case, one merely changes the wavelength of the incident x rays so as to sweep through the K edge of the metal ions.

Light scattering using PCS technique has been used to study the hydrodynamic properties of association colloids.[31–33] This technique relies on measurement of the mutual diffusion coefficient D of the colloidal particles. In the dilute limit, where interactions between colloidal particles can be neglected, D approaches the self-diffusion coefficient D_s of the aggregates. The Stoke-Einstein relation[23] can then be used to relate D_s to the hydrodynamic radius R_H, i.e.,

$$D_s = k_B T/6\pi\eta R_H, \qquad (16.32)$$

provided one knows the solvent viscosity.

Growth and polydispersity of ionic micelles of sodium dodecyl sulfate (SDS) near CMC and at high salt aqueous solution have been studied by this technique.[11] In an ionic micellar solution, it turns out that this noninteracting limit is hard to reach even at low concentration of the amphiphile molecules and at high salt levels.[10] Therefore measurement of R_H by this technique is subject to unknown errors. However, for reverse micelles such as AOT in oil, where charge neutrality is maintained in the aggregates, the interaction is not expected to be important at low amphiphile concentrations. Zulauf and Eicke[33] have used PCS to study reverse micelles of AOT in various oil media successfully.

16.2.2. SANS Experiment—Data Reduction and Normalization

16.2.2.1. The Instrument.
A modern small-angle scattering facility is characterized by a well-calibrated, two-dimensional multidetector controlled by a dedicated computer, which performs the on-line data acquisition and storage. A complete software package for data processing is usually available.

Figure 3 depicts a schematic arrangement of the SANS apparatus at the National Center for Small-Angle Scattering Research (NCSASR), Oak Ridge National Laboratory (ORNL), aside from its computer system. The neutron source plane is located at the high flux reactor core from which a thermal neutron beam emerges. The beam passes through a beryllium filter followed by a set of graphite monocromator crystals that select a wavelength $\lambda = 4.75$ Å with a spread $\Delta\lambda/\lambda = 6\%$. A nuclear reactor provides a neutron beam with a large cross-sectional area, thus specimens of large cross sections

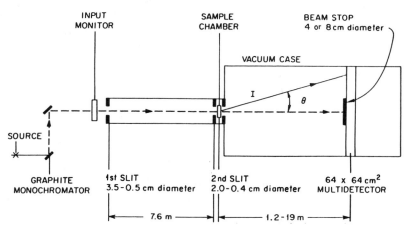

FIG. 3. The schematic arrangement of the 30-m SANS apparatus at the National Center for Small-Angle Scattering at ORNL.

can be used to increase the scattered intensity. The incident beam flux Φ_0 is continuously monitored by a low-efficiency neutron proportional counter before entering a double-slit evacuated flight path of length $L_1 = 7.6$ m. The entrance slit S_1 has a circular diameter ranging from 3.5 to 0.5 cm, and the sample slit S_2 has a diameter that can be varied from 2.0 to 0.4 cm. The two-slit geometry defines a spot of the transmitted beam at the center of the detector, which is covered by a beam stop (diameter = 4 or 8 cm). Neutrons that are scattered at an angle 2θ are detected by the area detector elements situated on a ring at a distance d from the center of the detector. The multi-detector consists of 4096 detector elements covering an area of 64×64 cm^2. For a large sample-to-detector distance L_2, the Bragg wave number can be written approximately as

$$Q \cong \frac{2\pi}{\lambda} \tan \theta = \frac{2\pi}{\lambda} \frac{d}{L_2}; \qquad (16.33)$$

L_2 can be varied from 1.2 to 19 m in order to span a Q range from 0.005 to 0.3 Å^{-1}.

A variation of this design is used at the high flux reactor of the Brookhaven National Laboratory (BNL). This SANS instrument is designed for measurements of solution samples containing macromolecules or biological macromolecules such as proteins or nucleic acids. It has also been used successfully for samples of association colloids. The instrument is situated at the exit of a beam port equipped with a cold moderator and thus has a supply of neutrons in the wavelength range of 4.0 to 8.0 Å. A multilayer monochromator is used to select a range of neutrons with wavelength spread $\Delta\lambda/\lambda = 10\%$. The corresponding neutron flux at 4.75 Å is somewhat higher (about 2×10^6 n/cm^2/s) than that of the instrument at NCSASR. The BNL instrument uses a two-dimensional detector, which has an area 17×17 cm^2 with an individual pixel area of 1.6×1.6 mm^2. Since the detector is smaller, the instrument has correspondingly smaller L_1 and L_2. However, the attainable Q range is the same as at NCSASR.[34]

16.2.2.2. Choice of Scattering Geometry. At constant S_2, the flux at the sample is approximately proportional to the area S_1. However, a large S_1 can be used only in combination with a small L_2 because of the finite size of the beam stop. Furthermore, the Q resolution $\Delta Q/Q$ has to be kept at a reasonably small value (e.g., 5%). This means that Q should be kept relatively large. The net result is that it is always preferable to work at the largest possible Q consistent with the size of the particle, namely $0.5 < QR < 5$. In this way, L_2 can be kept at the minimum and S_1 at the maximum value.

In general, the scattering geometry should be set up to suit the requirements that the transmitted beam is completely intercepted by the beam stop and that the size of the sample slit S_2 is approximately half that of the neutron source slit S_1.

16.2.2.3. *Data Collection, Correction, and Normalization.* Any SANS measurement from a sample in solution requires also the evaluation of the scattered intensity contributions from the solvent I_{so}, the empty container I_e, and the ambient background I_b when a sheet of cadmium (a strong neutron absorber) is blocking the beam. The transmission $T = (I_t/I_0)$ for each sample should also be determined to correct for the attenuation of the neutrons in traversing the sample. The transmission of the sample, T_s, should be kept above 80% to minimize the multiple scattering effects. The sample cells are usually flat quartz cells with 1-mm thick windows and path lengths of 1 to 10 mm. The empty cells should have a high transmission ($\geq 95\%$), meaning that their absorption and scattering cross sections are small. The measured intensity from the sample I_s is corrected for these various contributions.

For this correction, the number of detected neutrons at a pixel i, $I_{s,i}$, must first be normalized to a given number of counts (given incident neutron flux). Denoting the monitor counts for the sample, empty cell, and background runs as M_s, M_e, and M_b, respectively, one computes

$$I_{s,i}^0 = \left(\frac{I_{s,i}}{M_s} - \frac{I_{b,i}}{M_b}\right) - \frac{T_s}{T_e}\left(\frac{I_{e,i}}{M_e} - \frac{I_{b,i}}{M_b}\right) \quad (16.34)$$

for every sample, including the solvent runs. We shall call $I_{s,i}^0$ the corrected intensity. The factor, T_s/T_e is to correct for difference of attenuation of the incident and scattered beams in traversing the sample and the windows of the quartz cell.

The next step is to obtain the differential cross sections from these corrected intensities. One usually uses a 1-mm water sample as a standard scatterer to calibrate the detector.[35] The procedure is based on the following considerations.

Since the sample is supposed to be thin enough to have only single scattering, one can write

$$I_{s,i}^0 = \Phi_M T_s^0 t_s (d\Sigma_s/d\Omega)_i \Delta\Omega_i \varepsilon_i, \quad (16.35)$$

where Φ_M is the number of incident neutrons corresponding to the unit monitor count, T_s^0 the sample transmission (with empty cell part removed), t_s the sample thickness, $\Delta\Omega_i$ the solid angle subtended by the pixel i, ε_i the efficiency of the pixel i, and $(d\Sigma_s/d\Omega)_i$ the differential cross section per unit volume of the sample.

For the 1-mm water sample, the situation is slightly different, since the transmission T_W^0 is about 0.5693 at room temperature. With this transmission factor the multiple scattering contribution can be estimated to be about 30% of the total scattering and a formula such as Eq. (16.35) no longer holds. Fortunately, water is an isotropic scatterer, and the multiple scattering contribution is also expected to be isotropic. In this case we can write, to a good approximation,

$$I^0_{w,i} = f\Phi_M[(1 - T^0_w)/4\pi]\Delta\Omega_i \varepsilon_i. \tag{16.36}$$

This is plausible because $(1 - T^0_w)/4\pi$ is the fraction of the incident beam that is scattered isotropically into a unit solid angle. f is a factor greater than one, which corrects for deviation from isotropy of the scattering. In practice, for wavelength $\lambda = 5.3$ Å, this factor is ~ 1.3.[35]

If we now form a normalized intensity ratio, $N_{s,i}$, by dividing $I^0_{s,i}/T^0_s t_s$ by $I^0_{w,i}/T^0_w t_w$, we get

$$N_{s,i} = 4\pi t_w \left(\frac{d\Sigma_s}{d\Omega}\right)_i \frac{T^0_w}{1 - T^0_w} \frac{1}{f}. \tag{16.37}$$

From Eq. (16.37) we can get the absolute differential cross section per unit volume as

$$\left(\frac{d\Sigma_s}{d\Omega}\right)_i = N_{s,i} f \left(\frac{1 - T^0_w}{T^0_w}\right) \frac{1}{4\pi t_w}. \tag{16.38}$$

For a 1-mm water sample at room temperature the factor $(1 - T^0_w)/T^0_w/4\pi t_w$ is ~ 0.602 and is relatively constant for variation of a few degrees around 25°C. The major advantage of this method is that distilled water is readily available and the procedure is a relatively simple way of calibrating the unknown factor $\Phi_M \Delta\Omega_i \varepsilon_i$ for the instrument. This method is widely used and is expected to be accurate to within 10%. One must be able to count the water sample long enough so that each pixel detects at least a few thousand counts of neutrons to get good statistics. This is easy when the sample-to-detector distance (SDD) is relatively short so that the entrance slit S_1 can be opened up to gain intensity. This is possible when the particle sizes under study are relatively small, or when these particles interact strongly so that the interaction peak (see Section 16.4) is pushed out to a larger Q. For larger particles, one is forced to increase the SDD and concomitantly decrease the size of the entrance slit S_1 so as to avoid the direct beam from spilling out of the detector mask. One thus takes a tremendous loss in intensity due both to the smaller size of S_1 and to the increased SDD.

The reasons for the success of SANS studies of interacting micellar systems are that these micellar aggregates are relatively small and that they are strongly interacting so that the relevant information can be extracted from data taken at larger Q values. We shall see many examples of this kind in later sections.

For experiments involving larger particles and hence requiring longer SDD, one can use a well-characterized polymer sample or a radiation treated aluminum sample in which there are well-defined distributions of void sizes. For example, at NCSASR there is an Al-4 sample of thickness $t_A = 1.114$ cm, with a transmission $T_A = 0.954$. The calibrated values are $R_g = 202 \pm 5$ Å for the radius of gyration and $(d\Sigma_A/d\Omega)_0 = 129 \pm 10$ cm^{-1} for the extrapolated zero-angle scattering cross section. This is a strong scatterer that can be used to calibrate the detector at larger distances (SDD ~ 13 m) in a matter of 30 min.

In this case

$$I_{A,i}^0 = \Phi_M t_A T_A \left(\frac{d\Sigma_A}{d\Omega}\right)_i \Delta\Omega_i \varepsilon_i \qquad (16.39)$$

$$\Delta\Omega_i = \Delta a_i/(\text{SDD})^2, \qquad (16.40)$$

where a_i is the area of the ith pixel. By this sample one can determine the instrument parameter $K_N = \Phi_M \varepsilon_i \Delta a_i$, which can be used in Eq. (16.35) to determine $(d\Sigma_s/d\Omega)_i$.

16.2.3. The Basic Cross-Section Formula in SANS

We have stated that the measured Q-space scattered intensity distribution reflects the real-space particle density distribution of the sample through a Fourier transform. The characteristic size in real space is reciprocally related to the characteristic width of the measured scattered intensity distribution in Q space. For example, when a sample contains particles of 50 Å in diameter, one is interested in the intensity distribution over a Q range from zero to about two times $2\pi/50$ Å. This corresponds to scattering angles from zero to about 12° if $\lambda = 5$ Å.

When the masses of the molecules in the sample are far greater than the mass of a neutron, we may assume that the energy transfer between the neutron and the nucleus is negligibly small and use the so-called static approximation (SA) for the expression of the differential scattering cross section. This can be written as

$$\frac{d\Sigma}{d\Omega}(\mathbf{Q}) = \frac{1}{V}\left\langle \left|\sum_{l=1}^{N} b_l \exp(i\mathbf{Q}\cdot\mathbf{R}_l)\right|^2 \right\rangle, \qquad (16.41)$$

or more explicitly

$$\frac{d\Sigma}{d\Omega}(\mathbf{Q}) = \frac{1}{V}\left\langle \sum_{l=1}^{N}\sum_{l'=1}^{N} b_l b_{l'} \exp[i\mathbf{Q}\cdot(\mathbf{r}_l - \mathbf{r}_{l'})] \right\rangle, \qquad (16.42)$$

where b_l is the bound scattering length of the lth nucleus in the sample, $\mathbf{r}_l$ the position vector of the lth nucleus, and the bracket represents an average over all possible configurations. Placzek[36] has derived correction terms to the static approximation up to the first order in m/M, the mass ratio of neutrons to the scattering molecule. It can be shown that deviation from the static approximation is given by[37]

$$\frac{|\Delta(d\sigma/d\Omega)_{SA}|}{(d\sigma/d\Omega)_{SA}} = \left| -\frac{1}{2}\left(\frac{\hbar^2Q^2/2m}{E_0}\right)\left(\frac{m}{M}\right) - \left(\frac{Q}{k_0}\right)\left(\frac{m}{M}\right) \right.$$
$$- \frac{1}{4}\left(\frac{k_BT}{E_0}\right)\left(\frac{\hbar^2Q^2/2m}{E_0}\right)\left(\frac{m}{M}\right)$$
$$\left. + \frac{1}{2}\left(\frac{k_BT}{E_0}\right)\left[1 + \frac{3}{4}\left(\frac{Q}{k_0}\right)^2\right]\left(\frac{m}{M}\right) \right|, \quad (16.43)$$

where E_0 is the incident neutron energy. In a SANS experiment, Q is generally less than $0.2\,\text{Å}^{-1}$, E_0 is about 3.3 meV, and k_0 is $1.3\,\text{Å}^{-1}$ for $\lambda = 5\,\text{Å}$. For neutrons scattered from a typical surfactant molecule of molecular weight 500 at room temperature, the correction is about 0.7%. The correction might reach 20% when neutrons are scattered from small molecules such as H_2O, however.

Equation (16.42) is the result of the interference between the neutron waves scattered from each nucleus in the sample. In general, the scattering length b_l depends on the species of the nuclei as well as on the spin states. It would be more convenient to separate their differential scattering cross section into a coherent and an incoherent part. Assuming the isotopic and spin states of each nucleus are completely uncorrelated with its position, one obtains[24]

$$\frac{d\Sigma}{d\Omega}(\mathbf{Q}) = \left(\frac{d\Sigma}{d\Omega}\right)_c + \left(\frac{d\Sigma}{d\Omega}\right)_i, \quad (16.44)$$

where

$$\left(\frac{d\Sigma}{d\Omega}\right)_c = \frac{1}{V}\left\langle \left|\sum_{l=1}^{N} b_{l,c}\exp(i\mathbf{Q}\cdot\mathbf{R}_l)\right|^2 \right\rangle, \quad (16.45)$$

$$\left(\frac{d\Sigma}{d\Omega}\right)_i = \frac{1}{V}\sum_{l=1}^{N}(b_{l,i})^2, \quad (16.46)$$

$$b_{l,i} = (\overline{b_l^2} - \overline{b_l}^2)^{1/2}, \quad (16.47)$$

and

$$b_{l,c} = \overline{b_l}. \quad (16.48)$$

The average of the scattering length is taken over the spin and isotopic states for each species. Since the contribution from the incoherent part is simply a flat background, in the rest of the discussions the differential scattering cross sections will refer to the coherent part only.

A colloidal solution usually consists of particles of relatively stable structure that can be distinguished from the solvent. Thus it is more convenient to consider each colloidal particle as a scattering center. For this purpose the sample volume can be divided into N_p cells, each cell centered at a colloidal particle. The position vector of jth nucleus in the ith cell can be represented by a sum of $\mathbf{R}_i$ and $\mathbf{X}_j$, where $\mathbf{R}_i$ is the position vector of the center of mass of the particle i and $\mathbf{X}_j$ is the position vector relative to the center of mass of the particle i. Thus, we have

$$\frac{d\Sigma}{d\Omega}(\mathbf{Q}) = \frac{1}{V}\left\langle \left| \sum_{i=1}^{N_p} \exp(i\mathbf{Q}\cdot\mathbf{R}_i) \sum_{\text{cell } i} b_{ij} \exp(i\mathbf{Q}\cdot\mathbf{X}_j) \right|^2 \right\rangle. \quad (16.49)$$

Define the form factor of each cell as

$$F_i(\mathbf{Q}) = \sum_{\text{cell } i} b_{ij} \exp(i\mathbf{Q}\cdot\mathbf{X}_j). \quad (16.50)$$

In terms of the form factor, the differential cross section can be expressed as

$$\frac{d\Sigma}{d\Omega}(\mathbf{Q}) = \frac{1}{V}\left\langle \left| \sum_{i=1}^{N_p} F_i(\mathbf{Q}) \exp(i\mathbf{Q}\cdot\mathbf{R}_i) \right|^2 \right\rangle, \quad (16.51)$$

or more explicitly

$$\frac{d\Sigma}{d\Omega}(\mathbf{Q}) = \frac{1}{V}\left\langle \sum_{i=1}^{N_p} \sum_{i'=1}^{N_p} F_i(\mathbf{Q}) F_{i'}^*(\mathbf{Q}) \exp[i\mathbf{Q}\cdot(\mathbf{R}_i - \mathbf{R}_{i'})] \right\rangle. \quad (16.52)$$

We now rewrite the form factor $F_i(\mathbf{Q})$ in a Fourier integral form by introducing a scattering length density of ith particle $\rho_i(\mathbf{r})$ defined as

$$\rho_i(\mathbf{r}) = \sum_j b_{ij}\delta(\mathbf{r} - \mathbf{X}_j), \quad (16.53)$$

and a constant scattering length density of the solvent ρ_s. We assume that the solvent molecules are small enough so that we can treat the solvent as a continuum with a scattering length density equal to the sum of all scattering lengths in a unit volume of the solvent. The form factor can then be expressed as

$$F_i(\mathbf{Q}) = \int_{\text{particle } i} d^3r[\rho_i(\mathbf{r}) - \rho_s] \exp(i\mathbf{Q}\cdot\mathbf{r}) + \int_{\text{cell } i} d^3r \rho_s \exp(i\mathbf{Q}\cdot\mathbf{r}). \quad (16.54)$$

The second term is to a good approximation a delta function at $\mathbf{Q} = \mathbf{0}$. Thus, except at $\mathbf{Q} = \mathbf{0}$, we have

$$F_i(\mathbf{Q}) = \int_{\text{particle } i} d^3 r [\rho_i(\mathbf{r}) - \rho_s] \exp(i\mathbf{Q} \cdot \mathbf{r}). \tag{16.55}$$

This expression shows that the form factor actually depends only on the shape of the colloidal particle when the scattering density of the particle is uniform. Since the form factor depends on the relative scattering length density between the particle and the solvent, it implies that one may change the contrast by varying ρ_s to look at different parts of the particle if the particle has parts with different scattering length densities.

For a monodisperse spherical colloidal system, the form factor is identical for each particle and Eq. (16.52) can be written as

$$\frac{d\Sigma}{d\Omega}(Q) = \frac{N_p}{V} |F(Q)|^2 \frac{1}{N_p} \left\langle \sum_{i=1}^{N_p} \sum_{i'=1}^{N_p} \exp[i\mathbf{Q} \cdot (\mathbf{R}_i - \mathbf{R}_{i'})] \right\rangle. \tag{16.56}$$

Define now the quantities

$$n_p = N_p / V, \tag{16.57}$$

$$P(Q) = |F(Q)|^2, \tag{16.58}$$

$$S(Q) = \frac{1}{N_p} \left\langle \sum_{i=1}^{N_p} \sum_{i'=1}^{N_p} \exp[i\mathbf{Q} \cdot (\mathbf{R}_i - \mathbf{R}_{i'})] \right\rangle. \tag{16.59}$$

We can finally rewrite Eq. (16.56) in a compact form as

$$(d\Sigma/d\Omega)(Q) = n_p P(Q) S(Q). \tag{16.60}$$

We see that the differential cross section per unit volume is proportional to the particle density n_p, and depends on the product of the intraparticle structure factor $P(Q)$ and the interparticle structure factor $S(Q)$. The latter is essentially the particle center-to-center correlation function. The main theme in an experiment is therefore to determine the intraparticle structure factor $P(Q)$ and interparticle structure factor $S(Q)$ separately. Theoretically, the former can be computed from a structural model of the colloidal particle and the latter can be calculated from the interparticle interactions and statistical mechanical theory.

For a polydispersed system of n components consisting of spherical particles, the differential cross section takes a more complicated form:[38]

$$\frac{d\Sigma}{d\Omega}(Q) = \sum_{p=1}^{n} \sum_{q=1}^{n} \sqrt{\frac{N_p}{V}} \sqrt{\frac{N_q}{V}} F_p(Q) F_q^*(Q) S_{pq}(Q), \tag{16.61}$$

where N_p and N_q are the numbers of particles of pth and qth components, respectively, and $S_{pq}(Q)$ is the partial structure factor defined by

$$S_{pq}(Q) = \frac{1}{\sqrt{N_p}\sqrt{N_q}} \left\langle \sum_{i_p=1}^{N_p} \sum_{j_q=1}^{N_q} \exp[i\mathbf{Q} \cdot (\mathbf{R}_{i_p} - \mathbf{R}_{j_p})] \right\rangle. \quad (16.62)$$

In principle one may solve the multicomponent Ornstein–Zernike equation[39] to obtain the partial structure factor $S_{pq}(Q)$. So far this has been done only for hard sphere systems[40] and for a symmetric electrolyte in the mean spherical approximation (MSA).[41] For other situations one may try to use approximations that retain some important features of a multicomponent system and may use an averaged one-component structure factor. The partial structure factor $S_{pq}(Q)$ is related to the total correlation function $h_{pq}(r)$ between pth and qth components by[39]

$$S_{pq}(Q) = \delta_{pq} + \sqrt{\frac{N_p}{V}}\sqrt{\frac{N_q}{V}} \int_0^\infty \frac{\sin Qr}{Qr} h_{pq}(r) 4\pi r^2 \, dr. \quad (16.63)$$

Chen *et al.* have shown that Eq. (16.61) can be cast into a form[42]

$$\frac{d\Sigma}{d\Omega}(Q) = \frac{N}{V} \langle |F(Q)|^2 \rangle \bar{S}(Q) = n_p \langle P(Q) \rangle \bar{S}(Q), \quad (16.64)$$

where

$$\langle P(Q) \rangle = \langle |F(Q)|^2 \rangle = \sum_{p=1}^n \frac{N_p}{N} |F_p(Q)|^2, \quad (16.65)$$

$$N = \sum_{p=1}^n N_p. \quad (16.66)$$

The averaged structure factor $\bar{S}(Q)$ can be found by equating Eq. (16.61) to Eq. (16.64) and using the definition of $S_{pq}(Q)$ as given by Eq. (16.62). We then have

$$\bar{S}(Q) = 1 + \frac{\sum_{p=1}^{n_p} \sum_{q=1}^{n_q} F_p(Q) F_q^*(Q) (N_p/V)(N_q/V) \times \int_0^\infty (\sin Qr/Qr) h_{pq}(r) 4\pi r^2 \, dr}{(N/V) \langle |F(Q)|^2 \rangle}. \quad (16.67)$$

So far, by redefining some terms, we are able to put the multicomponent cross section of Eq. (16.61) in the form of Eq. (16.64), which is similar to the one-component result of Eq. (16.60). In order that $\bar{S}(Q)$ can be actually calculated from a microscopic theory, we approximate $\bar{S}(Q)$ by an effective one-component structure factor. The effective one-component system should have the same total particle number N and droplet volume fraction as the actual multicomponent system. Thus, we require that

$$\sigma^3 = \sum_{p=1}^{n} N_p \sigma_p^3 \bigg/ \sum_{p=1}^{n} N_p, \qquad (16.68)$$

where σ is the diameter of the particles of the effective one-component system. Such a model allows one to obtain the effective one-component total correlation function $h(r)$, and therefore the effective interaction potentials, from fitting the measured differential cross section. For a system with low polydispersity, or for a dilute system, this approximation should be a good one since it will reduce correctly to the one-component result. Even for a system with modest polydispersity, we expect that it will not deviate too much from the exact results, especially at low Q.

Another approach to deal with the polydispersity is to assume that the particle size and orientation are completely uncorrelated with the particle positions;[43] we can then write

$$\frac{d\Sigma}{d\Omega}(Q) = \frac{1}{V}\left\langle \sum_{i=1}^{N} \sum_{i'=1}^{N} \langle F_i(\mathbf{Q})F_{i'}^*(\mathbf{Q})\rangle \exp[i\mathbf{Q}\cdot(\mathbf{R}_i - \mathbf{R}_{i'})]\right\rangle. \qquad (16.69)$$

The term $\langle F_i(\mathbf{Q})F_{i'}^*(\mathbf{Q})\rangle$ can be decomposed into

$$\langle F_i(\mathbf{Q})F_{i'}^*(\mathbf{Q})\rangle = (\langle |F(\mathbf{Q})|^2\rangle - |\langle F(\mathbf{Q})\rangle^2|)\delta_{ii'} + |\langle F(\mathbf{Q})\rangle|^2. \qquad (16.70)$$

By defining

$$\beta(Q) = |\langle F(\mathbf{Q})\rangle^2|/\langle |F(\mathbf{Q})|^2\rangle, \qquad (16.71)$$

$$\bar{S}(Q) = \frac{1}{N}\left\langle \sum_{i=1}^{N} \sum_{i'=1}^{N} \exp[i\mathbf{Q}\cdot(\mathbf{R}_i - \mathbf{R}_{i'})]\right\rangle, \qquad (16.72)$$

and

$$S'(Q) = 1 + \beta(Q)[\bar{S}(Q) - 1], \qquad (16.73)$$

Eq. (16.69) can thus be put into a form

$$d\Sigma/d\Omega(Q) = N/V\langle P(Q)\rangle S'(Q). \qquad (16.74)$$

Here, $\bar{S}(Q)$ is also taken to be the structure factor of the effective one-component system. This approximation is expected to be good only for Q values near the first diffraction peak of $\bar{S}(Q)$. But it can be shown that at least for the case of polydispersed hard spheres at small Q values, it deviates significantly from the exact calculations of van Beurten and Vrij.[40]

For ellipsoidal particles, the form factor and the total correlation function, in principle, depend on the orientations of the particles. But in an isotropic fluid phase the total correlation function can be, in an average sense, approximated by a scalar quantity $h(r)$. If one regards different orientations as different components of a multicomponent system, one arrives at a similar expression, Eq. (16.64), for the differential cross section but in which $h_{pq}(r)$ in Eq. (16.67) is replaced by $h(r)$, the total correlation function of an effective one-component system. Thus the right-hand side of Eq. (16.67) reduces to the form of Eq. (16.73) with $\tilde{S}(Q)$ the structure factor the effective one-component system and the form factors that enter into the expression for $\beta(Q)$ averaged over the particle orientations. The final expression for the differential cross section is just Eq. (16.74) with $S'(Q)$ given by Eq. (16.73). It is not surprising that one obtains the same formula for the differential cross section as given by Eq. (16.74) since the fundamental assumptions are the same. Replacing $h_{pq}(r)$ by $h(r)$ in Eq. (16.67) is equivalent to the assumption that the particle positions are uncorrelated with the particle size and their orientations.

In short, when dealing with a monodispersed system, one uses Eq. (16.60) for analyzing the measured differential cross section. But for a polydispersed system one uses Eq. (16.64), and finally for monodispersed nonspherical systems of particles one may use Eq. (16.74).

The form factor depends on the structure of the scattering particle in the solution. In the simplest case of a monodispersed system of homogeneous spherical particles with a radius R, Eq. (16.59) becomes

$$F(Q) = V(\bar{\rho} - \rho_s)[3j_1(QR)/QR], \tag{16.75}$$

where $V = \frac{4}{3}\pi R^3$, $\bar{\rho}$ is the mean coherent scattering length density (csld) of the particle defined as

$$\bar{\rho} = \frac{1}{V} \int d^3 r \rho(\mathbf{r}), \tag{16.76}$$

and $j_1(x)$ is the first-order spherical Bessel function

$$j_1(x) = (\sin x - x \cos x)/x^2. \tag{16.77}$$

One can easily generalize this formalism to more complicated cases. For example, the form factor of a spherical particle with an internal core of radius R_1 and a csld ρ_1 surrounded by a shell with an outer radius R_2 and a csld ρ_2 is given by

$$F(Q) = \frac{4}{3}\pi R_1^3 (\rho_1 - \rho_2) \frac{3j_1(Q_1 R_1)}{QR_1} + \frac{4\pi}{3} R_2^3 (\rho_2 - \rho_s) \frac{3j_1(QR_2)}{QR_2}. \tag{16.78}$$

Applications to ellipsoidal particles will be discussed in Section 16.3.2.

16.3. Determination of Intraparticle Structures

16.3.1. Techniques of Contrast Variation and Selected Deuteration

In previous sections we have shown that the differential scattering cross section is related to the intraparticle structure, mainly through the term for the particle form factor. For a dilute solution or a weakly interacting suspension, the interparticle structure factor is very close to unity and the differential cross section for a monodispersed system is simply given by the product of the number density of the suspended particles n_p and the particle form (or structure) factor $P(Q)$:

$$\frac{d\Sigma}{d\Omega}(Q) = n_p P(Q). \tag{16.79}$$

The behavior of the scattered neutron intensity in the small-Q region depends weakly on the structure, size, and shape of the colloidal particle. The Guinier approximation,[21] which is valid in the small-Q region (or more explicitly for $QR < 1$ for a sphere of radius R) can be expressed as

$$P(Q) = P(0) \exp(-Q^2 R_g^2/3), \tag{16.80}$$

where R_g is the radius of gyration of the particle and is defined by

$$R_g^2 = \frac{\int r^2 [\rho(\mathbf{r}) - \rho_s] \, d^3 r}{\int [\rho(\mathbf{r}) - \rho_s] \, d^3 r}. \tag{16.81}$$

Thus a plot of the logarithm of the scattering intensity versus $Q^2/3$ will be a straight line in the small-Q region with a slope equal to negative square of the radius of gyration of the colloidal particle. The value for R_g can be given explicitly in terms of the geometrical parameters in a few simple cases:

(1) Sphere with radius R:

$$R_g^2 = 3R^2/5. \tag{16.82}$$

(2) Ellipsoid of revolution with semiaxis a, b, c:

$$R_g^2 = (a^2 + b^2 + c^2)/5. \tag{16.83}$$

(3) Circular cylinder with radius R and height h:

$$R_g^2 = \left(2R^2 + \frac{h^2}{3}\right)/4. \tag{16.84}$$

(4) Spherical shell with inner and outer radius R_1 and R_2:

$$R_g^2 = \frac{3}{5}\left[\frac{R_2^5 - R_1^5}{R_2^3 - R_1^3}\right]. \tag{16.85}$$

(5) Spherical particle with an inner core (radius R_1 and csld ρ_1) and an outer shell (outer radius R_2 and csld ρ_2):

$$R_g^2 = \frac{3}{5} R_2^2 \left\{ \frac{1 - [(\rho_2 - \rho_s)/(\rho_2 - \rho_1)]R_1^5/R_2^5}{1 - [(\rho_2 - \rho_s)/(\rho_2 - \rho_1)]R_1^3/R_2^3} \right\}. \qquad (16.86)$$

Besides R_g, the Guinier plot also gives $(d\Sigma/d\Omega)(0)$, which can be used in determining the aggregation number $\bar{n}$ of association colloids or the molecular weight of proteins.[35]

It is also worthwhile to note the asymptotic behavior of $P(Q)$ in the large-Q region, say $QR > 5$. For a spherical particle having a constant csld ρ and a radius R, one has

$$\lim_{Q \to \infty} \left[Q^4 \frac{d\Sigma}{d\Omega}(Q) \right] = 2\pi \frac{A}{V} (\rho - \rho_s)^2, \qquad (16.87)$$

where

$$A = n_p 4\pi R^2. \qquad (16.88)$$

Then, A/V is the total interface area per unit volume. Equation (16.87) shows that at large Q the asymptotic value of $Q^4(d\Sigma/d\Omega)(Q)$ is a constant. This is called the Porod's law in the literature.[44] Another interesting case is a thin shell of thickness d located at a distance R from the center. The intensity at large Q goes asymptotically as[45]

$$\lim_{Q \to \infty} \left[Q^4 \frac{d\Sigma}{d\Omega}(Q) \right] = 2\pi \frac{A}{V} (\rho - \rho_s)^2 Q^2 d^2. \qquad (16.89)$$

Thus for a particle having a core of radius R with csld ρ_1 surrounded by a thin interface film of thickness d and csld ρ_2, the asymptotic behavior is given by[45]

$$\lim_{Q \to \infty} \left[Q^4 \frac{d\Sigma}{d\Omega}(Q) \right] = 2\pi \frac{A}{V} [(\rho_1 - \rho_s)^2 + (\rho_2 - \rho_1)(\rho_2 - \rho_s) Q^2 d^2]. \qquad (16.90)$$

For any colloidal particle having a thin surface layer, $Q^4(d\Sigma/d\Omega)(Q)$ will not be a constant but a linear function of Q^2. This fact enables one to determine the layer thickness d by simply plotting $Q^4(d\Sigma/d\Omega)(Q)$ versus Q^2 up to the large-Q region.

A sufficient contrast between the solute and solvent can be provided by adjusting the H_2O and D_2O ratio for solutions containing water. The csld of an aqueous solvent can be varied by large amounts, from -0.56×10^{10} cm^{-2} to 6.34×10^{10} cm^{-2}, by changing from pure light water to pure heavy water. The variation of ρ's is large enough to cover the csld of both protonated and deuterated molecules of interest. The so-called external contrast variation method consists of measurements of a series of scattering patterns in the small-Q region from solutions in various H_2O/D_2O mixtures. Both the radius of gyration R_g and the scattered intensity at zero angle are then determined from these spectra. From Eq. (16.55) we have this relation for the external contrast variation measurements:

$$\left[\frac{d\Sigma}{d\Omega}(0)\right]^{1/2} = \sqrt{n_p} \left| \int_{particle} (\rho(\vec{r}) - \rho_s) \, d^3r \right|, \tag{16.91}$$

or in a different form:[35]

$$\left[\frac{d\Sigma}{d\Omega}(0)\right]^{1/2} = \sqrt{n_p} \left| \sum_{particle} b - \rho_s V_p \right|. \tag{16.92}$$

Here, V_p refers to the solvent-excluded volume of the particle and $\sum b$ is the total scattering length of the particle. One may define the csld of the particle averaged over the solvent excluded volume as

$$\bar{\rho} = \sum_{particle} b/V_p. \tag{16.93}$$

Denoting by α the fraction of D_2O in the solvent we have

$$\rho_s = \alpha \rho_{D_2O} + (1 - \alpha) \rho_{H_2O}. \tag{16.94}$$

A plot of the square root of $d\Sigma/d\Omega(0)$ versus α will result in a straight line. By knowing the total scattering length of the particle, one may then determine V_p and the number density of the particle n_p. Due to a large amount of incoherent scattering from H_2O in the solvent, it takes a much longer time to obtain precise experimental data when α is small. Thus one may turn to the so-called internal contrast variation method. Instead of varying csld of the solvent, we can keep it, for example, as D_2O and vary $\bar{\rho}$. We can do this by using partially deuterated molecules. In the case of association colloids, it is easier to vary $\bar{\rho}$ continuously by mixing deuterated monomers with undeuterated monomers. Denoting by β the fraction of the deuterated monomers, then

$$\bar{\rho} = \beta \bar{\rho}_D + (1 - \beta) \bar{\rho}_H. \tag{16.95}$$

Again a plot of $[d\Sigma/d\Omega(0)]^{1/2}$ versus β will yield a straight line. The same information can be obtained as in the case of the external variation measurements.

Besides the zero-intercept $d\Sigma/d\Omega(0)$, the Guinier plots yield also the radius of gyration R_g, which will in general depend on the contrast. As one changes the contrast either by deuteration of the solvent or the solute, values of R_g

will change accordingly. Using Eqs. (16.81) and (16.93), one can rewrite R_g as

$$R_g^2 = R_s^2 + \frac{\bar{\rho}}{\bar{\rho} - \rho_s} \frac{1}{V_p} \int d^3r \left(\frac{\rho(\mathbf{r}) - \bar{\rho}}{\bar{\rho}} \right) r^2, \quad (16.96)$$

where

$$R_s^2 = \frac{1}{V_p} \int d^3r \, r^2. \quad (16.97)$$

R_s is the radius of gyration associated with the shape of the particle, an asymptotic value of R_g when ρ_s approaches infinity, or when the particle has a uniform csld. The second term on the right-hand side of Eq. (16.96) is a contribution from the internal structure of the particle.

For a particle with a hydrophobic core (protonated hydrocarbons) surrounded by a hydrophilic region (molecular groups other than hydrocarbons), the factor in the integrand $[\rho(\mathbf{r}) - \bar{\rho}]/\bar{\rho}$ tends to be negative in the core region and positive in the outer layer. Since the integration is weighted by r^4, the net value of the second term tends to be positive in this case. Therefore a plot of R_g^2 versus $(\bar{\rho} - \rho_s)^{-1}$, the reciprocal of the contrast, would give a straight line with a positive slope. This is in fact the case for a globular protein myoglobin as was shown by the classic work of Ibel and Stuhrmann.[46]

Another important technique is the selected deuteration of a particular portion of the monomer in colloidal particles. By this technique one may selectively look at only that deuterated portion, namely, the scattering pattern due to that portion only. The idea is that the difference of the signal $[d\Sigma/d\Omega \, (Q)]^{1/2}$ between the deuterated and undeuterated samples is proportional to the Fourier transform of the spatial scattering length density distribution of the deuterated portion denoted by $\rho_x(r)$. This method is useful only when the deuterated portion distributes symmetrically with respect to the center of the particle. Denoting the difference by $A(Q)$, one has the equation[29]

$$A(Q) = \left[\frac{d\Sigma}{d\Omega}(Q) \right]_H^{1/2} - \left[\frac{d\Sigma}{d\Omega}(Q) \right]_D^{1/2}$$

$$= \sqrt{n_p} \int_0^\infty |\rho_x^D(r) - \rho_x^H(r)| \frac{\sin Qr}{Qr} 4\pi r^2 \, dr. \quad (16.98)$$

This equation is valid when polydispersity is small. Similar to the Guinier approximation, one can also obtain the radius of gyration from

$$A(Q) = A(0) \exp[-\tfrac{1}{6} Q^2 R_g^2]. \quad (16.99)$$

This radius of gyration R_g gives the second moment of the distribution of the deuterated portion of the monomers in a micelle.

We have shown in this section that one may extract information concerning the structure of colloidal particles from the zero-angle scattering intensities and from the intensity distribution in the small-Q region and in the large-Q region. In Section 16.3.2. we will show how the distribution of scattering intensities in the intermediate-Q region may be explained by actually modeling the structure of colloidal particles with the information obtained from the techniques described in this section as input.

16.3.2. Model Fitting of Spectra in the Intermediate Q Range

After one has done all the analyses of the scattering patterns according to the techniques described in the previous section, one should be able to determine the apparent molecular weight or, equivalently, the aggregation number of the molecular aggregate and to obtain a rough idea about the size of the particle from the values of R_g. Information about the detailed internal structure can also be obtained from selected deuterations. However, the exact shape of the particle or the degree of polydispersity can only be determined by examining the scattering over the entire Q range or, in practice, over the range of $Q\sigma \leq 10$, where σ is the characteristic size of the particle. This means that one would have to fit the measured spectrum $(d\Sigma/d\Omega)(Q)$ according to the formula

$$\frac{d\Sigma}{d\Omega}(Q) = n_p \langle P(Q) \rangle = n_p \langle |F(Q)|^2 \rangle, \tag{16.100}$$

when the interparticle structure factor is very close to unity. One may also incorporate the interparticle structure factor in the fitting if necessary. For spherical particles having a two-layer structure, the particle form factor $F(Q)$ is given by Eq. (16.78). For nonspherical particles it is necessary to average the form factor over the orientations of the particle. In the case of an ellipsoid having a uniform csld ρ, semimajor axis (the axis of symmetry) a, and semiminor axis b, the form factor is given by

$$\langle |F(Q)|^2 \rangle = (\rho - \rho_s)^2 \left(\frac{4\pi}{3} ab^2\right)^2 \left| \int_0^1 \frac{3 j_1(u)}{u} d\mu \right|^2 \tag{16.101}$$

where

$$u = Q[a^2\mu^2 + b^2(1 - \mu^2)]^{1/2}. \tag{16.102}$$

For a polydispersed system of particles, one should also average the form factor weighted by the distribution function of particle sizes as given by Eq. (16.65).

Figure 4 illustrates the extent to which the form factor depends on the a/b ratios. These curves are plotted as a function of a dimensionless quantity

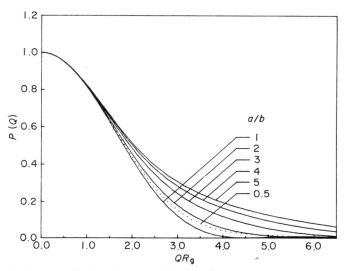

FIG. 4. The single particle form factor $P(Q)$, normalized to unity at $Q = 0$, for ellipsoidal particles with different ratios of semimajor axis to semiminor axis.

QR_g. The differentiation between shapes is possible at the tail part of the curve.[47] The effects of polydispersity on the form factor have been studied by van Beurten and Vrij[40] and by Kotlarchyk and Chen.[43] In general the polydispersity will change the form factor to a greater extent, and the radius of gyration determined from the Guinier plot will be larger than that of a monodispersed system.

A good example of the determination of the a/b ratio is the elongated globular protein bovine serum albumin. This protein can be obtained in a monodispersed form with a very small amount of dimer content. In an acetate buffer at the isoelectric pH = 5.1, a 0.3% protein solution in 0.03 M of LiCl can be well approximated by an interaction free system. Figure 5 shows the result of SANS measurements together with a thoretical fit by Bendedouch and Chen.[48] The results are well fitted by choosing $a = 70$ Å and $b = 20$ Å.

16.3.3. Structure of Ionic Micelles

Ionic micelles are spherical or rod-like aggregates formed by ionic surfactants in aqueous solutions. A model system formed from a lithium salt of laurel sulfate, called lithium dodecyl sulfate (LDS), shows a remarkably small degree of polydispersity and dependence of aggregate sizes on the detergent concentration, salt concentration, and temperature.[49] At room

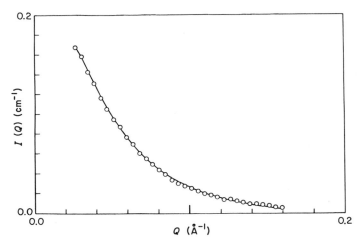

Fig. 5. Intensity distribution (circles) and theoretical fit (line) for a solution containing 0.3% BSA in acetate buffer pH 5.1, 0.03 M. The fit is obtained by assuming a prolate ellipsoidal shape for the solvated BSA ($a = 70$ Å and $b = 20$ Å) and $z = -1$. $S(Q)$, computed as explained in the text, turned out to be nearly unity rendering this system almost ideal.[48]

temperature, the CMC of LDS in 0.3 M LiCl solution is 0.85 mM, which is a negligible amount as compared to the commonly used 1% solution (43 mM) in the SANS study. Thus the system seems well suited for a detailed structural study. To illustrate the contrast variation technique, we will use the internal contrast variation data of Bendedouch et al.[50] Samples were prepared by mixing perdeuterated and protonated LDS solutions, each at 36.72 mM concentration in an aqueous D_2O solvent with 0.3 M LiCl. The internal contrast variation technique is easier to use and gives more accurate results than the external contrast variation technique, since the solvent can be kept as D_2O and one varies only the fraction of the perdeuterated monomers in the micelle. This way, the incoherent background from the solvent can largely be eliminated and the signal-to-noise ratio greatly enhanced.

With 0.3 M LiCl in the solution, the Coulomb interaction between micelles are expected to be weak due to the screening effect of the electrolyte. Thus the Guinier approximation could be applied to analyze small-Q data. We denote the average aggregation number by $\bar{n}$, the total scattering length of a monomer by b_m, the solvent-excluded volume per monomer by V_m, and the number density of monomers forming micelles by $n_1 = n_p \bar{n}$. We may then rewrite Eq. (16.92) as

$$\left[\frac{d\Sigma(0)}{d\Omega}\frac{1}{n_1}\right]^{1/2} = \sqrt{\bar{n}}\,|b_m - \rho_s V_m|. \qquad (16.103)$$

16. COLLOIDAL SOLUTIONS

For a given fraction β of the perdeuterated monomers, we have

$$b_m = \beta b_m^D + (1 - \beta)b_m^H. \tag{16.104}$$

By assuming that 70% of the LDS monomers in micelles are ionized[10] and hence lose their lithium ion, one may calculate the deuterated and undeuterated scattering lengths b_m^D and b_m^H, respectively, from known chemical formulas. We get $b_m^D = 2.712 \times 10^{-3}$ Å, and $b_m^H = 1.082 \times 10^{-4}$ Å. Substituting Eq. (16.104) into Eq. (16.103), we see that the zero-angle intensity [the quantity given by the square bracket on the LHS of Eq. (16.103)] is a linear function of β with the zero crossing at

$$\beta_0 = \left| \frac{b_m^H - \rho_{D_2O} V_m}{b_m^D - b_m^H} \right|, \tag{16.105}$$

and the slope m given by

$$|m| = \sqrt{\bar{n}}(b_m^D - b_m^H). \tag{16.106}$$

From these two relations, we can uniquely determine V_m and $\bar{n}$. Figure 6 shows the result obtained by Bendedouch et al.,[50] which gives $\beta_0 = 0.94$ and $|m| = 0.0230$ Å. From these values one obtains the dry volume of the LDS monomer $V_m = 402$ Å^3 and the aggregation number of the micelle at 37°C, $\bar{n} = 77.9$.

Guinier plots of small-Q data also give the same value of the radius of gyration $R_g = 15.4$ Å from $\beta = 0$ up to $\beta = 0.6$. A relatively constant R_g means that the scattering of neutrons is dominated by the tail parts of the LDS monomers forming the core of the micelle. If one assumes that the tails form a close-packed ellipsoidal core with semiminor axis b equal to the fully stretched tail length 16.7 Å derived from Eq. (16.10), the semimajor axis a is then determined to be 23.4 Å using 350 Å^3 as the steric volume of the hydrocarbon tail. The head group of the monomer should then occupy a volume equal to 402 Å^3 − 350 Å^3 = 52 Å^3, which corresponds to a sphere of diameter 4.6 Å. Thus the outer layer of the micelle can be assumed to be a shell of thickness 4.6 Å, consisting of 78 head groups plus solvent molecules. Such a model would predict that there are about 9.3 water molecules in the outer layer for every head group. This value is in substantial agreement with the magnitude of hydration of the head groups investigated by other techniques. The radius of gyration for the outer layer alone, denoted by R_{g2} is estimated to be 21.5 Å and R_{g1} the radius of gyration of the core alone is 14.9 Å calculated from Eq. (16.83). Thus the radius of gyration of the whole particle is given by

$$R_g = \left[\frac{R_{g1}^2 \int_1 (\bar{\rho}_1 - \rho_s) d^3r + R_{g2}^2 \int_2 (\bar{\rho}_2 - \rho_s) d^3r}{\int_1 (\bar{\rho}_1 - \rho_s) d^3r + \int_2 (\bar{\rho}_2 - \rho_s) d^3r} \right]^{1/2}, \tag{16.107}$$

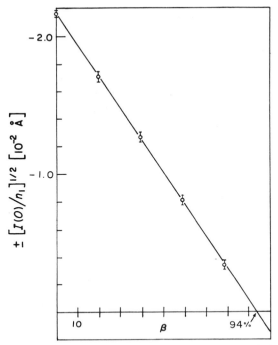

FIG. 6. Internal contrast variation measurements of LDS micelles in solutions of 0.3 M LiCl in D_2O. The degree of deuteration of the hydrocarbon chains that is necessary to match out the scattering length density of D_2O is given by $\beta = 0.94$, where β is the molar fraction of perdeuterated monomers.[50]

where the subscripts 1 and 2 stand for the core and the outer layer, respectively. The calculated values of R_g turn out to vary from 15.2 to 15.7 Å for $\beta = 0$ to $\beta = 0.6$. These values agree very well with the experimental value, which is about 15.4 Å. The scattering intensity profiles in the intermediate-Q range can also be fitted very well by this model.

It is generally assumed in the literature[4,51] that the micellar core is like a liquid hydrocarbon of the same alkane number as the tails. This means that the packing density of the core is like a liquid alkane. However, the conformation of the hydrocarbon tails cannot be completely like that in a liquid state because of a geometrical constraint that the head groups are distributed over the outer layer of the micelle. On the other hand, the familiar picture of a micellar core[52] as depicted in many textbooks also cannot be realistic. Namely, if the transconformation is the dominant feature as is commonly assumed, then there are too many open spaces where water can enter the hydrocarbon core. The hydrophobic energy would be too high in this case. Obviously the real situation is somewhere in between these two extreme pictures. Fortunately, a selective deuteration of the methyl groups at the end of each hydrocarbon tail can yield data that supply a partial answer to this question.

To study the spatial distribution of the end methyl groups, the scattering intensities of protonated and end-deuterated 36.72 mM LDS in 0.3 M LiCl D$_2$O solutions are compared in Fig. 7. Following the procedure described in Section 16.3.1, one obtains the scattering amplitude $A(Q)$, which is the Fourier transform of the methyl group radial distribution function as defined in Eq. (16.98). The $A(Q)$ in the Q range measured (i.e., $0 \leq Q \leq 0.25$ Å^{-1}) is very well fitted by the Guinier approximation with $R_g = 12.9$ Å. Comparing this to $R_g = 15.4$ Å, obtained for the protonated hydrocarbon core, we infer that the methyl groups have a higher probability of being at the inner part of the core and less probability of lying at the outer layer of the micelle. The zero Q limit obtained from extrapolation of the Guinier analysis gives $A(0) = (n_p \bar{n})^{1/2} b_{Me} = 0.128$ cm$^{-1/2}$ where $b_{Me} = |b_{CH_3} - b_{CD_3}|$, from which we extract $\bar{n} = 78.0$. This result is independent of the shape of the

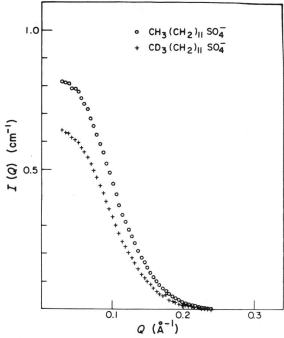

FIG. 7. Scattering intensity profiles for $36.72 \times 10^{-3} M$ protonated (open circles) and end-deuterated (crosses) LDS in 0.3 M LiCl in D$_2$O.[50]

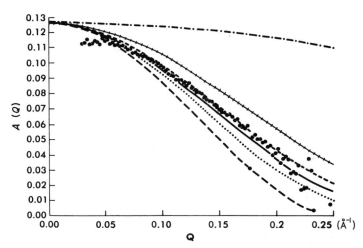

FIG. 8. Small-angle neutron scattering amplitudes $A(Q)$ against wave vector Q. Data from Bendedouch et al.,[50] corrected here for interparticle interactions. Predictions (radial shells are number i = 1, 2, 3, 4, toward the micellar center) are given by[53] (1) radial chain model, t_i = 0, 0, 0, 1 (–·–); (2) pincushion model, t_i = 0, 0.563, 0.375, 0.062 (++++); (3) interphase model, ω = 1, t_i = 0.234, 0.414, 0.289, 0.062 (–––); (4) interphase model, ω = 0.5, t_i = 0.326, 0.370, 0.241, 0.062 (——); (5) interphase model, $\omega \to 0$ (identical to Cabane model), t_i = 0.438, 0.312, 0.188, 0.062 (·····); (6) oil droplet model, t_i = 0.578, 0.297, 0.109, 0.016 (— —), where t_i is the fraction of chains whose termini are in layer i and ω represents the stiffness of chains from rigid to freely flexible chains by ω = 1 to $\omega \to 0$, respectively.

micelle and agrees well with the previous value $\bar{n}$ = 77.9 obtained from the internal contrast variation measurement.[50] Model analysis of $A(Q)$ has been carried out by Dill et al.[53] and the result is shown in Fig. 8. Agreement with the Dill–Flory model of the conformational distribution in the micellar core is seen to be good.[54] More detailed theoretical analyses and computer simulations were carried out by Gruen[55] and again the experimental data support a very reasonable model of micellar organization. Finally, Fig. 9 depicts a cross-sectional view of the micellar core. The dots are the methyl groups. This picture is schematic only, and a more realistic three-dimensional picture is presented in Dill et al.[53]

16.4. Determination of the Interparticle Structure

16.4.1. Ionic Micellar Solutions

It is an inherent characteristic of ionic micellar solutions that the intermicellar interaction always plays a significant role in determining their thermodynamic and microscopic structural properties. These systems can for

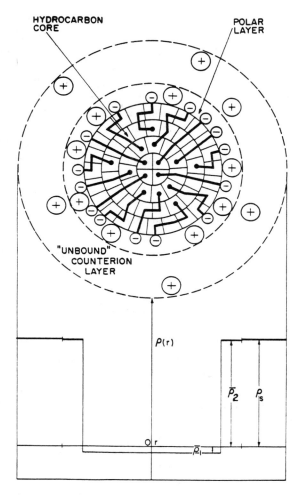

Fig. 9. Schematic structure of an ionic small micelle. The hydrocarbon core contains the tails that occupy completely the volume V_1 defined by the radius R_1. Our analysis shows that the core has a liquidlike density, but the chains have a degree of order between that of a completely disordered liquid state and a fully ordered all trans radial configuration. One possible close-packed structure is illustrated by a latticelike model. The polar shell has a density $\bar{\rho}_2$. It contains the sulfate head groups, the "bound" counter ions, the "bound" water (which together define the outer layer), some additional bulk water, and maybe a few protruding CH_2 groups. The scattering length density profile corresponding to the different regions is also shown in the bottom of the figure.

many purposes be regarded as consisting of charged hard spheres surrounded by a cloud of counter ions and of additional small ions if a salt has been added to the solution. The micelles are large compared to the small ions and to the molecules of the solvent. The role of a solvent (such as water) is to provide a dielectric constant ε for the charged particle interactions and to provide friction against the motion of the micelles. When there is no salt present in the solution, the intermicellar Coulomb interaction is such that it is already appreciable at CMC. One cannot conveniently add salt (electrolyte) to screen off the electrostatic interaction, as is the case for other macromolecules, because an addition of salt will induce micellar growth and thus change the particle structure factor $P(Q)$. It is also difficult to perform a dilution study and to extrapolate the scattering intensity to the limit of zero concentration because the micelles exist in solution only above CMC. Therefore in interpreting SANS data from micellar solutions, simultaneous model calculations of $P(Q)$ and $S(Q)$ are always necessary.

To model the interaction, one assumes the micelles to be charged hard spheres with the hard sphere diameter σ chosen to represent the excluded volume effect. When the charge is absent, the structure factor $S_{HS}(Q)$ can be, to a first approximation, calculated from an analytical solution of the Percus–Yevik equation.[56] When the charge is present, the additional charge interaction can be taken into account by treating it as a perturbation to the basic hard sphere interaction. The structure factor can again be calculated analytically using the mean spherical approximation (MSA). For ionic micelles the repulsive interaction is mainly due to the charged double layers forming around each particle. The strength of the attractive Hamaka interaction is only of the order of $k_B T$ and usually can be neglected except when a large amount of electrolyte (for example, more than 1 M for LiCl) is added to the solution.[57]

As discussed already in Section 16.1.3, the repulsive double layer potential can be well represented by a screened Coulomb form (or Yukawa form) of Eq. (16.27) for interparticle distances larger than σ. In the dimensionless form, it can be written (with $x = r/\sigma$) as

$$\frac{V_c(x)}{k_B T} = \begin{cases} \gamma e^{-k} \dfrac{e^{-k(x-1)}}{x}, & x > 1 \\ \infty & x \leq 1 \end{cases} \qquad (16.108)$$

where γe^{-k}, the dimensionless contact potential, can be related to the effective surface charge Z^*e of a micelle by

$$\gamma e^{-k} = \frac{4(Z^*e)^2}{\varepsilon\sigma(2 + k)^2} \bigg/ k_B T. \qquad (16.109)$$

The dimensionless screening constant $k = \kappa\sigma$ depends on the ionic strength I of the solution, which can be computed in terms of the molar concentration C of detergent monomers forming micelles and the fractional surface charge $\alpha = Z^*/\bar{n}$, when LiCl is added, by

$$I = \text{CMC} + [\text{LiCl}] + \tfrac{1}{2}\alpha C. \tag{16.110}$$

As long as the shape of the ionic micelle is ellipsoidal with a/b ratio not larger than 2, one may approximate the micellar core as a rigid sphere with a diameter σ given by

$$\sigma = 2[(ab^2)^{1/3} + 4.6 \text{ Å}], \tag{16.111}$$

where $(ab^2)^{1/3}$ is the mean radius of the hydrocarbon core and 4.6 Å is the thickness of the outer layer consisting of the hydrated head groups for the case of LDS micelles.

The interparticle structure factor for a monodisperse system is related to the total correlation function $h(r)$ in a way similar to Eq. (16.63):

$$S(Q) = 1 + \frac{N}{V} \int_0^\infty \frac{\sin Qr}{Qr} h(r) 4\pi r^2 \, dr. \tag{16.112}$$

The Ornstein–Zernicke equation[58] can be solved for the $h(r)$ in the mean spherical approximation.[59] The MSA assumes that the direct correlation function $c(r)$ outside the hard core is given by its asymptotic form

$$c(r) = -V_c(r)/k_B T, \qquad r > \sigma. \tag{16.113}$$

Using this Ansatz one can then solve the Ornstein–Zernicke equation to obtain $S(Q)$. A computer program, which was written in Fortran and developed by Hayter and Penfold,[60] is a compact and fast way of executing this solution. An alternative way to calculate $S(Q)$ is to follow a mathematical procedure developed by Baxer.[61] This method was adapted for this purpose by Blum and Hoye.[41]

To calculate $S(Q)$ according to this model, one needs two independent input parameters, namely, α and the aggregation number $\bar{n}$. Since the volume of the hydrocarbon core is related to $\bar{n}$ by $\bar{n}V_\text{tail}$, σ as given by Eq. (16.111) is now related to $\bar{n}$ for LDS micelles by

$$\sigma = 2\left[\left(\frac{3\bar{n}V_\text{tail}}{4\pi}\right)^{1/3} + 4.6 \text{ Å}\right]. \tag{16.114}$$

The volume fraction η is then equal to $(\pi/6)N_p\sigma^3$, where $N_p = N_A(C - \text{CMC})/1000/\bar{n}$.

By fitting the measured scattering spectra to the models for $P(Q)$ and $S(Q)$, one essentially determines $\bar{n}$ and α. Analysis of data from LDS micellar

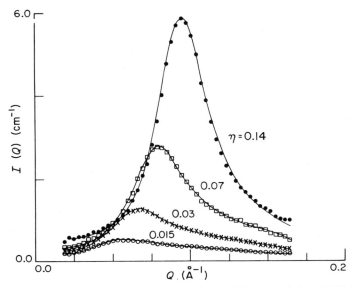

FIG. 10. Scattering patterns for 4 LDS concentrations: 0.037 (open circles), 0.074 (crosses), 0.147 (open squares), 0.294 M (closed circles) in pure D_2O at 37°C. The corresponding volume fractions η are indicated. The lines are theoretical fits and the symbols are data points. The building up of structure is clearly seen to follow the increased micellar concentration. The higher the concentration, the sharper the peak. The peak shifts to larger Q values when the LDS concentration increases, reflecting a decrease in the mean distance that results in a stronger interaction between particles in solution.[10]

systems with this kind of model shows that values of α are around 0.30 and that $\bar{n}$ increases with increasing concentrations of LiCl and LDS.[10] Figure 10 illustrates the sharpening and enhancement of the interparticle structure as the concentration of LDS is increased. The extracted $S(Q)$ and $P(Q)$ are shown in Fig. 11. For a strong repulsive interacting system, the structure factor at small Q is far less than unity, which means that the compressibility of the system is low. The prominent interaction peak in the SANS spectra is the key factor in being able to determine both parameters $\bar{n}$ and α. The double layer repulsion can be screened by adding more salts to the solution. The effectiveness of adding salts is shown clearly in Figs. 12 and 13. The scattering patterns were all very well fitted by this model, which supports the growth model of the micelle assumed and also the reliability of the extracted surface charge Z^*.

16.4.2. Charged Globular Protein Solutions

Many water soluble globular proteins tend to be negatively charged at a neutral pH = 7.0. As the solution pH is lowered, the surface charge gradually decreases until, at the so-called isoelectric pH (called also PI), the

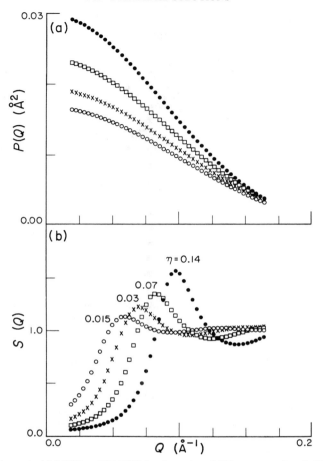

FIG. 11. Theoretical (a) $S(Q)$ and (b) $P(Q)$ for the various LDS concentrations in D_2O at 37°C corresponding to Fig. 10. The change in $P(Q)$ is due to the change in a/b ratio, which is only dependent on $\bar{n}$, according to our model (see text). In these low ionic strength conditions, $S(Q)$ shows a marked depression at low Q and a sharper and higher peak as the concentration increases.[10]

average surface charge is zero. For a pH below this value, the protein will acquire positive surface charge. Thus, water soluble globular proteins are interesting cases of macromolecular colloids for which the charge interaction can be adjusted by an external means. Except at PI it is impractical to perform experiments at a concentration low enough so that one can neglect the charge interaction. Since at low ionic strength, the double layer interaction is relatively long-ranged, the interaction has to be taken into consideration even at a concentration level of 1 gr/dl (1%).

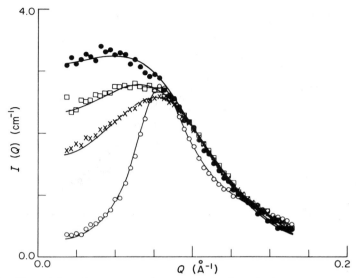

FIG. 12. Effect of electrolyte concentration on scattering intensity for 0.147 M LDS at 37°C. As the LiCl concentration is increased, the peak disappears gradually until, at the highest salt concentration, it becomes more like a shoulder;[10] 0.0 M LiCl (open circles), 0.2 (crosses), 0.4 (open squares), and 1.0 (closed circles).

A second motivation for studying a simple globular protein, such as bovine serum albumin (BSA), is that the purified protein is free of dimers, fairly monodispersed, and the shape of a native protein can be well approximated by an ellipsoid of revolution with an axial ratio $a/b = 3.5$.[62] We have therefore an ideal case for studying the interaction between elongated particles.

The structure factors $S'(Q)$ and $\bar{S}(Q)$ in Eq. (16.73) can be calculated in the same way as in Section 16.4.1. One assumes the pair potential between two protein molecules to consist of a hard core of diameter σ plus a screened Coulomb potential beyond the hard core diameter. The diameter σ is set equal to

$$\sigma = 2(ab^2)^{1/3}. \tag{16.115}$$

When the ratio a/b becomes larger than 3, a more accurate prescription is[43]

$$\sigma^3 = 2(f + 1)ab^2, \tag{16.116}$$

where

$$f = \frac{3}{4}\left(1 + \frac{\sin^{-1}(p)}{p(1-p^2)^{1/2}}\right)\left[1 + \frac{1-p^2}{2p^2}\ln\left(\frac{1+p}{1-p}\right)\right], \tag{16.117}$$

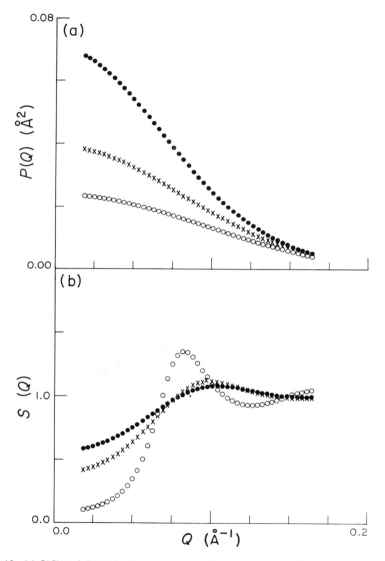

FIG. 13. (a) $S(Q)$ and (b) $P(Q)$ corresponding to Fig. 12 for 0.147 M LDS at 37°C, in pure D_2O (open circles), with 0.2 M LiCl (crosses) and with 1.0 M LiCl (closed circles). It is clear that even the highest salt concentration 1 M LiCl is not sufficient to screen completely the electrostatic repulsion between micelles.[10]

and

$$p^2 = (a^2 - b^2)/a^2, \qquad (16.118)$$

when $a \to b$, $f \to 3$ and Eq. (16.116) becomes identical to Eq. (16.115).

Using BSA as an example, one finds that the interparticle interaction is a sensitive function of the surface charge.[48] As the protein concentration is increased to 4% and above at neutral pH and with no salt added, the double layer interactions between protein molecules become so strong that the interparticle structure factor is significantly depressed in the small-Q region. This is manifested in the scattering patterns by a well-defined interaction peak with the curve plunging down steeply in the small-Q region. Figure 14

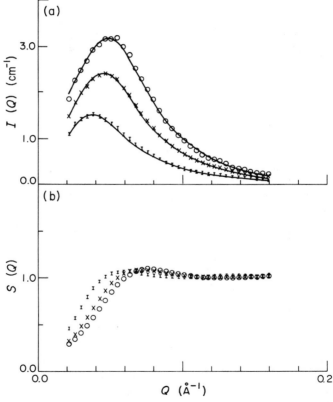

FIG. 14. (a) Intensity patterns for 4% (I), 8% (crosses), and 12% (open circles) BSA in D_2O solutions. The lines are theoretical fits. The sharpening of the interaction peak as the concentration increases is due to the building up of Coulombic repulsions between charged BSA molecules. (b) Corresponding $S(Q)$ calculated using $z = -8$ for all three concentrations.[48]

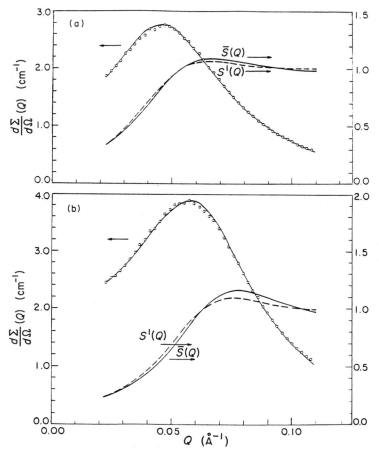

FIG. 15. SANS patterns and best fit curves for BSA solutions at (a) 8.73% BSA in D_2O and (b) 16.36% BSA in D_2O. The differences between $\bar{S}(Q)$ and $S'(Q)$ are also shown.[67]

shows the results of the analysis for 4, 8, and 12% BSA in pure D_2O, which gave a surface charge of $-8e$ and a surface potential of 3.4, 2.6, and 2.2 units of $k_B T$, respectively. The agreement between the theory and the experiment is very good.[48] Figure 15 shows the difference between $S'(Q)$ and $\bar{S}(Q)$ for the case of BSA. By carefully modeling the charge interaction, one can obtain a reliable measurement of the structural parameters.

A SANS experiment on BSA at the isoelectric pH = 5.1 has been carried out by Bendedouch and Chen.[57] In this case the double layer interaction is negligible and an effective attractive interaction was observed. By modeling this attractive interaction using the Hamaker interaction, an effective interaction parameter $A = -5.4 \, k_B T$ was obtained.

16.4.3. Microemulsions

The nature of the interactions in microemulsions is different from that of ionic micelles or charged proteins. A water-in-oil microemulsion has a structure similar to that shown in Fig. 1. In a microemulsion droplet, the water core is surrounded by a layer of detergent molecules with their hydrophilic head groups pointing inward toward the water core and the tails outward toward the oil. Because of the reversed micellar structure, the effective charge of the microemulsion droplet is zero and the mutual interaction beyond the hard core is thus expected to be attractive. This attractive interaction is considered to be responsible for the observed critical phenomena in different microemulsion systems.[63-66] At certain surfactant-to-oil ratios, upon increasing the temperature a microemulsion may become cloudy and separate into an upper and a lower phase that are different only in their number density of the microemulsion droplets.[28]

The AOT–water–oil system, where AOT stands for the ionic surfactant sodium di-2-ethylhexylsulfosuccinate, is a simple model system since it forms one of the few three-component microemulsions near room temperature. For example, at 22.5°C, the AOT/D_2O/n-decane system can be described by

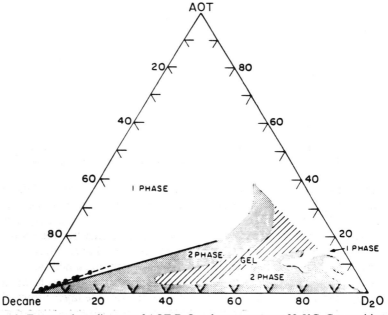

FIG. 16. Ternary phase diagram of AOT-D_2O-n-decane system at 22.5°C. Compositions are in volume percent. Solid points correspond to water-in-oil microemulsions with $x = [D_2O]/[AOT]$ kept at 40.8.

the phase diagram shown in Fig. 16, where at the oil corner (oil-rich phase), it shows a large transparent one-phase region. In this region the microemulsion droplets are just what we have depicted in Fig. 1. It is found that the average core radius $\bar{R}$ is linearly dependent on the parameter $X = [\text{D}_2\text{O}]/[\text{AOT}]$, the molar ratio of water to surfactant.[28] This linear relationship indicates that each surfactant molecule tends to subtend a constant head group area in forming the interfacial film at a given temperature. Analysis of SANS data shows that the droplets are, to a good approximation, spherical with a polydispersity index $p = \Delta R/\bar{R}$ of about 0.3.[28,67] Polydispersity is generally higher for microemulsions as compared with the case of ionic micelles. Thus in the analysis of SANS data from microemulsions, one has to deal with polydispersity, attractive interactions, and the divergence of scattering intensity near the critical temperature. For a polydisperse system, the averaged form factor can be calculated from Eq. (16.65), and the scattering intensities are analyzed according to Eq. (16.64). Since the hydrocarbon tails have a scattering length density similar to that of the oil, the form factor will be determined practically by the water core alone. The scattering length density within the water core is assumed to be uniform. It is generally assumed that the radius R of the water core has a Schultz-size distribution, i.e.,

$$f(R) = \left(\frac{Z+1}{\bar{R}}\right)^{Z+1} R^Z \exp\frac{[-(Z+1)R/\bar{R}]}{\Gamma(Z+1)}, \qquad (16.119)$$

where Z is the width parameter and is related to the polydispersity index p by

$$p = (\overline{R^2} - \bar{R}^2)^{1/2}/\bar{R} = (Z+1)^{-1/2}. \qquad (16.120)$$

For a continuous distribution of sizes, the averaged form factor is then calculated from

$$\langle |F(Q)|^2 \rangle = \int_0^\infty |F(Q, R)|^2 f(R)\, dR, \qquad (16.121)$$

where

$$F(Q, R) = \frac{4\pi}{3} R^3 \Delta\rho \frac{3(\sin QR - QR \cos QR)}{QR}, \qquad (16.122)$$

and $\Delta\rho$ is the difference between the scattering length densities of the water core and the oil. Given the volume fraction of AOT and D_2O, ϕ_{AOT} and $\phi_{\text{D}_2\text{O}}$, together with the two parameters $\bar{R}$ and Z, other parameters can be derived. The number of particles per unit volume is given by

$$\frac{N}{V} = \frac{\phi_{\text{D}_2\text{O}}}{\frac{4}{3}\pi\bar{R}^3} = \phi_{\text{D}_2\text{O}} \frac{(Z+1)^2}{(Z+2)(Z+3)} \frac{1}{\frac{4}{3}\pi\bar{R}^3}. \qquad (16.123)$$

As discussed in Section 16.2.3, the averaged structure factor $\bar{S}(Q)$ as defined by Eq. (16.67) is approximated by an effective one-component structure factor with a particle diameter σ defined by Eq. (16.68). The effective one-component system has the same volume fraction, $\phi = \phi_{AOT} + \phi_{D_2O}$, and the same number density of particles. Thus the parameter σ is given by

$$\sigma = \left[\frac{6\phi}{\pi(N/V)}\right]^{1/3}. \tag{16.124}$$

The final task is to model the attractive interactions. The long-range attractive potential function as given by Eq. (16.21) does not yield an analytical solution that is so essential in the fitting of data. One may, therefore, adopt a phenomenological potential function having a hard core plus a Yukawa tail that represents the attractive interactions. The two parameters k and ye^{-k} are then to be determined from data fitting. The critical behavior of such a model, the MSA for Yukawa fluids, has been studied by Cummings and Stell in detail.[68] In order to improve the accuracy of the MSA, especially in the small-Q region, which is important for an attractive system, one may use the optimized cluster theory (OCT),[69] which differs from the MSA by taking into account some contributions from the hard core to the direct correlation function for $r > \sigma$. The result of OCT for the pair correlation function is, when expressed as a linearized exponential approximation (LEXP),

$$g(r) = g_{HS}(r) \exp[h_{MSA}(r) - h_{HS}^{(PY)}(r)], \tag{16.125}$$

where HS stands for hard sphere system and (PY) stands for the Percus–Yevick approximation. One may further approximate $g_{HS}(r)$ by $g_{HS}^{(PY)}(r)$ and rewrite Eq. (16.125) in terms of the total correlation as[42]

$$h(r) = h_{MSA}(r) + h_{HS}^{(PY)}(r)[h_{MSA}(r) - h_{HS}^{(PY)}(r)]. \tag{16.126}$$

Thus we see that the LEXP in this formulation simply adds a correction term to the $h_{MSA}(r)$.

In terms of the structure factor, Eq. (16.126) gives

$$S(Q) = S_{MSA}(Q) + \frac{1}{(2\pi)^3(N/V)} \int [S_{HS}^{(PY)}(\mathbf{Q}') - 1][S_{MSA}(\mathbf{Q} - \mathbf{Q}') - S_{HS}^{(PY)}(\mathbf{Q} - \mathbf{Q}')] d\mathbf{Q}'. \tag{16.127}$$

Such a model allows one to obtain information about the average particle size, the polydispersity, and the strength and range of the attractive interactions ye^{-k} and k. This model works well up to volume fractions of 20%[42] as shown in Fig. 17 for volume fractions from 3.6 to 12.8%. Near the critical temperatures, the interaction peak is not visible for ϕ less than 20%, and

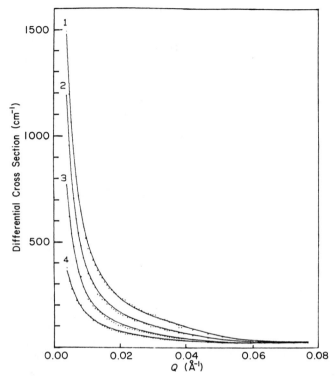

FIG. 17. SANS patterns for the AOT-D_2O-n-decane system with $[D_2O]/[AOT] = 40.8$ at 42.8°C. Solid curves are best fits to the experimental data using the linear exponential (LEXP) approximation, based on OCT, for the structure factor.[42] Volume fraction: 12.8% (1), 9.8% (2), 5.3% (3), and 3.6% (4).

the scattering intensities increase sharply in the small-Q region due to the attractive interactions. The differences between the LEXP and the MSA are best illustrated in Fig. 18, where the structure factors are plotted for different volume fractions. The importance of using LEXP is only significant in the region $Q < 2\pi/\sigma$. The corrections to $S(Q)_{MSA}$ are also more significant at higher volume fractions.

The correlation length ξ and isothermal compressibility χ_T are two parameters that characterize the critical behavior of a system. To extract these two parameters, we use another model for the interparticle structure factor. When a system is near the critical point, the structure factor for the critical scattering at small Q should take the Ornstein-Zernicke[58] form

$$S_{OZ}(Q) = \chi_T/(1 + Q^2\xi^2). \qquad (6.128)$$

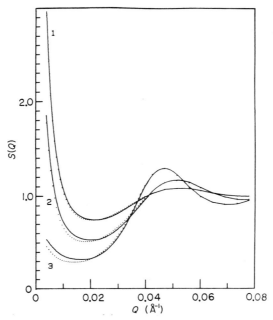

FIG. 18. Structure factors for $\phi = 5.3\%$ (1), 12.8% (2), and 21.3% (3) at 42.8°C. Solid curves are results obtained by using the LEXP approximation, and the dotted curves are obtained by using the MSA with the same parameters for the Yukawa tail as used in LEXP approximation.[42]

The peaking of the structure factor near the $Q = 0$ region is due to the attractive interaction between particles. Away from it the structure factor should be close to the hard sphere structure factor. Thus a good representation of the structure factor near the critical point should be[28,42]

$$S_{OZ+HS}(Q) = S_{HS}^{(PY)}(Q) + S_{OZ}(Q). \qquad (16.129)$$

This model has been applied successfully to the AOT-D_2O-n-decane system at $\phi = 0.075$ with $[D_2O]/[AOT] = 40.8$ from 30 to 42.6°C as shown in Fig. 19.[42] The temperature dependence of ξ and ψ_T can be described by the power laws $\xi(T) = \xi_0 \varepsilon^{-\nu}$ and $\chi_T = \chi_0 \varepsilon^{-\gamma}$, where $\varepsilon = (T_c - T)/T_c$. These parameters are determined to be $T_c = 43.12 \pm 0.05$°C, $\xi_0 = 12.8 \pm 0.3$ Å, $\nu = 0.62 \pm 0.01$, $\chi_0 = 0.058 \pm 0.017$, and $\gamma = 1.20 \pm 0.06$. The values of the critical exponents ν and γ are close to the Ising model values, and the phase transition belongs, therefore, to the universality class of the liquid–vapor phase critical point.

So far the effect of polydispersity on the interparticle structure is only taken into account in an approximate way. Further improvements on this as well as on the model for the attractive potential can be expected in the near future.

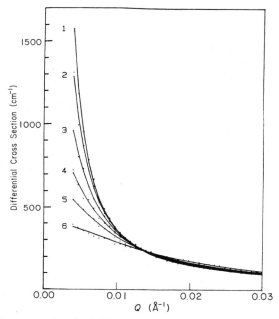

FIG. 19. SANS patterns for the AOT-D_2O-n-decane system with $[D_2O]/[AOT] = 40.8$ at 7.5% droplet volume fraction showing the temperature dependence. Solid curves are best fits using the OZ + HS model, Eq. (16.129), for the interparticle structure factor. Temperature dependence: 42.6°C (1), 41.7°C (2), 40.1°C (3), 381.1°C (4), 35.0°C (5), and 30.0°C (6).

16.5. Conclusion

Colloidal solutions are ideal systems to be studied by the SANS technique. This is especially true for the association colloids, such as micelles and microemulsions. The size range of these micellar systems ranges from 30 Å to about 100 Å, which, in terms of the scattering experiment, translates into an intensity distribution in the Q range of 0.01 to 0.2 Å^{-1}. This is the most easily accessible range of the SANS spectrometer for which the available neutron flux is also the highest. The fact that the surfactant tails tend to form a compact hydrophobic core implies that in a D_2O solvent the contrast between the micellar core and the solvent is very high and a micellar aggregate looks like a very simple object to the neutrons. The resulting simple particle structure factor $P(Q)$ in turn allows an accurate determination of the interparticle structure factor $S(Q)$ from the measurement. Thus questions such as the effective surface charge Z^* on micelles, or how micellar aggregation numbers grow as a function of concentration and salt, can be answered with little ambiguity. Knowledge of these quantities is vital to the future formulation of the statistical mechanical theory of micellar formation and growth. Likewise, the internal structure of the micellar core, its water content, and its hydrocarbon conformations can also be addressed once the SANS data are properly analyzed.

In microemulsions where one has an inverted micellar structure, the effective attractive interaction leads to the cloud point effect or the critical point phenomenon. Again, SANS is a unique tool for simultaneously measuring the structure of the aggregates and their mutual correlations. An important aspect in recent developments of the application of the SANS technique in colloids is the ability to take into account the interaction realistically. It seems that the traditional method of emphasizing the study of the small-Q part of the SANS patterns in dilute solutions would not have produced such a wealth of information. An obvious direction to pursue in the future is to explore more of the phase transitions and the structure of the dense phases of these surfactant solutions.[70]

Due to the lack of space we cannot, unfortunately, include a discussion of the dynamic measurement using SANS. The so-called neutron spin-echo spectroscopy (NSES) is a counterpart of the photon correlation spectroscopy in light scattering. NSES has the capability of measuring the density fluctuations in the larger Q range (0.02 to 2 $Å^{-1}$). Some interesting work in applying the NSES to micellar solutions has already been reported by Hayter and Penfold,[71] and a good introduction to the technique has been given by Hayter.[72]

In this chapter we have mainly illustrated the SANS techniques using results from our own group, since the chapter is not meant to be a review for all the SANS results on micelles and microemulsions. We would therefore like to apologize in advance for not presenting other results in a balanced manner. As an example, the structural determination of SDS ionic micelles by Cabane et al.[73] was very much along the line that we presented in Section 16.3. We can also cite a more recent work on the intermicellar interactions of a system of sodium alkylbenzenesulfonate micelles by Triolo et al.[74] Their analysis based on the DLVO potential in the mean spherical approximation yielded both the micellar aggregation number and the effective surface charge for a series of 0.055 M detergent solutions. These two parameters were shown to remain approximately constant as the fraction of D_2O in the solvent was varied. The method of analysis essentially parallels that given in Section 16.4, except in this work the structural model of the micelle was assumed rather than determined by an independent SANS contrast variation measurement.

Acknowledgment

This research is supported by the National Science Foundation and the American Chemical Society through the Petroleum Research Fund. T. L. Lin is grateful for the financial support from the T. J. Thompson Fellowship of the Nuclear Engineering Department of M.I.T. S. H. Chen owed a great deal to the late Dr. W. C. Koehler and to Dr. B. P. Schoenborn for their generous support in terms of the neutron beam time at NCSASR at ORNL and at the Biology Low-Angle Spectrometer in BNL.

References

1. P. S. Pershan, *Phys. Today* **35**(5), 34 (1982).
2. S. Friberg, *in* "Microemulsion Theory and Practice" (L. M. Prince, ed.), p. 133. Academic Press, New York, 1977.
3. D. J. Shaw, "Introduction to Colloid and Surface Chemistry," 2nd Ed. Butterworths, London, 1970.
4. C. Tanford, "The Hydrophobic Effect," 2nd Ed. Wiley, New York, 1980.
5. C. Tanford, *J. Phys. Chem.* **78**, 2469 (1974).
6. J. N. Israelachvili, "Physics of Amphiphiles: Micelles, Vesicles and Microemulsions" (V. Degiorgio and M. Corti, eds.), p. 24. North-Holland Publ., Amsterdam, 1985.
7. J. N. Israelachvili, D. J. Mitchell, and B. W. Ninham, *J.C.S. Faraday Trans. I* **72**, 1525 (1976).
8. D. J. Mitchell and B. W. Niham, *J.C.S. Faraday Trans. II* **77**, 601 (1981).
9. N. A. Mazer, M. C. Carey, and G. B. Benedek, *in* "Micellization, Solubilization and Microemulsions (K. L. Mittal, ed.), Vol. 1, p. 359. Plenum, New York, 1977.
10. D. Bendedouch, S. H. Chen, and W. C. Koehler, *J. Phys. Chem.* **87**, 2621 (1983).
11. P. Missel, N. A. Mazer, G. B. Benedek, C. Y. Young, and M. C. Carey, *J. Phys. Chem.* **84**, 1044 (1980).
12. J. Th. G. Overbeek, *in* "Colloid and Interface Science" (M. Kerker, R. L. Rowell, and A. C. Zettlemoyer, eds.), Vol. 1, p. 431. Academic Press, New York, 1977.
13. E. J. W. Verwey and J. Th. G. Overbeek, "Theory of the Stability of Lyophobic Colloids." Elsevier, Amsterdam, 1948.
14. W. H. Keesom, *Proc. Acad. Sci. Amsterdam* **18**, 636 (1915); **23**, 939 (1920).
15. P. Debye, *Z. Phys.* **21**, 178 (1920); **22**, 302 (1921).
16. F. London, *Z. Phys.* **63**, 245 (1930).
17. H. C. Hamaker, *Physica (Amsterdam)* **4**, 1058 (1937).
18. P. C. Hiemenz, "Principles of Colloid and Surface Chemistry," p. 418. Dekker, New York, 1977.
19. M. Medina-Noyola and D. A. McQuarrie, *J. Chem. Phys.* **73**, 6279 (1980).
20. O. Glatter and O. Kratky, "Small Angle X-Ray Scattering." Academic Press, New York, 1982.
21. A. Guinier and G. Fournet, "Small Angle Scattering of X-Rays. Wiley, New York, 1955.
22. B. Chu, "Laser Light Scattering." Academic Press, New York, 1974.
23. B. J. Berne and R. Pecora, "Dynamic Light Scattering." Wiley, New York, 1976.
24. S. H. Chen, B. Chu, and R. Nossal, eds., "Scattering Techniques Applied to Supramolecular and Non-equilibrium Systems," NATO ASI Series B73. Plenum, New York, 1981.
25. V. Degiorgio, M. Corti, and M. Giglio, eds., "Light Scattering in Liquids and Macromolecular Solutions." Plenum, New York, 1980.
26a. B. Jacrot, *Rep. Prog. Phys.* **39**, 911 (1976).
26b. S. H. Chen, *Ann. Rev. Phys. Chem.* **37**, 351–399 (1986).
27. J. Schelten and W. Schmatz, *J. Appl. Crystallogr.* **13**, 385 (1980).
28. M. Kotlarchyk, S. H. Chen, J. S. Huang, and M. W. Kim, *Phys. Rev. A* **29**, 2054 (1984).
29. D. Bendedouch, S. H. Chen, and W. C. Koehler, *J. Phys. Chem.* **87**, 153 (1983).

30. R. Zana, C. Picot, and R. Duplessix, *J. Colloid Interface Sci.* **93**, 43 (1983).
31. N. A. Mazer, G. B. Benedek, and M. C. Carey, *J. Phys. Chem.* **80**, 1075 (1978).
32. N. A. Mazer, G. B. Benedek, and M. C. Carey, *J. Phys. Chem.* **82**, 1375 (1978).
33. M. Zulauf and H. F. Eicke, *J. Phys. Chem.* **83**, 480 (1979).
34. D. K. Schneider and B. P. Schoenborn, *in* "Neutrons in Biology" (B. P. Schoenborn, ed.), p. 119. Plenum, New York, 1984.
34a. W. C. Koehler, *Physica B* **137**, 320 (1986).
35. B. Jacrot and G. Zaccai, *Biopolymers* **20**, 2413 (1981).
36. G. Placzek, *Phys. Rev.* **86**, 377 (1952).
37. P. A. Egelstaff, *Adv. Chem. Phys.* **53**, 1 (1983).
38. A. Vrij, *J. Chem. Phys.* **71**, 3267 (1979).
39. R. J. Baxer, *J. Chem. Phys.* **52**, 4559 (1970).
40. P. van Beurten and A. Vrij, *J. Chem. Phys.* **74**, 2744 (1981).
41. L. Blum and J. S. Hoye, *J. Phys. Chem.* **81**, 1311 (1977).
42. S. H. Chen, T. L. Lin, and M. Kotlarchyk, *Proc. Int. Symp. Surfactants Solutions, Bordeaux, Fr.* (K. L. Mittal and P. Bothorel, eds.), 7 (1987).
43. M. Kotlarchyk and S. H. Chen, *J. Chem. Phys.* **79**, 2461 (1983).
44. G. Porod, *Kolloid Z.* **124**, 83 (1951).
45. L. Auvray, J.-P. Cotton, R. Ober, and C. Taupin, *J. Phys. (Orsay, Fr.)* **45**, 913 (1984).
46. K. Ibel and H. B. Sturhmann, *J. Mol. Biol.* **93**, 255 (1975).
47. M. Kakudo and N. Kasai, "X-Ray Diffraction by Polymers." Elsevier, Amsterdam, 1972.
48. D. Bendedouch and S. H. Chen, *J. Phys. Chem.* **87**, 1473 (1983).
49. P. J. T. Missel, Ph.D. Thesis, MIT, Cambridge, Massachusetts, 1981.
50. D. Bendedouch, S. H. Chen, and W. C. Koehler, *J. Phys. Chem.* **87**, 153 (1983).
51. H. Wennerstrom and B. Lindman, *Phys. Rep.* **52**, 1 (1979); *Top. Curr. Chem.* **87**, 1 (1980).
52. C. R. Cantor and P. R. Schimmel, "Biophysical Chemistry." Freeman, San Francisco, California, 1980.
53. K. A. Dill, D. E. Koppel, R. S. Cantor, J. D. Dill, D. Bendedouch, and S. H. Chen, *Nature (London)* **309**, 42 (1984).
54. K. A. Dill and P. J. Flory, *Proc. Natl. Acad. Sci. U.S.A.* **78**, 676 (1981).
55. D. W. R. Gruen, *J. Phys. Chem.* **89**, 146, 153 (1985).
56. J. K. Percus and G. J. Yevik, *Phys. Rev.* **110**, 1 (1958).
57. D. Bendedouch and S. H. Chen, *J. Phys. Chem.* **88**, 648 (1984).
58. L. S. Ornstein and F. Zernicke, *Proc. Akad. Sci. (Amsterdam)* **17**, 793 (1914).
59. J. L. Lebowitz and J. K. Percus, *Phys. Rev.* **144**, 251 (1966).
60. J. B. Hayter and J. Penfold, Rep. No. 80HA07S. Inst. Laue-Langevin, Grenoble (1980).
61. R. J. Baxer, *Aust. J. Phys.* **21**, 563 (1968).
62. P. G. Squire, P. Moser, and C. T. O'Konsky, *Biochemistry* **7**, 4261 (1968).
63. J. B. Hayter and M. Zulauf, *Colloid Polym. Sci.* **260**, 1023 (1982).
64. J. S. Huang and M. W. Kim, *Phys. Rev. Lett.* **47**, 1462 (1981).
65. M. Kotlarchyk, S. H. Chen, and J. S. Huang, *Phys. Rev. A* **28**, 508 (1983).
66. C. Toprakcioglu, J. C. Dore, B. H. Robinson, A. Howe, and P. Chieux, *J.C.S. Faraday Trans.* **80**, 413 (1984).
67. M. Kotlarchyk and S. H. Chen, *J. Chem. Phys.* **79**, 2461 (1983).
68. P. T. Cummings and G. Stell, *J. Chem. Phys.* **78**, 1917 (1983).
69. H. C. Anderson and D. Chandler, *J. Chem. Phys.* **57**, 1918 (1972).
70. M. Kotlarchyk, S. H. Chen, J. S. Huang, and M. W. Kim, *Phys. Rev. Lett.* **53**, 941 (1984).
71. J. B. Hayter and J. Penfold, *J.C.S. Faraday Trans. I* **77**, 1851 (1981).

72. J. B. Hayter, *in* "Scattering Techniques Applied to Supramolecular and Non-Equilibrium Systems" (S. H. Chen, B. Chu, and R. Nossall, eds.), NATO ASI Series B73, p. 49. Plenum, New York, 1981.
73. B. Cabane, R. Duplessix, and T. Zemb, *in* "Surfactants in Solution" (K. L. Mittal and B. Lindman, eds.), Vol. 1, p. 373. Plenum, New York, 1984.
74. R. Triolo, J. B. Hayter, L. J. Magid, and J. S. Johnson, Jr., *J. Chem. Phys.* **79**, 1977 (1983).

17. LIQUID CRYSTALS

Jerzy A. Janik

Institute of Nuclear Physics
Krakow, Poland

Tormod Riste

Institute for Energy Technology
Kjeller, Norway

17.1. Introduction

17.1.1. Classification of Liquid Crystals: Structures

Liquid crystal phases show up in systems of molecules that are distinctly elongated or distinctly flat. The latter group (so-called discotic liquid crystals) will not be discussed here, as it has not been subject to neutron scattering studies so far. The former group includes, as typical examples, dimethoxyazoxybenzene

$$H_3CO-\phenyl-N_2O-\phenyl-OCH_3$$

within some temperature range, as well as linear polyethylene $(CH_2)_n$ and sodium palmitate $C_{15}H_{31}COONa$ within some temperature and solution concentration ranges. Again, the latter two systems (i.e., polymeric and lyotropic liquid crystals) will not be discussed here as they have only sporadically been studied by the neutron methods. Thus, what remain as a subject of this chapter are in some sense classic thermotropic liquid crystals, such as the already mentioned di-methoxy-azoxy-benzene [alias paraazoxyanisole (PAA)]. There exist many substances like PAA, which have only one phase, the *nematic* phase, between the *melting* point and the so-called *clearing* point. The behavior of PAA with increasing temperature is

$$\text{crystal} \xrightarrow{118°C} \text{nematic} \xrightarrow{135.5°C} \text{isotropic liquid.}$$

There also exist many substances that show a whole sequence of liquid crystal phases. Terephtal-dibutyl-aniline (TBBA) with its five different smectic phases and one nematic phase between the melting and the clearing points may be noted as an example. The behavior of TBBA with change of temperature is

$$Cr_{II} \underset{-33°C}{\longleftrightarrow} Cr_{I} \underset{113°C}{\longrightarrow} S_G \underset{144°C}{\longleftrightarrow} S_C \underset{172°C}{\longleftrightarrow} S_A \underset{200°C}{\longleftrightarrow} N \underset{236°C}{\longleftrightarrow} I$$

$$S_{VII} \underset{68°C}{\longleftrightarrow} S_H \updownarrow 84°C$$

Here S denotes various kinds of smectic phases (see below), N the nematic phase, and I the isotropic liquid.

In a crystal the molecules are ordered with respect to both the center of gravity positions (positional order) and to orientations (rotational order). The elongated form of the molecules causes a strong anisotropy of the potential for both molecular translations and rotations. Consequently, reorientational and translational jumps are liberated in various sequences, at various temperatures for various directions of translations, and for various axes of rotation. In this way one obtains various liquid crystal phases and finally the isotropic, normal liquid phase.

A certainly over-simplified picture of liquid crystal phases is this: very close to a crystal is the smectic E phase. The molecules form layers in this phase. The molecules are not only parallel to each other in the layer but form a two-dimensional array in it, as is presented in Fig. 1. The transition from

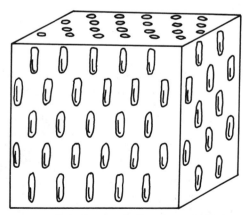

Fig. 1. Molecular arrangements in the smectic E and smectic B phases. It is possible that in the S_E phase the rotation around long axes is restricted to angular jumps by angle π; if so, the molecules seen from above should be represented by ellipses for S_E, and by circles of S_B. A nematic order parameter $S = 1$ is chosen for simplification of the figure.

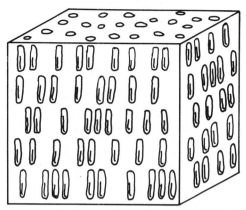

Fig. 2. Molecular arrangement in the smectic A phase. A nematic order parameter $S = 1$ is chosen for simplification of the figure.

a crystal to a smectic E phase is connected with a liberation of rather rare translations and rare reorientations (around the long axes) and with at least a partial loss of positional correlations between layers.

Figure 1 may also serve for the presentation of the smectic B phase in which the molecules probably perform a much less restricted reorientation around the long axes and on a much shorter time scale than in smectic E.

The smectic A phase shows even less restricted reorientation around the long axes and, what is more important, a much less restricted translation in the layers, which leads to a loss of the two-dimensional order in these layers, as seen in Fig. 2. With further increase in temperature, the molecules lose the

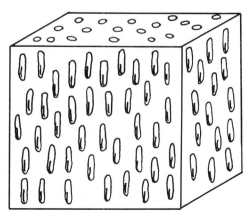

Fig. 3. Molecular arrangement in the nematic phase. A nematic order parameter $S = 1$ is chosen for simplification of the figure.

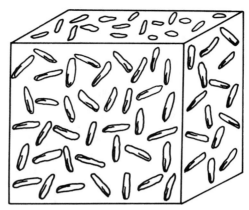

FIG. 4. Molecular arrangement in the isotropic liquid phase.

translatory correlation in the direction perpendicular to the layers, but maintain the reorientational order as in the smectic A phase. One then obtains the nematic phase shown in Fig. 3. A further transition—to the *isotropic* liquid—takes place when reorientations around other molecular axes are liberated, as shown in Fig. 4. It should be added that in addition to the group of smectics E, B, and A, there exists an analogous group of smectics H, G, and C, in which the long molecular axes are tilted with respect to the normal to the smectic layers, as shown for instance in Fig. 5.

We should remark here that the names of these phases are traditional. They were introduced when the miscibility criterion was applied for the classification of smectics.[1] Existing attempts to find an appropriate generalization of

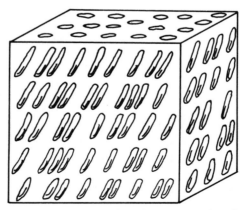

FIG. 5. Molecular arrangement in the smectic C phase. A nematic order parameter $S = 1$ is chosen for simplification of the figure.

the group theoretical classification valid for liquid crystals are not quite satisfactory (see, e.g., de Vries[2]). One should also remember that there are more phases than those listed here. There exist, for instance, the "chiral" nematics known as *cholesterics* and also smectic I, smectic D, and smectic F. Some phases were identified only in one substance, as for instance the smectic VII in TBBA. Finally there are such "exotic" phases as the "blue" phase or the "cubic" phase. For a review of these matters the reader is referred to Ref. 3.

17.1.2. Order Parameters

So far, we have been assuming an idealized picture of perfect orientational order of the long molecular axes, that is, we have assumed that in smectics as well as in nematics the long axes are parallel to each other. In reality this is not true. Therefore, we introduce an order parameter to characterize the deviations of orientations of the long molecular axes from an average direction given by the unit vector **n**, which in what follows will be called the *director*. Consequently, we must keep in mind that the realistic molecular arrangements are not those shown in Figs. 1–5. For instance, the corrected picture of the nematic phase (which has to be compared with the idealized Fig. 3) is presented in Fig. 6.

The way of introducing the nematic order parameter might be considered as conventional. It was originally defined[4] as

$$S = \tfrac{1}{2}\langle 3\cos^2\beta - 1\rangle, \tag{17.1}$$

where β is the angle of deviation of the long molecular axis from the director **n**, and the averaging is over the whole molecular assembly. It is worth

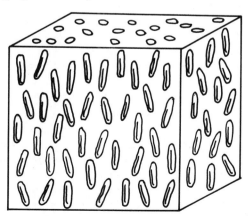

FIG. 6. Molecular arrangement in the nematic phase. A realistic value of the order parameter S is chosen.

noticing that a form equivalent to Eq. (17.1) is

$$S = \langle P_2(\cos \beta) \rangle, \qquad (17.2)$$

where P_2 is the Legendre polynomial of second order.

Now, taking into account the fact that the Legendre polynomials form a complete set, we may introduce complementary order parameters based upon P_4, P_6, so that the complete system of order parameters will be

$$S_1 = \langle P_2(\cos \beta) \rangle, \quad S_2 = \langle P_4(\cos \beta) \rangle, \quad S_3 = \langle P_6(\cos \beta) \rangle, \quad \text{etc.} \qquad (17.3)$$

Only even l's are present in P_l in these formulas, as we do not distinguish between the two molecular ends in the definition of the order parameters.

As is easily seen, the S_1 order parameter values are in the [0, 1]-interval. Its value just above the crystal–nematic (or smectic–nematic) transition, as measured for many liquid crystals, is ~ 0.7, and that just below the clearing point is ~ 0.3. Hence, the phase transition nematic–isotropic liquid is of the first order, as the order parameter jumps from ~ 0.3 to zero.

The nematic order parameters just introduced characterize the angular deviations from **n** for the smectics as well as for the nematics. For smectics, however, it is necessary to introduce besides the S_l some additional parameters connected with the two-dimensional order as well as with the lack of perfection in the smectic layers.

17.1.3. Dynamics

First of all, we must distinguish between the single-molecule and the collective motions in liquid crystals. Self-diffusion and stochastic reorientations of molecules are examples of the former. Incoherent neutron scattering has been used in their study, which is the subject of Section 17.2. Collective motion is expected in both the orientational and positional degrees of freedom and may in principle be investigated by coherent neutron scattering, as discussed in Section 17.3.

In Section 17.1.1, when introducing the classification of various smectic phases, it was natural to assume that the molecules may perform a stochastic reorientation around the long axes. A relatively high value of the order parameter indicates, on the other hand, that such a reorientation around the other (short) axes is much more restricted. Therefore it is natural to expect that the correlation times for the reorientations around the long axes will be relatively short, whereas the correlation times for reorientations around the short axes will be relatively long. Hence the molecular reorientation in liquid crystals should be strongly anisotropic, whereas above the clearing point, we may expect a much more isotropic behavior.

In a similar way, we may expect that the translatory self-diffusion in smectic liquid crystals should be different along **n** than perpendicular to it. Hence, we may expect an anisotropy of the self-diffusion coefficient and, moreover, it may happen that it will be of the opposite sign for smectics than for nematics. A macroscopic indication that such an anisotropy of diffusion does exist is the anisotropy of the viscosity (see, e.g., Miesowicz[5]).

The molecular reorientation may also (and even in a more direct way) be reflected in macroscopic measurements. As a matter of fact, many molecules that are liquid crystal constituents have electric dipole moments. In general these moments form an angle of 0 to ~80° with the long molecular axes. An example of such a molecule is methoxy-phenyl-azoxy-butyl-benzene

$$CH_3O-\langle\bigcirc\rangle-N_2O-\langle\bigcirc\rangle-C_4H_9.$$

Figure 7(a) shows, in the so-called Cole–Cole representation, the results of the dielectric relaxation measurements for the nematic phase of this substance.[6] The measurements performed in a direction perpendicular to **n**, leading to values $\varepsilon''_\perp$ and $\varepsilon'_\perp$, give a Debye relaxation region (a little deformed at the very high-frequency end) with a large dielectric increment,

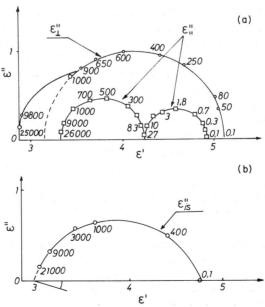

FIG. 7. Dielectric relaxation measurements for methoxy-phenyl-azoxy-butyl-benzene in the Cole–Cole representation. Numbers denote frequencies in megahertz. (a) Nematic phase and (b) isotropic phase. (From Parneix et al.[6])

which is associated with the large value of the component of the dipole moment perpendicular to the long axis. The (macroscopic) dielectric relaxation time for this region is 1.61×10^{-10} s.

The measurements performed in the direction parallel to **n**, leading to values $\varepsilon''_\parallel$ and $\varepsilon'_\parallel$, give two Debye regions: a low-frequency one, with a much smaller dielectric increment than before (as the component of the dipole moment parallel to the axis is much smaller than the perpendicular one) and a high-frequency one, with a small increment also, as this region is now seen only because the order parameter is smaller than one. The dielectric relaxation time for the high-frequency region is approximately the same as before. The low-frequency region, on the other hand, gives a dielectric relaxation time equal to 8.84×10^{-8} s.

In view of the geometry of measurements it is natural to connect the low-frequency relaxation region with the reorientation of molecules around their short axes and to connect the high-frequency region with reorientation around their long axes. Thus we have obtained evidence of the expected anisotropy of molecular reorientations in the nematic phase, as the relaxation time connected with reorientation around the long axis turned out to be two orders of magnitude smaller than that connected with reorientation around the short axes.

Figure 7(b) shows the dielectric relaxation results for the isotropic phase of the same substance. There is only one (high-frequency) relaxation region; the Cole–Cole diagram, however, is not a semicircle anymore, being now significantly lowered. This indicates an incomplete merging of the two relaxation regions, which were so different in the nematic phase. Hence, we may conclude that although the molecular reorientation in the isotropic phase is much more isotropic than in the nematic phase some anisotropy remains, as does some short-range orientational order.

It is worth noting here that the deformation observed with the geometry perpendicular to **n** at very high frequencies is connected with the so-called Poley absorption band in the far infrared. This Poley absorption indicates that the reorientation around the long axes is not a uniaxial rotational diffusion *sensu stricto*, but it is accompanied by anharmonic librations of the molecules around these long axes.

The dielectric relaxation results in methoxy-phenyl-azoxy-butyl-benzene described earlier may be considered as a typical example. Practically all nematic liquid crystals with polar molecules show the existence of one relaxation region around 10^{-10} s and another around 10^{-8} s. Only for molecules with relatively short end groups, as for instance PAA, are these relaxation times ca. 10^{-11} s and 10^{-9} s.

Hence, it seems that because of the power of the dielectric method we have a complete knowledge of both the anisotropy and the time scale of molecular reorientations in liquid crystals. We should, however, remember that dielectric measurements are macroscopic and do not directly concern single molecular motions. As a matter of fact, if we intend to obtain the single molecule correlation time, we must make a transition from the macroscopic to the microscopic response and next a transition from the polymolecular to the monomolecular response. In the case of liquids with simple polar molecules, these two transitions give correction factors not very different from 1. Therefore, for simple polar liquids, we may say that the dielectric relaxation time is approximately equal to the reorientational correlation time (connected with the first spherical harmonic), i.e., $\tau_D \approx \tau_1$.[7,8] For liquid crystals such an approximation is perhaps doubtful. We often pragmatically use it, however, when saying that not only the dielectric relaxation times but also the typical reorientational correlation times in liquid crystals are 10^{-10} s and 10^{-8} s for reorientation around the long and short molecular axes, respectively. It is, however, possible that the disagreements between the correlation times derived from the dielectric and neutron methods, which are discussed in Section 17.2.4, originate at least partly from this approximation.

17.2. Single-Molecule Properties: Incoherent Scattering

17.2.1. Scattering Functions as Derived from Various Models

The information given in Section 17.1.3 leads to the following conclusions concerning the application of incoherent neutron scattering methods to the problems of molecular motions in liquid crystals.

(1) It is practically impossible to observe the reorientation of molecules around their short axes with the neutron methods, because the corresponding correlation times are too long (for detection by neutrons).

(2) We may expect that the much faster reorientation of molecules around the long axes will be observable by quasi-elastic neutron scattering (QNS), giving a characteristic quasi-elastic component besides the elastic one.

(3) We may expect an additional quasi-elastic broadening due to translatory self-diffusion; especially well visible should be the self-diffusion broadening of the elastic component.

(4) The evaluation of the adequacy of models of molecular motions achieved by fitting the model scattering function to the experimental one may be improved when the liquid crystal sample is aligned and when some well-defined "geometries" (for instance, $\mathbf{Q} \parallel \mathbf{n}$ and $\mathbf{Q} \perp \mathbf{n}$) are used.

(5) As the molecules have movable parts (e.g., the end groups), which in addition reorient independently of the reorientation of the molecules as a

whole, it is advisable to use the method of partial deuteration to mask those parts that do not interest us in the problem in question.

In connection with this last point, it is worth noticing that a comparison of the experimental scattering functions in the QNS region for deuterated and nondeuterated end groups shows that the results are identical or almost identical. Hence, we may state that the motions of the molecules as a whole are mainly responsible for the QNS spectra.

Among these motions the main candidate for model fitting is the *uniaxial reorientation of molecules around the long axes*. Various models of this reorientation may be taken into consideration. It should, however, be noted that the single-parameter models of uniaxial reorientation are not sufficient in comparisons with the experiments and therefore, some additional motions have to be taken into account. The scattering functions of these motions have to be convoluted with that for the uniaxial reorientation. There are two candidates for these additional motions: fluctuations of the long molecular axes with respect to the **n** direction and, in the case of smectic phases, the strongly anharmonic translatory oscillations of the whole molecules perpendicular to smectic layers. We shall present the neutron scattering functions for several models of uniaxial reorientations as well as a phenomenological treatment of the two additional motions.

17.2.1.1. Rotational Diffusion Around One Axis. The incoherent neutron scattering function of a proton, moving on a circle of radius **r**, is[9]

$$S_i(\mathbf{Q}, \omega) = J_0^2(\mathbf{Q} \cdot \mathbf{r}) \delta(\omega) + \frac{2}{\pi} \sum_{l=1}^{\infty} J_l^2(\mathbf{Q} \cdot \mathbf{r}) \frac{1/\tau_l}{(1/\tau_l)^2 + \omega^2}, \quad (17.4)$$

where J_l stands for the cylindrical Bessel functions, and $\tau_l = \tau_1/l^2$. Instead of the correlation time τ_1, the rotational diffusion constant $D_r = 1/\tau_1$ is often used. There is one fitting parameter in this model, namely, τ_1.

17.2.1.2. Uniaxial Reorientation via Instantaneous Angular Jumps Between N Equidistant Equilibrium Positions on a Circle. The corresponding scattering function is[10]

$$S_i(\mathbf{Q}, \omega) = F_0(\mathbf{Q} \cdot \mathbf{r}) \delta(\omega) + \frac{1}{\pi} \sum_{l=1}^{N-1} F_l(\mathbf{Q} \cdot \mathbf{r}) \frac{\Gamma_l}{\Gamma_l^2 + \omega^2}, \quad (17.5)$$

where

$$F_l(x) = \frac{1}{N} \sum_{m=1}^{N} J_0\left(2x \sin \frac{\pi m}{N}\right) \cos \frac{2\pi l m}{N}$$

and

$$\Gamma_l = \frac{1}{\tau} \frac{\sin^2(\pi l/N)}{\sin^2(\pi/N)}.$$

There is again one fitting parameter in this model, i.e., the average residence time between instantaneous jumps, τ. In the $N \to \infty$ limit, Eq. (17.5) transforms into the uniaxial rotational diffusion formula (17.4).

A generalization of the case now discussed is the case of a uniaxial reorientation in the cosinusoidal potential with N minima $V_N(\phi) = -\frac{1}{2} V_N \cos N\phi$. The corresponding scattering function is very complicated[11] and contains V_N as a fitting parameter besides the rotational diffusion constant D_r. All these expressions must be subjected to an averaging when used. Such an averaging gives different results for a nonoriented ("polycrystalline") sample than, for instance, for $\mathbf{Q} \parallel \mathbf{n}$ or $\mathbf{Q} \perp \mathbf{n}$ cases when the sample is aligned.

We present two expressions for a nonoriented sample as useful examples. One is for uniaxial rotational diffusion in a nonoriented sample[12] where

$$S_i(Q, \omega) = \langle J_0^2(\mathbf{Q} \cdot \mathbf{r}) \rangle \delta(\omega) + \frac{2}{\pi} \sum_{l=1}^{\infty} \langle J_l^2(\mathbf{Q} \cdot \mathbf{r}) \rangle \frac{1/\tau_l}{(1/\tau_l)^2 + \omega^2} \quad (17.6)$$

[the averaging is over all orientations; the notations are as in Eq. (17.4)]. The other example is for uniaxial instantaneous angular jumps by an angle π in a nonoriented sample:[13]

$$S_i(Q, \omega) = \frac{1}{2}\left(1 + \frac{\sin 2Q \cdot r}{2Q \cdot r}\right) \delta(\omega) + \frac{1}{2\pi}\left(1 - \frac{\sin 2Q \cdot r}{2Q \cdot r}\right) \frac{2/\tau}{(2/\tau)^2 + \omega^2}, \quad (17.7)$$

where τ is the average residence time between the instantaneous jumps.

Analogous formulas for an aligned sample, when the director $\mathbf{n}$ forms an angle α with $\mathbf{Q}$, are for uniaxial rotational diffusion in an aligned sample[14]

$$S_i(Q, \omega) = J_0^2(Q \cdot r \sin \alpha) \delta(\omega) + \frac{1}{\pi} \sum_{l=1}^{\infty} J_l^2(Q \cdot r \sin \alpha) \frac{1/\tau_l}{(1/\tau_l)^2 + \omega^2}, \quad (17.8)$$

and for uniaxial angular jumps by angle π in an aligned sample[14]

$$S_i(Q, \omega) = \frac{1}{2}[1 + J_0(2Q \cdot r \sin \alpha)] \delta(\omega)$$

$$+ \frac{1}{2\pi}[1 - J_0(2Q \cdot r \sin \alpha)] \frac{2/\tau}{(2/\tau)^2 + \omega^2}. \quad (17.9)$$

In real, oriented samples the order parameter is, of course, smaller than 1, which introduces some complication in Eqs. (17.8) and (17.9). This can easily be taken into account as a static contribution to the fluctuations of the axis.

In some liquid crystal phases, especially in those with a lower order (smectic A or nematic), neutron experiments give an excess of the quasi-elastic component (over the elastic one) for $\mathbf{Q} \perp \mathbf{n}$. This additional quasi-elastic component is small as compared with the main one arising from reorientations around the long molecular axes. It is often interpreted as caused by *dynamic* fluctuations of the long axes with respect to the direction of **n**. Protons performing these fluctuations move perpendicular to **n** and therefore these motions are visible in the $\mathbf{Q} \perp \mathbf{n}$ and not in the $\mathbf{Q} \parallel \mathbf{n}$ geometry. If the sample is not aligned and if we are unable to distinguish clearly between the fluctuational and reorientational quasi-elastic components, we may phenomenologically take the fluctuations into account by writing[5]

$$S_i(Q, \omega) = pA_0(Q)\delta(\omega) + \frac{1}{\pi}[A_1(Q) + (1-p)A_0(Q)]\frac{1/\tau_1}{(1/\tau_1)^2 + \omega^2}$$
$$+ \frac{1}{\pi}\sum_{n>1} A_n(q)\frac{1/\tau_n}{(1/\tau_n^2) + \omega^2}. \qquad (17.10)$$

In this formula, the part of the term described by the Lorentz function with half-width $1/\tau_1$ is connected with fluctuations whose characteristic time parameter is $\tau_f \approx \tau_1$. The rest of this term and the other terms describe uniaxial rotational diffusion and should be made equal to the respective terms of Eq. (17.6). If the sample is aligned it is easy to derive similar phenomenological formulas; moreover, we can then abandon the $\tau_f \approx \tau_1$ approximation.

In a similar way we may detect another motion through an additional quasi-elastic component, this time visible in the $\mathbf{Q} \parallel \mathbf{n}$ geometry. This motion clearly shows up in highly ordered smectic phases (E or B). It is interpreted as a fluctuational translatory motion in which the molecules move perpendicular to the smectic layers. The phenomenological characteristic time parameter for this motion is τ_z[14] (where z is the direction perpendicular to smectic layer).

17.2.2. Reorientations and Phase Transitions

The experimental results concerning reorientations in the liquid crystal phases of molecules around the long axes will be discussed not in two examples, both performed at the Institute Laue Langevin reactor in Grenoble.

The first example is provided by Volino *et al.*[15] and concerns the already mentioned (Section 17.1.1) TBBA. Figure 8 presents the QNS spectra obtained for several smectic phases of TBBA. When interpreting these spectra we must take into account the fact that various hydrogen atoms in the molecule have various states with respect to the neutron scattering. One reason for this is connected with the fact that the end chains may move relative to the rings in addition to the reorientation of the molecule as a whole. Another reason is that the gyration radii of various protons

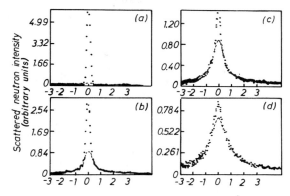

FIG. 8. Neutron quasi-elastic and elastic scattering results for TBBA with deuterated tails. The sample is not aligned. The incident neutron wavelength is 9.48 Å; $Q = 0.92$ Å^{-1}: (a) Solid, +70Ć; (b) smectic H, +82°C; (c) smectic G, +130°C; and (d) smectic C, +152°C. The dashed curve shows the quasi-elastic component deduced from the data. (From Volino et al.[15])

reorienting around the long axis are different. Figure 9 presents the molecule. As we see, the end chains were totally deuterated and thus practically invisible, giving only ~5% of the scattering, which, moreover, does not concentrate around $\omega = 0$. As regards the gyration radii, we must (as seen from Fig. 9) treat individual hydrogen atoms separately, ascribing to them scattering functions (17.5) that are formally the same but with different gyration radii for different atoms. These gyration radii are between 1.52 Å and 2.33 Å.

Now, we must check whether reorientation around the long axes is the only process that should be taken into consideration when fitting the model to the QNS experiment. Therefore we shall present the experimental data in a form convenient for this purpose, which is the ratio of the elastic component to the total intensity versus the momentum transfer Q (i.e., the elastic incoherent scattering form factor EISF versus Q). For determining EISF values from

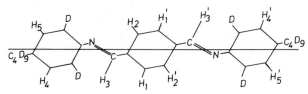

FIG. 9. Partly deuterated TBBA molecules. The axis of rotation is also shown. Protons labelled from 1 to 5 have different radii of trajectories when reorienting around this axis. (From Volino et al.[15])

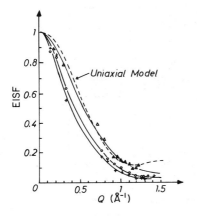

FIG. 10. Elastic incoherent scattering form (EISF) factor versus Q for TBBA at various temperatures. Open triangle, smectic G, 119°C; open circle, smectic G, 137°C; crosses, smectic C, 151°C. Nonaligned sample. (From Volino et al.[15])

data such as those in Fig. 8, we will not use Eq. (17.5) but the more general phenomenological Eq. (17.10). The results are shown in Fig. 10. We may also see in this figure the theoretical curve that corresponds to uniaxial rotational diffusion, i.e., to Eq. (17.6). We note that the experimental values of EISF obtained for three temperatures, that is, those corresponding to three values of the order parameter $S_1 = 0.92$, 0.75, and 0.61, lie below the uniaxial rotational diffusion curve. That may be interpreted as evidence of an *additional* motion (besides the uniaxial rotational diffusion), which is the *dynamic fluctuation of the long axis with respect to the director* **n**. In the experiments under discussion, this interpretation could not be confirmed by an argument based on the direction of the proton motion, as the experiment was carried out with a nonaligned sample.

Let us now return to the main topic, that is, to the problem of how uniaxial reorientation occurs and what its time scale is in the various phases of TBBA. A comparison of form factors $A_n(Q)$ of Eq. (17.10) with those given by Eq. (17.5), after averaging over the orientations, makes it possible to obtain, by fitting, both the number of equilibrium positions N in the rotational motion of the molecule around the long axis and the time parameter τ_1 of the model. In the most ordered smectic VII phase of TBBA, there exists no reorientational motion, as is evident from Fig. 8(a). In the next phase (S_H),[†] N is approximately equal to 2, hence the reorientation occurs via jumps by an angle π. In all the subsequent phases (S_G, S_C, S_A) the model fitting is unable to distinguish motion with N equilibrium positions from a uniaxial rotational diffusion. The time parameter τ_1, which in the S_H phase means the average

---

[†] We use here a now accepted notation, which differs from that of Volino et al.[15,16] In particular: phase VI of Volino et al.[15] is now denoted by S_H, and phase H of Volino et al.[15] is now denoted as S_G.

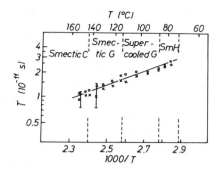

FIG. 11. Correlation time τ_1 for TBBA reorientation around long molecular axes. (From Volino et al.[15])

residence time between jumps by an angle π and in all other phases stands for the τ_1 correlation time of the uniaxial rotational diffusion, is presented in Fig. 11. We can see that τ_1 decreases with an increase of temperature, but it does not show any discontinuities at the phase transitions. The τ_1 values extrapolated to the nematic phase temperatures are surprisingly small, of the order of several picoseconds.

It should be noticed that the previously discussed fluctuational motion of the long molecular axes in relation to **n** suggested by the EISF versus Q dependence evidently occurs on approximately the same time scale of 10^{-11} s, since it does not give rise to a clearly separable quasi-elastic component.

The second example of a QNS experiment with liquid crystals comes from Richardson et al.[14] and concerns iso-butyl-phenyl-benzylidene-amino-cinnamate (IBPBAC) whose chemical formula is

$$\bigcirc\!-\!\bigcirc\!-\!CH\!=\!N\!-\!\bigcirc\!-\!CH\!=\!CHCOOCH_2CH(CH_3)_2.$$

This substance has smectic E, B, and A phases and a nematic phase. Most of the remarks made above in connection with the TBBA studies, those concerning end-chain deuteration and different gyration radii of different protons, are also valid here. In addition, let us stress these facts:

(1) Some of the QNS experiments with IBPBAC were made at low Q values (see Section 17.2.3), and they provided the translational diffusion coefficients $D_\parallel$ and $D_\perp$; these data were used for subtraction of the translational contribution from the total QNS spectra and thus for the isolation of the reorientational quasi-elastic and elastic components.

(2) The IBPBAC samples were aligned in this experiment; the orientation was induced by a magnetic field in the nematic phase, and it remained in the smectic phases without the field, giving a much better possibility of

identification and isolation of various motions than in the experiments with the nonaligned TBBA described before.

(3) In this experiment several neutron spectrometers installed at the Grenoble reactor were used. The spectrometers have different resolutions, and so measurements on one of them led to the identification of the fastest motions, not distinguishing the slower ones from the elastic component, whereas the measurements on another spectrometer (with a better resolution) did not distinguish the fastest motions from the broad background, thus making it possible to identify the slower components.

(4) It is worthwhile to comment here on the model of uniaxial reorientation of molecules around the long axes performed via angular jumps between N equilibrium positions. In Section 17.2.1 we discussed this case with an approximation that the jumps are instantaneous, which gave Eq. (17.5). A more general case also mentioned there is that the motion takes place in a cosinusoidal potential with N minima which, as has been said, leads to a very complicated formula. In this formula the potential amplitude V_N is of course present, as is the rotational diffusion coefficient D_r, because we assume that the equation for the self-correlation function $G_s(\phi, t)$ for the particle is now

$$\frac{1}{D_r} \frac{\partial G_s}{\partial t} = \frac{\partial}{\partial \phi} \left[\frac{\partial G_s}{\partial \phi} + \frac{NV_N(\phi)}{2k_B T} \sin N\phi \cdot G_s \right]. \qquad (17.11)$$

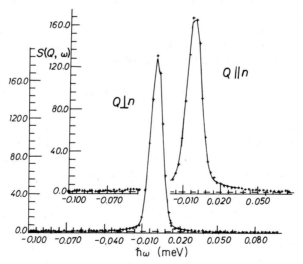

FIG. 12. QNS results for the smectic E phase of IBPBAC obtained with a resolution of ~ 10 μeV. Incident neutron wavelength 12 Å. $Q = 0.75$ Å^{-1}. The sample is aligned. Solid curves represent the model of reorientation around long axes in a cosinusoidal potential with two minima. The dashed curve represents a flat inelastic background. (From Richardson et al.[14])

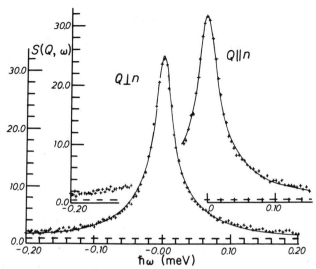

FIG. 13. QNS results for the smectic A phase of IBPBAC obtained with a resolution of ~20 μeV. Incident neutron wavelength 10 Å. $Q = 0.83$ Å^{-1}. The sample is aligned. Solid curves represent the model of uniaxial rotational diffusion. Dashed lines represent a flat inelastic background. (From Richardson et al.[14])

The scattering function for this case will not be given here; it is available in Volino et al.[15] This formula was used for fitting to the experimental results of IBPBAC.

Figure 12 presents an example of QNS results obtained for the smectic E phase with a spectrometer resolution ~10 μeV. The best fit was obtained for a potential with two minima ($N = 2$) with amplitude $V_2 = 1.2 \cdot 2k_B T$, and a uniaxial rotational diffusion coefficient $D_r = 0.32 \times 10^{10}$ rad^2 s^{-1}. In other smectic phases (B and A) the best fits were obtained for the normal uniaxial rotational diffusion model, with D_r increasing with increasing temperature (e.g., $D_r = 4.7 \times 10^{10}$ rad^2 s^{-1} for smectic phase A). Figure 13 shows, as an example, the results for the smectic A phase. Figure 14 shows the dependence of D_r on temperature (as well as phase) or, using another

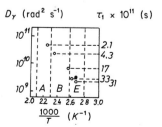

Fig. 14. Uniaxial rotational diffusion constant and correlation time τ_1 for IBPBAC reorientation around long molecular axes. (From Richardson et al.[14])

convention, it shows the dependence of the uniaxial reorientational correlation time τ_1 on temperature. As for TBBA, the τ_1 values extrapolated to the nematic phase region are surprisingly small, of the order of several picoseconds. Nevertheless, it seems evident that the IBPBAC molecules perform a reorientation around their long axes in all the liquid crystal phases; however, in the S_E phase this reorientation is a relatively rare event (τ_1 is several hundred picoseconds) and occurs via jumps by an angle π.

Another similarity to TBBA is connected with the occurrence of motions *other* than reorientation around the long axes. Evidence of their occurrence can be obtained from the EISF versus Q dependence (as in TBBA); also in the case of IBPBAC experiments now discussed, it was possible to separate the respective components by applying different energy resolutions and $\mathbf{Q} \parallel \mathbf{n}$ and $\mathbf{Q} \perp \mathbf{n}$ geometries. An example of such an attempt is shown in Fig. 15. It can be seen that the QNS component connected with the reorientation around the long axis is now very broad, giving a kind of background; on the other hand a narrower component shows up and is interpreted as caused (in this case) by the axis fluctuations with respect to the $\mathbf{n}$ direction (as in TBBA). The final conclusions from all these experiments are that the long-axis fluctuations about $\mathbf{n}$ start to appear (on the neutron time scale) in the smectic B phase, where their effective correlation time τ_f is ~ 400 ps, so it is one order of magnitude longer than the τ_1 time. In the smectic A

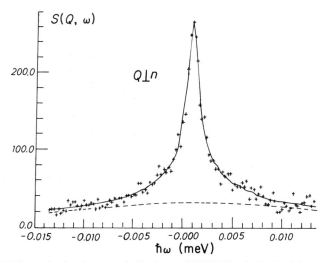

FIG. 15. QNS results for the smectic B phase of IBPBAC obtained with a resolution of ~ 1 μeV. Incident neutron wavelength 6.28 Å. $Q = 0.83$ Å^{-1}. The sample is aligned. The solid curve represents the phenomenologically introduced model of long-axis fluctuations. The dashed line separates the broader component due to uniaxial reorientational motion and a flat inelastic background. (From Richardson et al.[14])

phase τ_f is of the same order as τ_1 (20 ps). Moreover, in all smectic phases, there exists a localized translation motion perpendicular to the smectic layers; the effective correlation time τ_z of this motion is between 20 and 40 ps.

These two examples of QNS experiments with liquid crystals (TBBA and IBPBAC) are typical and so far the best ones. References 17-27 provide information on other work of this type, also with other substances. The conclusions from all these studies are similar to those given in the preceding paragraph.

17.2.3. Translational Diffusion and Phase Transitions

The neutron scattering method has been used for many years in studies of self-diffusion of simple molecules in liquids and of hydrogen in metals. With liquid crystals this technique acquires a new methodological aspect because we may apply it not only with nonoriented samples, which leads to the average translational coefficient $\bar{D}$, but also with aligned samples and $\mathbf{Q} \parallel \mathbf{n}$ and $\mathbf{Q} \perp \mathbf{n}$-geometries, which provides $D_\parallel$ and $D_\perp$ components. For *continuous* translational diffusion, the neutron scattering function is very simple—a Lorentz function with half-width

$$\Gamma = \hbar Q^2 \bar{D}, \qquad (17.12)$$

$$\Gamma_\parallel = \hbar Q^2 D_\parallel, \qquad (17.13)$$

and

$$\Gamma_\perp = \hbar Q^2 D_\perp. \qquad (17.14)$$

The frequently used formula

$$\bar{D} = \tfrac{1}{3}D_\parallel + \tfrac{2}{3}D_\perp$$

is, strictly speaking, not true although it may be considered as a fair approximation. The correct formula[28]

$$\bar{D} = \sqrt{(D_\parallel - D_\perp)D_\perp} \arctan \sqrt{(D_\parallel - D_\perp)/D_\perp} \qquad (17.15)$$

is impractical and advisable only when the measurements are very accurate.

The continuous diffusion model is correct only for low Q values. We may thus apply it for neutron scattering studies with low Q's, where we are not disturbed by reorientations, which give only a very weak QNS component at low Q's. Consequently, we have to deal in our treatment only with the elastic components, now broadened by translational diffusion. As a result we obtain the $D_\parallel$, $D_\perp$, and $\bar{D}$ values.

The first such measurement was made for PAA[29] in the nematic phase, in two situations: with a nonoriented sample and with an aligned one so that $\mathbf{Q} \perp \mathbf{n}$. Some results are shown in Fig. 16. The D values obtained were $\bar{D} = 4.1 \times 10^{-6}\,\text{cm}^2\,\text{s}^{-1}$ and $D_\perp = 3.4 \times 10^{-6}\,\text{cm}^2\,\text{s}^{-1}$, from which

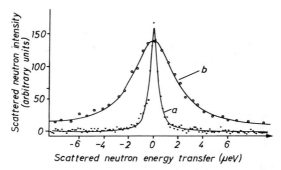

FIG. 16. Low-Q quasi-elastic neutron scattering results for PAA. (a) Solid PAA data giving the resolution function. (b) Nematic PAA data showing broadening by translational diffusion. The sample is aligned, $\mathbf{Q} \perp \mathbf{n}$. (From Töpler et al.[29])

$D_\parallel = 5.5 \times 10^{-6}$ cm^2 s^{-1} was deduced. A clear anisotropy showed up—the diffusion of molecules is easier in the $\mathbf{n}$ direction than perpendicular to $\mathbf{n}$.

Similar measurements carried out for nematic TBBA[30] gave the same order of magnitude of the diffusion coefficients: $D_\perp = 5.6 \times 10^{-6}$ cm^2 s^{-1}, $D_\parallel = 11.4 \times 10^{-6}$ cm^2 s^{-10}, and a marked anisotropy. The slightly larger values of the coefficients as compared to PAA, in spite of the fact that the molecule is more complicated than PAA, can be understood if one takes into account that the temperatures of the nematic phase in TBBA are around 220°C whereas those of the nematic phase in PAA are around 125°C.

It should be noticed that in the nematic phase of some substances the anisotropy vanishes, as for instance for

$$CH_3COO-\bigcirc-CH=N-\bigcirc-CH=CHCOOC_2H_5,^{24}$$

where $D_\perp = 1.4 \times 10^{-6}$ cm^2 s^{-1} and $D_\parallel = 2.0 \times 10^{-6}$ cm^2 s^{-1}. In some cases the anisotropy is even opposite as, for instance, for the nematic IBPBAC where $D_\perp = 4.4 \times 10^{-6}$ cm^2 s^{-1} and $D_\parallel = 3.6 \times 10^{-6}$ cm^2 s^{-1}.[14]

For smectic phases with low orientational order (S_A or S_C phases), the diffusion coefficients are somewhat smaller than in the neighboring nematic phases, but there is no distinct jump at the nematic–smectic A (or C) transition.[19] The anisotropy of diffusion is small. For instance, in the case of the smectic A phase of IBPBAC, $D_\perp = 1.5 \times 10^{-6}$ cm^2 s^{-1} and $D_\parallel = 1.3 \times 10^{-6}$ cm^2 s^{-1}.[19] In TBBA for the same phase $D_\perp$ is practically equal to $D_\parallel$ and the common value is 2.8×10^{-6} cm^2 s^{-1}.[31] When passing to more ordered smectic phases (S_B or S_G), one obtains a jump of D to values of the order of 10^{-7} cm^2 s^{-1}. Phases S_E or S_H have even smaller diffusion coefficients that cannot be measured by the neutron method.

17.2.4. Remarks Concerning Comparison with Other Methods

The results of QNS measurements concerning the reorientation of liquid crystal molecules around their long axes, discussed in Section 17.2.2, led to these conclusions:

(1) The reorientation around the long axis exists in almost all liquid crystal phases; moreover, it is the process that dominates the quasi-elastic scattering at not too low values of Q.

(2) Except for the most ordered smectic phases, this reorientation has the character of uniaxial rotational diffusion.

(3) Reorientational correlation times for phases with a low orientational order are of the order of 10 ps.

Here we should note (see Section 17.1.3) that the dielectric relaxation times in liquid crystals, which correspond to reorientation around the long axis, are of the order of 100 ps. On the other hand, for the substances discussed in Section 17.2.2 studied by QNS, there exist no dielectric relaxation data. Therefore a comparison is impossible.

Dielectric and QNS studies have been carried out so far for three substances. These are the three members of the PAA series, whose general formula is

$$H_{2n+1}C_nO-\langle\bigcirc\rangle-N_2O-\langle\bigcirc\rangle-OC_nH_{2n+1},$$

with $n = 1, 2,$ and 7. Table I shows a comparison of the QNS determined correlation time τ_1 with those obtained from the dielectric relaxation measurements, with the use of the $\tau_D \approx \tau_1$ approximation (see Section 17.1.3). It seems evident that the τ_1 values as obtained from these two different methods are different, although it is difficult to estimate how large

TABLE I. Comparison of QNS and Dielectrically Measured τ_1 Correlation Times

Substance	QNS $\tau_1{}^a$ (ps)	Dielectric $\tau_1{}^b$ (ps)	References
PAA, $n = 1$, nematic, 125°C	4	22	32, 33
PAP, $n = 2$, nematic, 140°C	6	30	34, 35
HOAB, $n = 7$, nematic, 105°C	12	110	36, 37
HOAB, $n = 7$, smectic C, 86°C	12	200	36, 37

[a] Determined from uniaxial diffusion model fitting to QNS.
[b] Assumes that $\tau_1 = \tau_0$.

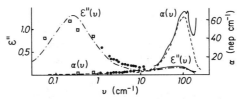

FIG. 17. Dielectric and far infrared results for nematic PAA. Open square, microwave dielectric data in ε'' representation (upper points) and absorption coefficient representation (low points). Closed circle, carcinometric technique data in ε'' (upper points) and α (lower points) representation. Solid lines, FIR data in ε'' (lower line) and α (upper line) representation. Dashed line, model of interrupted librations around long axes + rotational diffusion around the same axes in the α representation. The dash–dot line—the same model in the ε'' representation. (From Janik et al.[38])

a correction would be required if we gave up the $\tau_1 \approx \tau_D$ approximation. One cannot exclude the existence of short-time and long-time components in the molecular reorientational motion around the long axes. The short-time component could be connected with the time of duration of molecular librations around the long axis, whereas the long-time component would correspond to the time of duration of molecular reorientations around the same axis, the two processes appearing alternately.

Such a picture is not unreasonable in view of the existence of a so-called Poley absorption band in the far infrared for many liquid crystals. Figure 17 shows an example of such a band for PAA. The Debye relaxation region is also shown in this figure. As we see, the results (ε'' versus frequency or α versus frequency) for both the Debye and Poley regions can be described by a theoretical model in which one assumes that the molecule performing librations around the long axis during time τ is situated in a cage, which in turn performs a uniaxial rotational diffusion with the correlation time τ_D. By fitting this model to the data,[38] one obtains for τ the value 2 ps in the case of nematic PAA and the value 11 ps in the case of nematic HOAB (assuming for τ_D the dielectric values 22 ps and 100 ps, respectively). It can be seen that the τ values are not far from the QNS determined times (see also Table I).

There also exists a possibility that the molecular moieties (consisting of terminals coupled with benzene rings) may reorientate around the benzene para-axis. If so, this motion could be responsible for the QNS results, whereas the reorientation of the whole molecule could be responsible for the dielectric results.

However, it is still too early to conclude that a revision of the QNS interpretations is necessary. One should first carry out systematic studies of many liquid crystals via the QNS, dielectric, and spectroscopic methods.

The matter looks a little better when comparing the translational diffusion coefficients obtained by low QNS measurements with the coefficients

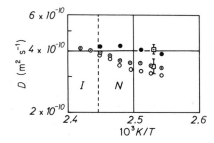

FIG. 18. Comparison of translational self-diffusion coefficients obtained with QNS and MTR methods for PAA. ⊙, nonaligned sample, MTR; ○, $D_\perp$ data, MTR method; ●, $D_\parallel$ data, MTR method;[40] □, nonaligned sample, QNS method; $D_\perp$ data, QNS method.[29]

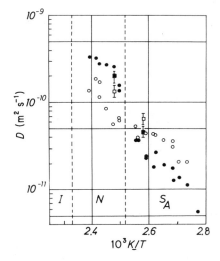

FIG. 19. Comparison of translational self-diffusion coefficients obtained with QNS and NMR methods of EABAC. Open circle, $D_\perp$ data, NMR method. Closed circle, $D_\parallel$ data, NMR method. Open square, $D_\perp$ data, QNS method. Closed square, $D_\parallel$ data, QNS method. (From Richardson et al.[41])

obtained by other methods, such as NMR or a direct measurement of the mass transport (MTR).

Figure 18 shows such a comparison between the QNS and MTR results for PAA.[39] Figure 19 shows a comparison between the QNS and NMR results for EABAC (ethyl-acetoxy-benzylidene-amino-cinnamate).[39] In the latter case the agreement is not bad, but it can be made better if one assumes that the translational diffusion is not just continuous but is performed in a cosinusoidal potential, which introduces some corrections to the previously obtained $D_\perp$ and $D_\parallel$ values.[42]

17.3. Collective Properties: Coherent Scattering

It appears from the previous sections that the collective properties to study in liquid crystals are their orientational and positional order and the collective motions in the corresponding degrees of freedom. The number of published papers in this field is still not very large, compared to other fields of application of neutron scattering, the main reason being the difficulty in obtaining fully deuterated samples.

17.3.1. Nematics

According to de Gennes[43] and Pynn,[44] the coherent cross section for neutrons scattered by a nematic can be divided into two parts: a single molecule term denoted $(d\sigma/d\Omega)_s$ and an interference term $(d\sigma/d\Omega)_d$. It turns out that part of the diffraction pattern, for $Q > 2.5$ Å in PAA, is quite well explained through the first term alone. For rigid molecules it may be written in the form[44]

$$(d\sigma/d\Omega)_s = \sum_l (-1)^l I_{2l}(Q) P_{2l}(\cos \theta_Q) \langle P_{2l} \rangle (4l + 1), \qquad (17.16)$$

where

$$\langle P_{2l} \rangle = \int_0^\pi f(\beta) \sin \beta \, P_{2l}(\cos \beta) \, d\beta, \qquad (17.17a)$$

$$I_{2l}(Q) = \sum_{jj'} a_j a_{j'} J_{2l}(Qv)_{jj'} P_{2l}(\cos \theta_{jj'}), \qquad (17.17b)$$

and

$$\cos \theta_{jj'} = v_{jj'}^z / |\mathbf{v}_{jj'}|; \qquad \mathbf{v}_{jj'} = \mathbf{u}_j - \mathbf{u}_{j'}, \qquad (17.17c)$$

where $v_{jj'}^z$ is the component of $\mathbf{v}_{jj'}$ parallel to the molecular axis.

In the foregoing equations $J_l(x)$ and $P_l(x)$ are, respectively, a spherical Bessel function and a Legendre polynomial, θ_Q is the angle between $\mathbf{Q}$ and the nematic axis, and a_j is the bound, coherent scattering length of the jth nucleus of a molecule and $\mathbf{u}_j$ is the position (defined in the molecular coordinate system) of that nucleus with respect to the molecular center of mass. In Eq. (17.16) the dependence of the cross section on scattering geometry (the θ_Q term), molecular structure [the $I_{2l}(Q)$ term], and nematic order (the $\langle P_{2l} \rangle$ term) are clearly separated.

The first coherent neutron diffraction pattern measured by Pynn et al.[45] was for fully deuterated para-azoxyanisole (PAA), see Fig. 20. In the Q range measured, the liquid pattern consisted of two peaks centered at approximately 1.8 Å^{-1} and 3 Å^{-1}. From the response to an applied magnetic field (i.e., the θ_Q dependence), these peaks give information about density correlations in directions perpendicular ($\mathbf{H} \perp \mathbf{Q}$) and parallel ($\mathbf{H} \parallel \mathbf{Q}$) to the long molecular axis, respectively. The nematic order parameter $\langle P_{2l} \rangle$ may be derived from the temperature variation of the intensity of either of the two peaks. The fact that they give the same result for the order parameter is surprising, since the analysis only makes use of $(d\sigma/d\Omega)_s$. The nematic order parameter is actually approximated by $S = \langle P_2 \rangle$. In principle higher terms of $\langle P_{2l} \rangle$ should

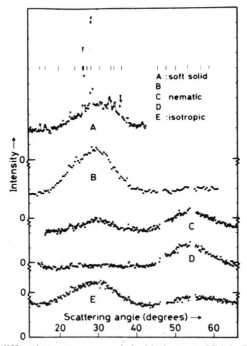

FIG. 20. Neutron diffraction patterns recorded with deuterated PAA. Part B: $\mathbf{Q} \perp \mathbf{H}$; part D: $\mathbf{Q} \parallel \mathbf{H}$; parts C and E–H = 0. Part A was obtained with solid PAA after the sample has been field cooled in the $\mathbf{Q} \perp \mathbf{H}$ configuration and at a temperature 2° below the melting. (From Pynn et al.[45] Copyright 1972, Pergamon Press Ltd.)

also be included, and they can be derived from the neutron data. The recipe for this has been given by Kohli et al.[46] For the second diffraction peak in Fig. 20, the observed intensity $I(\theta_Q)$ was fitted to an expression of the form

$$I(\theta_Q) = a_0 + a_2 P_2(\cos\theta_Q) + a_4 P_4(\cos\theta_Q) + a_6 P_6(\cos\theta_Q) + \cdots. \quad (17.18)$$

The results of two such fits to order $P_6(\cos\theta_Q)$ are displayed in Fig. 21 together with the experimental data. The fitted coefficients were further used to derive the ratio $\langle P_4 \rangle / \langle P_2 \rangle$, which was compared with mean-field theory. The result is displayed in Fig. 22. A procedure for obtaining $\langle P_2 \rangle$ was also given,[46] and the result is shown in Fig. 23. The width of the cross-hatched region represents the uncertainty, due to background scattering, in establishing an absolute value for the order parameter.

In the uppermost part of Fig. 20, the positions of the most intense Bragg peaks of a room-temperature, polycrystalline sample are indicated by vertical bars. The diffraction pattern of the nematic region evolves quasicontinuously

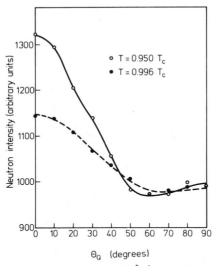

FIG. 21. Measured neutron intensity at $Q = 2.84$ Å^{-1} plotted as a function of θ_Q for two temperatures in the nematic phase of deuterated PAA. The data were obtained without energy analysis of the scattered neutron beam and with an incident neutron energy of 50 meV. Statistical errors in the measurements are less than the size of the plotted symbols. (From Kohli et al.[46])

out of the pattern of the solid. In a so-called "soft-solid" region, which sets in some 5° below the melting transition, sharp Bragg peaks coexist with the wide nematic peaks. Riste and Pynn[47] studied the pretransitional effect associated with the melting. They measured the temperature variation of the Bragg peaks (see Fig. 24) and fitted an apparent "critical exponent"

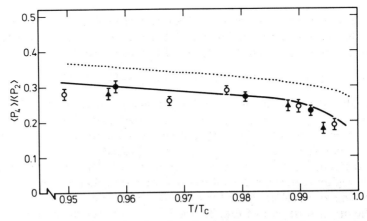

FIG. 22. The ratio $\langle P_4 \rangle / \langle P_2 \rangle$ as a function of reduced temperature for deuterated PAA. (······) mean-field theory; (———), calculations. (From Kohli, et al.[46])

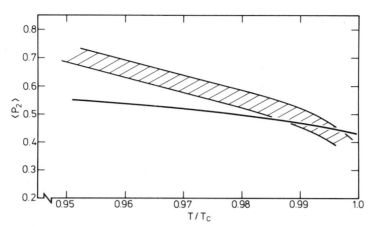

FIG. 23. Order parameter $\langle P_2 \rangle$ for deuterated PAA. The cross-hatched region represents experimental measurements, while the solid curve has been obtained from mean-field theory. (From Kohli et al.[46])

$\beta = 0.12 \pm 0.03$ to the curve. Pynn, as a result of calculations,[44] suggested an explanation for the observed pretransitional effects. Molecules were considered to be mutually parallel and to execute uncorrelated torsional oscillations about the long axes. A softening of the oscillations provides a qualitative explanation both of the evolution of the diffuse peaks and of the

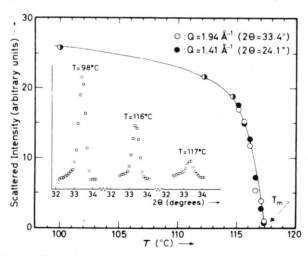

FIG. 24. Variation of integrated Bragg intensity with temperature for two Bragg peaks during the melting of PAA: the curve through the experimental points is intended only as a guide to the eye. The insert shows profiles of one of the Bragg peaks at three temperatures. (From Riste and Pynn.[47] Copyright 1973, Pergamon Press Ltd.)

behavior of the Bragg intensity. The melting of a nematogenic substance probably bears some relation to the progressive melting of smectogenic crystals, to be discussed in Section 17.3.2. In the latter case single crystals were used, which obviously would be needed also for putting the data on PAA on a more firm basis.

No neutron scattering study has yet been reported for the relaxational, collective excitations of the orientational degrees of freedom in the nematic state. An attempt has been made by Conrad et al.[48] to search for collective excitations in the disordered, positional degrees of freedom. On the basis of constant-angle observations by the time-of-flight technique, they suggest the existence of propagating, phononlike modes. These observations are challenging, but probably not conclusive, and their validity has been questioned.[49]

17.3.2. Smectics

Coherent scattering of radiation from smectics has been studied for two reasons: first, as part of the investigation of the smectic state and its relation to the solid and the other liquid crystalline states. Here neutron scattering plays an important role, especially in the determination of the positional order within the smectic layers and in the measurement of the lattice dynamics. Second, smectics provide a testing ground in which low-dimensional phenomena (strong fluctuations, melting, etc.) can be studied in a three-dimensional sample. The main merit of the diffraction techniques for the latter problems has been in distinguishing Bragg-type and Peierls-type scattering for the density wave normal to the smectic layers. Here only x rays[50] and synchrotron radiation[51] give sufficiently good angular resolution.

One of the more interesting smectic systems is TBBA (terephtal-di-butyl-aniline) whose phase diagram is given in Section 17.1. Hervet et al.[52,53] measured the molecular tilt angle in the S_C phase and its temperature dependence when approaching the S_C–S_A transition. This was done by measuring the angular distance between two peaks on the rocking curve.

Solid TBBA melts at 113.5°C to smectic G, which supercools and transforms at 84°C to smectic H. Both S_G and S_H have some positional order, not only out of, but also within, the smectic layers. Extensive studies by x rays and neutron scattering of these ordered smectic states have been performed by Doucet et al.[54,55] The molecules are arranged on a quasi-hexagonal lattice with their long axes tilted about 30° from c^*. Observations of reflections (hkl) with $l \neq 0$ allowed them to conclude that there are positional correlations between the planes. This rules out one of the models of the S_G phase,[56] according to which there should be no interlayer coupling.

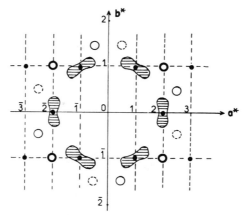

FIG. 25. (a^*, b^*) reciprocal plane of the smectic phase S_G. ○, Diffuse spot type 1; o, diffuse spot type 2; ⊙, diffuse spot type 3; •, Bragg peak; ⊜, zone of scattering. The type 1 diffuse spots become Bragg peaks in the smectic S_H phase. (From Levelut et al.[55])

Figure 25 shows reflections recorded by x rays in the (a^*, b^*) reciprocal plane of S_G. In addition to Bragg peaks, surrounded by a wide zone of diffuse scattering, there are three types of diffuse spots. Type 1 becomes Bragg spots in the S_H phase, but the other two are visible through part of this phase. The diffuse spots in S_G correspond to domains of short-range, orientationally ordered molecules.

Gane et al.[57] have studied the development of three-dimensional order and of tilted structures in other smectics. Using a neutron spectrometer with an area detector they were able to study the modulation of the smectic layers as a precursor to the tilted structure.

The three-dimensional, positional order of the smectic phases is expected to involve propagating phonons. Levelut et al.[55] have measured some phonons in the solid and in the smectic phases of the TBBA. The samples were monocrystalline and nearly fully deuterated. The experiments in the smectic phases were made difficult by the strong quasi-elastic scattering associated with the molecular motion.

Figure 26 shows the reciprocal plane (a^*, c^*) of the solid phase. Figure 27 indicates that the longitudinal acoustic mode along a^* changes only little from the solid to the smectic phases. Another mode, that was tentatively ascribed to a librational motion of the phenyl rings, vanished at the melting transition to the S_G. Such a transition to free rotations seems to support the results obtained for the solid nematic transition of PAA.[44,47] However, lacking so far is clear evidence of transverse phonons propagating along c^*. Such a mode would correspond to shear oscillations of the layers of orientationally ordered molecules.

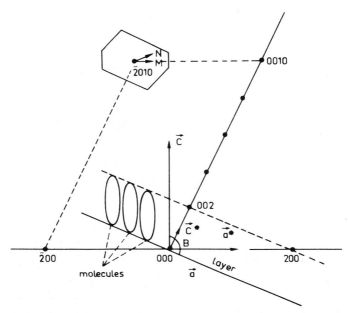

Fig. 26. Reciprocal plane (**a***, **c***) of the solid phase of TBBA. (From Levelut et al.[55])

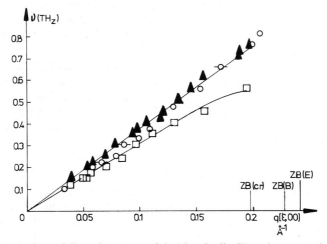

Fig. 27. Comparison of dispersion curves of the "longitudinal" mode propagating along **a*** for the three phases: solid phase (▲); S_H phase (○); and S_G phase (□). The values of q are given in Å^{-1}. Zone boundaries of the different phases are indicated. B and E correspond to S_G and S_H in our notation, respectively. (Adapted from Levelut et al.[55])

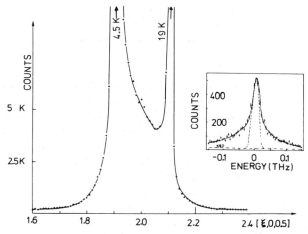

FIG. 28. Elastic scan along the **a*** direction in the smectic B_C phase showing the mosaic spread of [2, 0, 0] and [2, 0, 1] Bragg reflections in addition to large scattered intensity. In the insert an energy scan at the point [2.2, 0, 0.5] proves that this intensity is mostly quasi-elastic. The broken curve gives the resolution function. (Adapted from Levelut et al.[55])

Figure 28 shows an example of a scan that reveals the strong quasi-elastic scattering referred to above. Similar scans (both in **q** and ω) were performed at other positions in the (**a***, **b***) plane. This gave information about the correlation lengths and relaxation times of the domains in the layers of orientationally ordered molecules. Such measurements have so far been made at only one temperature, in the S_G phase. It would, of course, be desirable to have such data through the different phases, from the solid crystalline to the S_A phase.

17.4. Convective Instabilities

The flow pattern in a liquid layer, heated from below, changes character as the vertical temperature difference (ΔT) between the bottom and top plates of the container is increased. At small ΔT heat is transported by conduction and the liquid remains quiescent. When ΔT exceeds a well-defined threshold value ΔT_C, convection rolls are observed. By further increase of ΔT one ultimately reaches a turbulent region. At present the interest is concentrated on the temporally turbulent region, rather than on the spatially turbulent one. The former is reached as the system, for increasing ΔT, passes through a sequence of higher instabilities, each adding some complexity to the frequency spectrum of the flow. This development of irregular, time-dependent flow from an initial steady flow is just one example of a wider problem area in science known as the development of chaos. According to our present understanding, chaos may develop along different routes[58] of which period doubling may be mentioned. The interest among physicists in convection instabilities is due also to the possibility they offer for studying nonequlibrium phase transitions.

17.4.1. Why Use Neutrons and Liquid Crystals?

In equilibrium phase transition studies, neutrons have become an indispensable tool for studying order and fluctuations. In convection the order parameter is the velocity amplitude of a convection roll. To observe this order parameter by neutron scattering one needs a liquid crystal. The elongated molecules are anisotropic scatterers, and the intensity of a liquid diffraction peak depends strongly on the orientation of the director with respect to the scattering vector, as seen from Fig. 20. There is a coupling between orientation and flow in a liquid crystal,[59] and changes in the flow are detected through the corresponding changes in the neutron intensity. Fully deuterated nematic PAA has been used in such experiments.

It turns out that a nematic liquid crystal offers some general advantages over isotropic liquids in the study of convective instabilities. The solutions of the hydrodynamic equations in the different flow regimes can be given in the space of the dimensionless parameters R (the Rayleigh number, proportional to ΔT) and P (the Prandtl number: the ratio between the viscosity and the thermal conductivity). Using a nematic sample one can conveniently vary P by varying the strength of the applied magnetic field. This field also controls the intensity of orientational director fluctuations at the "microscopic" level. These fluctuations couple to and influence the macroscopic behavior of the liquid, thus making a nematic sample perhaps the only candidate in which to observe critical fluctuations near ΔT_C.[60] The dependence of the order parameter on both H and ΔT also opens up the possibility to study multicritical behavior.

It was shown (Fig. 20) that for a horizontally aligned nematic the intensity of the first diffraction peak is a minimum. It turns out that, for $\Delta T > \Delta T_C$, the intensity is proportonal to the order parameter v for a given convection roll. A change in neutron intensity therefore signals a change in v or in the roll configuration. The roll configuration may be measured by a direct diffraction method, by scanning across the cell with a narrow beam and observing the intensity variation.[61] Temporal changes of the roll structure or of the velocity amplitude are macroscopic and are measured in real time rather than through the conventional inelastic scattering technique.

17.4.2. The Real-Time Neutron Scattering Technique

In conventional neutron scattering one measures $S(Q, \omega)$, the Fourier transform of $G(r, t)$. This technique is applicable when the neutron passage time through a correlation range is longer than the period, or the relaxation time, of the fluctuation. The real-time technique is complementary to the conventional one, it works when the passage time is shorter. Hence, with the latter technique, temporal changes in a physical state are measured not by one and the same neutron but by neutrons passing successively through the scatterer. This is possible only when the dynamic changes affect the total intensity of the scatterer, that is, when the wavelength of the fluctuation is comparable to the dimension of the sample.

When the above conditions are satisfied, the technique consists of measuring $I(t)$, the coherently scattered intensity of neutrons as a function of time. For linear oscillations even the raw data may give the information needed. Usually one will have to calculate the autocorrelation function $C(\tau)$ and its Fourier transform, the power spectrum $P(\omega)$. The function $C(\tau)$ is defined as $\langle [I(t) - \bar{I}][I(t + \tau) - \bar{I}]\rangle$, where $\bar{I}$ denotes the time average. Problems regarding resolution, frame overlap, and false frequencies (aliasing) have been treated in Otnes and Riste.[62]

If the power spectrum consists of one line only, the phase space trajectory or phase portrait $[I(t), \dot{I}(t)]$ is a closed orbit, a limit cycle. For more complicated spectra, the phase portrait gives information not contained in $P(\omega)$, since the phases of the different spectral components are preserved in the phase portrait. The phase portrait depicts the attractors in phase space, their orbits, singular points, and strangeness.

17.4.3. Order Parameters and Attractors

Experiments by Riste et al.[61] on nematic PAA in a vessel of dimension $(38 \times 38 \times 5)$ mm^3 in a horizontal magnetic field H gave Fig. 29. The short dimension was horizontal and perpendicular to H. In this configuration one expects a single roll, with axis along the short edge, to develop as ΔT exceeds ΔT_C. Theory predicts[63]

$$\Delta T_C(H) = \Delta T_C(0)[(1 + H^2)/H_C^2], \qquad (17.19)$$

where H_C is the threshold field for alignment of the director. In the given geometry they measured $H_C = 25$ Oe, and this value was used in Fig. 30 when fitting Eq. (17.19) to the data of Fig. 29. The data at $H \geq 160$ Oe were then all plotted versus the reduced gradient $\varepsilon = (\Delta T - \Delta T_C)/\Delta T_C$. There is a reasonable collapse of all data to a single curve, as required from scaling theory. Fitting to $A\varepsilon^{1/2}$ (i.e., to a single roll) reproduces the data only at small ε. Fitting to $\Sigma A_n \varepsilon^{n/2}$, where A_n denotes the amplitude of the nth harmonic of the fundamental roll, gives a considerable improvement of the fit. The critical exponent $n/2$ is predicted by Landau theory. This theory also gives a prediction for the time behavior of the order parameter as the system evolves from one stationary state to another. It predicts an exponential growth at

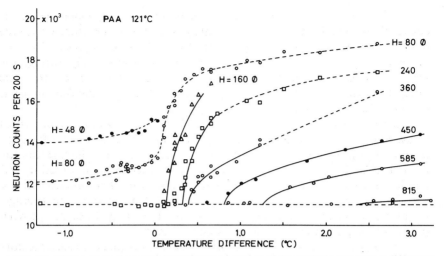

FIG. 29. Time-averaged neutron intensity as a function of the vertical temperature difference across the PAA sample cell. Values of horizontal magnetic field are given, fully drawn parabolic curves are least-square fitted to data. (From Riste et al.[61])

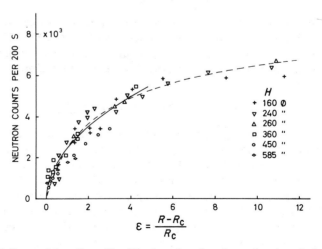

FIG. 30. Collapse of data (from Fig. 29) when plotted against reduced vertical temperature difference. A constant high-field background has been subtracted. The broken curve has been calculated from mean-field theory with allowance for admixture of higher harmonic rolls. The fully drawn curve is parabolic. (Adapted from Riste et al.[61])

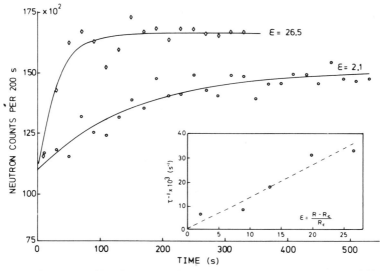

FIG. 31. After suppression of the convective flow by a strong field, the exponential buildup of the flow in a weak field (48 Oe) is observed by neutron scattering. The inset shows the inverse of the time constant τ versus ε, the reduced temperature difference. (From Riste et al.[61])

a rate proportional to ε (i.e., a critical slowing down). Figure 31 shows such a behavior.

All the above data refer to intensities that are long-time averages. In horizontal fields only higher instabilities are predicted to induce time-dependent flow, for which real-time counting will be needed. For a vertical field, however, theories by Lekkerkerker[64] and others[65] predict steady flow to be preceded by a time-dependent flow regime. The width (in ΔT) of the latter regime is predicted to shrink at increasing H and to disappear above a critical field H_1. This prediction was tested[66,67] on PAA contained in a cell with side walls of poor heat conductivity (stainless steel), in order to reach both flow regimes. For $H \geq 1.1$ kOe only one flow regime was seen. Figure 32 shows data for $H = 0.6$ kOe together with a broken curve for $H = 1.2$ kOe.

At the lower field there is a multistable regime in which the time-averaged intensity may reside in a number of levels. To take it from one level to another, one has to disturb the system by sudden changes of the field or of the gradient. Having settled at one level, however, one may move smoothly along it by slow variation of ΔT. Intensity profiles across the cell verified the expectation that each level corresponds to a certain spatial configuration and sense of rotation of the rolls. Each level is stable within a certain regime of ΔT. At the higher end, the intensity falls to the lowest level before joining the curve of the steady flow regime, as seen from Fig. 32. At low fields

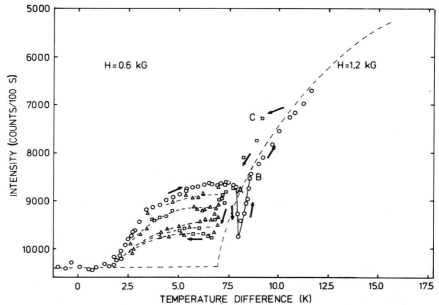

FIG. 32. Neutron intensity versus temperature difference (ΔT) from PAA in a vertical field $H = 0.6$ kG. Circular and square points are for a complete cycle of increasing and decreasing ΔT, respectively. Triangular points are measured after additional variation of H. Notice intensity dip between A and B. The broken curve obtained at $H = 1.2$ kG is also indicated. (From Otnes and Riste.[66])

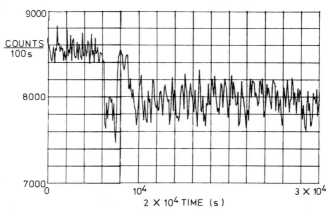

FIG. 33. Time record of neutron intensity at constant applied vertical field ($H = 0.12$ kG) and temperature difference, showing transition between states of different stability (K. Otnes and T. Riste, to be published).

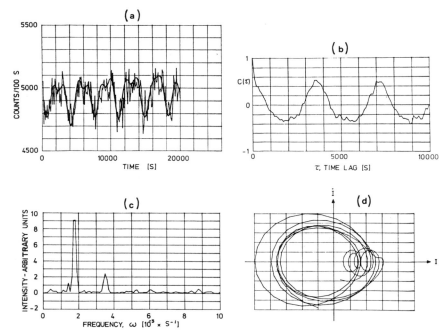

FIG. 34. (a) Portion of a time record $I(t)$ of scattered neutron intensity for a quasi-periodic state; (b), (c), and (d) are, respectively, the autocorrelation function, the power spectrum, and the phase space trajectory of $I(t)$. The heavy, smooth line in (a) is fitted and used in the calculation of (d). (From Otnes and Riste.[62])

(~0.1 kOe) the hysteretic region is wide, and the intensity may show spontaneous jumps from one level to another, see Fig. 33. As seen from this figure the levels represent states of different stability. An example[62] of a quasi-periodic state is given in Fig. 34, which gives the time series, the autocorrelation function, the power spectrum, and the phase portrait.

On the basis of Fig. 32 and similar curves at other values of the magnetic field, one arrives at the phase diagram of Fig. 35. (This diagram gives the extreme values of the threshold gradients at any one field only, and neglects the level structure.) According to the classification of equilibrium phase transitions, the diagram represents a bicritical system. In this diagram the lower flow regime has been denoted multistable rather than time periodic, although states of such a time character were seen in some parameter range. A full mapping of the temporal stability of the states in both regimes is a formidable task. Also, for a horizontally applied magnetic field there are time-dependent states, as first reported by Møller and Riste.[68] The oscillatory states may even have a soft-mode behavior.[69] It seems difficult to reconcile these observations with presently available theories.

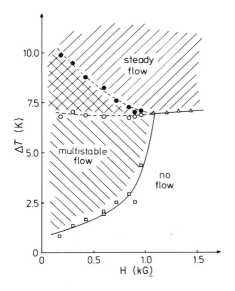

FIG. 35. Phase diagram showing confluence of first-order (----) lines and second-order (——) lines at a multicritical point. First-order lines are upper and lower stability limits of multistable and steady-flow regimes, respectively. (Adapted from Riste and Otnes.[67])

17.5. Concluding Remarks

It seems fair to state that neutron scattering has not yet been used to its full power in the study of liquid crystals. This is especially true in regard to the use of coherent, inelastic scattering that could shed light on the liquid crystalline phases and their relation to the other phases of condensed matter. The lack of deuterated samples is partly responsible for this. Another reason is the large mass of the constituent molecules, which makes inelastic scattering very difficult. Possibly the use of the real-time technique could be explored further, for example, in the study of novel materials with liquid-crystalline character, such as ferrofluids, holes in ferrofluids, and magnetically doped liquid crystals.

References

1. D. Demus and H. Sackmann, *Z. Phys. Chem. (Leipzig)* **238**, 215 (1968).
2. A. de Vries, *J. Chem. Phys.* **70**, 2705 (1979).
3. *Phys. Today, Spec. Issue: Liq. Cryst.* May (1982).
4. V. Tsvetkov, *Acta Physicochim. (URSS)* **16**, 132 (1942).
5. M. Miesowicz, *Nature* **17**, 261 (1935); *Nature (London)* **158**, 27 (1946).
6. J. P. Parneix, A. Chapoton, and E. Constant, *J. Phys. (Ursay, Fr.)* **36**, 1143 (1975).
7. P. Bordewijk, *Z. Naturforsch. A* **35A**, 1207 (1980).
8. E. Kluk, *Acta Phys. Pol.* **A58**, 51 (1980).
9. F. Volino, A. J. Dianoux, and H. Hervet, *Mol. Phys.* **30**, 1131 (1975).
10. J. D. Barnes, *J. Chem. Phys.* **58**, 5193 (1973).

11. A. J. Dianoux, H. Hervet, and F. Volino, *J. Phys. (Orsay, Fr.)* **38**, 809 (1977).
12. J. A. Janik, J. M. Janik, K. Otnes, and K. Rosciszewski, *Physica B + C (Amsterdam)* **83B + C**, 259 (1976).
13. I. Svare, B. O. Fimland, K. Otnes, J. A. Janik, J. M. Janik, E. Mikuli, and A. Migdal-Mikuli, *Physica B + C (Amsterdam)* **106B + C**, 195 (1981).
14. R. M. Richardson, A. J. Leadbetter, and J. C. Frost, *Mol. Phys.* **45**, 1163 (1982).
15. F. Volino, A. J. Dianoux, and H. Hervet, *J. Phys. (Orsay, Fr.)* **37**, 73 (1976).
16. F. Volino, *Mol. Phys.* **34**, 1263 (1977).
17. H. Hervet, F. Volino, A. J. Dianoux, and R. E. Lechner, *J. Phys., Lett. (Orsay, Fr.)* **35**, L-15 (1974).
18. H. Hervet, F. Volino, A. J. Dianoux, and R. E. Lechner, *Phys. Rev. Lett.* **34**, 451 (1975).
19. A. J. Dianoux, F. Volino, H. Hervet, and A. Heidemann, *J. Phys., Lett. (Orsay, Fr.)* **26**, L-257 (1975).
20. F. Volino, A. J. Dianoux, and H. Hervet, *Mol. Cryst. Liq. Cryst.* **38**, 125 (1977).
21. F. Volino and A. J. Dianoux, *Phys. Rev. Lett.* **39**, 763 (1977).
22. F. Volino, *J. Phys. (Orsay, Fr.)* **40**, 181 (1979).
23. R. M. Richardson, A. J. Leadbetter, and J. C. Frost, *Ann. Phys. (Paris)* **3**, 177 (1978).
24. A. J. Leadbetter and R. M. Richardson, *Mol. Phys.* **35**, 1191 (1978).
25. A. J. Leadbetter, R. M. Richardson, and J. C. Frost, *J. Phys. Colloq. (Orsay, Fr.)* **40**, C-3 (1979).
26. R. M. Richardson, A. J. Leadbetter, C. J. Carlile, and W. S. Howells, *Mol. Phys.* **35**, 1697 (1978).
27. D. H. Bonsor, A. J. Leadbetter, and F. P. Temme, *Mol. Phys.* **36**, 1805 (1978).
28. K. Rosciszewski, *Acta Phys. Pol.* **A41**, 549 (1972).
29. J. Töpler, B. Alefeld, and T. Springer, *Mol. Cryst. Liq. Cryst.* **26**, 297 (1973).
30. F. Volino, A. J. Dianoux, and A. Heidemann, *J. Phys., Lett. (Orsay, Fr.)* **40**, 583 (1979).
31. A. J. Dianoux, A. Heidemann, F. Volini, and H. Hervet, *Mol. Phys.* **32**, 1521 (1976).
32. J. A. Janik, J. M. Janik, K. Otnes, J. Krawczyk, and K. Rosciszewski, *Physica B + C (Amsterdam)* **92B + C**, 351 (1977).
33. S. Wrobel, J. A. Janik, J. Moscicki, and S. Urban, *Acta Phys. Pol.* **A48**, 215 (1975).
34. K. Chledowska, B. Janik, J. Krawczyk, J. A. Janik, J. M. Janik, and K. Otnes, *Liq. Cryst.* **1**, 127 (1986).
35. S. Urban, S. Wrobel, K. Chledowska, J. Chrusciel, J. A. Janik, and H. Kresse, *Mol. Cryst. Liq. Cryst.* **100**, 57 (1983).
36. J. A. Janik, J. M. Janik, J. Krawczyk, and K. Otnes, *Mol. Cryst. Liq. Cryst.* **89**, 171 (1982).
37. X. P. Nguyen, S. Urban, S. Wrobel, and H. Kresse, *Acta Phys. Pol.* **A54**, 617 (1978).
38. J. A. Janik, M. Godlewska, T. Grochulski, A. Kocot, E. Sciesinska, J. Sciesinski, and W. Witko, *Mol. Cryst. Liq. Cryst.* **98**, 67 (1983).
39. G. J. Krüger, *Phys. Rep.* **82**, 230 (1982).
40. C. K. Yun and A. G. Fredrickson, *Mol. Cryst. Liq. Cryst.* **12**, 73 (1970).
41. R. M. Richardson, A. J. Leadbetter, D. H. Bonsor, and G. J. Krüger, *Mol. Phys.* **40**, 741 (1980).
42. F. Volino, and A. J. Dianoux, *Mol. Phys.* **36**, 389 (1978).
43. P. G. de Gennes, *C. R. Hebd. Seances Acad. Sci., Ser. B* **274**, 142 (1972).
44. R. Pynn, *J. Phys. Chem. Solids* **34**, 735 (1973).
45. R. Pynn, K. Otnes, and T. Riste, *Solid State Commun.* **11**, 1365 (1972).
46. M. Kohli, K. Otnes, R. Pynn, and T. Riste, *Z. Phys. B* **24**, 147 (1976).
47. T. Riste and R. Pynn, *Solid State Commun.* **12**, 409 (1973).
48. H. M. Conrad, H. H. Stiller, and R. Stockmeyer, *Phys. Rev. Lett.* **36**, 264 (1976).
49. C. A. Pelizzari and T. A. Postol, *Phys. Rev. Lett.* **38**, 573 (1977).

50. J. Als-Nielsen, J. D. Litster, R. J. Birgeneau, M. Kaplan, C. R. Safinya, A. Lindegaard-Andersen, and S. Mathiesen, *Phys. Rev. B* **22**, 312 (1980).
51. J. Als-Nielsen, F. Christensen, and P. S. Pershan, *Phys. Rev. Lett.* **48**, 1107 (1982).
52. H. Hervet, S. Lagomarsini, F. Rustichelli, and F. Volino, *Solid State Commun.* **17**, 1533 (1975).
53. G. Albertini, S. Lagomarsini, F. Rustichelli, and F. Volino, *Solid State Commun.* **27**, 607 (1978).
54. J. Doucet, M. Lambert, A. M. Levelut, P. Porquet, and B. Dorner, *J. Phys. (Orsay, Fr.)* **39**, 173 (1978).
55. A. M. Levelut, F. Moussa, J. Doucet, J. J. Benattar, M. Lambert, and B. Dorner, *J. Phys. (Orsay, Fr.)* **42**, 1651 (1981).
56. P. G. de Gennes and P. G. Sarma, *Phys. Lett. A* **38A**, 219 (1972).
57. P. A. C. Gane, A. J. Leadbetter, and P. G. Wrighton, *Mol. Cryst. Liq. Cryst.* **66**, 247 (1981).
58. For review see, e.g., P. H. Coullet and C. Vanneste, *Helv. Phys. Acta* **56**, 813 (1983).
59. P. G. de Gennes, "The Physics of Liquid Crystals," p. 151. Oxford Univ. Press (Clarendon), London and New York, 1974.
60. R. Graham, in "Fluctuations, Instabilities and Phase Transitions" (T. Riste, ed.), p. 215. Plenum, New York, 1975.
61. T. Riste, K. Otnes, and H. B. Møller, in "Neutron Inelastic Scattering," Vol. 1, p. 511. IAEA, Vienna, 1978.
62. K. Otnes and T. Riste, *Atomenerg. Kerntech. Suppl.* **44**, 753 (1984).
63. For review see E. Guyon, p. 295 in Ref. 60.
64. H. N. W. Lekkerkerker, *J. Phys., Lett. (Orsay, Fr.)* **38**, L277 (1977).
65. E. Guyon, P. Pieranski, and J. Salan, *J. Fluid Mech.* **93**, 65 (1979); M. G. Verlarde and I. Zuniga, *J. Phys. (Orsay, Fr.)* **40**, 725 (1979); E. Dubois-Violette and M. Gabay, *J. Phys. (Orsay, Fr.)* **43**, 1305 (1982).
66. K. Otnes and T. Riste, *Helv. Phys. Acta* **56**, 837 (1983).
67. T. Riste and K. Otnes, *In* "Multicritical Phenomena" (R. Pynn and A. T. Skjeltorp, eds.), p. 451. Plenum, New York, 1984.
68. H. B. Møller and T. Riste, *Phys. Rev. Lett.* **34**, 996 (1975).
69. K. Otnes and T. Riste, *Phys. Rev. Lett.* **44**, 1490 (1980); **45**, 1463 (1980).

INDEX

A

Adsorbed monolayers, 1
  experimental, 6
  inelastic scattering, 53
Adsorption isotherms, 5
Aggregation number, 493
Amorphous alloys, 255
  metal–metal, 260
  metal–metalloid, 255
AOT–water–oil system, 534
Aqueous solutions, 471
Argon, liquid, 466
Association colloids, 490
Attenuation corrections, 421

B

Bhatia–Thornton formalism, 251
Blocking factor, 149
Bose condensation, 331
Bovine serum albumin, 532
BSA, *see* Bovine serum albumin

C

Catalytic surfaces, 63
Chaos, 575
Chemisorbed systems, 63
Chudley–Elliott model, 132, 151, 201
Clustering, 99
  kinetics, 103
CMC, *see* Critical micellar concentration
Colloidal solutions, 489
  interparticle structures, 524
  intraparticle structures, 514
Commensurate–incommensurate transitions, 36
Contrast matching, 502
Convolution approximation, 151, 202
Correlation function, space-time, 245
Correlations between jumps, 149
Creep, 117
Critical micellar concentration, 490
Cu–Ni alloys, 102

D

Debye Waller factor
  fast ions, 197
  hydrogen, 138
Defect clusters, 226
Defects, 85
  dynamic properties, 120
  instrumention, 93
  point, 91
    isolated, 96
Derjaguin–Landau–Verwey–Overbeek theory, *see* DLVO theory
Diffuse scattering, 197
  elastic, 97, 107
Diffusion
  hydrogen in metals, 132
  nondilute, 150
Diffusion coefficient, 142, 483
  chemical, 152
Di-methoxy-azoxy-benzene, *see* PAA
Dislocations, two-dimensional, 46
DLVO theroy, 495
Dynamic form factor, *see* Scattering functions
Dynamic susceptibilities, 308

E

Elastic incoherent structure factor, 557
Electrochemical cells, 193

F

Faber–Ziman formalism, 251
Faraday transition, 191
Fast ion conductors, 115, 187
  anharmonicity, 211
  diffraction, 205
  inelastic scattering, 233
  neutron scattering methods, 189
  quasi-elastic scattering, 223
Fatigue, 117
FeP glasses, phonon density of states, 294
Feynmann, R. P., 330

INDEX

Floating solid phase, 38
Fluctuations
 concentration, 253
 number density, 253
Fluids, 405
 computer simulation, 455
 data analysis, 419
 experimental techniques, 414
 molecular, 408
 normal modes, 461
 pair approximation, 405
 scattering function, 411, 450
 structure factor, 408, 445
Fluorite structures, 204, 226, 230
Form factor, cell, 509
Fourier synthesis, 208

G

Germanate anomaly, 278
Glasses, 243
Globular protein solutions, 528
Gorski relaxation method, 152
Graphite, 4
 exfoliated, 4
Graphite intercalation compounds, 70
Guinier approximation, 514
Guinier–Preston zones, 114, 125

H

Halperin–Nelson theory, 46
Hamaker constant, 496
Helium, 303
 phase diagrams, 303
Helium-3, liquid, 370
 dynamic susceptibility, 374
 interactions, 377
 kinetic energy, 383
 paramagnon model, 380
 spin fluctuations, 370
 zero sound, 370
Helium-4, liquid, 326
 adsorbed on graphite, 59
 anomalous dispersion, 337
 condensate, 331, 350
 elementary excitations, 328, 337
 kinetic energy, 355
 linewidths, 342
 Monte Carlo calculation, 357, 366
 multiphonon scattering, 348
 pair correlations, 359
 scattering function, 338
 singular behavior, 354
 structure factor, 359
 superfluidity, 304, 326
 two-fluid model, 342
Helium 3–4 mixtures, 385
 excitations, 388
 scattering function, 387
Helium-3, solid, magnetic ordering, 325
Helium-4, solid, 309
 anharmonicity, 324
 interference terms, 321
 dynamics, 310
 kinetic energy, 322
 phonons, 317
 scattering function, 315
 short-range correlations, 314
 single-particle excitations, 321
Hexatic phase, 46
Huang scattering, 94, 98
Hydrides, acoustic branches, 137
Hydrocarbons, adsorbed on graphite, 29, 54, 62
Hydrodynamic regime, 464
Hydrogen, adsorbed on graphite, 27
Hydrogen in amorphous metals, 175
Hydrogen in metallic glasses, 284
 local vibrations, 289
Hydrogen in metals, 131
 anharmonicity, 161
 dilute, 132
 isotope effects, 165
 nondilute, 149
 potentials, 160
 precipitation, 174
 self-trapping, 168
 shear constant anomalies, 158
 superconductivity, 177
 tunneling states, 172
 vibrations, 153
  acoustic, 153
  local, 170
  optic, 156
  table, 166
Hydrogen storage, 147

I

IBPBAC, 559
Impulse approximation, 322, 350
Incommensurate phases, 17, 42
Inelasticity corrections, 428, 472

INDEX

Inhomogeneities, 93
Intercalated compounds, 1, 69
Intermediate-range fluctuations, 270
Ionic conductivity, 188
Ionic solutions, 471
Iso-butyl-phenyl-benzylidene-amino-
 cinnamate, see IBPBAC
Isotopes, table, 476
Isotopic substitution, 474

J

Jump diffusion, 230
 activated, 160
Jump sequences, 135

K

Kinetic energy, 322
Kosterlitz–Thouless transition, 45
Krypton, dense gas, 406, 461

L

Landau theory, 377
Landau–Khalatnikov theory, 336
Landau–Pomaranchuk model, 386
Landau, L., 326
LDS, see Lithium dodecyl sulfate
LEED, 1
Light scattering, 499
Liquid crystals, 545
 attractors, 577
 cholesterics, 549
 coherent scattering, 567
 convective instabilities, 575
 dielectric relaxation, 551
 diffusion, 563
 dynamic fluctuations, 556, 558
 dynamics, 550
 incoherent scattering, 553
 nematic phase, 545, 563
 order parameters, 577
 phase transitions, 556, 563
 pretransitional effect, 570
 quasi-elastic scattering, 553
 reorientations, 550, 556
 rotational diffusion, 554
 scattering functions, 553
 self-diffusion, 551
 smectic phases, 546, 572
Lithium dodecyl sulfate, 519
 LiCl solutions, 527

Lithium ion conductors, 228, 232
Localized modes, 124
Local moments, 120
 relaxation, 121
London dispersion force, 496
Long-wavelength fluctuations, 93, 109
 magnetic, 113
Low-energy electron diffraction, see LEED

M

Melting, two-dimensional, 42
Metallic glasses, phonon density of states, 293
Methane
 adsorbed on graphite, 17, 37, 43
 solid, 108
Method of differences, 474
MgZn glasses, correlated motions, 296
Micellar solutions, ionic, 524
Micelles, 490
Microemulsions, 534
Molecular dynamics, computer simulation, 238
Mori formalism, 381
Multilayers, 51
Multiple scattering corrections, 419, 422

N

NiB glasses, 259
 phonon density of states, 294
Ni–Cu alloys, see Cu–Ni alloys
NiTi metallic glasses, 261
NiZr metallic glasses, 264
Nitrogen, adsorbed on graphite, 22

O

Order parameter, nematic, 549
Oxide glasses, 267
Oxygen, adsorbed on graphite, 24

P

PAA, 545, 568, 577
Pair correlations
 anion–water, 478
 cation–water, 477
 ion–ion, 479
Pair density function, 246
 partial, 252
Pair distribution function, 471

INDEX

Partial structure factor, 472
PdGe glasses, 259
PdSi glasses, 255
Phase transitions, fast ion conductors, 191
Physisorbed systems, 2
   phase diagrams, 16
Placzek corrections, *see* Inelasticity corrections
Placzek expansion, 433
Polarization potential theory, 334
   liquid helium-3, 379
Porod's law, 515
Potts model, 34
Precipitation, 111

Q

Quasi-elastic scattering, 120, 132, 223
   coherent, 151, 224
   incoherent, 229

R

Radial distribution function, 256
Radius of gyration, 516, 556
Rare gases, adsorbed on graphite, 28, 41, 57
Registered phases, 17
Resolution function, 199
Resonance modes, 124
Rotational correlations, 483
Rotational disorder, 108
Rotons in liquid helium-4, 326
Rubidium, liquid, 458

S

Scattering cross sections, 87, 194
   one-phonon coherent, 155
   polarized neutron, 89
Scattering functions, 86, 194, 306
   intermediate, 195
   spin-dependent, 307
Selective deuteration, 502, 514
Self-consistent phonon theory, 312
Short-range order in alloys, 99
Silver ion conductors, 203, 228, 232
SiN glasses, 270
SiO_2, 274
   correlated motions, 296
Small-angle scattering, 116, 499
   cross section, 507
   technique, 503
Small-angle x-ray scattering, 499
Solid electrolytes, 190
Spin-echo spectrometer, 124, 344

Spin glasses, 107, 123
   ferromagnetic, 119
Spinodal decomposition, 111
Sputtered glasses, 274
Static approximation, 246
Structural anisotropy in glasses, 270
Structure factors, 245, 408
   interparticle, 510
   intraparticle, 510
Substitutional defects, magnetic, 99
Sum rules, 308
Superfluidity, 326
Superionic conductors, *see* Fast ion conductors
Surfaces, 1
   magnetization, 67
Synchrotron radiation, 3

T

TBBA, 546, 556, 564, 572
TeO–BaO glasses, 281
Terephtal-dibutyl-aniline, *see* TBBA
Thermal diffuse scattering, 206
Total reflection, 67
Total scattering spectrometer, 247
Trapping impurities, 139
Truncation of Fourier transform, 249
Tunneling, hydrogen in metals, 172, 179
Two-dimensional structures, 10
   finite, 15
   line shape, 13
Two-roton bound states in liquid helium-4, 348

V

Vanadium measurements, 416, 442
Van der Waals interactions, 495
Van Hove, L., 245
Vessicles, 490
Vineyard, G. H., 202

W

Warren line shape, 15, 22
Warren–Cowley short-range order parameter, 102, 254
Water exchange, 481
Wetting transition, 51

Z

ZrNiD glasses, 285

CONTENTS OF VOLUME 23, PART A

1. Introduction to Neutron Scattering
   DAVID L. PRICE AND KURT SKÖLD

2. Neutron Sources
   JOHN M. CARPENTER AND WILLIAM B. YELON

3. Experimental Techniques
   COLIN G. WINDSOR

4. Neutron Optics
   SAMUEL A. WERNER AND ANTHONY G. KLEIN

5. Chemical Crystallography
   ARTHUR J. SCHULTZ

6. Lattice Dynamics
   C. STASSIS

7. Molecular Dynamics and Spectroscopy
   G. STUART PAWLEY

Appendix. Neutron Scattering Lengths and Cross Sections
   VARLEY F. SEARS

   INDEX

CONTENTS OF VOLUME 23, PART C

18. Phase Transitions
    R. A. COWLEY

19. Magnetic Structures
    J. ROSSAT-MIGNOD

20. Magnetic Excitations
    WILLIAM G. STIRLING AND KEITH A. MCEWEN

21. Nuclear Magnetism
    H. GLÄTTLI AND M. GOLDMAN

22. Polymers
    JULIA S. HIGGINS AND ANN MACONNACHIE

23. Neutron Crystallography of Proteins
    N. V. RAGHAVAN AND ALEXANDER WLODAWER

24. Molecular Biology
    HEINRICH B. STRUHRMANN

25. Industrial Applications
    MICHAEL T. HUTCHINGS AND COLIN G. WINDSOR

    INDEX